Optimization and Applicability of Bioprocesses

Hemant J. Purohit • Vipin Chandra Kalia
Atul N. Vaidya • Anshuman A. Khardenavis
Editors

Optimization and Applicability of Bioprocesses

Springer

Editors
Hemant J. Purohit
Environmental Biotechnology and Genomics Division, CSIR – National Environmental Engineering and Research Institute (CSIR-NEERI)
Nagpur, Maharashtra, India

Atul N. Vaidya
Solid and Hazardous Waste Management Division, CSIR – National Environmental Engineering and Research Institute (CSIR-NEERI)
Nagpur, Maharashtra, India

Vipin Chandra Kalia
Microbial Biotechnology and Genomics
CSIR – Institute of Genomics and Integrative Biology
New Delhi, India

Anshuman A. Khardenavis
Environmental Biotechnology and Genomics Division, CSIR – National Environmental Engineering and Research Institute (CSIR-NEERI)
Nagpur, Maharashtra, India

ISBN 978-981-10-6862-1 ISBN 978-981-10-6863-8 (eBook)
https://doi.org/10.1007/978-981-10-6863-8

Library of Congress Control Number: 2017960179

Printed on acid-free paper

This Springer imprint is published by Springer Nature
The registered company is Springer Nature Singapore Pte Ltd.
The registered company address is: 152 Beach Road, #21-01/04 Gateway East, Singapore 189721, Singapore

Dedicated to our mentors
and
inspiration –
The respected Mr. Dashrath Manjhi:
The Mountain Man

Preface

Natural resources are limited and their consumption rates are very high. A systematic approach is required for sustainable management of resources. It primarily involves integration of green innovative biotechnological strategies and ecoengineering. The issue revolves around the interaction and adaptability of plants, animals, microbes, and their adaptability in soil, air, and water. Plants need to enhance their tolerance toward hydrocarbons, petrochemical pollutants and soil salinity, temperature, and drought. These strategies are expected to enhance biomass and bioenergy production under suboptimal conditions, resulting in sustainable agriculture, forestry, protection of landscape, and biodiversity. Biological processes can be optimized by integrating bioinformatics tools, providing great perspectives to sustainable development. An integral part of the interaction of different organisms is to monitor them at a single cell level. Modern molecular techniques such as fluorescence in situ hybridization (FISH), highly sensitive catalyzed reporter deposition (CARD)-FISH, and in situ DNA-hybridization chain reaction (HCR) and methods for detecting mRNA and/or functional genes have been described. These techniques have been supplemented with metagenomics analysis, which reveals that a large proportion of microorganisms are as yet to be identified and also that they play an important and necessary role in establishing bioprocesses. In the book entitled *Optimization and Applicability of Bioprocesses*, we will be presenting the status of the diverse possibilities and the strategies for their application for human welfare. The book provides strategies for systematic approaches for sustainable management of resources and insights into integration of green innovative biotechnological strategies and ecoengineering. The well-illustrated chapters have been written by the experts in the area. It provides information on thrust scientific R&D areas for prospective researchers and graduate students. We are sincerely indebted to all the contributors, whose efforts have brought this book to fruition.

New Delhi, India Vipin Chandra Kalia

Contents

About the Editors

Hemant J. Purohit is head of the Environmental Biotechnology and Genomics Division, CSIR-National Environmental Engineering and Research Institute, Nagpur. He completed his Ph.D. at Nagpur University in 1986 with support from CSIR Research Fellowships (JRF and SRF). He has training in microbial biochemistry from the Department of Biochemistry, University of Hull, UK (Commonwealth Fellowship), and also from the Laboratory of Molecular Biology, NINDS, NIH, Bethesda (Fogarty International Exchange Program Fellowship). He also developed the environmental genomics program at CSIR-NEERI. His key area of interest is decision-making by the bacterial/microbial community. He has been involved in designing a strategy for capturing microbial diversity by interfacing culturable and DNA fingerprinting tools; developing genomics-based monitoring tools for EIA and bioremediation processes; studying stress-dependent microbial response using dynamic gene expression and its application in bioprocess optimization; and developing insights into microbial capacities for utilization of organics through genome sequence analysis, etc. He has been project coordinator for a number of high-value projects and has 225 publications to his credit. Dr. Purohit is an associate editor of *Applied Biochemistry and Biotechnology* and editor of the *Indian Journal of Microbiology* and *PLoS One*.

Vipin Chandra Kalia is an emeritus scientist and ex-chief scientist at Microbial Biotechnology and Genomics, CSIR-Institute of Genomics and Integrative Biology, Delhi. He obtained his M.Sc. and Ph.D. degrees in genetics from the Indian Agricultural Research Institute, New Delhi. He has been elected as fellows of (1) the Association of Microbiologists of India (AMI), (2) the National Academy of Sciences (NASc), and (3) the National Academy of Agricultural Sciences (NAAS). His main areas of research are microbial biodiversity, genomics and evolution, bioenergy, biopolymers, antimicrobials, quorum sensing, and quorum quenching. He has published over 100 papers in scientific journals such as (1) *Nature Biotechnology*, (2) *Biotechnology Advances*, (3) *Trends in Biotechnology*, (4) *Critical Reviews in Microbiology*, (5) *Bioresource Technology*, (6) the *International Journal of Hydrogen Energy*, (7) *PLoS ONE*, (8) *BMC Genomics*, (9) *Gene*, and (10) the *Annual Review of Microbiology*. He has an h index of 32 and an i10 index of 71. He has edited three books: *Quorum Sensing Vs Quorum Quenching: A Battle with No*

End in Sight (2015, Springer India) and *Microbial Factories* Vol 1 and 2 (2015). He is currently the editor in chief of the *Indian Journal of Microbiology* and editor of (1) *PLoS ONE*, (2) the *Journal of Microbiology and Biotechnology* (Korea), (3) *Applied Biochemistry and Biotechnology* (USA), (4) *International Scholarly Research Notices* (Energy), (5) *Dataset Papers in Science* (Microbiology), and (6) the *Journal of Molecular and Genetic Medicine*. He is a life member of the following scientific societies: (1) the Society of Biological Chemists of India; (2) the Society for Plant Biochemistry and Biotechnology, India; (3) the Association of Microbiologists of India; (4) the Indian Science Congress Association; (5) the Bioenergy Society of India; and (6) the Biotech Research Society of India (BRSI). He is also a member of the American Society for Microbiology.

Atul N. Vaidya is head of the Solid and Hazardous Waste Management Division, National Environmental Engineering Research Institute (CSIR), Nagpur. He is also a professor at AcSIR (Academy of Scientific and Innovative Research), New Delhi. Since completing his Ph.D. at Nagpur University in 2001, he has been involved in developing strategies for managing municipal solid waste through sustainable methods, the application of biotechnological processes and cleaner technologies for treating industrial wastewaters, and the restoration of the environment through recovery of value-added products such as biogas from the organic fraction of municipal solid waste. He has been a project leader in a number of high-value projects and has 63 publications to his credit. He is a member of the scientific advisory panels of several national and international committees and has been recognized for his work as member of the Expert Committee for Proper MSW Management at Srinagar, Jammu and Kashmir, by the National Green Tribunal.

Anshuman A. Khardenavis is a senior scientist at the Environmental Biotechnology and Genomics Division, National Environmental Engineering Research Institute (CSIR), Nagpur. He is also an asst. professor at AcSIR (Academy of Scientific and Innovative Research), New Delhi. He completed his Ph.D. at Nagpur University in 2011 and has been involved in developing methods and strategies for the recovery of value-added products from organic solid waste, such as keratinases from feather waste, and polyhydroxyalkanoates (PHAs) from activated sludge biomass; the design and operation of reactors for managing organic waste through biomethanation; studying the role of substrate-specific microbial community plasticity in the efficiency of anaerobic digestion; and the evaluation of microbial capacities for the utilization of organics through genome sequence analysis. He has been a project leader in a number of high-value projects sponsored by government agencies and industries. He has 24 publications to his credit. He has received a number of prestigious awards, including a Lien Environmental Fellowship, Singapore, and CSIR-UGC Research Fellowships (JRF and SRF).

Recent Advances in Optimization of Environmental Bioprocesses

1

Anshuman A. Khardenavis, Atul N. Vaidya,
Vipin Chandra Kalia, and Hemant J. Purohit

Abstract

In the scenario of increasing population and concomitant pressure on the depleting fossil fuel resources, the focus of current research is on the development of innovative options for achieving sustainability and eco-efficiency in industrial and environmental bioprocesses. Further, there is a need for managing the ever-increasing waste generated as a result of human and industrial activities in an efficient manner. Of the currently available tools, environmental processes based on biomass utilization hold the highest potential owing to the abundance and renewable nature of such biomass. For achieving the highest efficiencies, these processes need to be optimized. In the current chapter, the underlying functioning of the various bioprocesses active in the environment has been discussed with particular reference to their optimization in bioremediation of solid and liquid waste. The chapter further discusses the tools for achieving higher efficiencies in such bioprocesses in addition to the value addition through energy generation in order to achieve the twin objectives of profitability and sustainability.

Keywords

Environmental bioprocesses • Eco-efficiency • Waste management • Bioremediation

A.A. Khardenavis (✉) • H.J. Purohit
Environmental Biotechnology and Genomics Division, CSIR – National Environmental Engineering and Research Institute (CSIR-NEERI), Nehru Marg, Nagpur 440020, Maharashtra, India
e-mail: aa_khardenavis@neeri.res.in

A.N. Vaidya
Solid and Hazardous Waste Management Division, CSIR – National Environmental Engineering and Research Institute (CSIR-NEERI), Nehru Marg, Nagpur 440020, Maharashtra, India

V.C. Kalia
Microbial Biotechnology and Genomics, CSIR – Institute of Genomics and Integrative Biology, New Delhi, India

H.J. Purohit et al. (eds.), *Optimization and Applicability of Bioprocesses*,
https://doi.org/10.1007/978-981-10-6863-8_1

1.1 Introduction

Most industrial processes which are seen as a symbol of affluence in a country are heavily dependent on the fossil fuel deposits. In a developing country like India, where the population is expected to surpass China, there is a heavy burden on the fossil fuel deposits in the country. As of now, the global crude oil reserve is estimated to be around 1656 billion barrels which is depleting fast (http://instituteforenergyresearch.org/topics/encyclopedia /fossil-fuels/). Hence, there is a need for focusing on organic biomass-based economy by developing and optimizing technologies and processes which target effective utilization of available biomass in a sustainable manner. Among the different biomass options available for exploitation in the country, those sourced from plants and algae, solid organic waste, industrial wastes, agricultural wastes, etc. occupy prime position owing to their abundance and renewable nature, which can serve as raw material for value addition to industrial and environmental bioprocesses (Kuhad 2012; Ferreira-Leitão et al. 2017).

Profitability and eco-efficiency are the main prerequisites for any bioprocess to be effective and sustainable in the long run (Kumar and Prasad 2011). The current quest for achieving global sustainability has targeted various aspects of bioprocesses including the biodiversity, provision for clean ecosystem, etc. which need to be considered in an integrated manner for critical understanding of socioeconomic and environmental linkages (Bennett et al. 2015; Liu et al. 2015). Purohit et al. (2016) have extensively reviewed the different strategies which could be employed for surpassing the challenges of profitability, eco-efficiency, and sustainability in aerobic and anaerobic bioprocesses, viz., (i) knowledge about global biogeochemical cycles as in ocean ecosystem (Jiang and Jiao 2016), agricultural and fallow lands (Sharma and Singh 2017), and desert ecosystems (Agarwal et al. 2014), (ii) efficient biomass utilization (Patel et al. 2016; Soares et al. 2016), (iii) microbial bioprocesses (Nair et al. 2016; Gupta et al. 2017; Patel et al. 2017), (iv) metabolic engineering (Nigam et al. 2017). These solutions are all based on exploiting the versatility of microbes and their metabolic pathways (Kalia 2015; Kalia et al. 2016).

1.2 Tools for Optimization of Environmental Bioprocesses

Despite rapid advances in the above technologies, significant gaps exist in the level of efficiency which can be achieved and the economic and environmental parameters which need to be addressed to, especially in industries such as those on bio-based economy which have gained importance in current scenario as sustainable alternatives to petrochemicals. In the last decade or so, a lot of research has been carried out for developing innovative methodologies for optimizing bioprocesses in order to increase the efficiency. Among the various tools, response surface methodology is the basic and conventional tool adopted by researchers for optimizing bioprocesses for economically important compounds such as L-DOPA (3,4-dihydroxyphenyl-l-alanine) (Gurme et al. 2013). Application of laser stimulation stands out as the most innovative approach which could find application in optimization of natural processes such as in sewage and wastewater treatment, in bioremediation of pollutants, and in improving the growth rate, biomass productivity, and adaptability of plants to pollutants such as heavy metals. Pioneering work in this area was carried out by Dobrowolski et al. (2012) who showed significant increase in resistance of experimental plants belonging to different genera with improvement in bioremediation efficiency of contaminated environmental niches on exposure to laser. The optimum efficiency achieved in the above processes has been attributed to the optimal phenotypic expression achieved on account of laser irradiation (Tursunov and Dobrowloski 2015). Very recently, photostimulation by semiconductor laser was demonstrated to result in improvement in seed germination potential in *Silene vulgaris* from heavy metal-contaminated site and serpentine waste dump site (Szajsner et al. 2014). Researchers have integrated the modeling approaches with the conventional approaches to develop a comprehensive **Mu**lti-scale framework for modeling **S**ustainable **I**ndustrial **C**hemicals production (MuSIC). MuSIC framework was applied for producing precursors of polymers from agricultural feedstocks by utilizing 66 pathways in *Escherichia coli* and *Saccharomyces cerevisiae* which allowed the user to explore various trade-offs and interactions between economy-scale objectives (Zhuang and Herrgard 2015). Similar integrated approach was applied for rapid evaluation of economic and environmental parameters in bioprocess involved in biosynthesis of polyhydroxyalkanoates (PHAs) through the coupling of cost analysis and life-cycle assessment which could facilitate screening for large-scale PHA production from various microbes such as *Bacillus* sp. (Kumar et al. 2014; Leong et al. 2016). The produced PHAs could find use in numerous biomedical applications tissue engineering, implants, and anticancer agents (Ray and Kalia 2017).

1.3 Optimization of Bioprocesses in Bioremediation

Further, in addition to the biosynthetic processes and tools facilitating improvement in their efficiencies, there is tremendous scope for applying such tools for optimization of process efficiencies of catabolic bioprocesses involved in the bioremediation of different ecological niches contaminated by various pollutants such as wastewater contaminated with polychlorinated biphenyls (PCBs), excess phosphorus, and terrestrial oil spills. Diverse approaches have been explored for bioremediation, biotransformation, and mineralization of pollutants, prominent among them being (i) through the formation of microbial biofilms, (ii) through bioaccumulation in microbial cells, (iii) by use of microbial enzymatic pathways, and (iv) through phytoremediation.

1.3.1 Biofilms and Bioremediation

Biofilm formation appears to be by far the most significant strategy adopted by microorganisms in wastewater treatment systems for providing protection from high concentration of toxic and hazardous pollutants. Conventionally, the microbial community is at a loss due to their inefficient removal capacity owing to their low abundance in treatment systems. To overcome this problem, different species of bacteria have evolved a strategy of forming biofilms by secreting extracellular polysaccharide (EPS) material, thus enabling them to reside in symbiotic harmony in a suitable microenvironment, which is under the regulation of certain signaling molecules also called quorum-sensing molecules or autoinducers (Donato et al. 2016; Mangwani et al. 2016). At the same time, the concentration of various groups of bacteria with diverse metabolic capacities at a single location provides the community the ability to metabolize different pollutants which was otherwise not possible for single bacterial species (Burmolle et al. 2014). A recent review has pointed out the limitations in efficient treatment of industrial wastewater carrying recalcitrant compounds such as persistent organic pollutants (POPs) and heavy metals (Edwards and Kjellerup 2013). Researchers have suggested that biofilm formation ability provided to the participating microorganisms an advantage of efficient nutrient and genetic exchange in addition to protection from surrounding environment, thereby aiding in the improvement of bioremediation potential for a variety of priority pollutants ranging from hydrocarbons to pharmaceutical and personal care products (PPCPs) (Burmolle et al. 2014). The role of EPS material in bioprocess involving heavy metal uptake and resistance was demonstrated in *Rhizobia*, which was also shown to exert a protective effect on non-EPS-producing heavy metal-sensitive strains (Nocelli et al. 2016). However, EPS secretion and subsequent biofilm formation itself does not solely guarantee efficient bioremediation. In addition to these conditions, there is a need for the presence of particular stresses which induce specific enzymatic pathways for the degradation of the pollutants. One such stress factor was suggested to be benzoate, which was essential for inducing genes of catechol 1,2-dioxygenase pathways in biofilm-forming *P. mendocina* EGD-AQ5 (Ghosh et al. 2017). Current advances in omics and mathematical modeling tools have further advanced our knowledge about the structural and functional aspects of the biofilm community (Turki et al. 2017). Random transposon mutagenesis library screening in *Shewanella oneidensis* was applied for getting a hyperadherent mutant strain which was shown to possess enhanced cohesiveness and higher bioremediation of Cr (VI) which was linked to the disruption of putrescine biosynthesis gene (Ding et al. 2014). One of the latest developments in the field of biofilms has been their application in bio-nanotechnology specifically with reference to the inhibition of biofilm formation and the significance of anti-biofilm molecules in the disruption of biofilms in drinking water and wastewater treatment systems (Kaur et al. 2014; Qureshi et al. 2015; Pal et al. 2016; Ahiwale et al. 2017). Certain molecules have been held responsible for the inhibition of biofilm formation and disruption of biofilms, which has been attributed to the quenching of quorum-sensing signals from bacterial population (Shang et al. 2014; Koul et al. 2016; Koul and Kalia 2017). Further prospects of biofilms have been envisaged for the recovery of precious metals from industrial wastes through reductive immobilization (Ng et al. 2015).

1.3.2 Bioremediation of Phosphorus Through Bioaccumulation

Among the different contaminants frequently encountered in our ecosystem, phosphorus (P) occupies an important position considering that it has both nutrient value and nuisance value. However, the nutrient value of phosphorus has been overshadowed by its nuisance value in domestic wastewater owing to its eutrophication potential which leads to depletion of oxygen and dead-zone formation in water bodies, thereby leading to loss of biodiversity due to death of normal flora and fauna. Numerous methods have been devised for phosphorus removal from wastewaters. One such method employed hydrated cement-based materials resulting in greater than 94% P removal at 20–1000 mg/L (Wang et al. 2014). Most commonly used technology for phosphorus sequestration in conventional wastewater treatment systems has been enhanced biological phosphorus removal system (EBPR), wherein for the first time, metabolic models were applied for describing the roles of polyphosphate-accumulating and glycogen-accumulating organisms (PAOs and GAOs) in biotransformation taking place in EBPR (Lanham et al. 2014). EBPR technology was further integrated with microalgal technology resulting in a modified EBP2R system which was effective in sequestering 75% of influent P such that the resultant effluent had required ratios of nitrogen (N) and P suitable for microalgal growth (Valverde-Pérez et al. 2015). Another innovative approach has been applied for phosphorus removal from wastewater and sludge, which involved the use of microalgae in a solar-driven P recycle, thus providing dual benefit of P removal from wastewater and generation of P-enriched algal biomass for application as soil fertilizer (Solovchenko et al. 2016; Yewalkar-Kulkarni et al. 2017). Genomic tools based on high throughput sequencing revealed the community structures in denitrifying and EBPR systems operating under different conditions which indicated the existence of highly diverse bacterial community in the reactor operated under alternating anaerobic-oxic conditions than in the reactor operated under anaerobic-anoxic conditions (Lv et al. 2014).

1.3.3 Microbial Enzymatic Pathways in Bioremediation

Several enzymatic pathways have been suggested to be responsible for performing the bioprocesses involved in bioremediation of both natural and anthropogenic compounds. Prominent among them which have been extensively explored for bioremediation of several pollutants involve reductases (chromium), dioxygenases and dehydrogenases (hydrocarbons), dioxygenases and hydrolases (PCBs), and oxidases, dehydrogenases, lyases, and reductases (amino acids, amines, nitriles, amides, and nitroaromatics) (Foster et al. 2014; Fuentes et al. 2014; Passatore et al. 2014; Thatoi et al. 2014; Bastida et al. 2016; Sharma et al. 2016; Kumari et al. 2017). In addition to the bacterial enzymatic pathways, numerous enzymes have been employed by fungi for bioremediation of recalcitrant compounds, prominent being the ones catalyzed by catalases, laccases, peroxidases, and cytochrome P450 (Deshmukh et al. 2016). Of the various pollutants which are a cause of concern to the environment, hydrocarbon contamination of ecological niches on account of oil spills (both terrestrial and marine) seems to have the most profound impact. Mesocosm studies with natural seawater spiked with up to 1000 ppm crude oil when subjected to biostimulation and bioaugmentation assays indicated that the total microbial abundance increased with inhibition of certain enzymes such as leucine aminopeptidase (LAP). The positive effect of bioaugmentation was clearly visible from the higher degradation percentage (95%) in comparison to biostimulation (Hassanshahian et al. 2014). The study highlighted the need for optimizing oil spill bioremediation bioprocesses by augmenting with suitable microbial cultures before the developed methodologies are applied to real-life scenarios. Studies on microbial community dynamics under influence of oil spills are extremely essential for providing an in-depth insight into the response and adaptation strategies of different microorganisms under such circumstances especially with reference to metabolic networks, mechanisms for tolerance/resistance, etc. (Fuentes et al. 2014; Bastida et al. 2016; Scoma et al. 2016). Modeling approach can accentuate the outcome of such bioremediation studies by answering open-ended questions with reference to assembly, functional diversity, abundances of microbial communities, mass flow and regulation of fluxes among species, applicability of generic rules for oil degradation, and possibility for engineering oil-degrading microbial communities (Rolling and van Bodegom 2014). Newer innovations in the methods for in vitro and field assessment of pollutants can further be of help in monitoring the biodegradation of such pollutants as reviewed by Kohli et al. (2017) in the application of compound-specific stable isotope method of analysis of hexachlorocyclohexane (HCH).

1.4 Bioprocess Optimization in Solid Waste Management

As has been discussed by numerous researchers, the solid waste generated from urban and rural areas is increasing at an alarming pace. Current means of disposal of solid waste in developing countries like India is dominated by open site dumping in landfills. With land coming at a premium, managing the solid waste especially the organic fraction in a decentralized manner can provide the benefit of reduction in the quantum of degradable waste reaching the landfill site. This has been shown to result in subsequent reduction in the environmental problems associated with leachate and the obnoxious odor and flammable gases generated during degradation of organic waste. Pandit et al. (2015) have reviewed the multiple means of disposal available for solid waste along with the limitations of such technologies and have suggested the option of renewable energy generation from the organic fraction of solid waste to be the most organized and attractive one. This can provide dual benefit of managing the organic fraction of solid waste in addition to the generation of biofuel in the form of methane.

Among the different components of solid waste generated in urban areas, waste-activated sludge (WAS) refers to the excess biomass generated from the aerobic treatment system of wastewater/sewage treatment plants (WWTPs/STPs) at the expense of nutrients/organics present in wastewater/sewage. The quantum of WAS generated is very large and presents a serious problem considering the land requirement for safe disposal. An estimated 11.2 million tons of WAS is reported to be generated in China annually, while the same was over 10 million tons in the European Union (EU), and its treatment accounted for nearly 50–60% of the maintenance costs in treatment plants (Sun et al. 2015). Various treatment strategies have been devised for managing WAS, which holds tremendous potential for value addition through energy generation in addition to the benefit of reduction in volume of sludge to be disposed. Cavinato et al. (2014) highlighted the economic importance of managing WAS to WWTPs through anaerobic co-digestion with other organic substrates under mesophilic, thermophilic conditions for the recovery of biogas and solid residue with fertilizer value. Subjecting the sludge to degradation in a microbial electrolysis cell was also shown to hold promise for volume reduction along with biomethane generation (Sun et al. 2015). Improvements in efficiency of the above environmental bioprocesses have been made through optimization of operating conditions and reactor designs. Leite et al. (2016) evaluated the performance of single- and two-phase reactors for anaerobic digestion of excess sludge under thermophilic conditions and observed a 15% gain in energy generation in two-phase systems over the single-phase ones. Further, the energetic sustainability of the process was established by the authors as the heat generated from such processes satisfied all the parameters for sustainability.

Though there is scope for converting food/kitchen and vegetable waste components of organic solid waste into biomethane, the anaerobic bioprocess involved needs to be optimized for achieving maximum efficiency. This can be achieved by an integrated approach consisting of information about different global nutrient recycle pathways for organic substrates such as cellulosic material, and the optimization of microbial bioprocesses through the use of novel reactor systems and analysis of the associated microbial community networks (Purohit et al. 2016). Khardenavis et al. (2013) demonstrated the effective conversion of food and vegetable waste into biogas through a hybrid anaerobic solid-liquid (HASL) reactor. Pandit et al. (2016) demonstrated that the efficiency of anaerobic bioprocesses was not only linked to the type of reactor system used but was also dependent on the substrate-specific differential enrichment of microbial community. Metagenomic tools were used by the authors to establish the dominance of lingo-hemicellulose degradation pathways and the abundance of genes for aromatic degradation, xylanase, and xylosidase in the rice straw anaerobic digestion metagenome owing to relative recalcitrance of rice straw in comparison to food and vegetable wastes. The performance of anaerobic digestion was further refined to achieve higher efficiency and optimum carbon mass balance during the treatment of vegetable market waste in a novel anaerobic baffled reactor (ABR) (Gulhane et al. 2016). This reactor system provided the advantage of superior performance which was attributed to the high stability to rapid changes in organic and hydraulic shock loads, biomass retention, and segregation of acidogenic and methanogenic bacterial communities along the length of the reactor. The authors showed that fine-tuning of operational condition in the form of recycling of effluent with feed was not only aimed at providing a method for maintaining alkalinity to neutralize the rapidly acidifying feed slurry but was also aimed at reducing the amount of effluent generated at the end of anaerobic digestion process. In yet another report, the authors were further able to establish the plasticity of microbial communities in different chambers under different operating conditions through the application of metagenomic tools (Gulhane et al. 2017). The study indicated that the segregation of the microorganisms in different chambers of ABR occurred based on their functional attributes in the methanogenic pathway. Such studies can go a long way in achieving the objectives of sustainability and eco-efficiency in industrial and environmental bioprocesses.

1.5 Perspectives

Conventional tools, which include RSM, and novel tools developed recently such as laser stimulation of plants in addition to those based on modeling approach (MuSIC framework) have helped in plugging the gaps in achieving the target of higher efficiencies in environmental bioprocesses. The combination of knowledge about the biogeochemical cycles operating in the different ecological niches and the underlying biological processes operating in such systems is essential for formulating mechanisms in order to achieve higher sustainability and profitability. This can be achieved in different environmental bioprocesses such as those operating in microbial biofilms and in anaerobic digestion, through detailed study about the enzymatic pathways involved in bioremediation and through information on the value addition which can be achieved from the recovery of energy in the form of biofuel.

Acknowledgment The authors are thankful to the Director of CSIR-National Environmental Engineering Research Institute Nagpur, and the Director of CSIR-Institute of Genomics and Integrative Biology New Delhi, for providing the necessary facilities for this work [KRC Manuscript No.: CSIR-NEERI/KRC/2017/JULY/EBGD/12].

References

Agarwal L, Qureshi A, Kalia VC, Kapley A, Purohit HJ, Singh RN (2014) Arid ecosystem: future option for carbon sinks using microbial community intelligence. Curr Sci 106:1357–1363

Ahiwale SS, Bankar AV, Tagunde S, Kapadnis BP (2017) A bacteriophage mediated gold nanoparticles synthesis and their anti-biofilm activity. Indian J Microbiol 57:188–194. doi:10.1007/s12088-017-0640-x

Bastida F, Jehmlich N, Lima K, Morris BEL, Richnow HH, Hernández T, von Bergen M (2016) The ecological and physiological responses of the microbial community from a semiarid soil to hydrocarbon contamination and its bioremediation using compost amendment. J Proteome 135:162–169. doi:10.1016/j.jprot.2015.07.023

Bennett EM, Cramer W, Ai B, Cundill G, Díaz S, Egoh BN, Geijzendorffer IR, Krug CB, Lavorel S, Lazos E (2015) Linking biodiversity, ecosystem services, and human well-being: three challenges for designing research for sustainability. Curr Opin Environ Sustain 14:76–85. doi:10.1016/j.cosust.2015.03.007

Burmolle M, Ren D, Bjarnsholt T, Sorensen SJ (2014) Interactions in multispecies biofilms: do they actually matter? Trends Microbiol 22:84–91. doi:10.1016/j.tim.2013.12.004

Cavinato C, Da Ros C, Pavan P, Cecchi F, Bolzonella D (2014) Treatment of waste activated sludge together with agrowaste by anaerobic digestion: focus on effluent quality. Water Sci Technol 69:525–531. doi:10.2166/wst.2013.736

Deshmukh R, Khardenavis AA, Purohit HJ (2016) Diverse metabolic capacities of fungi for bioremediation. Indian J Microbiol 56:247–264. doi:10.1007/s12088-016-0584-6

Ding Y, Peng N, Du Y, Ji L, Cao B (2014) Disruption of putrescine biosynthesis in *Shewanella oneidensis* enhances biofilm cohesiveness and performance in Cr (VI) immobilization. Appl Environ Microbiol 80:1498–1506. doi:10.1128/AEM.03461-13

Dobrowolski JW, Śliwka M, Mazur R (2012) Laser biotechnology for more efficient bioremediation, protection of aquatic ecosystems and reclamation of contaminated areas. J Chem Technol Biotechnol 87:1354–1359. doi:10.1002/jctb.3798

Donato PD, Poli A, Taurisano V, Abbamondi GR, Nicolaus B, Tommonaro G (2016) Recent advances in the study of marine microbial biofilm: from the involvement of quorum sensing in its production up to biotechnological application of the polysaccharide fractions. J Mar Sci Eng 4:34. doi:10.3390/jmse4020034

Edwards SJ, Kjellerup BV (2013) Applications of biofilms in bioremediation and biotransformation of persistent organic pollutants, pharmaceuticals/personal care products, and heavy metals. Appl Microbiol Biotechnol 97:9909–9921. doi:10.1007/s00253-013-5216-z

Ferreira-Leitão VS, Cammarota MC, Aguieiras ECG, de Sá LRV, Fernandez-Lafuente R, Freire DMG (2017) The protagonism of biocatalysis in green chemistry and its environmental benefits. Catalysts 7:9. doi:10.3390/catal7010009

Foster A, Barnes N, Speight R, Keane MA (2014) The repertoire of nitrogen assimilation in *Rhodococcus*: catalysis, pathways and relevance in biotechnology and bioremediation. J Chem Technol Biotechnol 89:787–802. doi:10.1002/jctb.4321

Fuentes S, Méndez V, Aguila P, Seeger M (2014) Bioremediation of petroleum hydrocarbons: catabolic genes, microbial communities, and applications. Appl Microbiol Biotechnol 98:4781–4794. doi:10.1007/s00253-014-5684-9

Ghosh S, Qureshi A, Purohit HJ (2017) Enhanced expression of catechol 1,2 dioxygenase gene in biofilm forming *Pseudomonas mendocina* EGD-AQ5 under increasing benzoate stress. Int Biodeterior Biodegrad 118:57–65. doi:10.1016/j.ibiod.2017.01.019

Gulhane M, Khardenavis AA, Karia S, Pandit P, Kanade GS, Lokhande S, Vaidya A, Purohit HJ (2016) Biomethanation of vegetable market waste in an anaerobic reactor: effect of effluent recirculation and carbon mass balance analysis. Bioresour Technol 215:100–109. doi:10.1016/j.biortech.2016.04.039

Gulhane S, Pandit P, Khardenavis AA, Purohit HJ (2017) Study of microbial community plasticity for anaerobic digestion of vegetable waste in plug flow bioreactor. Renew Energy 101:59–66. doi:10.1016/j.renene.2016.08.021

Gupta V, Capalash N, Gupta N, Sharma P (2017) Bio-prospecting laccases in the bacterial diversity of activated sludge from pulp and paper industry. Indian J Microbiol 57:75–82. doi:10.1007/s12088-016-0624-2

Gurme ST, Surwase SN, Patil SA, Jadhav SB, Jadhav JP (2013) Optimization of biotransformation of l-Tyrosine to ***l***-DOPA by *Yarrowia lipolytica*-NCIM 3472 using response surface methodology. Indian J Microbiol 53:194–198. doi:10.1007/s12088-012-0346-z

Hassanshahian M, Emtiazi G, Caruso G, Cappello S (2014) Bioremediation (bioaugmentation/biostimulation) trials of oil polluted seawater: a mesocosm simulation study. Mar Environ Res 95:28–38. https://doi.org/10.1016/j.marenvres.2013.12.010. http://instituteforenergyresearch.org/topics/encyclopedia /fossil-fuels/

Jiang X, Jiao N (2016) Vertical distribution of bacterial communities in the Indian Ocean as revealed by analyses of 16S rRNA and *nasA* genes. Indian J Microbiol 56:309–317. doi:10.1007/s12088-016-0585-5

Kalia VC (2015) Microbes: factories for bioproducts. In: Kalia VC (ed) Microbial factories biofuels, waste treatment, vol 1. Springer, New Delhi, pp 1–5. doi:10.1007/978-81-322-2598-0_1

Kalia VC, Prakash J, Koul S (2016) Biorefinery for glycerol rich biodiesel industry waste. Indian J Microbiol 56:113–125. doi:10.1007/s12088-016-0583-7

Kaur J, Niharika N, Lal R (2014) Biofilms: united we stand, divided we fall. Indian J Microbiol 54:246–247. doi:10.1007/s12088-014-0460-1

Khardenavis AA, Wang JY, Ng WJ, Purohit HJ (2013) Management of various organic fractions of municipal solid waste via recourse to VFA and biogas generation. Environ Technol 34:2085–2097. doi:10.1080/09593330.2013.817446

Kohli P, Richnow HH, Lal R (2017) Compound-specific stable isotope analysis: implications in Hexachlorocyclohexane in-vitro and field assessment. Indian J Microbiol 57:11–22. doi:10.1007/s12088-016-0630-4

Koul S, Kalia VC (2017) Multiplicity of quorum quenching enzymes: a potential mechanism to limit quorum sensing bacterial population. Indian J Microbiol 57:100–108. doi:10.1007/s12088-016-0633-1

Koul S, Prakash J, Mishra A, Kalia VC (2016) Potential emergence of multi-quorum sensing inhibitor resistant (MQSIR) bacteria. Indian J Microbiol 56:1–18. doi:10.1007/s12088-015-0558-0

Kuhad RC (2012) Microbes and their role in sustainable development. Indian J Micorbiol 52:309–313. doi:10.1007/s12088-012-0267-x

Kumar RR, Prasad S (2011) Metabolic engineering of bacteria. Indian J Microbiol 51:403–409. doi:10.1007/s12088-011-0172-8

Kumar P, Pant DC, Mehariya S, Sharma R, Kansal A, Kalia VC (2014) Ecobiotechnological strategy to enhance efficiency of bioconversion of wastes into hydrogen and methane. Indian J Microbiol 54:262–267. doi:10.1007/s12088-014-0467-7

Kumari S, Regar RK, Bajaj A, Ch R, Satyanarayana GNV, Mudiam MKR, Manickam N (2017) Simultaneous biodegradation of polyaromatic hydrocarbons by a *Stenotrophomonas* sp: characterization of *nid* genes and effect of surfactants on degradation. Indian J Microbiol 57:60–67. doi:10.1007/s12088-016-0612-6

Lanham AB, Oehmen A, Saunders AM, Carvalho G, Nielsen PH, Reis MAM (2014) Metabolic modelling of full-scale enhanced biological phosphorus removal sludge. Water Res 66:283–295. doi:10.1016/j.watres.2014.08.036

Leite WRM, Gottardo M, Pavan P, Filho PB, Bolzonella D (2016) Performance and energy aspects of single and two phase thermophilic anaerobic digestion of waste activated sludge. Renew Energy 86:1324–1331. doi:10.1016/j.renene.2015.09.069

Leong YK, Show PL, Lin HC, Chang CK, Loh HS, Lan JCW, Ling TC (2016) Preliminary integrated economic and environmental analysis of polyhydroxyalkanoates (PHAs) biosynthesis. Bioresources Bioprocess 3:41. doi:10.1186/s40643-016-0120-x

Liu J, Mooney H, Hull V, Davis SJ, Gaskell J, Hertel T, Lubchenco J, Seto KC, Gleick P, Kremen C, Li S (2015) Systems integration for global sustainability. Science 347, 1258832. doi:10.1126/science.1258832

Lv X-M, Shao M-F, Li C-L, Li J, Gao X-L, Sun F-Y (2014) A comparative study of the bacterial community in denitrifying and traditional enhanced biological phosphorus removal processes. Microb Environ 29:261–268. doi:10.1264/jsme2.ME13132

Mangwani N, Kumari S, Das S (2016) Bacterial biofilms and quorum sensing: fidelity in bioremediation technology. Biotechnol Genet Eng Rev 32:43–73. doi:10.1080/02648725.2016.1196554

Nair NR, Sekhar VC, Nampoothiri KM (2016) Augmentation of a microbial consortium for enhanced polylactide (PLA) degradation. Indian J Microbiol 56:59–63. doi:10.1007/s12088-015-0559-z

Ng CK, Mohanty A, Cao B (2015) Biofilms in bio-nanotechnology: opportunities and challenges. In: Singh OV (ed) Bio-nanoparticles: biosynthesis and sustainable biotechnological implications, 1st edn. Wiley, New York

Nigam VK, Arfi T, Kumar V, Shukla P (2017) Bioengineering of nitrilases towards its use as green catalyst: applications and perspectives. Indian J Microbiol 57:131–138. doi:10.1007/s12088-017-0645-5

Nocelli N, Bogino PC, Banchio E, Giordano W (2016) Roles of extracellular polysaccharides and biofilm formation in heavy metal resistance of Rhizobia. Materials 9:418. doi:10.3390/ma9060418

Pal S, Qureshi A, Purohit HJ (2016) Antibiofilm activity of biomolecules: gene expression study of bacterial isolates from brackish and fresh water biofouled membranes. Biologia 71:239–246. doi:10.1515/biolog-2016-0045

Pandit PD, Gulhane MK, Khardenavis AA, Vaidya AN (2015) Technological advances for treating municipal waste. In: Kalia VC (ed) Microbial factories biofuels, waste treatment, vol 1. Springer, New Delhi, pp 217–229. doi:10.1007/978-81-322-2598-0_13

Pandit PD, Gulhane MK, Khardenavis AA, Purohit HJ (2016) Mining of hemicellulose and lignin degrading genes from differentially enriched methane producing microbial community. Bioresour Technol 216:923–930. doi:10.1016/j.biortech.2016.06.021

Passatore L, Rossetti S, Juwarkar AA, Massacci A (2014) Phytoremediation and bioremediation of polychlorinated biphenyls (PCBs): state of knowledge and research perspectives. J Hazard Mater 278:189–202. doi:10.1016/j.jhazmat.2014.05.051

Patel SKS, Lee JK, Kalia VC (2016) Integrative approach for producing hydrogen and polyhydroxyalkanoate from mixed wastes of biological origin. Indian J Microbiol 56:293–300. doi:10.1007/s12088-016-0595-3

Patel SKS, Lee JK, Kalia VC (2017) Dark-fermentative biological hydrogen production from mixed biowastes using defined mixed cultures. Indian J Microbiol 57:171–176. doi:10.1007/s12088-017-0643-7

Purohit HJ, Kapley A, Khardenavis A, Qureshi A, Dafale NA (2016) Insights in waste management bioprocesses using genomic tools. In: Sariaslani S, Gadd GM (eds) Adv Appl Microbiol 97:121–170. https://doi.org/10.1016/bs.aambs.2016.09.002

Qureshi A, Pal S, Ghosh S, Kapley A, Purohit HJ (2015) Antibiofouling biomaterials. Int J Recent Adv Multidiscip Res (IJRAMR) 2:677–684

Ray S, Kalia VC (2017) Biomedical applications of polyhydroxyalkanoates. Indian J Microbiol 57(3):261–269. https://doi.org/10.1007/s12088-017-0651-7

Roling WFM, van Bodegom PM (2014) Toward quantitative understanding on microbial community structure and functioning: a modeling-centered approach using degradation of marine oil spills as example. Front Microbiol 5:Article 125

Scoma A, Barbato M, Hernandez-Sanabria E, Mapelli F, Daffonchio D, Borin S, Boon N (2016) Microbial oil-degradation under mild hydrostatic pressure (10MPa): which pathways are impacted in piezosensitive hydrocarbonoclastic bacteria? Sci Rep 6:23526. doi:10.1038/srep23526

Shang Z, Wang H, Zhou S, Chu W (2014) Characterization of *N*-acyl-homoserine lactones (AHLs)-deficient clinical isolates of *Pseudomonas aeruginosa*. Indian J Microbiol 54:158–162. doi:10.1007/s12088-014-0449-9

Sharma S, Singh DK (2017) Temporal variations in diazotrophic communities and nifH transcripts level across the agricultural and fallow land at Jaipur, Rajasthan, India. Indian J Microbiol 57:92–99. doi:10.1007/s12088-016-0634-0

Sharma NK, Thakur S, Thakur N, Savitri BTC (2016) Thermostable xanthine oxidase activity from *Bacillus pumilus* RL-2d isolated from Manikaran thermal spring: production and characterization. Indian J Microbiol 56:88–98. doi:10.1007/s12088-015-0547-3

Soares LCSR, Chandel AK, Pagnocca FC, Gaikwad SC, Rai M, da Silva SS (2016) Screening of yeasts for selection of potential strains and their utilization for in situ microbial detoxification (ISMD) of sugarcane bagasse hemicellulosic hydrolysate. Indian J Microbiol 56:172–181. doi:10.1007/s12088-016-0573-9

Solovchenko A, Verschoor AM, Jablonowski ND, Nedbal L (2016) Phosphorus from wastewater to crops: an alternative path involving microalgae. Biotech Adv 34:550–564. doi:10.1016/j.biotechadv.2016.01.002

Sun R, Zhou A, Jia J, Liang Q, Liu Q, Xing D, Ren N (2015) Characterization of methane production and microbial community shifts during waste activated sludge degradation in microbial electrolysis cells. Bioresour Technol 175:68–74. doi:10.1016/j.biortech.2014.10.052

Szajsner H, Koszelnik-Leszek A, Nowosad K (2014) Influence of laser light on the germination parameters and morphological characteristics of seedlings of selected *Silene vulgaris* ecotypes. Pol J Natur Sc 29:5–16

Thatoi H, Das S, Mishra J, Rath BP, Das N (2014) Bacterial chromate reductase, a potential enzyme for bioremediation of hexavalent chromium: a review. J Environ Manag 146:383–399. doi:10.1016/j.jenvman.2014.07.014

Turki Y, Mehri I, Lajnef R, Rejab AB, Khessairi A, Cherif H, Ouzari H, Hassen A (2017) Biofilms in bioremediation and wastewater treatment: characterization of bacterial community structure and diversity during seasons in municipal wastewater treatment process. Environ Sci Pollut Res 24:3519–3530. doi:10.1007/s11356-016-8090-2

Tursunov O, Dobrowolski J (2015) Addressing environmental issues and risks in Uzbekistan. Integer J Engg Res Tech 2:62–69

Valverde-Pérez B, Ramin E, Smets BF, Plósz BG (2015) EBP2R – an innovative enhanced biological nutrient recovery activated sludge system to produce growth medium for green microalgae cultivation. Water Res 68:821–830. doi:10.1016/j.watres.2014.09.027

Wang X, Chen J, Kong Y, Shi X (2014) Sequestration of phosphorus from wastewater by cement-based or alternative cementitious materials. Water Res 62:88–96. doi:10.1016/j.watres.2014.05.021

Yewalkar-Kulkarni S, Gera G, Nene S, Pandare K, Kulkarni B, Kamble S (2017) Exploiting phosphate-starved cells of *Scenedesmus* sp. for the treatment of raw sewage. Indian J Microbiol 57:241–249. doi:10.1007/s12088-016-0626-0

Zhuang KH, Herrgard MJ (2015) Multi-scale exploration of the technical, economic, and environmental dimensions of bio-based chemical production. Metab Eng 31:1–12. doi:10.1016/j.ymben.2015.05.007

2 Visualization of Microorganisms in Bioprocesses

Tsuyoshi Yamaguchi and Kengo Kubota

Abstract

Fluorescence in situ hybridization (FISH) is currently a standard technique for detecting, identifying, and enumerating microorganisms. The method allows for visualization of cell morphology as well as in situ localization of microorganisms at single-cell resolution. Although FISH has been used for more than 20 years, many challenges remain with regard to improving the technique for understanding microbial ecology and physiology. This chapter describes the recent developments in this field, such as improved sensitivity by catalyzed reporter deposition (CARD)-FISH and in situ DNA-hybridization chain reaction (HCR). Highly sensitive methods for detecting mRNA and/or functional genes are also described. Moreover, methods combining isotope probing to reveal microbial metabolic activities are introduced.

Keywords

Microorganisms • Ecology • Fluorescence • Metabolic activity • Phylogeny

T. Yamaguchi (✉)
Department of Civil and Environmental Engineering, National Institute of Technology, Matsue College, 14-4 Nishiikuma, Matsue, Shimane 690-8518, Japan
e-mail: yamaguchi@matsue-ct.jp

K. Kubota
Department of Civil and Environmental Engineering, Tohoku University, 6-06-6 Aoba, Aramaki, Aoba-ku, Sendai, Miyagi 980-8579, Japan

H.J. Purohit et al. (eds.), *Optimization and Applicability of Bioprocesses*,
https://doi.org/10.1007/978-981-10-6863-8_2

2.1 Introduction

Bioprocesses utilize microorganisms for waste and wastewater treatment, and understanding the microorganisms involved in a bioprocess is important for clarifying what occurs during the process. Conventional microbiological methods based on cultivation, morphological, metabolic, and genetic assays can provide information of microbial behavior in certain environments. However, most of the culture media used favor the growth of certain groups of microorganisms, and it is generally accepted that more than 99% of environmental microorganisms are yet-to-be cultured. These facts suggest that a majority of microorganisms that are important and necessary for establishing bioprocesses cannot be identified by culture-dependent methods.

With the development of molecular techniques, culture-independent approaches have been applied for understanding the microorganisms in complex microbial communities involved in bioprocesses. Indeed, the introduction of a ribosomal RNA (rRNA) approach expanded our view of environmental microorganisms, and it is currently recognized that the diversity of environmental microorganisms is much greater than previously recognized (Hug et al. 2016). Phylogenetic analysis based on the small subunit of ribosomal RNA (SSU rRNA) and/or on the SSU rRNA gene is a gold standard for describing a microbial community. These molecules have been selected because of the following reasons: (1) they are found in all living organisms; (2) they offer sufficient resolution for distinguishing from the domain/phylum level to the species level; (3) a high copy number of the transcripts is present (hundreds to thousands per cell); and (4) both conserved and variable regions exist (suitable for primer and probe design). Sequencing of SSU rRNA genes allows rapid identification of non-culturable microorganisms. However, SSU rRNA gene-based phylogenetic analysis is not capable of identifying the functional features of microorganisms, which are important for understanding bioprocesses.

The physiological properties of microorganisms are investigated by the sequencing functional genes (Hatamoto et al. 2011; Saito et al. 2015). Researchers are interested in determining the origin of genes, i.e., linking phylogeny and function, but this is generally a laborious task because there is typically a large distance between an rRNA gene and a functional gene in the genome. Recent advancements in sequencing technologies and bioinformatics have allowed for high-throughput and systematic analyses of microbial communities, resulting in new insight toward an understanding of microbial communities in bioprocesses (Nobu et al. 2014, 2015). The genome sequences of environmental microorganisms, including phyla involved in various bioprocesses (e.g., TM7, Caldiserica, OP9, Synergistetes, NKB19, Armatimonadetes, WS1, WWE1), have been reconstructed from metagenomic sequence datasets (Rinke et al. 2013). Although these sequencing data are informative, it is impossible to understand microorganism cell morphology and in situ localization, which is especially important for understanding biofilm ecosystems.

In this chapter, we describe a method for visualizing microorganisms in bioprocesses without cultivation at a single-cell resolution: fluorescence in situ hybridization (FISH). The first application of FISH was reported more than 20 years ago

(DeLong et al. 1989), ushering in a new era in environmental microbiology. After the proposal of a full-cycle rRNA approach (Amann et al. 1995), FISH became the essential tool for detecting, identifying, and enumerating microorganisms. To expand its applicability, FISH has since been further improved by many researchers. This technique is currently capable of detecting not only rRNA but also mRNA and functional genes on chromosomes. In addition, combining FISH with stable/radioactive isotope probing is a powerful tool for deciphering in situ physiology of microbes involved in bioprocesses.

2.2 Phylogenetic Identification of Microorganisms at Single-Cell Resolution

2.2.1 Fluorescence In Situ Hybridization

The detection, identification, and visualization of microbial cells are important for understanding microbial community structures in bioprocesses. FISH allows phylogenetic identification of microorganisms at a single-cell resolution. The principle of conventional FISH is shown in Fig. 2.1a. For the application of FISH, an oligonucleotide probe of approximately 16–25 bases targeting an rRNA is designed for the group of interests. Oligonucleotide probes can be found in the literature and databases (Loy et al. 2003) but can also be designed using such software as ARB (Ludwig et al. 2004). The hybridization efficiency of a designed oligonucleotide probe can be predicted using math FISH based on a thermodynamic model of hybridization (Yilmaz and Noguera 2004; Yilmaz et al. 2011).

The oligonucleotide probes used for FISH are labeled with a fluorescent dye such as FITC, Cy3, and Cy5. A series of Alexa Fluor dyes is also often used. To apply FISH, microorganisms are first fixed with either paraformaldehyde and/or ethanol, which permeabilizes the cells. Additional cell permeabilization processes are sometimes necessary for Gram-positive bacteria and some archaeal cells because they have thick cell walls. As there is no gold standard protocol for permeabilization of all microbial cells, cell treatment methods should be appropriately selected and optimized for the targets of interest. Enzymatic treatments are often employed; treatment with lysozyme is popular because it can digest bacterial peptidoglycan.

Hybridization of oligonucleotide probes is carried out in a buffer that typically contains 0.9 M NaCl, 20 mM Tris-HCl [pH 7–8], and 0.01% SDS (sodium dodecyl sulfate). Tris is used as the buffer, salt is added to increase the melting temperature (Tm) of oligonucleotides, and SDS is added to improve probe penetration and to make the target region of rRNA accessible for the probe (Behrens et al. 2003). In general, the hybridization is carried out in a humid chamber at 46 °C in the dark. To control for hybridization stringency, formamide is added to the buffer; in the case of a DNA-RNA hybrid, 1% formamide decreases the Tm by 0.5 °C. Because each probe has a different dissociation temperature, a different formamide concentration is added to the buffer to maintain specific hybridization. A blocking reagent is sometimes included in the buffer to prevent nonspecific oligonucleotide binding.

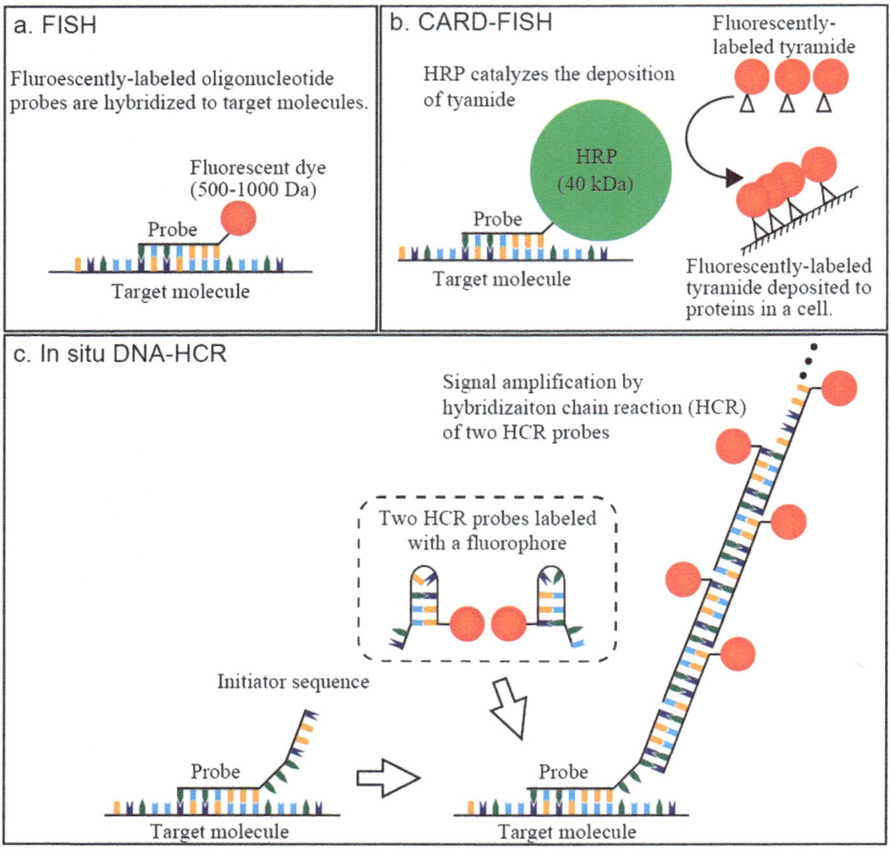

Fig. 2.1 Schematic diagrams of FISH (**a**), CARD-FISH (**b**), and in situ DNA-HCR (**c**)

The incubation time is an important factor for effective hybridization, and maximum hybridization efficiency can be obtained at different incubation times (Yilmaz et al. 2006). To accelerate hybridization of the oligonucleotides, the addition of formamide to the buffer is recommended (Yilmaz et al. 2006). Hybridization is generally carried out for more than 2 h. Excess probe is then removed by immersing the samples in a buffer at 48 °C for 15–20 min. The stringency of the washing buffer is adjusted by decreasing the salt concentration and not by adding formamide, using the formula 16.6 log M, where M is the concentration of NaCl (e.g., the addition of 10% formamide is equivalent to a 5 °C decrease in Tm, which corresponds to a decrease in NaCl concentration from 0.9 to 0.45 M). After counterstaining with DAPI to visualize all microorganisms, the samples are subjected to observation under an epifluorescence microscope.

To date, complex microbial communities in various bioprocesses have been analyzed by employing FISH. The well-established upflow anaerobic blanket (UASB) process is implemented in many industries to treat organic-rich wastewater. The formation of granular sludge in a UASB reactor allows separate control of hydraulic retention time and sludge retention time, and thus high organic loading can be achieved. Nevertheless, although many microorganisms should be involved in treatment of organic matter in an anaerobic environment (e.g., syntrophic associations), the microbial community structure of granular sludge is considered a black box. The research team of Sekiguchi attempted to clarify the microbial composition as well as the in situ localization of selected microorganisms by FISH and found that *Bacteria* were more abundant at the outer layer of a granule, whereas *Archaea* were more frequent in the inner portion (Sekiguchi et al. 1999). Syntrophs such as propionate-oxidizing bacteria were found close to hydrogenotrophic methanogens (Imachi et al. 2000). A member of the phylum *Nitrospirae* found in various UASB granular sludges was also detected (Iguchi et al. 2009). With an expectation of microbial metabolism and substrate diffusion to granular sludge, Sekiguchi proposed a model for predicting microbial function after localization of the microbial habitat by FISH (Sekiguchi 2006). The team has also intensively investigated the microbial dynamics of sludge bulking in UASB reactors. By applying FISH, these authors found that members of *Chloroflexi* or KSB3 are involved in sludge bulking (Sekiguchi et al. 2001; Yamada et al. 2005, 2007, 2011).

2.2.2 Sensitive FISH Techniques: CARD-FISH and In Situ DNA-HCR

One problem of FISH is the detection of microorganisms with low rRNA contents. Microorganisms in oligotrophic environments are often difficult to detect by FISH. Although microorganisms in bioprocess generally possess sufficient rRNA for FISH detection, more sensitive approaches are necessary in some cases. FISH combined with catalyzed reporter deposition (CARD), also known as tyramide signal amplification (TSA), was reported 20 years ago (Schönhuber et al. 1997). The principle of CARD-FISH is shown in Fig. 2.1b, and reviews have been published elsewhere (Amann and Fuchs 2008; Kubota 2013). In brief, instead of a fluorophore horseradish peroxidase (HRP) is conjugated to an oligonucleotide probe. HRP mediates the linkage of tyramide to cellular proteins in the presence of hydrogen peroxide. As a result, a large number of tyramide molecules are deposited around HRP. Fluorophores can also be conjugated with tyramide; thus, CARD-FISH is reported to show 20–40 times greater signal intensity than FISH (Schönhuber et al. 1997; Hoshino et al. 2008). Application of CARD-FISH to bioprocess samples has been reported, e.g., the detection of *Thiobacillus denitrificans* in a UASB reactor (Fernández et al. 2006, 2008).

Despite the high sensitivity of CARD-FISH, there are some drawbacks for its application. Endogenous peroxidase activities need to be inactivated prior to application of CARD-FISH, as they may cause false-positive signals. However, inactivation of these activities is not simple because a high concentration of hydrogen peroxide, which is effective for inactivation, also causes nucleic acid degradation, resulting in a lack of signal. Thus, the concentration is generally kept low (approximately 0.1–0.5%) (Ishii et al. 2004), but in some cases, it has to be much higher, e.g., 30% for anammox bacteria in a bioreactor (Pavlekovic et al. 2009). Another drawback of CARD-FISH is penetration of the HRP-labeled probe due to a large HRP molecule (approximately 40 kDa) (Thiele et al. 2011). Therefore, permeabilization steps are generally required for CARD-FISH application (Kubota 2013). Many researchers use enzymatic cell treatment such as lysozyme or proteinase K. Lysozyme treatment is the permeabilization most often used, as it can digest bacterial peptidoglycan. In contrast, archaeal cells have a completely different cell wall structure, and lysozyme treatment is not effective. Instead, proteinase K is used for permeabilizing archaea. PeiW treatment is also effective for the detection of methanogens, which are important components in anaerobic bioprocesses (Kubota et al. 2008).

Recently, a novel sensitive FISH, in situ DNA-hybridization chain reaction (HCR), was reported as overcoming the problems of CARD-FISH (Yamaguchi et al. 2015b) (Fig. 2.1c). First reported in 2004, HCR uses two stable hairpin oligonucleotides, which only assemble after exposure to an initiator oligonucleotide (Dirks and Pierce 2004). In situ DNA-HCR has two advantages over CARD-FISH. First one is that inactivation of endogenous peroxidase activity can be ignored as the method does not utilize any enzymes. Another advantage of in situ DNA-HCR is that penetration of the probes used is much easier than HRP-labeled probes. This is more effective for detection of *Archaea*. Yamaguchi et al. reported that the high penetrability of the probes used for in situ DNA-HCR into *Archaea*. As a result, greater detection rates were observed for anaerobic granular sludge and digester sludge samples compared to FISH and CARD-FISH (Fig. 2.2). An accelerated and intensified protocol of in situ DNA-HCR (quick HCR-FISH) was published very recently (Yamaguchi et al. 2015a). Although the signal intensity of in situ DNA-HCR is not as intensive as CARD-FISH, it has sufficient sensitivity to detect microorganisms with low rRNA contents in marine and sediment samples. Overall, in situ DNA-HCR can be a powerful tool for detecting environmental microorganisms and is an alternative to CARD-FISH.

2.3 Understanding the Physiology of Microorganisms at Single-Cell Resolution

2.3.1 Detection of mRNA and/or Genes Encoding Key Enzymes

Phylogenetic identification of microorganisms based on SSU rRNA gene sequence information does not allow for determination of physiological properties, whereas detection of mRNA and/or functional genes encoding key enzymes is useful (Table 2.1). Nevertheless, the copy numbers of mRNA and/or functional genes are low in a single microbial cell. Regarding the detection limit of FISH using fluorescently labeled oligonucleotide probes, approximately 1,400 target molecules are necessary (Hoshino et al. 2008). This detection limit can be further improved by CARD-FISH, though it is still as low as 30–60 target molecules in a single microbial cell. The mRNA copy number ranges from 10^0 to 10^2 copies per cell and that of

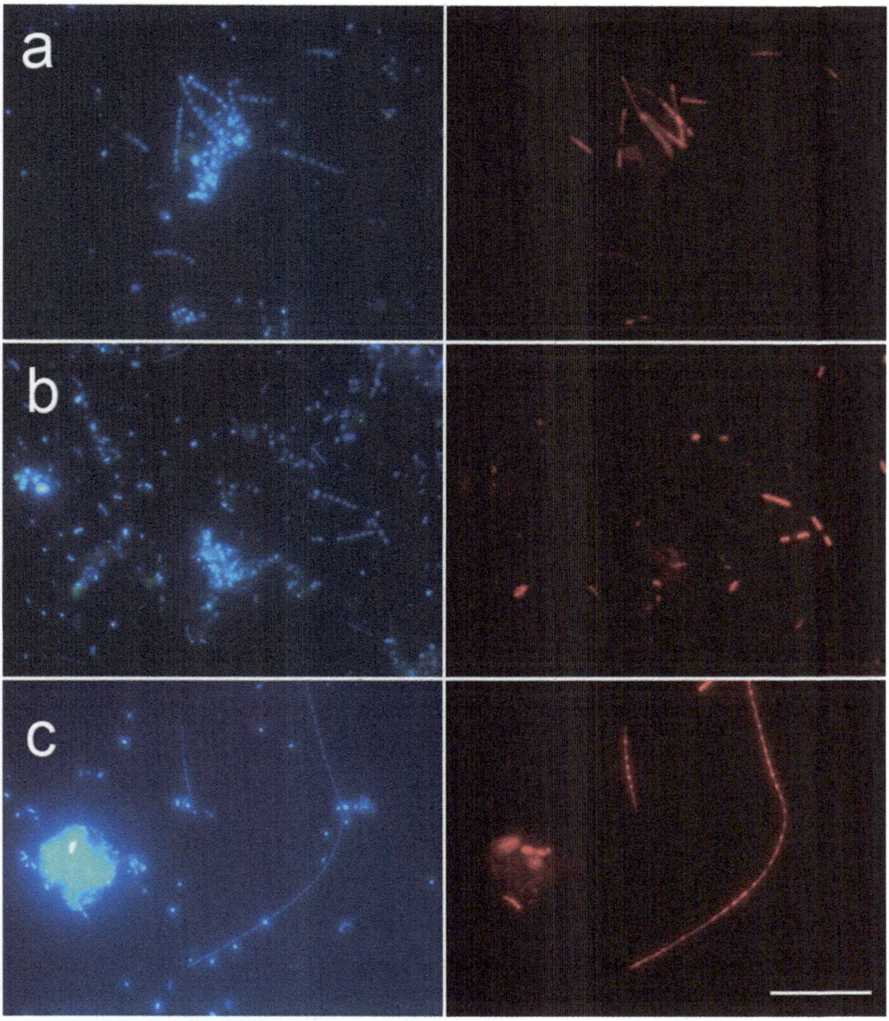

Fig. 2.2 Detection of *Archaea* in an anaerobic sludge sample using FISH (**a**), CARD-FISH (**b**), and in situ DNA-HCR (**c**). Photomicrographs of DAPI-stained cells (left panels) and epifluorescence (right panels) of the identical fields. Exposure times were adjusted for each method (CARD-FISH < in situ DNA-HCR < FISH). The bar represents 20 μm

functional genesis, generally less than 10 (many are single copy on the chromosome). Therefore, more sensitive methods are necessary for detection. Zwirglmaier et al. reported a method termed recognition of individual genes-FISH (RING-FISH) (Zwirglmaier et al. 2004). In RING-FISH, a large amount of RNA polynucleotide probes is incubated with fixed cells, during which the probes are hybridized to each other and generate a network by anchoring to a target gene on the chromosome. RING-FISH has been applied for detection of the *nirK* gene in an activated sludge sample (Pratscher et al. 2009). Another approach is two-pass TSA-FISH, which

Table 2.1 List of functional genes used for analyzing microbial community in bioprocess

Gene name		Target microorganisms	References used for sequencing-based analysis	References used for FISH analysis
mcrA	(Methyl coenzyme M reductase)	Methanogenic and anaerobic methane-oxidizing archaea	Steinberg and Regan (2008)	Kubota et al. (2006)
			Saito et al. (2015)	Kawakami et al. (2012)
amoA	(Ammonia monooxygenase)	Ammonia-oxidizing prokaryotes	Araki et al. (2004)	Hoshino et al. (2003)
			Limpiyakorn et al. (2010)	
pmoA	(Particulate methane monooxygenase)	Methanotrophic bacteria	Luesken et al. (2011)	–
			Hatamoto et al. (2011)	
nirK	(Copper nitrite reductase)	Denitrifying bacteria	You (2005)	Pratscher et al. (2009)
			Kim et al. (2011)	
			Herbert et al. (2014)	
nirS	(Cytochrome cd1 nitrite reductase)	Denitrifying bacteria	Kim et al. (2011)	Hoshino and Schramm (2010)
			Herbert et al. (2014)	Mota et al. (2012)
apsA	(Adenosine-5′-phosphosulfate reductase)	Sulfate-reducing and sulfur-oxidizing prokaryotes	Sumino et al. (2007)	Kawakami et al. (2012)
			Ben-Dov et al. (2007)	
dsrA	(Dissimilatory sulfate reductase)	Sulfate-reducing bacteria	Luo et al. (2011)	–
			Biswas et al. (2014)	

involves an additional amplification step in CARD-FISH (Kubota et al. 2006). The two-pass TSA-FISH method first involves a TSA reaction with DNP-labeled tyramide (Kubota et al. 2006). After HRP-labeled anti-DNP-antibody is immunologically bound to the deposited DNP molecules, a second TSA reaction with fluorophore-labeled tyramide was carried out. Using this approach, *mcrA* mRNA in methanogens was successfully detected, with strong and reliable signals (Fig. 2.3). Kawakami et al. later extended application of this method for detecting functional genes in granular sludge samples (Kawakami et al. 2010, 2012).

Linking physiology and phylogeny via simultaneous detection of mRNA/functional genes and rRNA is an approach for understating microbial ecosystems (Pernthaler and Amann 2004; Moraru et al. 2010). Combining mRNA detection with flow cytometry for cell sorting is another option. To reveal the diversity of denitrifying bacteria in a bioprocess, Mota et al. detected *nirK* mRNA followed by phylogenetic analysis of sorted cells (Mota et al. 2012).

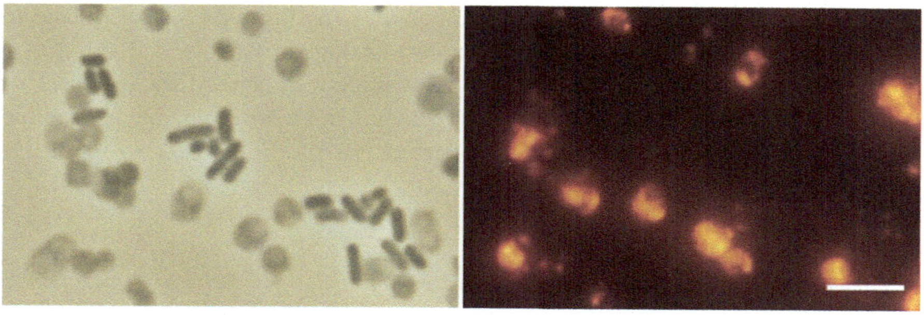

Fig. 2.3 Two-pass TSA-FISH detection of *mcrA* mRNA in *Methanococcus vannielii* in an artificial mixture with *Escherichia coli*. Photomicrographs of phase contrast (left panel) and epifluorescence (right panel) of the identical field. The bar represents 5 μm

2.3.2 Isotope Probing of Microorganisms: A Combination with Microautoradiography or NanoSIMS

FISH methods can be further combined with isotope probing, which can provide information about the metabolic activities of microbial cells. FISH combined with microautoradiography (MAR-FISH) has been successfully applied to identify the phylogeny and metabolic activities of microorganisms (Okabe et al. 2004). In MAR-FISH, a sample of interest is incubated with a radioactive isotope-labeled substrate such as ^{14}C, ^{3}H ^{35}S, or ^{33}P. Microorganisms assimilating the substrate can then be visualized after the formation of silver grains. At the same time, FISH is performed for phylogenetic identification, with both MAR and FISH signals observed under a microscope. MAR-FISH has been successfully applied to several bioprocess samples. Indeed, the polyphosphate-accumulating ability of *Tetrasphaera* in an activated sludge sample was confirmed using a^{33}P-labeled substrate (Nguyen et al. 2011). The ability of *Nitrospira* to take up pyruvate was also demonstrated by MAR-FISH (Daims et al. 2001). Moreover, ecophysiological interaction between nitrifiers and heterotrophic bacteria was investigated MAR-FISH (Kindaichi et al. 2004).

Because MAR-FISH uses radioactive isotopes, direct visualization of nitrogen is not possible. The recently introduced nanoscale secondary ion mass spectrometry (NanoSIMS) overcomes this problem because NanoSIMS has high spatial and mass resolutions, allowing subcellular-scale analysis and measurement of nitrogen as CN^-. However, as NanoSIMS is not equipped with a fluorescent microscope, phylogenetic staining of microorganisms for NanoSIMS analysis is achieved in different ways. As halogen atoms have a high ionization efficiency in NanoSIMS and are less abundant in biomass, microbial cells are labeled with halogen atoms such as iodine, fluorine, and bromine by CARD-FISH (Behrens et al. 2008; Musat et al. 2008). Indeed, fluorophores containing halogen atoms, such as 5-carboxy-2′,4′,5′,7′-tetrabromosulfone fluorescein (containing four bromine atoms), Oregon Green 488-X (containing two fluorine atoms), and BODIPY TMR-X (containing two fluorine atoms), are labeled with tyramide. During NanoSIMS analysis, these halogen atoms are simultaneously detected, and metabolic activities can then be linked to phylogeny. Gold can be used as an alternative to halogens. Kubota et al. developed Gold-ISH for phylogenetic identification compatible with NanoSIMS and successfully detected lactate-assimilating microorganisms in an anaerobic sludge sample under a sulfate-reducing condition (Kubota et al. 2014).

2.4 Conclusion

Microbial reactions in bioprocesses are still considered as a black box. However, accumulating fundamental knowledge, the basic questions of how the process can be maintained and what is the trigger for process deterioration can be answered. In situ visualization of microorganisms provides direct evidence for understanding microbial physiology and ecology. Although FISH has been applied to environmental microbiology for more than 25 years, it is still a leading-edge technique, and many researchers are working on improving this method. Indeed, microbial community analysis is changing with publication of an increasing number of papers related to "omics," with large amounts of data and informatics, thus providing new discoveries for understanding bioprocesses. Regardless, visualization of microorganisms at single-cell resolution is essential for elucidating the microbiological phenomena that occur in bioprocesses.

References

Amann RI, Fuchs BM (2008) Single-cell identification in microbial communities by improved fluorescence in situ hybridization techniques. Nat Rev Microbiol 6:339–348. https://doi.org/10.1038/nrmicro1888

Amann RI, Ludwing W, Schleifer KH (1995) Phylogenetic identification and in situ detection of individual microbial cells without cultivation. Microbiol Rev 59:143–169

Araki N, Yamaguchi T, Yamazaki S, Harada H (2004) Quantification of amoA gene abundance and their amoA mRNA levels in activated sludge by real-time PCR. Water Sci Technol 50(8):1–8

Behrens S, Fuchs BM, Mueller F, Amann RI (2003) Is the in situ accessibility of the 16S rRNA of *Escherichia coli* for Cy3-labeled oligonucleotide probes predicted by a three-dimensional structure model of the 30S ribosomal subunit? Appl Environ Microbiol 69:4935–4941. https://doi.org/10.1128/AEM.69.8.4935-4941.2003

Behrens S, Losekann T, Pett-Ridge J, Weber PK, Ng WO, Stevenson BS, Hutcheon ID, Relman DA, Spormann AM (2008) Linking microbial phylogeny to metabolic activity at the single-cell level by using enhanced element labeling-catalyzed reporter deposition fluorescence in situ hybridization (ELFISH) and nanoSIMS. Appl Environ Microbiol 74:3143–3150. https://doi.org/10.1128/AEM.00191-08

Ben-Dov E, Brenner A, Kushmaro A (2007) Quantification of sulphate reducing bacteria in industrial wastewater by real-time polymerase chain reaction (PCR) using *dsrA* and *apsA* genes. Microb Ecol 54:439–451. https://doi.org/10.1007/s00248-007-9233-2

Biswas K, Taylor NW, Jurner SJ (2014) *dsrAB*-based analysis of sulphate-reducing bacteria in moving bed biofilm reactor (MBBR) wastewater treatment plants. Environ Biotechnol 98:7211–7222. https://doi.org/10.1007/s00253-014-5769-5

Daims H, Nielsen JL, Nielsen PH, Schleifer KH, Wagner M (2001) In situ characterization of Nitrospira-like nitrite-oxidizing bacteria active in wastewater treatment plants. Appl Environ Microbiol 67:5273–5284. https://doi.org/10.1128/AEM.67.11.5273-5284.2001

DeLong EF, Wickhan GS, Pace NR (1989) Phylogenetic stains: ribosomal RNA-based probes for the identification of signal cells. Science 243:1360–1363. https://doi.org/10.1126/science.2466341

Dirks RM, Pierce NA (2004) Triggered amplification by hybridization chain reaction. Proc Natl Acad Sci U S A 101:15275–15288. https://doi.org/10.1073/pnas.0407024101

Fernández N, Gomez R, Amils R, Sierra-Alvarez R, Field JA, Sanz JL (2006) Microbiological and structural aspects of granular sludge from autotrophic denitrifying reactors. Water Sci Technol 54:11–17. https://doi.org/10.2166/wst.2006.480

Fernández N, Sierra-Alvarez R, Field JA, Amils R, Sanz JL (2008) Microbial community dynamics in a chemolithotrophic denitrification reactor inoculated with methanogenic granular sludge. Chemosphere 70:462–474. https://doi.org/10.1016/j.chemosphere.2007.06.062

Hatamoto M, Miyauchi T, Kindaichi T, Ozaki N, Ohashi A (2011) Dissolved methane oxidation and competition for oxygen in down-flow hanging sponge reactor for post-treatment of anaerobic wastewater treatment. Bioresour Technol 102:10299–10304. https://doi.org/10.1016/j.biortech.2011.08.099

Herbert RB, Winbjork H, Hellman M, Hallin S (2014) Nitrogen removal and spatial distribution of denitrifier and anammox communities in a bioreactor for mine drainage treatment. Water Res 66:350–360. https://doi.org/10.1016/j.watres.2014.08.038

Hoshino T, Schramm A (2010) Detection of denitrification genes by in situ rolling circle amplification-fluorescence in situ hybridization to link metabolic potential with identity inside bacterial cells. Environ Microbiol 12:2508–2517. https://doi.org/10.1111/j.1462-2920.2010.02224.x

Hoshino T, Tsuneda S, Hirata A, Inamori Y (2003) In situ PCR for visualizing distribution of a functional gene "amoA" in a biofilm regardless of activity. J Biotechnol 105:33–40. https://doi.org/10.1016/S0168-1656(03)00171-8

Hoshino T, Yilmaz LS, Noguera DR, Daims H, Wagner M (2008) Quantification of target molecules needed to detect microorganisms by fluorescence in situ hybridization (FISH) and catalyzed reporter deposition-FISH. Appl Environ Microbiol 74:5068–5077. https://doi.org/10.1128/AEM.00208-08

Hug LA, Baker BJ, Anantharaman K, Brown CT, Probst AJ, Castelle CJ, Butterfield CN, Hernsdorf AW, Amano Y, Ise K, Suzuki Y, Dudek N, Relman DA, Finstad KM, Amundson R, Thomas BC, Banfield JF (2016) A new view of the tree of life. Nat Microbiol 2016:16048. https://doi.org/10.1038/NMICROBIOL.2016.48

Iguchi A, Terada T, Narihiro T, Yamaguchi T, Kamagata Y, Sekiguchi Y (2009) In situ detection and quantification of uncultured members of the phylum *Nitrospirae* abundant in methanogenic wastewater treatment systems. Microbes Environ 24:97–104. https://doi.org/10.1264/jsme2.ME08562

Imachi H, Sekiguchi Y, Kamagata Y, Ohashi A, Harada H (2000) Cultivation and in situ detection of a thermophilic bacterium capable of oxidizing propionate in syntrophic association with hydrogenotrophic methanogens in a thermophilic methanogenic granular sludge. Appl Environ Microbiol 66:3608–3615. https://doi.org/10.1128/AEM.66.8.3608-3615.2000

Ishii K, Mußmann M, MacGregor BJ, Amann RI (2004) An improved fluorescence in situ hybridization protocol for the identification of bacteria and archaea in marine sediments. FEMS Microbiol Ecol 50:203–212. https://doi.org/10.1016/j.femsec.2004.06.015

Kawakami S, Kubota K, Imachi H, Yamaguchi T, Harada H, Ohashi A (2010) Detection of single copy genes by two-pass tyramide signal amplification fluorescence in situ hybridization (two-pass TSA-FISH) with single oligonucleotide probes. Microbes Environ 25:15–21. https://doi.org/10.1264/jsme2.ME09180

Kawakami S, Hasegawa T, Imachi H, Yamaguchi T, Harada H, Ohashi A, Kubota K (2012) Detection of single-copy functional genes in prokaryotic cells by two-pass TSA-FISH with polynucleotide probes. J Microbiol Methods 88:218–223. https://doi.org/10.1016/j.mimet.2011.11.014

Kim YM, Cho HU, Lee DS, Park D, Park JM (2011) Influence of operational parameters on nitrogen removal efficiency and microbial communities in a full-scale activated sludge process. Water Res 45:5785–5795. https://doi.org/10.1016/j.watres.2011.08.063

Kindaichi T, Ito T, Okabe S (2004) Eco-physiological interaction between nitrifying bacteria and heterotrophic bacteria in autotrophic nitrifying biofilms as determined by MAR-FISH. Appl Environ Microbiol 70:1641–1650. https://doi.org/10.1128/AEM.70.3.1641-1650.2004

Kubota K (2013) CARD-FISH for environmental microorganisms: technical advancement and future applications. Microbes Environ 28:3–12. https://doi.org/10.1264/jsme2.ME12107

Kubota K, Ohashi A, Imachi H, Harada H (2006) Visualization of *mcr* mRNA in a methanogen by fluorescence in situ hybridization with an oligonucleotide probe and two-pass tyramide signal amplification (two-pass TSA-FISH). J Microbiol Methods 66:521–528. https://doi.org/10.1016/j.mimet.2006.02.002

Kubota K, Imachi H, Kawakami S, Nakamura K, Harada H, Ohashi A (2008) Evaluation of enzymatic cell treatments for application of CARD-FISH to methanogens. J Microbiol Methods 72:54–59. https://doi.org/10.1016/j.mimet.2007.10.006

Kubota K, Morono Y, Ito M, Terada T, Itezono S, Harada H, Inagaki F (2014) Gold-ISH: a nanosize gold particle-based phylogenetic identification compatible with NanoSIMS. Syst Appl Microbiol 37:261–266. https://doi.org/10.1016/j.syapm.2014.02.003

Limpiyakorn T, Sonthiphand P, Rongsayamanont C, Polprasert C (2010) Abundance of amoA genes of ammonia-oxidizing archaea and bacteria in activated sludge of full-scale wastewater treatment plants. Bioresour Technol 102:3694–3701. https://doi.org/10.1016/j.biortech.2010.11.085

Loy A, Horn M, Wagner M (2003) probeBase: an online resource for rRNA targeted oligonucleotide probes. Nucleic Acids Res 31:514–516. https://doi.org/10.1093/nar/gkg016

Ludwig W, Strunk O, Westram R, Richter L, Meier H, Buchner YA, Lai T, Steppi S, Jobb G, Forster W, Brettske I, Gerber S, Ginhart AW, Gross O, Grumann S, Hermann S, Jost R, Konig A, Liss T, Lussmann R, May M, Nonhoff B, Reichel B, Strehlow R, Stamatakis A, Stuckmann N, Vilbig A, Lenke M, Ludwig T, Bode A, Schleifer KH (2004) ARB: a software environment for sequence data. Nucleic Acids Res 32:1363–1371. https://doi.org/10.1093/nar/gkh293

Luesken FA, Zhu B, van Alen TA, Butler MK, Rodriguez DM, Song B, Op den Camp HJM, Jetten MSM, Ettwig KF (2011) *pmoA* primers for detection of anaerobic methanotrophs. Appl Environ Microbiol 77:3877–3880. https://doi.org/10.1128/AEM.02960-10

Luo JF, Lin WT, Guo Y (2011) Functional gene based analysis of sulphur oxidizing bacteria community in sulfide removing bioreactor. Appl Microbiol Biotechnol 90:769–778. https://doi.org/10.1007/s00253-010-3061-x

Moraru C, Lam P, Fuchs BM, Kuypers MMM, Amann RI (2010) GeneFISH-an in situ technique for linking gene presence and cell identity in environmental microorganisms. Environ Microbiol 12:3057–3073. https://doi.org/10.1111/j.1462-2920.2010.02281.x

Mota CR, So MJ, de los Reyes FL (2012) Identification of nitrite-reducing bacteria using sequential mRNA fluorescence in situ hybridization and fluorescence-assisted cell sorting. Microb Ecol 64:256–267. https://doi.org/10.1007/s00248-012-0018-x

Musat N, Halma H, Winterhollerb B, Hoppe P, Peduzzi S, Hillion F, Horreard F, Amann RI, Jørgensen BB, Kuypers MM (2008) A single-cell view on the ecophysiology of anaerobic phototrophic bacteria. Proc Natl Acad Sci U S A 105:17861–17866. https://doi.org/10.1073/pnas.0809329105

Nguyen HTT, Le VQ, Hansen AA, Nielsen JL, Nielsen PH (2011) High diversity and abundance of putative polyphosphate-accumulating *Tetrasphaera*-related bacteria in activated sludge systems. FEMS Microbiol Ecol 76:256–267. https://doi.org/10.1111/j.1574-6941.2011.01049. x

Nobu MK, Tamaki H, Kubota K, Liu WT (2014) Metagenomic characterization of '*Candidatus* Defluviicoccustetraformis strain TFO71', a tetradforming organism, predominant in an anaerobic-aerobic membrane bioreactor with deteriorated biological phosphorus removal. Environ Microbiol 16:2739–2751. https://doi.org/10.1111/1462-2920.12383

Nobu MK, Narihiro T, Rinke C, Kamagata Y, Tringe SG, Woyke T, Liu WT (2015) Microbial dark matter ecogenomics reveals complex synergistic networks in a methanogenic bioreactor. ISME J 9:1710–1722. https://doi.org/10.1038/ismej.2014.256

Okabe S, Kindaichi T, Ito T (2004) MAR–FISH—an ecophysiological approach to link phylogenetic affiliation and in situ metabolic activity of microorganisms at a single-cell resolution. Microbes Environ 19:83–98. https://doi.org/10.1264/jsme2.19.83

Pavlekovic M, Schmid MC, Schmider-Poignee N, Spring S, Pilhofer M, Gaul T, Fiandaca M, Löffler FE, Jetten M, Schleifer KH, Lee NM (2009) Optimization of three FISH procedures for in situ detection of anaerobic ammonium oxidizing bacteria in biological wastewater treatment. J Microbiol Methods 78:119–126. https://doi.org/10.1016/j.mimet.2009.04.003

Pernthaler A, Amann RI (2004) Simultaneous fluorescence in situ hybridization of mRNA and rRNAin environmental bacteria. Appl Environ Microbiol 70:5426–5433. https://doi.org/10.1128/AEM.70.9.5426-5433.2004

Pratscher J, Stichternoth C, Fichtl K, Schleifer K-H, Braker G (2009) Application of recognition of individual genes-fluorescence in situ hybridization (RING-FISH) to detect nitrite reductase genes (*nirK*) of denitrifiers in pure cultures and environmental samples. Appl Environ Microbiol 75:802–810. https://doi.org/10.1128/AEM.01992-08

Rinke C, Schwientek P, Sczyrba A, Ivanova NN, Anderson IJ, Cheng J-F, Darling A, Malfatti S, Swan BK, Gies EA, Dodsworth JA, Hedlund BP, Tsiamis G, Sievert SM, Liu WT, Eisen JA, Hallam SJ, Kyrpides NC, Stepanauskas R, Rubin EM, Hugenholtz P, Woyke T (2013) Insights into the phylogeny and coding potential of microbial dark matter. Nature 499:431–437. https://doi.org/10.1038/nature12352

Saito Y, Aoki M, Hatamoto M, Yamaguchi T, Takai K, Imachi H (2015) Presence of a novel methanogenic archaeal lineage in anaerobic digesters inferred from *mcrA* and 16S rRNA gene phylogenetic analyses. J Water Environ Technol 13:279–289. https://doi.org/10.2965/jwet.2015.279

Schönhuber W, Fuchs B, Juretschko S, Amann RI (1997) Improved sensitivity of whole-cell hybridization by the combination of horseradish peroxidase-labeled oligonucleotides and tyramide signal amplification. Appl Environ Microbiol 63:3268–3273

Sekiguchi Y (2006) Yet-to-be cultured microorganisms relevant to methane fermentation processes. Microbes Environ 21:1–15. https://doi.org/10.1264/jsme2.21.1

Sekiguchi Y, Kamagata Y, Nakamura K, Ohashi A, Harada H (1999) Fluorescence in situ hybridization using 16S rRNA-targeted oligonucleotides reveals localization of methanogens and selected uncultured bacteria in mesophilic and thermophilic sludge granules. Appl Environ Microbiol 65:1280–1288

Sekiguchi Y, Takahashi H, Kamagata Y, Ohashi A, Harada H (2001) In situ detection, isolation, and physiological properties of a thin filamentous microorganism abundant in methanogenic granular sludges: a novel isolate affiliated with a clone cluster, the green nonsulfur bacteria, subdivision I. Appl Environ Microbiol 67:5740–5749. https://doi.org/10.1128/AEM.67.12.5740-5749.2001

Steinberg LM, Regan JM (2008) Phylogenetic comparison of the methanogenic communities from an acidic, oligotrophic fen and an anaerobic digester treating municipal wastewater sludge. Appl Environ Microbiol 74:6663–6671. https://doi.org/10.1128/AEM.00553-08

Sumino H, Takahashi M, Yamaguchi T, Abe K, Araki N, Yamazaki S, Shimozaki S, Nagano A, Nishio N (2007) Feasibility study of a pilot-scale sewage treatment system combining an up-flow anaerobic sludge blanket (UASB) and an aerated fixed bed (AFB) reactor at ambient temperature. Bioresour Technol 98:177–182. https://doi.org/10.1016/j.biortech.2005.10.020

Thiele S, Fuchs BM, Amann RI (2011) Identification of microorganisms using the ribosomal RNA approach and fluorescence in situ hybridization. In: Wilderer P (ed) Treatise on water science. Elsevier, Oxford. ISBN: 0444531998, pp 171–189. https://doi.org/10.1016/B978-0-444-53199-5.00056-7

Yamada T, Sekiguchi Y, Imachi H, Kamagata Y, Ohashi A, Harada H (2005) Diversity, localization, and physiological properties of filamentous microbes belonging to *Chloroflexi* subphylum I in mesophilic and thermophilic methanogenic sludge granules. Appl Environ Microbiol 71:7493–7503. https://doi.org/10.1128/AEM.71.11.7493-7503.2005

Yamada T, Yamauchi T, Shiraishi K, Hugenholtz P, Ohashi A, Harada H, Kamagata Y, Nakamura K, Sekiguchi Y (2007) Characterization of filamentous bacteria, belonging to candidate phylum KSB3, that are associated with bulking in methanogenic granular sludges. ISME J 1:246–264. https://doi.org/10.1038/ismej.2007.28

Yamada T, Kikuchi K, Yamauchi T, Shiraishi K, Ito T, Okabe S, Hiraishi A, Ohashi A, Harada H, Kamagata Y, Nakamura K, Sekiguchi Y (2011) Ecophysiology of uncultured filamentous anaerobes belonging to the phylum KSB3 that cause bulking in methanogenic granular sludge. Appl Environ Microbiol 77:2081–2087. https://doi.org/10.1128/AEM.02475-10

Yamaguchi T, Fuchs BM, Amann RI, Kawakami S, Kubota K, Hatamoto M, Yamaguchi T (2015a) Rapid and sensitive identification of marine bacteria by an improved in situ DNA hybridization chain reaction (quickHCR-FISH). Syst Appl Microbiol 38:400–405. https://doi.org/10.1016/j.syapm.2015.06.007

Yamaguchi T, Kawakami S, Hatamoto M, Imachi H, Takahashi M, Araki N, Yamaguchi T, Kubota K (2015b) In situ DNA-HCR: a facilitated in situ hybridization chain reaction system for the detection of environmental microorganisms. Environ Microbiol 17:2532–2541. https://doi.org/10.1111/1462-2920.12745

Yilmaz LS, Noguera DR (2004) Mechanistic approach to the problem of hybridization efficiency in fluorescent in situ hybridization. Appl Environ Microbiol 70:7126–7139. https://doi.org/10.1128/AEM.70.12.7126-7139.2004

Yilmaz LS, Okten HE, Noguera DR (2006) Making all parts of the 16S rRNA of Escherichia coli accessible in situ to single DNA oligonucleotides. Appl Environ Microbiol 72:733–744. https://doi.org/10.1128/AEM.72.1.733-744.2006

Yilmaz LS, Parnerkar S, Noguera DR (2011) mathFISH, a web tool that uses thermodynamics-based mathematical models for in silico evaluation of oligonucleotide probes for fluorescence in situ hybridization. Appl Environ Microbiol 77:1118–1122. https://doi.org/10.1128/AEM.01733-10

You SJ (2005) Identification of denitrifying bacteria diversity in an activated sludge system by using nitrite reductase genes. Biotechnol Lett 27:1477–1482. https://doi.org/10.1007/s10529-005-1314-z

Zwirglmaier K, Ludwig W, Schleifer KH (2004) Recognition of individual genes in a single bacterial cell by fluorescence in situ hybridization-RINGFISH. Mol Microbiol 51:89–96. https://doi.org/10.1046/j.1365-2958.2003.03834.x

Integrated Innovative Biotechnology for Optimization of Environmental Bioprocesses and a Green Economy

3

Jan W. Dobrowolski, Dawid Bedla, Tomasz Czech, Florian Gambuś, Krystyna Górecka, Waldemar Kiszczak, Tomasz Kuźniar, Robert Mazur, Agata Nowak, Malgorzata Śliwka, Obid Turunov, Aleksandra Wagner, Jerzy Wieczorek, and Magdalena Zabochnicka-Świątek

Abstract

A systematic approach to sustainable management of natural resources incorporates integration of innovative biotechnologies and eco-engineering. Here we review complementary eco-innovations for sustainable development in different regions. One of the newer biotechnologies is laser photostimulation of different species of plants and microorganisms to increase their adaptability to xenobiotics in soil, air, and water. Empirically selected algorithms of laser irradiation significantly increase biodegradation of hydrocarbons, phytoremediation of trace metals by willow *Salix sp.*, elephant grass *Miscanthus x giganteus*, Virginian mallow *Sida hermaphrodita,* and to increase tolerance of different species of plants to petrochemical pollutants and efficiency of reclamation of contaminated areas and tolerance to salinity of soil and suboptimal temperatures as well as water deficiency. This biotechnology is also useful for biomass enhancement and

J.W. Dobrowolski (✉) • R. Mazur • M. Śliwka • O. Turunov • A. Wagner
AGH University of Science and Technology, Kraków, Poland
e-mail: dobrowol@agh.edu.pl; awagner@agh.edu.pl

D. Bedla • T. Czech • F. Gambuś • T. Kuźniar • J. Wieczorek
University of Agriculture in Krakow, Krakow, Poland

K. Górecka • W. Kiszczak
Research Institute of Horticulture in Skierniewice, Skierniewice, Poland
e-mail: krystyna.gorecka@inhort.pl

A. Nowak
Poznań University of Life Sciences, Poznań, Poland

M. Zabochnicka-Świątek
Institute of Environmental Engineering, Czestochowa University of Technology, Częstochowa, Poland

H.J. Purohit et al. (eds.), *Optimization and Applicability of Bioprocesses*,
https://doi.org/10.1007/978-981-10-6863-8_3

bio-energy production under suboptimal conditions for more efficient vegetative multiplication of some vegetables and development of sustainable agriculture, forestry, protection of the landscape, and biodiversity. Integration of transdisciplinary cooperation and application of complementary biotechnologies and innovative methods of environmental pollutant bioremediation (such as wastewater treatment) and reclamation, supported by the implementation of a neural network, can optimize bioprocesses that are useful for a better quality of life, globally. Long-term research-developing studies are supplemented by problem-solving training and case studies in different regions (including ecotourism, recreation, and promotion of ecological culture), long distance education and life-long education for the common action of experts and knowledge-based society, promoting sustainable development that is based on integrated biological sciences and sustainable models of consumption.

Keywords
Sustainable development • Bio-based economy • Climate change • Adaptation • Laser biostimulation • Microalgae biomass • Bio-energy • Bioremediation • Biodiversity • Environmental education

3.1 Introduction

Sustainable management of biological resources is a key factor in both improving the quality of human life and achieving more efficient protection of biodiversity on a regional and global scale. The crucial factor is attaining proper homeostasis in relation to the function of every living organism and all ecosystems. A biocybernetic mechanism of a negative feedback system is the basis of homeostasis, from a single cell to the entire biosphere. We want to propose a common action for integration of complementary research-developing studies in this field with promotion of the concept of integration of biomimetic processes with ecological engineering. Such a systematic approach is a promising tool for protection of homeostasis as well as promotion of a bio-based green economy and the creation of a sustainable labor market.

The main task of contemporary human civilization is to change from the paradigm of unlimited use of natural resources to a sustainable society based on knowledge. Mankind should move from a destructive and wasteful way of life to working with the natural environment. The key factor for sustainability is replacement of energy- and resource-consuming technologies by a biology-based green economy that follows physiological processes that work with ecological engineering.

The depletion of fossil fuels and natural resources will come sooner or later, and has been prognosticated for many years now, but our society should not wait until then; we need to develop and discover eco-technologies that are able to provide the necessary amount of energy and products for our society in the spirit of conservation of natural resources.

The progress of basic biological and research-developing studies is an inspiration for the new model of civilization and sustainable management of natural resources focused on green economy.

Research-developing studies aimed toward the improvement of biotechnology and sustainability (Lee et al. 2016) and optimization of bioreactors (Mandenius 2016) and biological treatment of solid wastes (Rada 2016) are providing greater insight and perspective into bio-based sustainable development.

3.2 Laser Biotechnology for Bio-based Sustainable Development

Laser biotechnology is a new tool that is useful for optimization of biological processes involved both in improvement of the quality of the natural environment and in biomass and bio-energy production. The basic idea of this technology is photostimulation of selected species of microorganisms and plants to intensify bioenergetic processes that are associated with the acceleration of growth and biomass production in a suboptimal environment. The purpose of this innovative biotechnology is optimization of the phenotype without changing the genotype.

The most sustainable and globally used energy source is biologically-based photosynthesis for conversion of the sun's energy into useful products both as a source of human food as well as biomass-based renewable source of energy. According to J. W. Dobrowolski, for ethical reasons, new and more effective biotechnologies are recommended for better management of unused areas and for the production of biomass as a source of bio-energy, but not instead of food production (Dobrowolski 2016b). Another promising area for future integration is environmental engineering and sustainable development of many regions all over the world. There are research-developing studies that are focused on the integration of management of organic wastes with energy production.

Taking into account the growth of the human population and even greater increase in consumers' demands, it is very useful to apply eco-innovations integrating better environmental management (sewage treatment, waste management) with the increase in biomass (e.g., food, wood, etc.) and bio-energy production. Extensive research studies strongly support development of more efficient sources of bio-energy and bio-fuel production (Junginger et al. 2006; Larson 2006; Obernberger et al. 2008; Bain 2007; Bauen et al. 2009; Gnansounou et al. 2009; Fischedick et al. 2011; Chum et al. 2014; Zwickel et al. 2014). Introduction on a wide scale of new and more efficient methods of biomass and bio-energy production may be especially promising and the most ecologically friendly method for satisfying these requirements. This also represents a creative contribution to the promotion of sustainable development in terms of cost-benefit analysis. There are massive areas that are out of use in many developed and developing countries. Thus the application of already positively tested laser photostimulation of seeds, seedlings, or cuttings of plants before their cultivation could markedly increase biomass production under suboptimal conditions. Introduction of this biotechnology could also support reclamation of degraded areas (Dobrowolski 1986; Dobrowolski 2000a, b, 2001a, b; Dobrowolski and Różanowski 1998 Dobrowolski and Zielinska-Loek 2002; Dobrowolski et al. 2004; 2012a, b; Jakubiak and Śliwka 2009). This type of reclamation also makes sense from an economic point of view. Low-power lasers are recommended for short-time irradiation of selected species of plants. The total costs associated with biostimulation are comparatively lower than the economic value of biomass produced as a result of photostimulation (Dobrowolski 2000a, b).

Stimulation with coherent light of high-energy density of seedlings of rapidly growing bushes and trees could massively increase wood production in areas after reclamation as well as in energy plantations. Overexploitation of primary forest could be eliminated as a result of enhancement of wood production. An empirically selected algorithm of laser stimulation is recommended for better adaptation of different plant species to an environment that has deteriorated as a result of the combination of anthropogenic pressure and the impact of climate change. The process of adaptation to environmental stress factors always requires additional energy, so the appropriate supplementation of organisms with energy increases their adaptability. Therefore, laser biotechnology could be recommended for more efficient reclamation of degraded areas, enhancement of wastewater treatment, and biomass production in various energy plantations and cultivation of industrial plants under suboptimal conditions.

Laser biotechnology has also been applied for more efficient reclamation of regions that have degenerated, enhancement of biomass followed by gas production, and for better bioremediation of soil contaminated with highly toxic trace metals as well as acceleration of biodegradation of petrochemical pollutants in contaminated soil and aquatic ecosystems. Results of the recent but not yet published experimental studies on the influence of laser light on different plant species open new perspectives on biomass production in areas currently not used. Application of a proper algorithm of coherent light (of high-energy density) in relation to the seeds of the cress plant (*Cardamine parviflora)* and summer lupin (*Lupinus hartwegi)* cultivated in soil polluted by hydrocarbons from an oil well could significantly increase biomass production (Dobrowolski 2016b). Especially promising for the future study is enhancement of the growth rate of lupin as a symbiotic plant with nitrogen-fixing bacteria, which are very important for stimulating the reclamation of degraded areas and sustainability of biomass production (Dobrowolski 2016b). These recent results also indicate the possibility of more efficient reclamation of areas contaminated by very high amounts of toxic trace metals like lead (Pb) and cadmium (Cd) as a result of irradiation of sunflower (*Helianthus sp.*) and the great pumpkin (*Cucurbita maxima)* seeds (Dobrowolski 2016b) by empirically selected laser light algorithms. New enzymological and physiological criteria are recommended for assessing the germinability of cultivated plants (Badek et al. 2016).

New possibilities for significant increases in biomass production as a source of biofuel are now being presented as a result of increases in laser photostimulation of the seeds of *Sorghum vulgare variety Sucrosorgo 506* with regard to resistance to low temperatures during the critical period of germination. This is a climate-limiting factor in Central European countries for the introduction of this cereal for widescale cultivation (Dobrowolski 2016b).

In support of Dobrowolski and co-workers' new application of laser photostimulation of the inoculum of *Trichophyton mentagrophytes var. granulosum* for much more efficient biodegradation of petrochemical pollutants including polycyclic aromatic hydrocarbons (PAH), recently carried out experiments achieved very successful application of empirically selected algorithms of laser irradiation of whole *biocenosis* from the area of the oldest oil well for much more efficient reclamation of areas contaminated by hydrocarbons (Dobrowolski 2016a). New complementary methods for enhancement of biodegradation of hydrocarbons in oil-contaminated soil are based on fortification of the soil with nitrogen and phosphorus (Jia et al. 2016). Application of a consortium of bacteria adapted to an oil-contaminated environment immobilized on a sponge bioreactor could stimulate biodegradation of aromatic hydrocarbons (El-Borai et al. 2016). Detailed information about the complementary environmentally friendly application of laser biotechnology has been published as an international e-book (Dobrowolski 2016a, b).

Laser biostimulation also increases the assimilation of carbon dioxide by the irradiated plants. Thus, one of the reasons for recommending this eco-innovation is the contribution to primary prevention of climate change through the promotion of low-carbon bio-energy and sustainable development in different regions and in different continents (Dobrowolski et al. 2012a, b). Another perspective of the progress in bio-energy production, which is reasonable in terms of cost-benefit analysis, involves innovative management of biodegradable wastes (Horsek and Hrebicek 2014) or municipal solid waste (MSW) (Tursunov 2014). Municipal solid waste is primarily waste that is produced by households, but also includes some industrial and commercial waste that has been deposited in municipal landfills, causing a wide array of harmful effects on the environment and human health. An ongoing increase in the production of waste results in loss of materials and energy and contributes to the deterioration of the natural environment. Currently, recycling of MSW is an important issue in many developed and developing countries. Some of the effective thermo-chemical techniques for MSW recycling and biomass-to-energy production are pyrolysis and gasification (Tursunov 2014; Tursunov et al. 2011, 2015a, b; Maoyun et al. 2010; Chum et al. 2014; Tursunov and Dobrowolski 2015). Bio-energy from municipal waste is produced by gasification, and pyrolysis is recommended for more efficient production of synthetic gas (syngas) from wood. The products (yields) of gasification are hydrogen (H_2), carbon monoxide (CO), and methane (CH_4) (Chum et al. 2014; Tursunov and Dobrowolski 2015; Tursunov et al. 2015a, b). Gasification is a process that converts biomass into gaseous products and bio-oil or bio-fuel (Tursunov 2014). However, the byproducts of gasification and pyrolysis processes are tar and char. Many research-developing studies have focused on the improvement of catalyst activity for tar cracking and char yield reduction (Shen and Yoshikawa 2013; Tursunov et al. 2015b). Calcined dolomite and nickel-based catalysts were introduced as the most effective catalysts in biomass gasification and pyrolysis for tar cracking and reforming (Tursunov 2014; Tursunov et al. 2015a, b).

There are therefore real opportunities for supplementation of renewable energy sources using laser stimulation for biomass production in areas currently out of use, including reclamation of landfills and application of clean thermal technologies for large-scale production of bio-energy. This eco-innovation may be a key factor for increasing efficiency of use of biomass and organic wastes for bio-energy production. Moreover, a linkage between more efficient management of the natural environment with progress in production of bio-energy from renewable sources could contribute to the development of a sustainable labor market and to enhancement of the quality of the environment. Therefore, the introduction of this approach through joint pilot projects is recommended in several countries.

3.3 The Application of Laser Biostimulation on Growth of Algae *Chlorella vulgaris*

3.3.1 Biological Properties of Microalga *Chlorella vulgaris*

Chlorella vulgaris are the smallest and oldest algae in the world. They are unicellular green algae (Zabochnicka-Świątek 2010), i.e., the simplest organisms, that consist of a nucleus, a chloroplast, mitochondria, and a cell wall. *C. vulgaris.* contains up to 4% chlorophyll—it has the highest concentration of chlorophyll of any plant. Such a high content of chlorophyll enables production of ten times more oxygen than other plants and the capacity to absorb ten times more carbon dioxide. *C. vulgaris* is able to reproduce itself very quickly. A single stem cell is divided in 16–20 h into four new cells, which over the next 16–20 h multiply in a similar way. The structure and composition of chlorella makes it irreplaceable in the environment, and a fully-fledged food. In relation to dry matter it contains more than 50% protein, and is rich in polyunsaturated fatty acids, minerals, fiber, vitamins, and, as mentioned, chlorophyll. *C. vulgaris* contains the highest amount of chlorophyll in dry matter of all foods (about 4%). It does not produce any toxic products from decomposition or metabolism. Previous experimental research using *C. vulgaris* strains indicates that *C. vulgaris* can be used as a biosorbent for heavy metal removal (Zabochnicka-Świątek et al. 2012, 2014, Zabochnicka-Świątek 2013).

Lasers are sources of coherent monochromatic light with a high density of energy. The possibility of applying laser to many biological materials has been investigated (Dobrowolski 2016a, b), and it has been proposed that laser biostimulation could have a positive effect on plant growth (Dobrowolski 2000a, b; Dobrowolski et al. 1996); an increase in plant biomass and greater uptake of some elements from contaminated water and soils has been found (Dobrowolski 2006a; Dobrowolski et al. 2012a, b).

Because of their absorption and oxygen-releasing properties, algae can be used for carbon capture in wastewater treatment technology (Rogulska et al. 2011), and laser stimulation can be used for more efficient sewage treatment bioprocessing (Dobrowolski et al. 2012b; Dobrowolski 2016b).

The mass cultivation of algae using lasers for production of biofuels, pharmaceuticals, and chemicals is a promising technology. Algae can grow and divide successfully under specific wavelength laser illumination (Kuwahara et al. 2011).

The overall goal of this study was to evaluate the influence of laser biostimulation on the growth of *C. vulgaris.* Two phases of experiments were conducted: With reference to the results obtained from the first phase, further experiments were conducted with a chosen exposition time for laser stimulation, and the cultivation period was extended from 11 to 33 days. The study was carried out under static conditions (batch tests) in a climatic chamber at 29 °C during the day and 25 °C at night.

3.3.2 Material and Methods

3.3.2.1 Experimental Methods

The strain of *C. vulgaris* used in this study was obtained from the Culture Collection of Baltic Algae (CCBA), Institute of Oceanography, University of Gdansk, Poland (http://ccba.ug.edu.pl/pages/en/home.php).

The inoculum was grown in ten 250-ml Erlenmayer flasks for 4 weeks. A cultivation medium (Bioflorin) was used for maintenance and inoculum preparation.

The experimental groups were exposed to varying parameters of laser stimulation: type of diode, length of wave, time, and power of radiation. In every experiment, a control group of a non-irradiated strain of *C. vulgaris* was also prepared.

Experimental groups of *C. vulgaris:*

- K. Control group: non-irradiated,
- DL. Laser diode: exposition time 3 × 1 s, 3 × 3 s, 3 × 9 s, light wavelength: 470 nm
- LM. Impulse medical laser: exposition time 3 × 1 s, 3 × 3 s, 3 × 9 s, light wavelength: 670 nm. After laser stimulation, samples were stored in darkness for 12 h.

The initial pH was 7.65 and turbidity 120 NTU. The experiments were conducted under a 12/12-h day/night cycle in a climatic chamber with a temperature of 29 ± 2 °C during the day and 24 ± 2 °C at night for 2 weeks (the first phase) and then for 5 weeks (the second phase). The experiments were performed in triplicates.

Two phases of experiments were conducted. During the first phase of the research, the inoculum was placed in 50-ml flasks. During the second phase, the inoculum was placed in a higher level of cultivation medium: 250 ml, and two samples were placed, as during the first phase, in 50-ml flasks: LM3(2) and DL3(2).

With reference to the results obtained from the first phase, further experiments were conducted with a chosen exposition time to laser stimulation, and the cultivation period was extended from 11 to 33 days.

3.3.2.2 Methods of Analysis

During cultivation, observations at equal time intervals on pH, turbidity, dissolved oxygen (DO), productivity of algal biomass, chlorophyll content, and organic and inorganic carbon concentration were made.

The pH was measured using a pH-meter, turbidity was carried out using a densitometer, and DO determination was done using an oxygen meter. Quantification of chlorophyll (Chl) was spectrophotometrically determined with the extinction coefficients in acetone and calculated using appropriate formulae. Organic and inorganic carbon concentrations were determined by a Multi N/C 2100 analyser.

Measurements of physicochemical parameters performed during the second phase in three different time intervals within the light phase made it possible to evaluate the influence of photosynthetic phases on those particular parameters. Productivity of algal biomass was determined as dry matter content. Dry matter was measured according to Polish standards (PN-R-04013:1988).

3.3.3 Results and Discussion

3.3.3.1 pH

Figures 3.1 and 3.2 present changes in pH values with time during cultivation of *C. vulgaris*.

During the first 5 days of the research the pH value remained at levels of 7.58–7.64 in all cases. After the sixth day up till the end of the experiment, the pH value depended on the exposition time of laser stimulation. The lowest level of pH was observed for the LM3 sample and the highest for the DL3 sample.

During the second phase of the experiment pH values for LM3 and DL3 samples were the lowest. The lower volume of cultivation medium (45 ml) resulted in a decrease in pH from 7.68 to the minimum value of 6.9. Cultivation of *C. vulgaris* in 250-ml flasks resulted in higher pH values.

3.3.3.2 Turbidity

Figures 3.3 and 3.4 illustrate the variation in turbidity with time measured by means of a turbidimeter during the first and second phases of the experiment.

The results achieved in turbidity measurements with the turbidimeter confirmed the increase in the algal population over the course of the experiment.

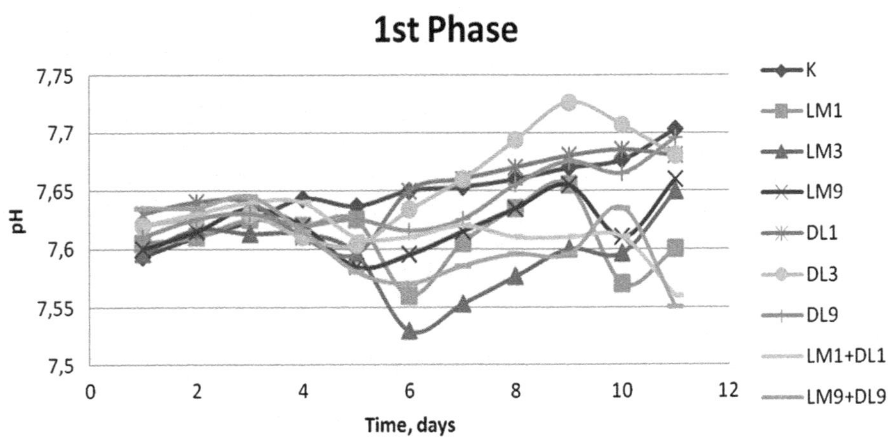

Fig. 3.1 pH values observed during the first phase of the experiment. K: non-irradiated sample; exposition time: 1 – 3 × 1 s, 3 – 3 × 3 s, 9 – 3 × 9 s. Values are expressed as means ± s.d. (n=3)

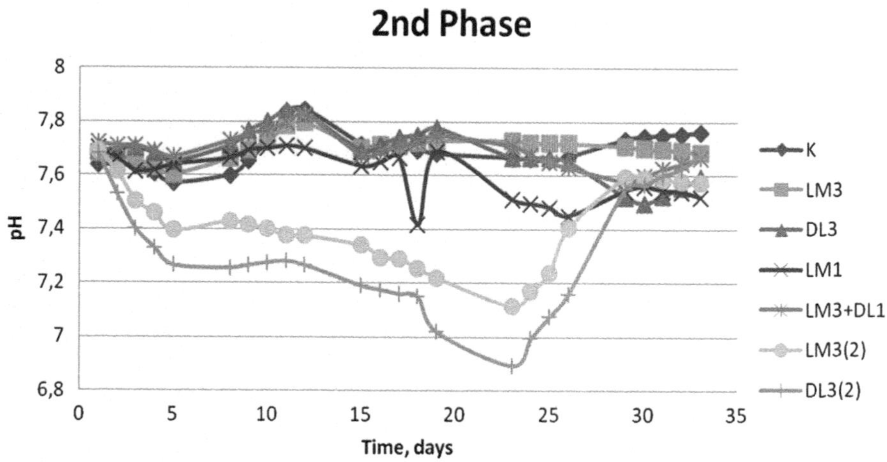

Fig. 3.2 pH values observed during the second phase of the experiments. K: non-irradiated sample; exposition time: 1 – 3 × 1 s, 3 – 3 × 3 s. Values are expressed as means ± s.d. (n=3)

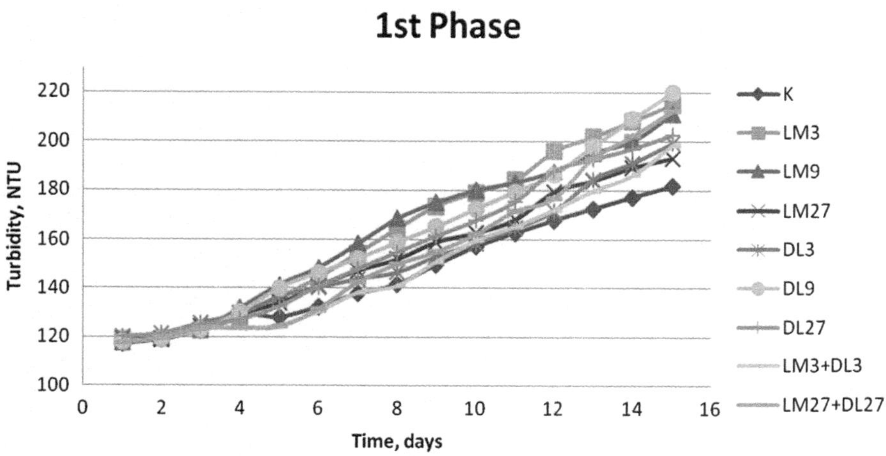

Fig. 3.3 Turbidity values observed during the first phase of the experiment. K: non-irradiated sample; exposition time: 1 – 3 × 1 s, 3 – 3 × 3 s, 9 – 3 × 9 s. Values are expressed as means ± s.d. (n=3)

An increase in turbidity with time was also observed during that phase. The highest turbidity was observed for LM1 (325 NTU). Cell density varied with the lower volume of cultivation medium (45 ml). This decrease could have been affected by the lower efficiency of photosynthesis in the lower level of cultivation medium.

3.3.3.3 Chlorophyll Content

Figures 3.5 and 3.6 present chlorophyll a content changes with time in the first and second phases of experiment.

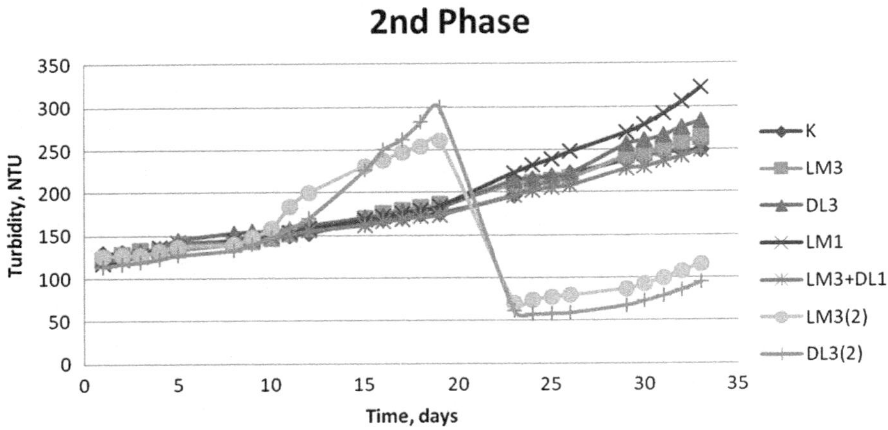

Fig. 3.4 Turbidity values observed during the second phase of the experiments. K: non-irradiated sample; exposition time: 1 – 3 × 1 s, 3 – 3 × 3 s. Values are expressed as means ± s.d. (n=3)

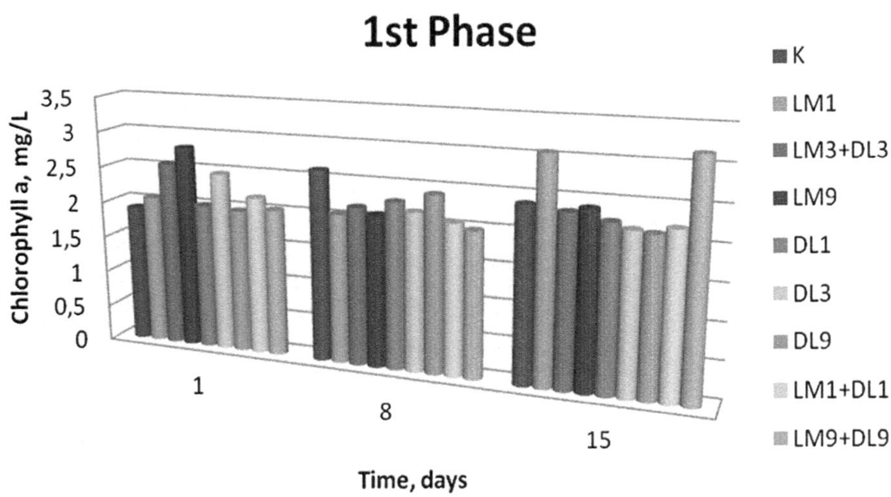

Fig. 3.5 Chlorophyll a content observed during the first phase of the experiment. K: non-irradiated sample; exposition time: 1 – 3 × 1 s, 3 – 3 × 3 s, 9 – 3 × 9 s. Values are expressed as means ± s.d. (n=3)

During the first phase of the research the chlorophyll a content remained at levels from 1.8 to 3.2 mg/L in all cases. The highest chlorophyll levels were observed on the 15th day for LM1 (3.0 mg/L) and LM9+DL9 (3.2 mg/L).

Chlorophyll a content increased with time during the second phase of the study. The highest levels of chlorophyll content were observed on the 29th day for the LM1 sample (4.4 mg/L), DL3 (3.75 mg/L), and LM3 (3.5 mg/L). The lowest levels were found for DL3(2) (2.53 mg/L) and LM3(2) (2.75 mg/L). This decrease could have been due to the lower volume of the cultivation medium.

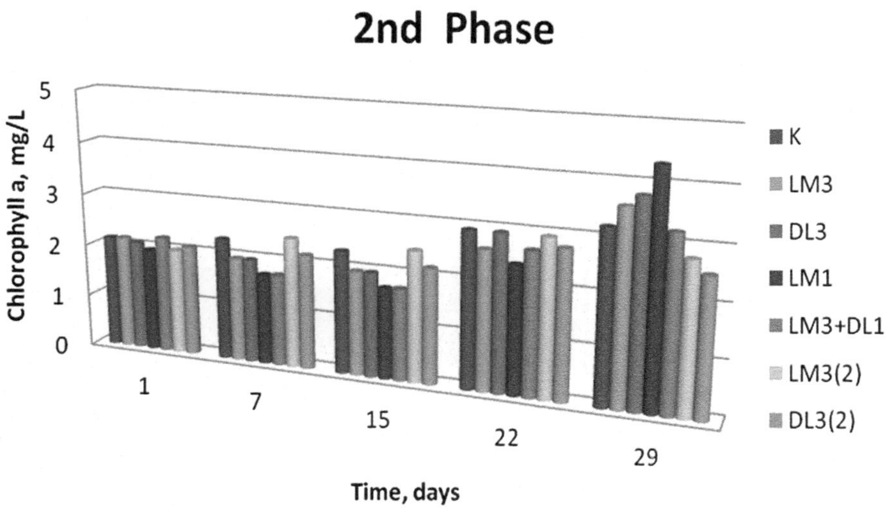

Fig. 3.6 Chlorophyll a content observed during the second phase of the experiment. K: non-irradiated sample; exposition time: 1 – 3 × 1 s, 3 – 3 × 3 s. Values are expressed as means ± s.d. (*n*=3)

Figures 3.7 and 3.8 illustrate chlorophyll b content changes with time in the first and second phases of experiment.

During the first phase of the research, chlorophyll b content decreased with time. The highest chlorophyll levels were found on day 1 for LM9 and on day 15 for LM9+DL9.

Chlorophyll b content increase with time during the second phase of the study. The highest chlorophyll value was observed for LM1 (3.9 mg/L). A higher volume of cultivation medium used in the second phase of 250 ml resulted in an increase in chlorophyll b content for the LM1 sample.

3.3.4 Dissolved Oxygen (DO) and Total Organic Carbon (TOC) Content

Figure 3.9 shows changes in dissolved oxygen values with time during the cultivation of *C. vulgaris* in the second phase of the experiment.

Laser stimulation of *C. vulgaris* resulted in a DO increase by the fifth day for all samples except those samples cultivated in a lower volume of cultivation medium, LM3(2) and DL3(2), in which DO was at the lowest level. The achieved results of DO measurements for the LM1 sample confirmed the increase in the algal population during the experimental course for the LM1 sample.

The growth of a microalgae culture occurs in successive phases: the adaptation phase (lag or inductive phase), the exponential phase (logarithmic phase), the phase of relative declining growth, the stationary phase, and the death phase (Zabochnicka-Świątek 2012; Kozieł and Włodarczyk 2011). The lowest level of DO was recorded on the 15th day of research, which could be a consequence of the death phase after which the second exponential phase started.

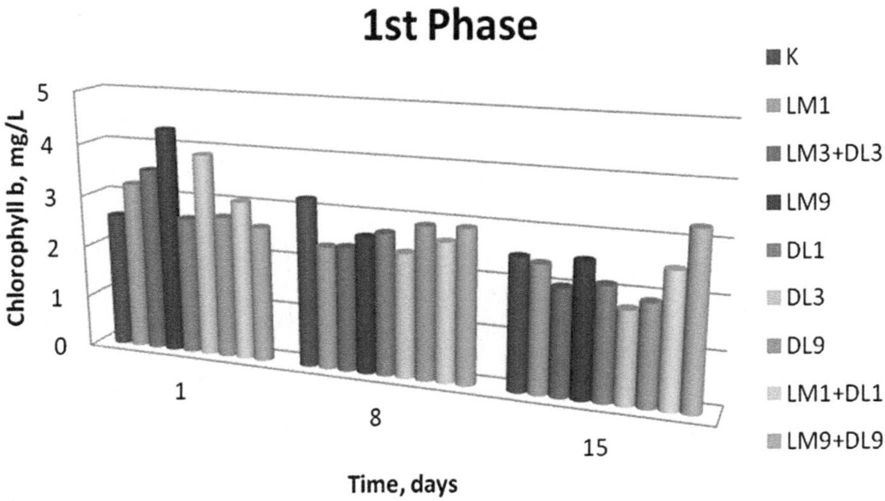

Fig. 3.7 Chlorophyll b content observed during the first phase of the experiment. K: non-irradiated sample; exposition time: 1 – 3 × 1 s, 3 – 3 × 3 s, 9 – 3 × 9 s. Values are expressed as means ± s.d. (n=3)

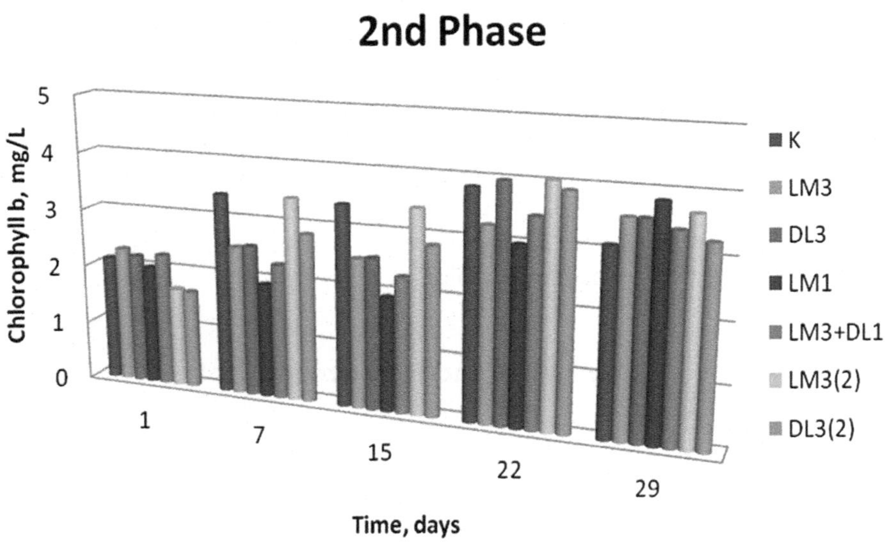

Fig. 3.8 Chlorophyll b content observed during the second phase of the experiment. K: non-irradiated sample; exposition time: 1 – 3 × 1 s, 3 – 3 × 3 s. Values are expressed as means ± s.d. (n=3)

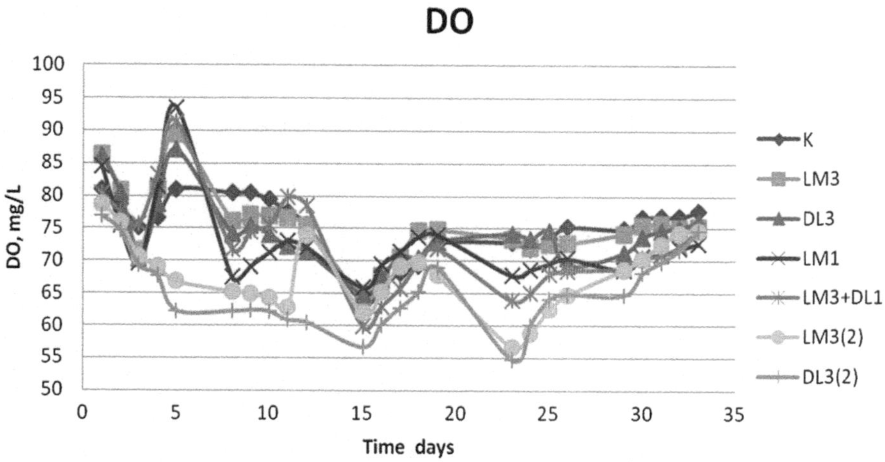

Fig. 3.9 Dissolved oxygen values observed during the second phase of the experiments. K: non-irradiated sample; exposition time: 1 – 3 × 1 s, 3 – 3 × 3 s. Values are expressed as means ± s.d. (n=3)

3.3.5 Organic and Inorganic Carbon Content

Figure 3.10 shows the total organic carbon (TOC) changes with time in the second phase of the experiment.

In the second phase of experiment, the TOC values were determined at the first, eighth, 15th, 22nd, and 29th days of the research. In the case of TOC, an increase in its concentrations was determined during the experiment. The highest TOC content was found for LM1 on the 29th day. This could have been affected by the higher efficiency of photosynthesis and higher biomass production after laser stimulation.

Figure 3.11 shows inorganic carbon (IC) changes with time in the second phase of the experiment.

The IC concentration varied with time. The highest level of IC was found for the LM3+DL1 sample.

3.3.6 Biomass Concentration

The growth dynamics and cell divisions are determined by cell-specific metabolism and physicochemical processes such as photosynthesis, respiration, ionic assimilation, etc. (Kozieł and Włodarczyk 2011).

Figure 3.12 presents the changes in the quantity of produced biomass of *Chlorella vulgaris* measured by the Thoma counting chamber during the second experimental phase.

There were no significant changes with cultivation of *C. vulgaris* in 250 ml. Cultivation of *C. vulgaris* in 45-ml flasks resulted in a decrease in biomass concentration from the 20th to the 25th days of the experiment.

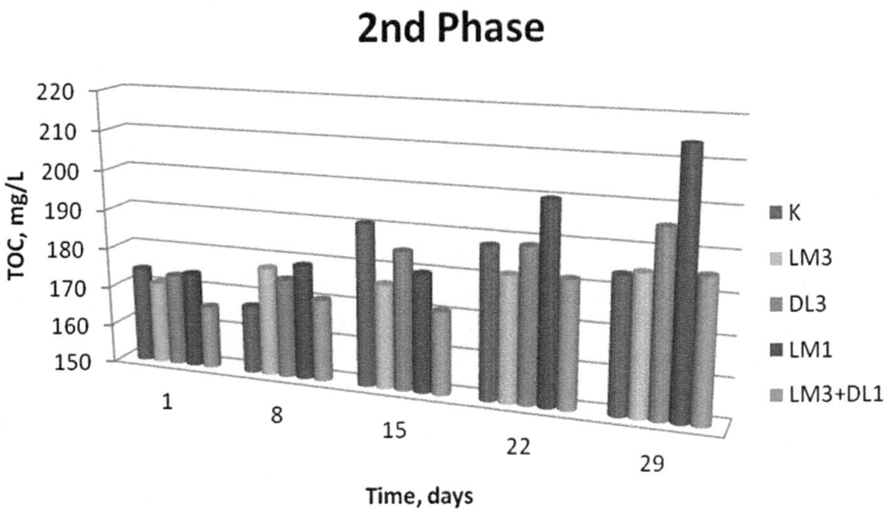

Fig. 3.10 Total organic carbon values observed during the second phase of the experiment. K: non-irradiated sample; exposition time: 1 – 3 × 1 s, 3 – 3 × 3 s. Values are expressed as means ± s.d. (n=3)

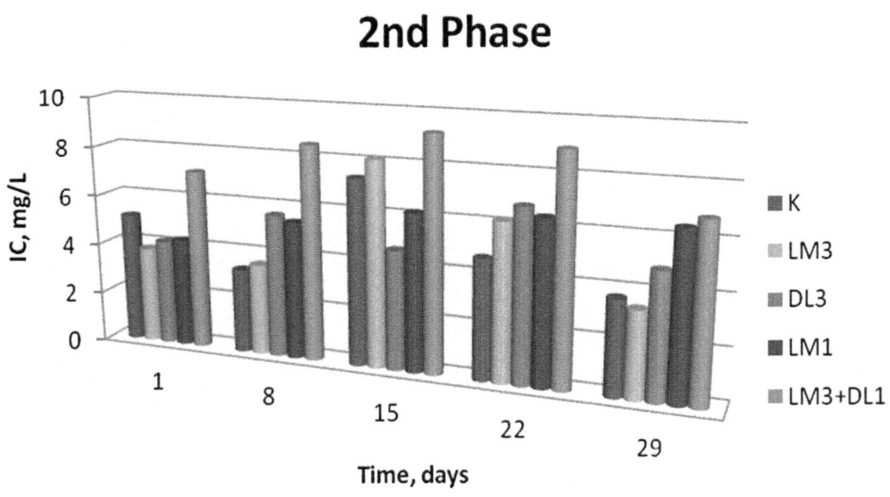

Fig. 3.11 Inorganic carbon changes observed during the second phase of the experiment. K: non-irradiated sample; exposition time: 1 – 3 × 1 s, 3 – 3 × 3 s. Values are expressed as means ± s.d. (n=3)

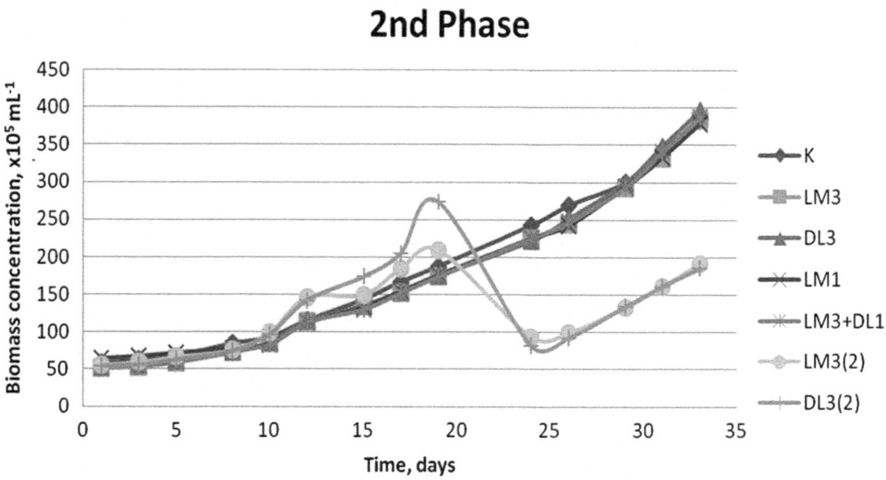

Fig. 3.12 The changes in produced algae biomass during the second phase of the experiment. K: non-irradiated sample; exposition time: 1 – 3 × 1 s, 3 – 3 × 3 s. Values are expressed as means ± s.d. (n=3)

3.3.7 Conclusions of the Experiments on Algae

The study yielded the following conclusions:

1. Laser biostimulation has a positive effect on algal growth.
2. Cultivation of *C. vulgaris* in 250-ml flasks resulted in higher levels of pH.
3. The turbidity varied with less cultivation medium (45 ml). This decrease could have been affected by the lower efficiency of photosynthesis in the smaller cultivation medium environment.
4. The highest level of chlorophyll a content was observed for the LM1 sample and the lowest level was found for LM3(2) and DL3(2). This decrease could have been due to a lower cultivation medium environment.
5. The higher volume of cultivation medium used in the second phase of 250 ml resulted in an increase in chlorophyll b content for the LM1 sample.
6. The dissolved oxygen level varied with time during the experiment.
7. The lowest level of DO as well as biomass concentration was found for LM3(2) and DL3(2).
8. In the case of total organic carbon, an increase in its concentrations was determined during the experiment. The inorganic carbon concentration varied with time.
9. The results obtained indicate that application of laser biostimulation can have a positive effect on cultivation of *C. vulgaris*. The positive effect occurs after selecting the best biostimulation parameters.

Optimization of the wavelength of low-power laser and time of irradiation in relation to different species of algae may be a subject for future studies. Algal biomass could be used for carbon capture and/or heavy metal and nutrient removal from the contaminated environment.

The results of this study support the prospect of introducing laser biotechnology for the optimization of chlorella cultures, including the possibility of application of innovative biotechnology for more efficient wastewater treatment by algae. Innovative biotests based on image analysis would be useful for evaluation of efficiency of different forms of biotechnology for wastewater treatment.

Following the concept of eco-engineering is the perspective of construction of innovative hydrobotanic wastewater treatment plants (constructed wetlands), using laser-stimulated algae as well as different species of plants, e.g., duck weed, reed, willow, etc. There is the real possibility of achieving more efficient prevention of eutrophization and water-borne diseases by integration hydrobotanic plants with a network of underground sewage treatment plants using the highly efficient (with a reduction value of up to 96.5 BOD5, 89.5 COD, 89.5 TOC) and cheap Japanese biotechnology "Bando." There have been very good results from long-term test studies in different regions of Japan and the first in Europe, in Poland (Amaya et al. 1990).

3.4 Biotechnology in Lake Restoration

Biological methods play a very important role in water treatment for organic pollution, but most of these methods are based on natural processes that take place in water ecosystems. Constructed wetlands and natural wetlands provide suitable conditions for wastewater treatment, and result in self-purification processing of water. Sometimes the load of pollutant runoff to a water reservoir is too high, and under these conditions this process is not very effective, with water quality drastically deteriorating.

Environmental engineering and biotechnology offer many possibilities for development of effective tools that can ensure proper water quality. One of the ideas is to transfer the operating principle of wastewater biotechnology from wastewater treatment plants directly to the polluted water reservoir. Classical aeration processes provide many advantages in the way of organic pollutant reduction, but this technology requires a proper infrastructure enabling a proper aeration process.

New methods of innovative technology are offered that are dedicated to small- and mid-size water reservoirs (up to 3 ha). These technologies are based on MBBRs (Moving Bed Biofilm Reactors). Within the project dealing with bio-aero-treatment, the construction of MBBRs was carried out to remediate small water bodies in a stationary mode. Hydraulic tests were carried out in the Water Laboratory at the Department of Hydraulic and Sanitary Engineering, Section of Rural Water Supply and Sanitation, University of Life Science in Poznań.

Bioreactors are characterized by having a mobile construction, and thus they are easy to transport and quick to assemble in the conditions surrounding lakes and small water bodies. In addition, they have cartridge constructions leading to the water bodies containing effective microorganisms. Additional dosing of microorganisms during the process of purification significantly speeds up the process and diminishes the time for formation of the activated sludge. Bioreactors work in a flow mode and can be moved around within the lake, depending on the length of the aeration ducts.

The designed MBBR uses a process of proliferation of microorganisms on the biofilm, as a result of forced flow of wastewater, which is caused by a diffuser. Water–air mixing occurs, allowing the substrate to flow in the bioreactor, additionally creating oxygen conditions for biological biofilm (McQuarrie and Boltz 2011). The improvement in efficiency of the biological treatment of sewage is supported by the progress in environmental microbiology linked with ecological engineering (Ivanov 2016).

The direction of flow of the polluted water is a result of the difference in pressures between the aeration chamber and the external environment. The flow goes through the strainer, and then with the feed line φ – 160 mm reaches the aeration zone.

In the aeration zone, the water–air mixture has a smaller density and reaches the upper part of the bioreactor. In this zone, a biofilm disc diffuser is located (Fig. 3.13). Aerated oxygen-rich mixture flows through the moving biofilter that consists of

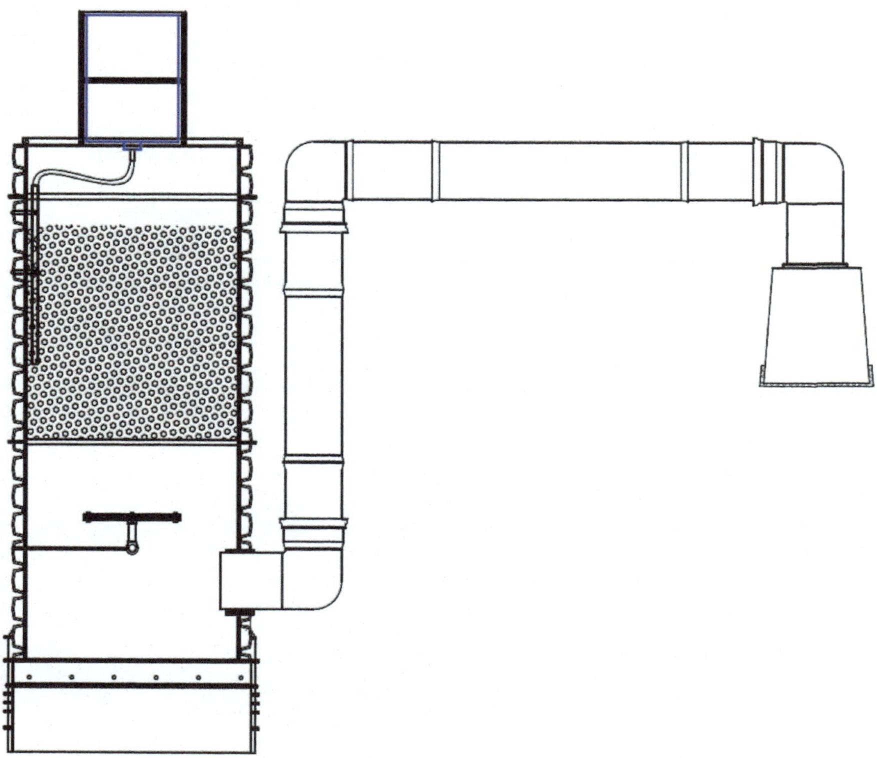

Fig. 3.13 The Moving Bed Biofilm Reactor (MBBR) model

emptying fittings and goes beyond the bioreactor in the upper zone of the reactor. The biofilter of the fittings in the system is in the limited aeration space isolated by a plastic mesh (∅ 10 mm) to prevent the loss of fittings.

It has been estimated that filling the bed from 50% to 70% slows the circulation through fittings and achieves the proper process of cleaning through the biological membrane (Fig. 3.14).

The reactor works in a flow system (Fig. 3.14). Microorganisms are used in the purification process, and cultures of effective microorganisms are administered to the bed by the cartridge of the drip applicator (Fig. 3.15). A modified construction of the cartridge with a de-aerator allows a continuous and long-lasting action of supplementing the purifying microorganisms. The diffuser, which is installed above

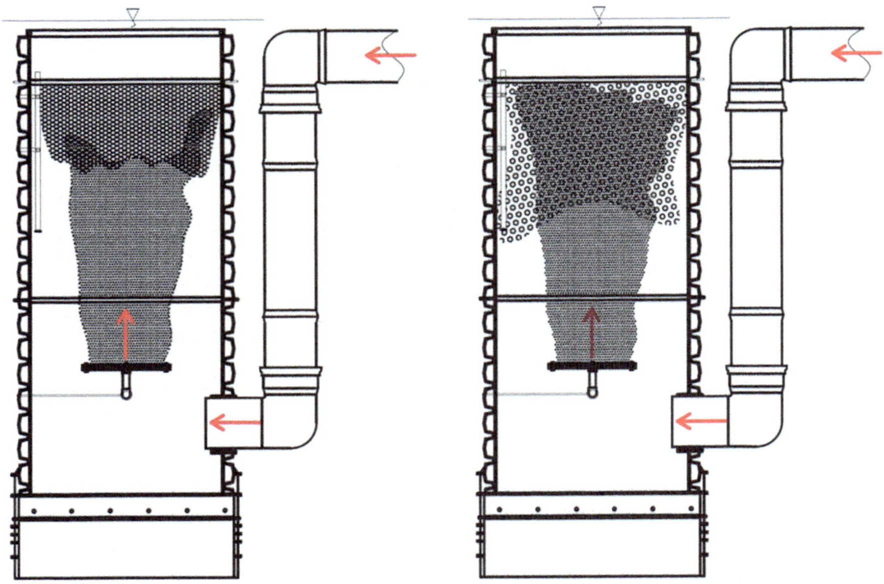

Fig. 3.14 The influence of filling quantity of the reactor with the bed on air and water flow, left: 70% full, right: 50% full

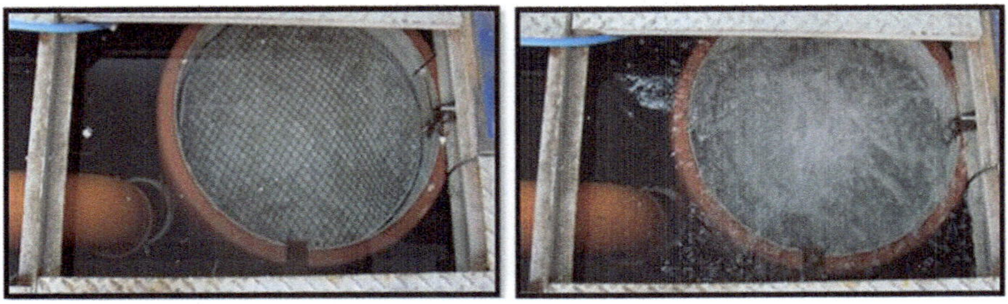

Fig. 3.15 Small-bubble aeration—formation of water–air mixture (bioreactor with the bed during the hydraulic tests). Phases with aeration and without aeration

the suction port, facilitates proper aeration and creates a water–air mixture, fueling the bubbling system. Four air ducts (hose ½-in.) are connected to the suction port and blowers (side-channels or biofilms). Fueling of the device is achieved with a mobile generator from the local power grid.

The bio-reagent is introduced at the regulated flow below the diffuser level and then uniformly distributed in the bed of the bioreactor with the water–air mixture. This system facilitates the presence of microorganisms from the bio-reagent into the bed. This allows rapid introduction of biofilm into the bed and an appropriate structure of the biocoenosis of microorganisms, depending on the specifics of the wastewater treatment station (Fig. 3.16).

In lakes, the use of this type of bioreactor with an aeration line, provides an efficient process of purification of these types of water bodies that have been degraded and polluted by large amounts of organics (Fig. 3.17.).

Fig. 3.16 The infusion applicator for bio-reagents used in the cartridge construction for containers of V = 20 L

Fig. 3.17 Remediation of lakes applying the system of Moving Bed Biofilm Reactors (MBBRs) and linear aeration

The next stage in developing the remediation project is based on advanced technology of reclamation with MBBRs, building mobile platforms fueled from renewable energy sources. Units of this type will be fully independent from stationary sources and will be able to continue the purification process in an on-line mode in any part of a water body.

Potential risk factors in areas of shale gas exploitation require on-line monitoring of flowback water from biological sewage treatment plants (Bartoszewicz et al. 2016).

3.5 Micropropagation, Androgenesis *in vitro*, and Their Integration with Laser Stimulation for Healthy and Nutritious Food

Another application of the optimization of bioprocesses is associated with research-developing studies that are focused on qualitative and quantitative improvement of vegetable production. Modern methods of *in vitro* cultures of plants are very promising and can be used in two ways: vegetative multiplication resulting in the production of identical plants in unlimited quantities, and the production of new varieties of plants. Multiplication *in vitro*, often called micropropagation, is used in the production of ornamental and fruit plants. With vegetable gardening, the micropropagation method can be used as an auxiliary tool for other areas of horticultural sciences, such as breeding, seed production, and protection. In the Research Institute of Horticulture a technique of releasing horseradish plants (*Cochlearia armoracia*) from the turnip mosaic virus (TuMV), which poses a threat for this species, and intense multiplication of healthy material using *in vitro* cultures has been developed (Górecka 1992). A method of cauliflower reproduction for the modern F_1 hybrid breeding has been devised (Górecka 1985). Effective ways of propagating tomatoes through a combination of decapitation of sterile seedling techniques with multiplication of leaf fragments and section shoots have been developed (Górecka et al. 1993, 1997).

In another use of *in vitro* cultures, the Institute of Horticulture conducted research on gametic embryogenesis of vegetables, mainly for breeding purposes.

The germ cells in *in vitro* cultures under the influence of applied external stimuli change their development from gametophyte to sporophyte and produce haploid embryos (Guha and Maheshwari 1964; Segui-Simarro 2016). The diploid plants are obtained through chromosome doubling of haploid plants, regenerated from these embryos in a spontaneous or induced way. Homozygous lines are obtained from these plants; the lines are necessary to produce heterosis hybrids.

The use of androgenesis or gynogenesis contributes to shortening the time needed to achieve homozygosity in relation to conventional breeding methods. A particular highlight is that those lines that are obtained by gametic embryogenesis *in vitro* are fully homozygous, which is not possible to achieve with traditional breeding methods. The Research Institute of Horticulture developed technologies of homozygous lines using *in vitro* androgenesis for head cabbage (*Brassica oleracea* var. *capitata*) (Górecka 1998), Brussels sprouts (*Brassica oleracea* var *gemmifera*) (Krzyżanowska and Górecka 2004), and carrots (*Daucus carota*) (Górecka et al. 2009). Plants obtained during the tests have been used for breeding purposes, which resulted in achieving the first three already registered varieties of head cabbage: Julia F_1, Justyna F_1, and Jowita F_1, and two varieties of Brussels sprouts: President F_1 and Apetite F_1. In Poland, these are the first and so far only varieties of vegetables from plants produced by androgenesis using an *in vitro* method.

The combination of biotechnological methods is very valuable. An example would be a study of the effect of laser-light stimulation on plant life processes in the *in vitro* culture of some vegetables.

Particularly promising results have been obtained in the preliminary study of the application of selective algorithms of particular laser irradiations on the stimulation of proliferation for androgenetic multiplication of carrots and gynogenetic multiplication of red beet plants.

Low-power laser-emitting red light was the most efficient in the stimulation of the production of unrooted and rooted rosettes and growth of the androgenic rosettes for the cultivation of carrots.

For multiplication of the red beet from gynogenetic embryos, the best results were obtained by combining the integration of photostimulation with the red-light laser followed by green coherent light. Previous experimental studies of different species of cultivated plants, including vegetables, resulted in a very significant increase in the accumulation of selenium and other trace elements as a result of laser treatment of seeds (Dobrowolski 1986, 1996; Dobrowolski et al. 1987).

The results clearly indicate the biostimulating nature of the applied laser light, manifested by increasing the capacity for *in vitro* regeneration of more plants per embryo. Rybiński and Garczyński (Rybiński and Garczyński 2004) have demonstrated that red light emitted by a helium-neon laser had a positive effect on barley doubled haploid plants. They noted an increase in leaf area, which results in increased photosynthesis and transpiration. Hawkins and Abrahams (2006) reported that despite the lack of full knowledge about the bio-stimulating nature of the laser radiation, the vast majority of published studies indicate that laser radiation changes the cellular processes in a non-thermal way, depending on the wavelength. Low-energy laser irradiation alters cellular functions by interacting with protein synthesis, cell growth and differentiation, and change in membrane potential (Hawkins and Abrahamse 2006).

Recently, we have developed the Yacon (*Smallanthus sonchifolius*) micropropagation method. This is an extremely valuable plant that is well suited as a nutraceutical food and also as a valuable source of nutrients and medicines, useful in cancer treatment. It contains many valuable elements including a large amount of inulin (Cazetta et al. 2005). The human body does not have the enzyme to hydrolyze inulin, therefore this sugar passes through the alimentary tract in a non-hydrolized form. This means that the Yacon plant is low-calorie and contains large amounts of fiber (Milella et al. 2011). Due to the lack of other effective methods of reproduction it is very important to develop a method of intensive propagation of this plant *in vitro*.

We started with red beet gynogenesis in order to speed up the breeding of new varieties of this vegetable. Another branch of our research has been an attempt to fortify the *in vitro* culture of carrots with zinc (Zn), a deficit of which is the cause of many diseases. Zinc deficiency due to geochemical reasons poses a health hazard for the inhabitants of many countries (Vohora and Dobrowolski 1990).

3.6 New Application of Laser Biostimulation in Biological Plant Protection

Integrated biology following an ecologically-based system approach takes into consideration very different and complementary biotechnology adapted both to protecting ecological balance and environmental health and to sustainable food production.

Biological methods for crop protection products are an alternative to the widely used chemical preparations. Biological preparations very often have similar effectiveness to chemical preparations, and, importantly, they do not pose a threat to human health and life. Living organisms used in insecticides (e.g., bacteria, fungi, nematodes), have the ability to penetrate the victim's body, reproduce in it, and even actively seek out the victim. Insecticidal fungi, as an active ingredient of biological preparations, are used in many countries. Their effectiveness depends on the species and strain of fungi, the size of the population of target insects, the climate conditions, and the form and method of application. Among the limitations are: slow response (compared to chemical preparations), sensitivity to environmental conditions, pollution (e.g., heavy metals), and difficulties with the storage of the biological preparations. From the point of view of protection of both the natural environment and consumers' health, it is very important to increase the effectiveness of biological preparations for crop protection. Among the fungi used in preparations for biological plant protection are *Isaria fumosorosea*, *Bauveria bassian,* and *Metarhizium anisopliae*, which are common in the soil environment (Shi, Feng 2004). The mechanism of action of entomopathogenic fungi is based on infecting the insect's body by means of spores, and its destruction through deprivation of water, nutrients, mechanical damage, and release of toxins. A very important factor that affects sporulation of insecticidal fungi is light (Avery et al. 2004). Research shows that constant illumination affected the increase in number of spores generated and sporulation, and that irradiation with blue light also has an effect on increasing sporulation (Sanchez-Murillo et al. 2004). As a result of the specific properties of light emitted by the lasers, and the observed number of interesting interactions between laser light and living organisms (Dobrowolski 1986, 1996; 2000b; Dobrowolski and Różanowski 1998; Dobrowolski and Zielinska-Loek 2002; Dobrowolski et al. 2012a; Jakubiak and Sliwka 2009), observations were carried out on the effects of laser light on spores of entomopathogenic fungi.

Kuźniar and Śliwka referring to Dobrowolski's (2000a, b, 2001a, 2006a) concept of application of laser biotechnology for sustainable management of biological resources and successful laser photostimulation of some pathogenic fungi

(Dobrowolski et al. 1995) conducted an experimental study on the influence of laser light on sporulation of the mould *Isaria fumosorosea.* This species of fungus is common in soil and can be easily propagated (Sahayaraj and Namasivayam 2008). It is therefore easy to produce a large number of the spores that can be used for the production of insecticides (Zimmermann 2007). Spores of *Isaria fumosorosea* in suspension were stimulated with coherent light of different wavelengths of 670 nm (red light emitted by Laser D68-1, MARP Elektronic) or 473 nm (blue light from the laser diode, Changchun New Industries Optoelectronics Tech Co.). Experimental groups were segregated in terms of the time and manner of exposure. After 24 h the length of the hyphae was rated and after 7 days of cultivation, circles were prepared from mycelium. Circles were irradiated with laser light again, with time and manner of exposure differing again for the groups as for the first part of the experiment.

After 3 weeks of cultivation under the same conditions, germination capacity and pathogenicity of spores were determined. Control groups (not irradiated with laser) were prepared and cultivated under the same conditions. Results of the experiments showed a significant effect of coherent red light (672 nm) in stimulating the growth of fungal hyphae and inhibition of the growth of fungal hyphae as a result of irradiation with blue laser light (473 nm). Exposure time of mycelium had a significant impact on the numbers of spores produced (Kuźniar 2012).

3.7 Innovative Technologies for Optimization of the Contribution of Agriculture to a Green Economy

One of the biggest future world challenges as noted by the Food and Agriculture Organization of the United Nations (FAO) (Alexandratos and Bruinsma 2012) is to provide adequate quantities of food to feed a constantly increasing global population. In this document the authors estimated that we need to increase food production by 60% between 2005/2007 and 2050 (Alexandratos and Bruinsma 2012; Van Dijk and Meijerink 2014). The current system of food production needs to implement methods that incorporate a reduction in fertilizer and pesticide use, while maintaining the same or higher yield and quality, but at the same time having lower production costs and environmental benefits (i.e., decreasing the impact on the environment). The only solution for this global problem is to propagate in the current agricultural sector sustainable farming and crop production, and introduce into it precision or climate smart agriculture (CSA) (Zhang et al. 2002; Long et al. 2016). Sustainable agriculture should be understood as being able conduct all activities related to animal husbandry and cultivation of plants while taking into account respect for the environment. Therefore, this concept (precision agriculture) is convergent with the vision of FAO (Alexandratos and Bruinsma 2012), but to achieve these goals we need to incorporate advanced techniques such as artificial intelligence, machine learning, concept-internetting, and many other innovative methods and products into the agricultural production process. A series of techniques and new issues used in precision agriculture are currently being developed.

The first devices to be described are artificial neural networks (ANNs). These methods are one of the main artificial intelligence approaches to forecasting, often providing quicker and more precise information than methods previously used to estimate the impact of toxic substances in the environment or estimate transfer of nutrients in the environmental system. Precise characteristics of ANNs (classification algorithm and multilayer perceptron (MLP)) have been described in many scientific papers and have found wide application in different fields (Bagheri et al. 2015; Carlon et al. 2004; Cigizoglu 2004; Fernando et al. 2012; Inn-Sil et al. 2002; Llanos et al. 2013; Long et al. 2016; Singh et al. 2012).

ANN models have two major applications: classification and forecasting (Hu Yi-Chung 2010; Gazzaz et al. 2012). In a paper by Czech et al. (2014), application of ANN was used for classification and forecasting (classification algorithm and MLP). These methods have been applied to predict the transfer of lead (contamination class) in the soil-plant-animal system. In this paper another method, namely MLP (Multilayer Neural Networks), was tested for developing a prediction model for determining the concentration of lead in tissue and organs of rodents based on selected environmental properties; the authors obtained promising results. The functional principle of such tools was described in detail by Hu and Weng (2009) and Sadrzadeh et al. (2008). These methods can be used to develop new farming systems and improve activities in precision agriculture.

There is no precision agriculture without innovative devices and innovative means of production. Many research teams continuously focus on reuse of waste to create new products, and good results have been achieved with implementing those products in the market. Therefore, waste plant materials derived from food production can be successfully used for manufacturing a biopolymer that is a mixture of ingredients that are acceptable and bioavailable in the natural environment. The composition could have a reduced share of synthetic polymer to allow use of more natural ingredients, which will make this polymer more favorable to nature. The composition could be made up of natural resources available in the region. Short-term use articles can also be produced, for instance by injection moulding for pots for seedlings (Gazzaz et al. 2012). Using these materials achieves multiple environmental benefits, such as for example using these materials after their primary use in production for composting with household waste, meaning the farmer will not be faced with a decision about whether it is worth supplying compost into a soil for which manufacturing was determined not by the value of the fertilizer, but rather by the necessity of having to manage biodegradable fractions of waste (Gambuś et al. 2014).

Another innovative sector of the economy that is developing fast in Poland is a fertilizer sector. Many companies work on creating more efficient fertilizers for increasing the yield of cultivated plants, and increase the rate of nutrient utilization, but often without thinking about the impact on the environment. Compared with other European countries, environmental conditions for agricultural production in Poland are, as a rule, worse [9]. The predominance of very light and light soil with low natural fertility and a relatively short vegetation period are the reason for this situation. It is thus important to care about soil fertility and maintain appropriate soil parameters that determine fertility, such as: pH, content of organic matter, content of available nutrients, etc. However, an even worse situation in Poland lies in the soil composition, because most soils in Poland have an acidic or very acidic composition, which can considerably limit crop production. Thus Czech et al. (2013), Gambuś et al. (2013) and Wieczorek et al. (2016) are working toward developing a new innovative fertilizer for soil liming, and for this purpose they use waste material from hard coal burning and also a dolomite-based deposit from the "Jozefka" mine (Wieczorek et al. 2016). In summary, many researchers in Poland see a great need for using innovative techniques for creating a more efficient agriculture system, especially to help face the biggest future global challenge, an increase in world-scale crop production of 60% by 2050.

Biological monitoring is very useful for selection of proper rural areas for the production of pollutant-free food and promotion of healthy recreation that is in touch with nature.

3.8 The Role of Biodiversity of Anthropogenic Areas in the Development of Nature-Based Tourism and Ecological Education

Application of the results of the studies on biological processes is also important for more efficient protection of the proper functioning of ecological systems. There are two complementary aspects to environmentally friendly activities. The first one includes research-developing studies of environmental biotechnology, and the second aspect, a complementary one, is education of all of society for active contribution to sustainable development. Both aspects are related to a green economy and the establishment of green jobs.

Very promising for the future is the development of qualified tourism. This includes ecotourism and agrotourism. In selecting areas that are appropriate for agrotourism it is very important to apply new methods of biological monitoring. Proper selection of rural regions is needed for the production of pollutant-free food (vegetables in particular), as well as for providing proper conditions for physical activities that are in line with nature. In this respect it is also important to provide appropriate conditions for aquaculture, which is useful for the production of pollutant-free food and the promotion of an ecological culture among anglers. Another field of environmentally friendly tourism could be in an anthropogenic landscape after renaturization. The residents of big cities often look afar for nature experiences, not realizing that these experiences are available to them close by. In any kind of nature-based tourism as well as in ecological education the key issue is biodiversity.

The concept of biodiversity is relatively new. The term "biological diversity" was first used by Lovejoy (de Andrade Franco 2013) and was primarily referred to incorporation of different species. Later it was extended into genetic variability and diversity of ecosystems. Biodiversity is usually associated with remote areas that have minimum human involvement. An example of making such areas accessible to tourists is the Whale and Seabird Cruises in Canada. This is also an example of combining scientific research, commercial activities and environmental education

(Brier Island, Nova Scotia, Canada, September 2014, personal communication with the crew of the boat).

Nevertheless, cultural landscapes (cities, suburban areas, agricultural land, post-industrial areas) can also be rich in biodiversity. Paradoxically, in an anthropogenic environment some species might find it easier to thrive than in the wild. They are often easier to observe in an anthropogenic environment, which is important in the development of nature-based tourism. In some cases, the populations of endangered species adapt well to a human-made environment and their demographic situation is better than in the wild habitat. An example is the population of Barbary macaques (*Macaca sylvanus*)—a species endangered in their natural environment in Morocco and Algeria, but having a well-established semi-natural population in Gibraltar (Maibeche et al. 2015; Cortes and Shaw 2006).

The processes of synathropization (adaptation of organisms to life in a human-made environment) and synurbization (adaptation to an urban environment) have intensified over the last several decades. Thus studies on the urban fauna and flora have been intensifying and their role in maintaining biodiversity on a local, regional, and global scale is being more and more appreciated (Adamski et al. 2005; Kudłek et al. 2005; Moretti et al. 2012; Orlewicz-Musiał and Wagner 2012; Wagner and Orlewicz-Musiał 2011, 2014).

One of the examples of an anthropogenic environment being rich in biodiversity, is man-made water bodies. They are also places that are attractive for tourists, especially in summer, when swimming is possible. Apart from swimming, other activities can be carried out there, like water sports and angling. In Kraków there are two artificial water bodies officially recognized as swimming resorts. Both are borrow pits formed after the excavation of sand and gravel. One, called Bagry (with the area of 30 ha), is situated in the southern part of the city, on the right bank of the Vistula River. The other is in the eastern part of the city, on the left bank of the Vistula, and is called Przylasek Rusiecki. It is a complex of 11 ponds of a total area of 82 ha. Both places provide habitats for wildlife, especially bird species, including species that were included in the Annex I of the Bird Directive (Directive 2009/147/EC), which lists particularly threatened species. At Bagry the common tern (*Sterna hirundo*) has been found (Wagner and Orlewicz-Musiał 2014) and in Przylasek there is a nesting place of the white stork (*Ciconia ciconia*) (Adamski et al. 2005; Kudłek et al. 2005). The other bathing resorts used by the residents of Kraków are the borrow pits Zalew na Piaskach (formerly known as Kryspinów) in the commune of Liszki (west of Kraków), which is the water body most frequently visited by the residents of Kraków (Wagner 2007), and the borrow pit of Zabierzów Bocheński—east of Kraków in the commune of Niepołomice. The latter place is very rich in bird life. Apart from the common tern (*Sterna hirundo*), the *Sterna albifrons* has also been observed (Dariusz Baran, personal communication, 1996). Another species included in the Directive is the whooper swan (*Cygnus cygnus*). The whooper swan has also been observed in the Nowa Huta (the quarter of Kraków) in the water body called Zalew Nowohucki (Wagner 2012). This place is a 7-ha water reservoir, which was initially developed as a water reservoir for the purposes of industry, but later used for various recreational purposes; however, at present bathing is officially forbidden there. The function of this water body is now as an angling resort.

Two water bodies in Kraków are places officially recognized as environmentally useful areas (in Polish "*użytek ekologiczny*") by the City Council of Krakow. These are the ponds of Dąbie (Uchwała nr XC/1202/10 XC/1202/10) and the pond of Kaczeńcowa Street (Uchwała nr XXXI/405/07). The first one is situated near the city center and is the habitat for the Amur bitterling (*Rhodeus sericeus*), a species included into the Habitats Directive (Council Directive 92/43/EEC).

The presence of people does not seem to be very harmful to the fauna. The escape distance seems to be small, which makes good conditions for bird watching. Nevertheless, something that is lacking is the information about the plant and animal species living there. Only recently (2017 - observation by A. Wagner) information boards about one bird spicies and some plant species were put in Bagry area. Even in environmentally useful areas like Dąbie and Kaczeńcowa, the available information is very general. Information should be placed in posters like the ones in another ecologically useful area in Kraków— "Łąki Nowohuckie" (Orlewicz-Musiał and Wagner 2014), where there is a didactic path and information; posters regarding habitats as well as plant and bird species. Water bodies have been studied in the international project of Terre de Rivieres. This project was given the task of comparing the existing water environments in different European countries. It brought forth many ideas related to educational activities connected to the protection of water areas (Dobrowolski and Wagner 2006).

An open question referring to biodiversity in the anthropogenic landscape is the invasion of alien species. In 2009, 16 locations were studied in and around the water bodies in Kraków and vicinity, focusing on alien plant species (Wagner and Hruševar 2015). The authors found 52 alien taxa (20.9% of all the species found in the studied area). Within this number there were 30 neophyte taxa and 18 archaeophytes. Four taxa have uncertain status. Tokarska-Guzik et al. (2012), studying the entire Polish territory, reported a higher percentage of alien species (27.4%), but a lower percentage of archaeophytes. The presence of alien species is acceptable in an urban environment, such as in parks and other organized green areas, where they can be controlled. With regard to alien animal species things are more complicated. Sometimes introduced species contribute to biodiversity and tourist attractiveness (as with the abovementioned Barbary macaques in Gibraltar), but sometimes they can dangerously compete with the native species (like gray squirrels in Britain or feral parrot populations in Europe (Menchetti and Mori 2014)). In Kraków, like in many cities worldwide, the best known feral species is the rock pigeon (*Columba livia var. domestica*), which has been present since Medieval times and its impact on the urban environment has been widely discussed (Blechman 2007).

A particular case of an anthropogenic environment already mentioned is the agricultural environment. An example of such an environment is the National Park of Cinque Terre (Italy). In 2005 a Smart History Project was carried out there, with the idea to move the enormous amount of tourism they experience from the coastal area to the vineyard area situated in the mountains. A project that delivered more adequate tourist information, including electronic devices, was embarked on ("From Smart History towards common European Heritage..." 2006, Dobrowolski et al. 2006).

3.9 Bio-based Training and Life-Long Education for Cooperation Experts and a Knowledge-Based Society for Sustainable Management of Biological Resources

Sustainable development is linked to a balance in efficient protection of life and a logical balance in the entire biosphere and good economy based on low resources and low energy, environmentally-friendly technologies, i.e., eco-innovation. The most sustainable management of natural resources as well as renewable sources of energy is a bio-based green economy. This concept is also associated with a sustainable model of consumption and sustainable society in the future.

Sustainable development is recommended by the European Union (EU) and the United Nations (e.g., by UNs GA in October 2016). The turning point in understanding the importance of sustainable management of natural resources was the Report of the former UN Secretary General U'Thant's Report of the Club of Rome and Bruntland's Team Report "Our Common Future."

The concept of sustainable management of natural resources facing the needs of the future generation was introduced by former president (rector) of the AGH University of Science and Technology in Krakow, Prof. Walery Goetel, over 60 years ago. GA of the IUCN in UK adopted his concept in 1956, and changed the name to the International Union for Conservation of Nature and Natural Resources. Prof. Goetel also recommended interdisciplinary cooperation of experts in natural, technical, and social sciences within so-called sozology and sozotechnology focused on human-oriented activity. He also introduced open-for-all seminars on the protection of nature and sustainable management of natural resources for the future about half a century ago.

Representatives of three generations took an active part in the related activities. The leading idea of the series of our International Conference from 1989 to 2012 is the common action of experts and knowledge-based society for sustainable management of natural resources (including energy) focused on improvement of the quality of the human environment and quality of life for all.

The key factor for better efficiency seems to be integration of problem-solving training of future experts with progress in science and technology (based on interdisciplinary case studies) as well as with implementations in selected regions as a result of cooperation with local stakeholders and decision-makers.

What is promising for future activity is integration of a common action in different regions and countries with an International Network for Cooperation focused on common global problems, e.g., adaptation to climate change, ozone layer depletion, and prevention of deterioration of the natural environment. The crucial issue is more efficient protection of natural resources for a sustainable society in the near future.

Let us hope that it is still possible to change the trends of overexploitation of biological resources (as a result of the positive feedback system of the common consumption model and "cheap" but polluting technologies) into environmentally-friendly models of consumption based on clean biotechnologies as a way toward a green economy. This basic change in the model of civilization is necessary for mankind to progress into a negative feedback system of maintaining a homeostatic balance in the entire biosphere.

The top priority of a bio-based green economy is the direction that the international scientific community would like to support. There are good examples of sustainability—for example, the Chinese economy, which for centuries was based on biotechnologies connected to rice, silk, tea, vegetables, and other food products. Our hope lies not only with great progress in new fields of technology as a result of learning from nature-like biomimetics (bionics), biomaterial engineering—including bio-nanotechnology, photobiology, applications of bio-mining, and extremophiles in environmental engineering—but also in development on an international scale of ecological engineering as initiated by the Polish professor Henryk Zimny in Warsaw more than 50 years ago. It is a promising concept for the future to integrate study on bionanoparticles (Böker and van Rijn 2015; Junhui He 2016) and nanobiophysics (Karachevtsev 2016).

Widescale introduction of a new model of professional training and education for all that can face these challenges is necessary for the proper implementation of the lessons learnt from nature, and the successful cooperation of experts (in sozology, sozotechnology, and ecological engineering) with knowledge-based society and decision-makers.

In Poland there is a very long tradition of interdisciplinary research (beginning from Nicolaus Copernicus in 15th century, as well as E. Czyrniański and M. Skłodowska-Curie in 19th century) and education for sustainable management of natural resources linked with the humanization of technical subjects.

Following the inspiration of Prof. Walery Goetel, we have to recommend (e.g., within the Bologna process of modernization of universities) basic knowledge of sozology and sozotechnology as obligatory elements of fundmental education necessary for an efficient partnership between experts and knowledge-based society. This would be a promising way forward for creation of a sustainable society that would be a responsible community with regard to sustainable use of natural resources (focused on the protection of ecological balance in the whole biosphere) and proper reproduction of biological resources, related to nutrition for future generations and protection of biodiversity in particular.

Let us begin with the recommendation to introduce this kind of basic knowledge as being obligatory for all the students of AGH University of Science and Technology, as a starting point for dissemination of such a practice on an international scale. With regard to long-term tradition, let us also consider founding an International Centre of Sustainable Development and Eco-biotechnology (based on innovative environmental biotechnology and human ecology).

The great challenge is dissemination of good practice from different leading universities and cooperation with teams of top experts and with the World Academy of Arts and Science and the Royal Academy Institute of Spain, in particular.

Let us refer to some examples of good practice that have already been tested over time. Dobrowolski, as the chairman of this Conference (as well as several other interdisciplinary international meetings of experts focused on sustainable society since 1989), introduced 47 years ago the concept of problem-solving training in this field. This concept was based on a regional common action of experts (in natural, technical, social, and other disciplines) and local society.

The logo of our preparatory Seminar in 1967 and Polish Summer Schools on the Human Environment (since 1968) as well as International Summer Schools on Sustainable Development (since 1972) was the image of the European bison (*Bison bonasus*)—a symbol of the protection of nature—combined with images of the Earth and a spaceship. This was a starting point for the integration of global thinking with local (on a regional scale) activity for better use of new technical achievements, e.g. space technology, as being useful for better knowledge about limited natural resources on a global scale and the importance of protection of ecological balance (as in long-term manned outer space missions). The author, during his university studies, in a seminar discussion with Prof. W. Goetel and a well-known expert in science fiction Stanisław Lem—introduced the idea of interdisciplinary and international cooperation for better protection of the natural environment outside of Earth (including the Moon or Mars). This idea seems to be very up-to-date, especially if we take into consideration contamination of the upper part of the Earth's atmosphere with still more and more anthropogenic waste materials. There are some potential risk factors for life on Earth connected with possible contamination of Mars with some anaerobic bacteria during automatic and/or manned missions on this planet and the reintroduction of mutants adapted to a much more extreme environment to the biosphere.

We have to answer the question of how experts (scientists and practitioners) could contribute more in a common action, together with knowledge-based society, to make better use of the progress in different fields of science and technology for improvement of the quality of human life (facing daily problems).

There are at least two fields of integrated action of experts and society, namely common action on a local scale within the habitat (e.g., village, town, or a district of a large city) and a common action of inhabitants and visitors involved in qualified tourism linked with education about nature and cultural heritage and protection of the cultural landscape.

Our common hope for improvement of the quality of the human environment and changing the model of overexploitation of natural resources into sustainable management of natural resources and energy, can became reality if basic knowledge in this field is seen as an obligatory element of training for students of all subjects. This is the target group for knowledge-based society as well as the future decision-makers.

Sustainability of universities as well as daily habits (including campuses) and also promotion of a sustainable society by qualified tourism seems to be the appropriate way towards changing the contemporary model of civilization based on consumption. Our beneficial experiences of 44 years of voluntary activity from the students' scientific club (non-government organizations) indicated a real perspective involving all students (of natural, technical, social, and many other study subjects) into a common action in cooperation with both top experts and local inhabitants of different regions.

Complementary experiences from different countries should be gathered into common databases and supplemented by an effective system of data mining. As experts, we have to face a common expectation and interest from younger generations, e.g., in widescale application of information technology including games. Instead of very popular games promoting pathological behavior, we have to develop cooperation with teachers involved in all levels of education (from kindergartens to postgraduate courses) to motivate the younger generations (involved both in formal and in voluntary forms of education) to take an active part in the improvement of both the indoor environment and the natural environment in the area they are living in or visiting for recreation, which is connected with education and physical activity in a clean, healthy environment.

The keen interest of the younger generation in games may be a very valuable tool for getting them involved in improvement of the quality of their environment in a heuristic way. They have to understand the importance of widescale introduction of clean technologies as being good for the environment and health, as well as necessary for securing the proper quality and quantity of natural resources for the stabilization of development in the future (which is especially important for the younger generation). The Founding International Network of Training promoting training of staff focused on eco-innovation being important both for developed as well as developing countries (including training of experts in industrialized countries to help less developed regions of the world).

Dissemination of good practice in a common action of knowledge-based society and products (as a corporate social responsibility, e.g. in Scandinavian countries) seems to be very promising for a better future for all (including feedback systems between expectations of consumers forcing the introduction of low-resource and low-energy clean technologies).

The curricula of training of experts and education of the whole society should take into consideration up-to-date knowledge and ability in application of scientific and technical achievements for early detection of potential risk factors (e.g., for health, nature, culture heritage, as well as for ecological balance, climate, and natural resources for the future) as well as for a common action of experts and society as a whole, focusing on effective improvement of quality of the human environment and for better, sustainable management of natural resources, e.g., waste management, use of renewable clean sources of energy as well as dissemination of better biotechnologies for bioremediation and prevention of ecological catastrophe. We could introduce international guidelines based on good practice of effective environmental monitoring (including synergistic effects of different environmental factors and control of individual, personal exposure and related health hazards) as well as new cheap and simple biotechnology, e.g., laser biotechnology for better treatment of waste water, reclamation of degraded areas, and biomass and bio-energy production.

There are also good examples of international cooperation helping victims of ecological catastrophes, e.g., in the Minamata region in Japan, Bhopal region in India (Vohora and Dobrowolski 1990), Chernobyl region in Ukraine, and more recently in the Fukushima region in Japan. Let us start with training experts in providing more efficient help for the victims of these or similar catastrophes in the future as well as for better adaptation to green-house effects and related processes of desertification and more frequent and larger flood occurrences.

Let us also develop different forms of life-long education and intergenerational and international solidarity with dissemination of good practice on a wider scale. Let us refer for example to the open seminars introduced by Prof. Goetel, our AGH-UST Open University and a new concept of cooperation between the AGH-UST Open University with the Jawaharlal Nehru National Open University, focused on modern distance learning, for a common action of experts and contemporary information in society for promoting sustainable development in countries such as Poland and India. Complementary to these beneficial experiences of long-term activity of national networks of student scientific clubs (founded in 1972 in the Polish University Youth Committee of Environmental Management and Protection), some other nationwide environmental NGOs, e.g., the Polish Ecological Club, as well as international NGOs, founded 40 years ago the Youth Federation for Environmental Studies and Conservation (IYF) and 30 years ago Youth and Environmental Europe (YEE).

Creating a modern system for permanent exchange of useful information among experts from different fields as well as scientific clubs and environmental NGOs of young generation seems to be much needed for a more efficient common action integrating top-quality experts in bodies like WAAS and national academies with voluntary activity of university students' NGOs.

Let us focus on the problem of contamination of the natural environment and climate change. According to the report of the European Environment Agency of 2011, the Carbon Capture and Storage Technologies for an additional 15–25% more energy produced by burning more coal is required. Therefore, the application of these technologies will be associated with the increase of nitrogen oxides and particulate matter emission that already exceed permissible level in many regions. According to forecasting, it is also expected that there will be an increase in pollution by ammonia, as well as expansion of the eutrophication process already affecting ca. 70% of sensitive ecosystems in EU countries.

As a result of anthropogenic pollution more than 80% of the EU urban population is subject to health hazards linked to high exposure to particulate matter (less than 2.5 nm in size, as nanoparticles become more and more important as a new type of harmful pollutant for human health and ecosystems).

In this situation as extensive as possible dissemination of biological methods of fixation of CO_2 and biodegradation and bioremediation, as well as wider application of biomass (of organic wastes as well as biomass produced by algae and other water plants in hydrobotanic wastewater plants and in energy plantations in areas out of use), should be recommended as the best contribution to the prevention of pollution of the natural environment, disturbance of the ecological balance as well as the best contribution to the reduction of greenhouse gasses. These methods should be the best from both ecological and economical points of view.

Another crucial problem is the development of interdisciplinary studies and training focused on proper adaptation of environmental management and economic activity to climatic change on a global scale (as a network of cooperating regions).

Introduction of new curricula and an international network for new situational training activities at university level could also contribute to a very significant development of the labor market with regard to long-term positive experiences in this field for some developed countries. According to the forecasting of the Political Economic Research Institute of the University of Massachusetts, introduction of new technologies linked to new air clean rules could create a large number of new jobs, e.g., in the USA about 1.5 million jobs over 5 years. Development of global investments driven by promotion-sustainable development seems to be an important factor for reducing unemployment among graduates of universities. The crucial factor for optimal use of related financial sources is modernization of training activity at universities (by developing problem-solving interdisciplinary and innovative approaches) in combination with large-scale education of society for a common action focused on a better quality of life and proper management of the natural resources in the future.

This may entail a contribution to the new model of reasonable life instead of the overconsumption of the contemporary society, depending on computers, cars, and other machines. It is important that experts and decision-makers (politicians, owners, and managers of enterprises) can contribute to the development of the labor market (especially through dissemination of innovative eco-technologies, which are useful for sustainable management of natural resources and energy), as well as provide protection of the quality of the human environment for future generations.

There are new challenges for mankind that are connected with sustainable exploitation of resources under extreme conditions (e.g., in the depths of the oceans), introduction of new environmentally friendly technologies for production of new-generation materials in industrial centers on the Earth's orbit as well as in outer space man-made ecosystems (mines and industrial centers on the Moon, Mars, etc.), including training of experts in the protection of the natural environment and development of new technologies adopted to new needs of bioeconomy and human ecology in the future (Dobrowolski 2006b).

Let us start with the database (including cooperation on modern data mining and application of artificial intelligence for automatic knowledge discovering) and network for permanent international cooperation related to both the practical use of progress in science and biotechnology for everybody as well as training of experts for the future system for solving crucial problems based on a heuristic approach and dissemination of good practice, supplemented by internet-based training, international schools, and workshops, are steps in the right direction for these projects.

We already have positive effects from tests in practice for very long periods of implementation of bioeconomy; this active contribution is open for students and graduates of all subjects (e.g., in national and international summer schools and workshops under J. W. Dobrowolski's scientific leadership that took part in over 40 subjects of technical, natural, social, and economic studies from people active in culture, policy, etc.).

We also have positive experiences with university education for over 4000 university students (from Voluntary Scientific Clubs as NGOs) contributing to interdisciplinary studies and common action for improvement of the quality of life, together with local communities and visitors to regions of national parks, health resort areas for recreation, etc. in different regions in Poland, as well as in Spain and Italy (especially in Florence in cooperation with 10 years of activity of Workshops of the Del Blanco Foundation and new action Life beyond Tourism) as well as in the European Projects promoting a sustainable society like in the model area of Cinque Terre and within the UTdR International Network of River Regions promoting Sustainable Development (cooperation of 18 universities and local administrative bodies from Portugal, Spain, France, Italy, UK, Germany, Hungary, Slovakia, and Poland).

Other positive experiences are associated with a contribution from our team of scholars and interested university students in the European Project Smart History, which was focused on interdisciplinary cooperation for promotion of qualified tourism and sustainable management of areas of great natural and historical value. The pilot project was related to the region of the Italian Cinque Terre National Park. The output of the project and related workshop was sustainable design for protection of the cultural landscape and promotion of sustainable tourism in association with education related to personal topics of interest as well as international e-handbooks.

There is also a positive experience with long-term cooperation with the top quality experts in complementary fields from Japan, beginning at the First International Congress of Scientists for Better Environment in Kyoto in 1975, and common action in Poland (through the introduction of a Japanese method for the detection of air and water pollutants as risk factors for ecosystems, human health, and culture heritage) as well as for efficient and cheap biotechnology of wastewater treatment, and a common action helping victims of ecological disasters in the Minamata area in Japan. A world famous expert from the Minamata, Prof. M. Harada, was the main scientific adviser for a documentary film about the history of contamination of the biosphere with mercury and the etiology of Minamata disease. This educational film was given an UNEP award and was supported by the results of a long-term study of global action for elimination of the use of mercury, after the Minamata case study. Dobrowolski et al. also took an active part in cooperation with the Indian scientists Prof. S.B. Vohora, Prof. B. Belsare, et al. in 1983 in providing help for victims of the ecological disaster in Bhopal. Introduction of an innovative biotechnology was very useful for the reduction of environmental health hazards in both regions. Joint interdisciplinary and international studies were followed by publications and a new doctoral dissertation at the Faculty of Biotechnology at Bhopal University.

Training of expert staff in new complementary biotechnology (including more efficient wastewater treatment, waste management, and energy production as well as protection of the indoor environment against both toxinogenic fungi as well as air pollutants) is a promising way forward for reduction of global health risks connected with air and water pollutants. There are several million premature deaths annually, both in developing countries and in big cities in developed countries, as a result of the common use of out-of-date technologies. Replacement of high energy consuming technologies through the widescale application of environmentally friendly biotechnologies could significantly reduce emission of chemical and microbiological pollutants at their sources and contribute to a much longer lifetime expectancy as well as improve the quality of environmental health. Training of experts in this field should be supplemented by life-long education of society as a whole.

With reference to major progress in the activity of open universities in many countries all over the world and so-called Universities of the Third Generation in Europe, I want to briefly elaborate on good practice in voluntary education of knowledge-based sustainable society. We started in 1989 with a common reflection of Bruntland Report "Our Common Future" and we also have sound experience spanning 26 years of the activity of the AGH-UST Open University. This university is open to different age groups, beginning from teenagers, candidates for becoming academic students, to over 80-year-old men. Among the lecturers are well-known experts from the majority of Polish universities, the Polish Academy of Sciences, research institutes, scientists from other countries of Polish origin, as well as representatives of local and central self-governing bodies, administration (including mayors, MPs, ministers), as well as people holding cultural roles.

We hope to exchange new ideas, useful methodological experiences, and good practice as a creative contribution to the development of a permanent cooperation focused on better protection of the quality of human life, *making also our life more reasonable and useful*, according to the poem "*An hour*" J.W. Dobrowolski received from a Nobel Prize Winner Czesław Milosz in 1981:

Before the five senses were opened, and earlier than any beginning,
They waited, ready for all those who would call themselves mortals,
*So that they might praise, as I do,**life**, that is,**happiness**.*

3.10 Innovative Biotechnology Driven by Education for a Bio-based Green Economy

Following official recommendations of EU authorities, a bio-based green economy has been a top priority for high education. In referring to this, it is worth to mention the good practice of 47 years of problem-solving training based on eco-innovations and a green economy for sustainable development. Related methodological experiences seem to be useful for integration of training staff of experts with trends in the labor market (including foresight analysis) and cooperation with the knowledge-based society in different regions and countries in a global network for common action focused on improvement in the quality of the human environment and life (Dobrowolski 2014, 2016b).

Research and educational activity from Dobrowolski and former diploma and doctoral students were focused on the integration of biotechnology for early detection of environmental risk factors with new biotechnological methods for more efficient elimination of pollutants in air, water, soil, and the food chain, as well as the indoor environment as risk factors for human and animal health.

One subject of "the case study" in clusters of proliferative diseases in rural regions versus control areas was interdisciplinary research of differences in the level of different elements in the trophic chains of cattle suffering from leukemia and in healthy subjects, as well as in human blood from subjects suffering from proliferative diseases and control donors, and comparison of the frequency of the incidence of the toxinogenic fungi *Aspergillus flavus* and *Penicillium meleagrinum* in the indoor environment. Field studies were supplemented by laboratory investigations on experimental animals and *in vitro* cultures of human blood cells under the influence of mycotoxins produced by these moulds. Both studies were followed by a practical output of a new patent of mycostatic additives for protection against contamination by toxinogenic fungi, as well as by recommendations of nutritional prevention against health hazards to consumers by fortification of food with deficient trace elements in cancer/leukemia clusters.

Another field of problem-solving study and training is related to monitoring of individual exposure to air pollutants such as SO_2 and related monitoring of sulfur in single blood cells (using cell monitoring by SEM and X-ray microanalysis of elements) as well as evaluation of individual health hazards. Personal monitoring was followed by medical education regarding the health of people with a high concentration of toxic elements like Pb and Cd in their blood cells, or a deficiency of the elements necessary for proper function of the human body, i.e., iron, magnesium, zinc, selenium and iodine. (Dobrowolski 1986, 1996, 2016b Vohora and Dobrowolski 1990).

J. W. Dobrowolski and his students applied Amaya-Krochmal samplers to monitor changes in the exposure of nature and cultural heritage in different regions (historical cities, health resorts, national parks, etc.) over the long term (more than 30 years). The monitoring referred to the mean concentration in the air of some indicators of traffic output and other sources of low emission. These transdisciplinary case studies were supplemented by the author introducing laser photostimulation of different species of bushes for acceleration of formation of protective high hedges. This is a new biotechnology that is useful for limiting dissemination of traffic output and more efficient protection of biodiversity and environmental health.

In the areas of cancer clusters i.e., the areas with higher concentration of nitrosamine precursors, laser biostimulation of local water plants was applied in combination with ecological engineering for more efficient treatment of wastewater and protection of the animal and human food chains against contamination by carcinogenic nitrosamines (including synergistic effects with mycotoxins).

Modern management of biodegradable wastes can provide a substantial amount of renewable energy (Horsek and Hrebicek 2014). Wider application of organic wastes and bioenergy based on innovative technology was recommended by the president of the International Consortium of Clean Energy Prof. G.R. Grob at the 14th International Conference on Sustainable Development and Eco-innovation at AGH-UST in Krakow, Poland in 2012 (keynote report).

Integration of interdisciplinary case studies including human ecology, ecotoxicology, and innovative environmental biotechnology could be recommended for widescale application of a more efficient primary prevention of environmental health hazards as well as for better management of city environments and rural regions in particular (Dobrowolski 2001a, b).

These innovative research-developing studies were supplemented by a transfer of know-how from several countries during training activities of creative diploma and doctoral students in Poland and during scientific visits in different regions of the world, as well as 14 International Conferences on Sustainable Development and

Eco-innovation chaired by Dobrowolski from 1989 to 2012. For 26 years our staff contributed at the AGH-UST Open University in a cooperation among experts from complementary fields and inter-generational integration of a knowledge-based society. The leading subject of the training was integration of a common action of experts and knowledge-based society for more efficient protection of the biosphere and the improvement of human life.

In our old university city, Krakow, in Poland, the Interuniversity Team of Sustainable Development and Eco-innovation was founded for promotion of innovative interdisciplinary research and problem-solving training on a national and international scale.

The main goal of this model activity is the common action for improvement in the quality of the human environment and life on a local scale in combination with consideration of global problems and the contribution to the creation of new green jobs through the introduction of innovative biotechnologies, including the transfer of useful know-how to developing countries.

3.11 Conclusions

Complementary applications of laser biotechnology would be useful for better prevention against contamination of the natural environment, providing conditions for primary prevention against related health hazards as well as for proper reproduction of plants and animals, *ecological balance and biodiversity*.

There are real prospects of widescale application of laser irradiation of seedlings before their cultivation in energy plantations with supplementation of their wood with inexpensive inorganic catalysts for enhancement of syngas production (Tursunov and Dobrowolski 2015). A hybrid approach to increasing production of biogas from municipal waste is connected to the fortification by a cheap powder like dolomite (Tursunov et al. 2015a, b) as a complementary eco-innovation for promotion of a green economy in different regions.

Other environmentally friendly applications of laser biotechnology were also successfully tested resulting in three times the acceleration of the formation of tall hedges alongside main roads and car parks, in combination with a sustainable design of new living houses including a new generation of Eco-hotels, habitats, and areas for recreation (Dobrowolski et al. unpublished).

This study confirmed that biotechnological methods, including *in vitro* culture, contributed to the knowledge of the life processes of plants, especially with regard to vegetative and generative propagation.

Information that we gained led to the development of methods of very fast and intensive specific plant multiplication and technology output starting with materials for creating modern hybrids. Application of plants obtained during experiments led to the first varieties of vegetables produced using androgenesis to be registered in Poland.

3.12 Perspectives for the Future

Widescale application of *laser biotechnology* in combination with ecoengineering could contribute to *better adaptation of environmental management to climatic change (in particular in relation to sewage treatment and reuse of water, production of food, biomass, and waste management to bio-energy)* and also to the creation of many green jobs and a contribution to sustainable development as a bio-based *green economy* of many countries.

Possibilities of developing interdisciplinary study integrating laser photostimulation with the introduction of nanoparticles (NPs) containing Zn for enhancement of Zn bioaccumulation in some species of vegetables have already been tested. Therefore, more such research-developing studies are recommended.

Based on haploidization and micropropagation vegetable technologies, there is the opportunity to develop research on the effect of laser light on improving productivity of the technologies and increasing the content of necessary trace elements and biologically active substances in plants *in vitro*.

Studies on Yacon micropropagation open up the prospect of research-developing studies and the prospect of commercialization of using this plant for the protection of health in both preventive measures and supportive therapy.

Acknowledgements The scientific work presented in Sect. 3 was supported by the National Centre for Research and Development, as Strategic Project PS/E/2/66420/10 "Advanced Technologies for Energy Generation: Oxy-combustion technology for PC and FBC boilers with CO_2 capture". The support is gratefully acknowledged.

References

Adamski P, Barszcz P, Bąba W, Cierlik G, Jarkiewicz K, Kalemba A, Kjas ZJ, Kurzyński J, Mielnicka B, Mróz W, Partyka J, Perzanowska J, Ślizowski J, Tadel A, Zarzyka-Ryszka M (2005) Skarby Przyrody i kultury Krakowa i okolic. Ekologiczne ścieżki edukacyjne. IOP PAN, OSF WT PAT, INOŚ UJ, Wyd. WAM, Kraków. 495 p. ISBN: 83-7318-608-5

Alexandratos N, Bruinsma J (2012) World agriculture towards 2030/2050: the 2012 revision. ESA working paper no. 12-03. June 2012. Agricultural Development Economics Division. Food and Agriculture Organization of the United Nations, pp 1–154. www.fao.org/economic/esa

Amaya K, Tanuma K, Dobrowolski JW (1990) "Bando" biotechnology of sewage treatment and its application in Japan and Poland. In: Norwicz K, Rajpolt B, Dobrowolski JW (eds) Perspektywy zastosowania nowej metody oczyszczania ścieków w Polsce na tle wyników badań testowych japońskiej oczyszczalni "Bando". Zeszyty Naukowe AGH, Sozologia i Sozotechnika, z. 28:7–20. Pages 147, PL ISSN 0138-0923

Avery PB, Faull J, Simmonds MSJ (2004) Effect of different photoperiods on the growth, infectivity and colonization of Trinidadian strains of *Paecilomyces fumosoroseus* on the greenhouse whitefly, *Trialeurodes vaporariorum*, using a glass slide bioassay. J Insect Sci 4(38):10. https://doi.org/10.1080/09583157.2010.515299

Badek B, Romanowska-Duda Z, Grzesik M, Kuras A (2016) Physiological markers for assessing germinability of *Lycopersicon esculentum* seeds primed by environment-friendly methods. Pol J Environ Stud 25(5):1831–1838. https://doi.org/10.15244/pjoes/63065

Bagheri M, Mirbagheria SA, Ehteshamia M, Bagheriba Z (2015) Modeling of a sequencing batch reactor treating municipal wastewater using multi-layer perceptron and radial basis function artificial neural networks. Process Saf Environ 93:111–123. https://doi.org/10.1016/j.psep.2014.04.006

Bain RL (2007) World biofuels assessment, worldwide biomass potential: technology characterizations. National Renewable Energy Laboratory, Golden. NREL/MP-510-42467

Bartoszewicz M, Michalska M, Cieszynska-Semenowicz M, Czernych R, Wolska L (2016) The problem of wastewater in shale gas exploitation. The influence of fracturing flowback water on activated sludge at a wastewater treatment plant. Pol J Environ Stud 3(5):1839–1845. ISBN: 1230-1485

Bauen A, Berndes G, Junginger M, Londo M, Vuille F, Ball R (2009) Bioenergy – a sustainable and reliable energy source: a review of status and prospects. IEA Bioenergy. IEA Bioenergy ExCo:2009:06

Blechman A (2007) Pigeons-the fascinating saga of the world's most revered and reviled bird. University of Queensland Press, St Lucia, 256 p. ISBN 10:0702236411 ISBN 13:9780702236419

Böker A, van Rijn P (eds) (2015) Bio-synthetic hybrid materials and bionanoparticles. A biological chemical approach towards material science, Royal Society of Chemistry, Cambridge, p 314. ISBN: 978-1-84973-822-4

Carlon C, Dalla Valle M, Marconimi A (2004) Regression models to predict water-soil heavy metals partion coefficients in risk assessment studies. Environ Pollut 127:109–115. ISSN 0269-7491

Cazetta ML, Martins PMM, Monti R, Contiero J (2005) Yacon (*Polymnia sanchifolia*) extract as a substrate to produce inulinase by *Kluyveromyces marxianus* var. *bulgaricus*. J Food Process Eng 66(3):301–305. https://doi.org/10.1016/jjfoodeng.2004.03.022

Chum H, Faaij A, Moreira J, Berndes G, Dhamija P, Dong H (2014) Bioenergy. In: Edenhofer O, Pichs-Madruga R, Sokona Y, Seyboth K, Matschoss P, Kadner SJ. https://www.ipcc.ch/pdf/special-reports/srren/SRREN_FD_SPM_final.pdf

Cigizoglu HK (2004) Estimation and forecasting of daily suspended sediment data by multi-layer perceptrons. Adv Water Resour 27:185–195. ISSN: 0309-1708

Cortes J, Shaw E (2006) The Gibraltar macaques: management and future. In: Hodges JK, Cortes J (eds) The Barbary macaque: biology, management and conservation. Nottingham University Press, Nottingham, pp 199–210. https://doi.org/10.1007/s10764-008-9269-2

Council Directive 92/43/EEC of 21 May 1992 on the conservation of natural habitats and of wild fauna and flora

Culture Collection of Baltic Algae (CCBA) Institute of Oceanography, University of Gdansk, Poland. http://ccba.ug.edu.pl/pages/en/home.php (on-line access: October 5, 2017)

Czech T, Gambuś F, Wieczorek J (2013) Assessment of chemical composition of waste materials from hard coal burning in view of their agricultural and environmental applications. Ecol Eng 34:89–95. https://doi.org/10.12912/23920629/323

Czech T, Gambus F, Wieczorek J (2014) Mathematical forecasting methods for predicting lead contents in animal organs on the basis of the environmental conditions. Ecotoxicol Environ Saf 110:232–238. https://doi.org/10.1016/j.ecoenv.2014.09.006

de Andrade Franco JL (2013) The concept of biodiversity and the history of conservation biology: from wilderness preservation to biodiversity conservation. Historia 32(2). https://doi.org/10.1590/S0101-90742013000200003

Directive 2009/147/EC of the European Parliament and of the Council of 30 November 2009 on the conservation of wild birds (2009) Off J European Union L 20/7, 26/01/2010

Dobrowolski JW (1986) Laser biostimulation and nutritional prevention of deficiency of essential trace elements. Magazine of Hamdard Tibbi College, Hamdard University Press, pp 5–11

Dobrowolski JW (1996) The influence of laser photostimulation of plants on bioaccumulation of elements. In: Pais I (ed) Proceedings of the international symposium new perspectives in the research of hardly known trace elements. University of Horticulture and Food Industry, Budapest, pp 47–52

Dobrowolski JW (2000a) Perspectives of application of laser biotechnology in management of the natural environment. Pol J Environ Stud 10:7–9. ISSN: 1230-1483

Dobrowolski JW (2000b) Application of laser biotechnology in environmental management, The world congress on biotechnology, Section VI, Environmental biotechnology, vol 3. Society of Chemical Engineering and Biotechnology, Berlin

Dobrowolski JW (2001a) Perspectives of application of laser biotechnology in management of the natural environment. Pol J Environ Stud 10:7–12. ISSN: 1230-1483

Dobrowolski JW (2001b) Ecotoxicology, human ecology, laser biotechnology in primary prevention of environmental health hazard. Przegl Lek 58(7):1–4. ISSN: 0033-2240

Dobrowolski JW (2006a) Innovative biological monitoring and laser biotechnology in management of the natural environment. In: Sustainable development of rivers regions, La Sapienza University, Roma. ISBN-10: 88-6049-018-9, ISBN-13: 978-88-6049-018-6

Dobrowolski JW (2006b) Preface: the main purposes of the 11th international conference Euro-Eco 2006 interdisciplinary co-operation for the sustainable development, proposals for the future cooperation. Pol J Environ Stud 15(5C):1–3 and 196–197. ISSN: 1230-1483

Dobrowolski JW (2014) Chapter 4: University education on sustainable development as a contribution to the shared responsibility of experts and knowledgebased society. In: Mulej M, Dyck RG (eds) Social responsibility-sustainability, education and management, pp 96–118. https://doi.org/10.2174/97816080590411140201

Dobrowolski JW (2016a) Laser biostimulation to enhance biodegradation of hydrocarbons, bioremediation of trace metals, reclamation of deteriorated areas, biomass production and protection of biodiversity. In: Carioca JOB, Moura Barros L, Coelho AL (eds) Biotechnology for the development of a green economy. IUPAC, Universida de Federal Ceara, Fortaleza, pp 141–150. ISBN: 978-85-420-0788-6

Dobrowolski JW (2016b) Laser biotechnology for more efficient bioremediation and promotion sustainable development. In: Carioca JOB, Moura Barros L, Coelho AL (eds) Biotechnology for the development of a green economy. IUPAC, Universida de Federal Ceara, Fortaleza, pp 151–164. ISBN: 978-85-420-0788-6

Dobrowolski JW, Różanowski B (1998) In: Anke M et al (eds) The influence of laser light on accumulation of selected macro-trace and ultra elements by some plants. Menegeund Spurenelemente, vol 18. Friedrich-Schiller-Universitat, Jena, pp 147–156

Dobrowolski JW, Wagner AW (2006) Interdisciplinary co-operation for the sustainable development of historical cities and protected areas. Perspectives of sound tourism. Pol J Environ Stud 15(5c) ISSN: 1230-1483

Dobrowolski JW, Zielinska-Loek A (2002) In: Anke M et al (eds) The laser photostimulation of willow cuttings planted alongside main roads and change of concentration of elements in the willow's organs. Mengen und Spurenelemente, vol 21. Friedrich-Schiller-Universitat, Leipzig, pp 334–340

Dobrowolski JW, Borkowski J, Szymczyk S (1987) Laser stimulation for cummulation of selenium in tomato fruits. In: Jezowska-Trzebiatowska B, Kochel B, Slawinski J, Starek W (eds) Photon emission from biological systems. World Scientific, Singapore, pp 211–218

Dobrowolski JW, Budak A, Bogusz B (1995) Effect of laser irradiation on pathogenic fungi in vitro. In: Proceedings of the. international biomedical optics conference, San Jose

Dobrowolski JW, Wachalewski T, Smyk B, Różycki E, Barabasz W (1996) Experiments on the influence of laser light on some biological elements of the natural environment. Environ Manag Health 8(4):136–141

Dobrowolski JW, Rozanowski B, Zielinska-Loek A, Sliwka M, Gowin K, Mazur R (2004) Perspectives of application of laser biostimulation for more bioremediation of soil and wastewater. International conference on bioremediation of soil and groundwater, Politechnika Śląska, Krakow, pp. 133–148

Dobrowolski JW, Jachimski J, Hejmanowska B, Wagner A, Boroń A, Drzewiecki W, Mikrut S, Śliwka M, Mazur R, Jakubiak M, Patuła B (2006) Interdyscyplinarna współpraca w zakresie edukacji na rzecz ochrony krajobrazu kulturowego na przykładzie Parku Narodowego Cinque Terre (Włochy). [Interdisciplinary co-operation in the area of the protection of cultural landscape, focussed on the Cinque Terre National Park (Italy)], vol 12. (2/1, Geodezja, Akademia Górniczo-Hutnicza, pp 167–182. ISSN: 1234-6608

Dobrowolski JW, Budak A, Trojanowska D, Rymarczyk M, Macuda J (2012a) Laser stimulation of trichophyton mentagrophytes for the enhancement biodegradation of hydrocarbons. J Environ Eng Manag 11(10):1783–1788. ISSN: 1582-9596

Dobrowolski JW, Sliwka M, Mazur R (2012b) Laser biotechnology for more efficient bioremediation, protection of aquatic ecosystems and reclamation of contaminated areas. J Chem Technol Biotechnol 87:1354–1359. ISSN: 1097-4660

El-Borai AM, Eltaych KM, Mostafa AR, El-Assar SA (2016) Biodegradation of industrial oil-polluted wastewater in Egypt by bacterial consortium immobilized in different types of carriers. Pol J Environ Stud 25(5):1901–1909. ISSN: 1230-1483

Fernando HJS, Mammarella MC, Grandoni G, Fedele P, Di Marco R, Dimitrova R, Hyde P (2012) Forecasting PM10 in metropolitan areas: efficacy of neural networks. Environ Pollut 163:62–67. https://doi.org/10.1016/j.envpol.2011.12.018

Fischedick M, Schaeffer R, Adedoyin A, Akai M, Bruckner T, Clarke L (2011) Mitigation potential and costs. In: Edenhofer O, Pichs-Madruga R, Sokona Y, Seyboth K, Matschoss P, Kadner S, Zwickel T, Eickemeier P, Hansen G, Schlömer S, von Stechow C (eds) IPCC special report on renewable energy sources and climate change mitigation. Cambridge University, Cambridge, pp 791–864

From Smart History towards common European Heritage by preservation model of Cinque Terre National Park – Italy (2005) Joint international project handbook. In: De Naeyer A (ed) EU Progr. Culture 2000: SMART HISTORY, Parco Nazionale Cinque Terre, Riomaggiore http://home.agh.edu.pl/~zfiit/publikacje_pliki/Smart%20History%20Handbook.pdf . Accessed 21 Mar 2016

Gambuś F, Czech T, Wieczorek J (2013) Changes of soil reaction and consumption of lime fertilizers in Poland. 19 international conference "Reasonable use of fertilizers dedicated to calcium and liming", at the Czech University of Life Sciences Prague. nr 19. 79–84

Gambuś F, Wieczorek J, Czech T, Gorczyca O, Spałek I, Urbańska K, Babula J, Mierzwa-Hersztek M, Rydarowski H, Kopeć M (2014) Yield forming efect of application of composts containing polymer materials enriched in biocomponents. J. Ecol Eng 15(1):81–86. https://doi.org/10.12911/22998993.1084222

Gazzaz NM, Yusoff MK, Aris AZ, Juahir H, Ramli MF (2012) Artificial neural network modeling of the water quality index for Kinta River (Malaysia) using water quality variables as predictors. Mar Pollut Bull 64:2409–2420. https://doi.org/10.1016/j.marpolbul.2012.08.005

Gnansounou E, Dauriat A, Villegas J, Panichelli L (2009) Life cycle assessment of biofuels: energy and greenhouse gas balances. Bioresour Technol 100(21):4919–4930. https://doi.org/10.1016/j.biortech.2009.05.067

Górecka K (1985) Propagation of cauliflower from flower buds in tissue culture. Bull Pol Acad Sci Biol Sci 33(6):27–29. ISSN: 0867-1656

Górecka K (1992) In: Bajaj YPS (ed) Micropropagation of horseradish (*Cochlearia armoracia* L.). Biotechnology in agri-culture and forestry. Hightech and micropropagation III, vol 19. Springer-Verlag, Berlin, pp 58–71. ISBN: 978-3-642-76415-8

Górecka K (1998) Obtaining of homozygous lines of head cabbage (*Brassica oleracea* L. var. *capitata* L.) with aid of anther culture. Research Institute of Vegetable Crops, Habil. Thesis 14, pp 1–71

Górecka K, Krzyżanowka D, Górecki R, Glapś T (1993) Increasing of intensity in vitro tomato propagation with decapitation seedlings method. Material of polish symposium. Smal plants and technology in horticulture. Polish Society for Horticultural Science. Poznań University of. Life Sciences, pp 154–156 [in Polish]

Górecka K, Krzyżanowska D, Śmiech M, Hoser-Krauze J (1997) The use of anther culture for obtaining head cabbage cv. Kamienna Głowa homozygous lines. Advances in biometrical genetics. In: Proceedings of the tenth meeting of the EUCARPIA section biometrics in plant breeding, pp 123–130

Górecka K, Krzyżanowska D, Kiszczak W, Kowalska U (2009) Plant regeneration from carrot (*Daucus carota* L.) anther culture derived embryos. Acta Physiol Plant 31(6):1139–1145. ISSN: 1861-1664

Guha S, Maheshwari SC (1964) In vitro production of embryos from anthers of *Datura*. Nature 204:497. https://doi.org/10.1038/204497a0

Hawkins DH, Abrahamse H (2006) The role of laser fluence in cell viability, proliferation, and membrane integrity of wounded human skin fibroblasts following helium-neon laser irradiation. Lasers Surg Med 38:74–83. https://doi.org/10.1002/lsm.20271

Hawkins D, Abrahamse H (2005) Laboratory methods for evaluating the effect of low level laser therapy (L LLT) In Hu E. (2004) Zinc deficiency, DNA damage and cancer risk. J Nutr Biochem 15:572–578. https://doi.org/10.1016/j.jnutbio.2004.07.005

Junhui He (2016) Nanomaterials in energy and environmental applications. Pan Stanford Publications, 548 p. ISBN-10: 9814463787

Horsek Z, Hrebicek J (2014) Biodegradable waste management in the Czech Republic. A proposal for improvement. Pol J Environ Stud 23(6):2019–2025. ISSN: 1230-1483

Hu X, Weng Q (2009) Estimating impervious surfaces from medium spatial resolution imagery using the self-organizing map and multi-layer perceptron neural networks. Remote Sens Environ 113:2089–2102. https://doi.org/10.1016/j.rse.2009.05.014

Hu Y-C (2010) Pattern classification by multi-layer perceptron using fuzzy integral-based activation function. Appl Soft Comput 10:813–819. https://doi.org/10.1016/j.asoc.2009.09.011

Inn-Sil K, Tae-Soo C, Hyun-Min K, Nam-Il C, Jong-Sang K, Sung Cheol K, Sung-Kyu L, Yoo-Shin K (2002) Pattern recognition of the movement tracks of medaka (Oryzias latipes) in response to sub-lethal treatments of an insecticide by using artificial neural networks. Environ Pollut 120:671–668. https://doi.org/10.1016/S0269-7491(02)00183-5

Ivanov V (2016) Environmental microbiology for engineers. CRC Press. ISBN: 9781420092349 – CAT# 92340

Jakubiak M, Sliwka M (2009) Research on effects of laser light stimulation on selected strains of energetic willow. Pol J Environ Stud 18:123–127. ISSN: 1230-1483

Jia J, Zhao S, Hu L, Wang Y, Yao L, Liu Y, Yuan Z (2016) Removal efficiency and the mineralization mechanism during enhanced bioventing remediation of oil-contaminated soils. Pol J Environ Stud 25(5):1955–1963. ISSN: 1230-1483

Junginger M, de Visser E, Hjort-Gregersen K, Koornneef J, Raven R, Faaij A (2006) Technological learning in bioenergy systems. Energ Policy 34(18):4024–4041. https://doi.org/10.1016/j.enpol.2005.09.012

Karachevtsev VA (ed) (2016) Nanobiophysics. Fundamentals and application. Pan Stanford Publishing PTE. Ltd., Singapore, 417 p. ISBN: 978-981-4613-96-5

Kozieł W, Włodarczyk T (2011) Glony – produkcja biomasy. Acta Agrophysica 17:105–116. http://www.old.actaagrophysica.org/pl/polrocznik.html?stan=detail&paper=1407&vol=17&numer=1

Krzyżanowska D, Górecka K (2004) Characteristic of Brussels sprouts plants obtained by anther culture. Veg Crop Res Bull 60:25–31. PL ISSN 1506-9427

Kudłek J, Pępkowska A, Walasz K, Weiner J (2005) Koncepcja ochrony różnorodności biotycznej miasta Krakowa. Instytut Nauk o Środowisku UJ, Kraków. 169 p. http://www.eko.uj.edu.pl/przyrodakrakowa/download/koncepcja.pdf

Kuwahara SS, Cuello JL, Myhre G, Pau S (2011) Growth of green algae *Chlamydomonas reinhardtii* under red and blue lasers. Opt Lasers Eng 49:434–438. https://doi.org/10.1016/j.optlaseng.2010.11.015

Kuźniar T (2012) Use of the entomopathogenic fungi in crop plant protection against insects. Doctoral dissertation, Faculty of Agriculture and Economics, University of Agriculture in Krakow

Larson ED (2006) A review of life-cycle analysis studies on liquid biofuel systems for the transport sector. Energy Sustain Dev 10(2):109–126. ISSN: 0973-0826

Lee SY, Nielsen J, Stephanopoulos G (2016) Advanced biotechnology. In: Willadsen J (ed) Fundamental bioengineering, vol 1. Willey – VCH, Weinheim

Llanos J, Rodrigo MA, Cañizares P, Popa Furtuna R, Curteanu S (2013) Neuro-evolutionary modelling of the electrodeposition stage of a polymer-supported ultrafiltratione-electrodeposition process. for the recovery of heavy metals. Environ Model Softw 42:133–142. ISBN: 978-3-527-33674-6

Long TB, Blok V, Coninx I (2016) Barriers to the adoption and diffusion of technological innovations for climate-smart agriculture in Europe: evidence from the Netherlands, France, Switzerland and Italy. J Clean Prod 112:9–21. https://doi.org/10.1016/j.jclepro.2015.06.044

Maibeche Y, Moali A, Yahi N, Menard N (2015) Is diet flexibility an adaptive life trait for relictual and peri-urban populations of the endangered primate *Macaca sylvanus*? PLoS One, Genet Genomic Res 10:e0118596. https://doi.org/10.1371/journal.pone.0118596

Mandenius CF (2016) Bioreactors. Design, operation and novel. Willey VCH. ISBN: 978-3-527-33768-2

Maoyun H, Xiao B, Shiming L, Zhiquan H, Xianjun G, Siyi L, Fan Y (2010) Syngas production from pyrolysis of municipal solid waste (MSW) with dolomite as downstream catalysts. J Anal Appl Pyrolysis 87:181–187. ISSN: 0165-2370

McQuarrie JP, Boltz JP (2011) Moving bed biofilm reactor technology: process applications, design, and performance. Water Environ Res 83(6):560–575. https://doi.org/10.2175/106143010X12851009156286

Menchetti M, Mori E (2014) Worldwide impact of alien parrots (Aves Psittaciformes) on native biodiversity and environment: a review. Ethol Ecol Evol 26(2-3):172–194. https://doi.org/10.1080/03949370.2014.905981

Milell L, Martelli G, Salava J, Fernandez E, Ovesná J, Greco I (2011) Total phenolic content, RAPDs, AFLPs and morphological traits for the analysis of variability in Smallanthus sonchifolius. Genet Resour Crop Evol 58(4):545–551. https://doi.org/10.1007/s10722-010-9597-x

Moretti M, Bontadina F, Bauer N, Duelli P, Della Bruna P, Gloor S, Hunziker M, Obrist M, Wagner S (2012) BiodiverCity – ecological and social value of urban nature: tools to identify, maintain and improve biodiversity and its acceptance in urban areas. Nat Res Program 54 project BiodiverCity, on-line: http://www.biodivercity.ch/. Accessed 28 Mar 2016

Obernberger I, Thek G, Reiter D (2008) Economic evaluation of decentralized CHP applications based on biomass combustion and biomass gasification. BIOS Bioenergiesysteme GmbH, Graz. Online: http://bios-bioenergy.at/uploads/media/Paper-Obernberger-Cost-assessment-CHPBM-comustion-gasification-2008-05-30.pdf. Accessed 22 Nov 2016

Orlewicz-Musiał M, Wagner A (2012) Rola terenów zieleni w zrównoważonym rozwoju miasta na przykładzie Błoń Krakowskich i Parku im. Dr Henryka Jordana w Krakowie. In: Kosmala M (ed) Zieleń a klimat społeczny miasta, Toruń, pp 147–158. ISBN: 978-83-931293-8-6

Orlewicz-Musiał M, Wagner A (2014) Przeobrażenia terenów zieleni miejskiej w związku z rozwojem infrastruktury sportowo-rekreacyjnej na przykładzie dzielnic Nowej Huty w Krakowie. In: M Kosmala (ed) Kierunki zmian terenów zieleni w miastach, Toruń, pp 241–252. ISBN: 9788393574049

Polish Standards, PN-R-04013:1988 – Chemico-agricultural analysis of plants – determination of air-dry and dry matter (1988)

Rada EC (2016) Biological treatment of solid waste. Enhancing sustainability. Apple Academic Press, CRC Press. ISBN: 978-1-77188-279-8. URL: https://www.crcpress.com/Biological-Treatment-of-Solid-Waste-Enhancing-Sustainability/Rada/p/book/9781771882798

Rogulska M, Grzybek A, Szlachta J, Tys J, Krasuska E, Biernat K, Bajdor K (2011) Powiązanie rolnictwa i energetyki w kontekście realizacji celów gospodarki niskoemisyjnej w Polsce. Pol J Agron 7:92–101. ISSN: 1230-1483, article-9caad312-de6b-427a-809d-4fd85508fe8f

Rybiński W, Garczyński S (2004) Influence of laser light on leaf area and parameters of photosynthetic activity in DH lines of spring barley (*Hordeum vulgare* L.) Int. Agrophysics 18:261–267. www.Jpaan.lublin.pl/Int-agrophysics

Sadrzadeh M, Mohammadi T, Ivakpour J, Kasiri N (2008) Separation of lead ions from wastewater using electrodialysis: comparing mathematical and neural network modeling. Chem Eng J 144:431–441. https://doi.org/10.1016/j.cej.2008.02.023

Sahayaraj K, Namasivayam SKR (2008) Mass production of entomopathogenic fungi using agricultural products and by products. Afr J Biotechnol 7(12):1907–1910. ISSN 1684–5315

Sanchez-Murillo RI, Torre-Martinez M, Aguirre-Linares J, Herrera-Estrella I (2004) Light-regulated asexual reproduction in *Paecilomyces fumosoroseus*. Microbiology 150:311–319. https://doi.org/10.1099/mic.0.26717-0

Segui-Simarro JM (2016) Androgenesis in *Solanaceae*. Methods Mol Biol 1359:209–244. https://doi.org/10.1007/978-1-4939-3061-6_9

Shen Y, Yoshikawa K (2013) Recent progresses in catalytic tar elimination during biomass gasification or pyrolysis – a review. J Renew Sustain Energy Rev 21:371–392. https://doi.org/10.1016/j.rser.2012.12.062

Shi WB, Feng MG (2004) Lethal effect of *Beauveria bassiana*, *Metarhizium anisopliae*, and *Paecilomyces fumosoroseus* on the eggs of Tetranychus cinnabarinus (Acari: Tetranychidae) with a description of a mite egg bioassay system. Biol Control 30:165–173. https://doi.org/10.1016/j.biocontrol.2004.01.017

Singh A, Imtiyaz M, Isaac RK, Denis DM (2012) Comparison of soil and water assessment tool (SWAT) and multilayer perceptron (MLP) artificial neural network for predicting sediment yield in the Nagwa agricultural watershed in Jharkhand, India. Agric Water Manag 104:113–120. https://doi.org/10.1016/j.agwat.2011.12.005

Tokarska-Guzik B et al (2012) The plants of the alien origin in Poland, focus on invasive species Rośliny obcego pochodzenia w Polsce ze szczególnym uwzględnieniem gatunków inwazyjnych. Generalna Dyrekcja Ochrony Środowiska, Warszawa. ISBN: 978-83-62940-34-9

Tursunov O (2014) A comparison of catalysts zeolite and calcined dolomite for gas production from pyrlolysis of municipal solid waste (MSW). Elsevier Appl Sci Sci Direct J Ecol Eng 69:237–243. https://doi.org/10.1016/j.ecoleng.2014.04.004

Tursunov O, Dobrowolski JW (2015) A brief review of application of laser biotechnology as an efficient mechanism for the increase of biomass for bio-energy production via clean thermo-technologies. Am J Renew Sustain Energy 1(2):66–71

Tursunov O, Isa K, Ong S (2011) Review paper of catalyst (dolomite) analysis for MSW pyrolysis (gasification). International Postgraduate Conference on Engineering (IPCE), Perlis

Tursunov O, Dobrowolski J, Klima K, Kordon B, Ryczkowski J, Tylko G, Czerski G (2015a) The influence of laser biotechnology on energetic value and chemical parameters of rose multiflora biomass and role of catalysts for bio-energy production from biomass: case study in Krakow-Poland. World. J Environ Eng 3(2):58–66. https://doi.org/10.12691/wjee-3-2-5

Tursunov O, Dobrowolski JW, Nowak W (2015b) Catalytic energy production from municipal solid waste biomass: case study in Perlis, Malaysia. World J Environ Eng 3(1):7–14. https://doi.org/10.12691/wjee-3-1-2

Uchwała nr XC/1202/10 XC/1202/10 Rady Miasta Krakowa z dnia 13 stycznia 2010 r. w sprawie ustanowienia użytku ekologicznego 'Staw Dąbski' [Resolution no. XC/1202/10 of the City Council of Krakow of 13th January 2010 on establishing the ecologically useful area 'The Pond of Dąbie'] [Online]. http://www.infor.pl/akt-prawny/U80.2010.045.0000302,uchwala-nr-xc120210-rady-miasta-krakowa-w-sprawie-ustanowieniauzytku-ekologicznego-staw-dabski.html

Uchwała nr XXXI/405/07 Rady Miasta Krakowa z dnia 19 grudnia 2007 r. w sprawie ustanowienia użytku ekologicznego 'Staw przy Kaczeńcowej' [Resolution no. XXXI/405/07 of the City Council of Krakow of 19th December 2007 on establishing the ecologically useful area 'The Pond of the Kaczeńcowa Street'] [Online]. https://www.bip.krakow.pl/?dok_id=167&sub_dok_id=167&sub=uchwala&query=id%3D17225%26amp%3Btyp%3Du

Van Dijk M, Meijerink GW (2014) A review of global food security scenario and assessment studies, Results, gaps and research priorities. Global Food Secur 3:227–238. ISSN: 2211-9124

Vohora SD, Dobrowolski JW (eds) (1990) New horizon of trace elements and health. Hamdard University, New Delhi

Wagner A (2007) Nature-based tourism in anthropogenic landscape: focus on water bodies in the Krakow area. In: Environmental protection into the future, eds. January Bień, Wojciech Nowak, Częstochowa, Wydawnictwo Politechniki Częstochowskiej, pp 462–463. ISBN: 978-83-7193-340-0

Wagner A (2012) Management of the industrial area of Nowa Huta (Kraków, Poland), focus on water bodies. Pol J Environ Stud 21(5A):435–439. ISSN: 1230-1483

Wagner A, Hruševar D (2015) Contribution to the knowledge of plant diversity in the Malopolska region. Int J Adv Life Sci 7(3, 4):164–165. ISSN:1942-2660

Wagner A, Orlewicz-Musiał M (2014) Functions and dysfunctions of tourism and recreation in and how they influence aquatic environments. Pol J Environ Stud 23(3):1045–1050. ISSN: 1230-1483

Wagner A, Orlewicz-Musiał M (2011) Zbiorniki rekreacyjne na terenie Krakowa na przestrzeni dziejów. [Water bodies in Krakow over the history]. In: Kosmala M (ed) Miasta wracją nad wodę [Cities are coming back towards water]. PZIiTS, Toruń, pp 197–206. 978-83-931293-5-5

Wieczorek J., Czech T, Koncewicz-Baran M., (2016) Assessment of the value of fertilizers prepared on the basis of dolomite from "Jozefka" mine – research report. Department of Agricultural and Environmental Chemistry. University of Agriculture in Krakow, pp 1–56

Zabochnicka-Świątek M (2010) Algae – feedstock of the future. Arch Combust 30:225–236

Zabochnicka-Świątek M (2012) Effect of air and CO2 supply on growth of microalgae *Chlorella vulgaris*. In: Proceedings of the 37th international technical conference on clean coal and fuel systems, pp 183–189. 03–07.06.2012, Clearwater, Florida

Zabochnicka-Świątek M (2013) Utilization of *Chlorella vulgaris* and sediments after N-NH4 removal containing clinoptilolite for sorption of heavy metals from wastewater. Rocz Ochr Sr 15:324–347

Zabochnicka-Świątek M, Gyliene O, Cederkvist K, Holm PE (2012) Adsorption processes. In: Best practice guide on metals removal from drinking water by treatment. IWA Publishing, London, pp 61–69. ISBN: 9781843393849

Zabochnicka-Świątek M, Malińska K, Krzywonos M (2014) Removal of biogens from synthetic wastewater by microalgae. Environ Prot Eng 40:87–104. https://doi.org/10.5277/epe140207

Zhang N, Wang M, Wang N (2002) Precision agriculture/a worldwide overview. Comput Electron Agric 36:113–132. https://doi.org/10.1016/S0168-1699(02)00096-0

Zimmermann G (2007) The entomopathogenic fungi *Isaria farinosa* (formerly *Paecilomyces farinosus*) and the *Isaria fumosorosea* species complex (formerly *Paecilomyces fumosoroseus*): biology, ecology and use in biological control. Biocontrol Sci Tech 18(9):865–901

Zwickel T, Eickemeier P, Hansen G, Schlömer S, von Stechow C (eds) (2014) IPCC special report on renewable energy sources and climate change mitigation. Cambridge University, Cambridge, pp 209–332

Bioprocess for Solid Waste Management

4

Radhika Deshmukh, Anshuman A. Khardenavis, and Hemant J. Purohit

Abstract

Among the diverse methods of solid waste management, bioprocess is the most reasonable one pertaining to its cost, potential, and generation of non-toxic products. The present era of population explosion demands fast and safe ways to manage the waste, and bioprocess is the best solution for that. A multitude of microbes can be used in bioprocessing, and a lot of research has to be done to fully explore their degradation potential. Moreover, bioprocesses can be combined with other waste treatment technologies to enhance the efficacy of waste management. Process network technologies can help in proper landfill management. A consolidative view of established as well as amateur techniques of bioprocessing being used for solid waste management will give an overall picture of the current status and will serve for finer waste administration.

Keywords

Bioprocess • Solid waste management • Bioremediation • Pre-treatment • Biofuel production

4.1 Introduction

A bioprocess is the process involving the use of a living cell or a cell component to obtain a desired product (Moser 2012). It basically involves the transport of mass and energy in the systems (Harun et al. 2010). The bioprocess technology is being

R. Deshmukh • A.A. Khardenavis • H.J. Purohit (✉)
Environmental Biotechnology and Genomics Division, CSIR – National Environmental Engineering and Research Institute (CSIR-NEERI), Nehru Marg, Nagpur 440020, Maharashtra, India
e-mail: hj_purohit@neeri.res.in; hemantdrd@hotmail.com

H.J. Purohit et al. (eds.), *Optimization and Applicability of Bioprocesses*,
https://doi.org/10.1007/978-981-10-6863-8_4

used in a multitude of areas like manufacturing of food products, life care solutions and even waste management (Kim et al. 2008; Mussatto et al. 2011; Brandelli et al. 2015; Kalia 2016).

Municipal solid waste (MSW) is mainly produced by human activities, which includes biodegradable as well as toxic and recalcitrant waste and pharmaceutical and industrial by-products. Almost 90% MSW is dumped on open land without taking proper care of the associated health issues and other sanitation problems (Pandey et al. 2016). Large-scale industrialization due to population explosion has given rise to pollution as well as MSW accumulation (Sharholy et al. 2008). The current annual global MSW production of 1.3 billion metric tons is expected to increase to 2.2 billion metric tons by 2025 (Kawai and Tasaki 2016). Around 700 million tons of waste is generated in urban parts of India per year (Pan et al. 2012). More the sophisticated products generated for mankind, more will be the complex waste produced and tougher it will be to device eco-friendly ways for managing this misplaced resource.

The MSW management methods include recycling, degradation, disposing or using the waste for energy generation (Subramanian 2000; Khardenavis et al. 2013; Paleologos et al. 2016). Integrated solid waste management (ISWM) provides a holistic approach combining all the above methods for prevention as well as management and reuse of the waste which can be effectively achieved through life cycle assessment (Buttol et al. 2007; Astrup et al. 2015; Quirós et al. 2015). Resource recovery is considered as a very important aspect of any waste management programme (Zaman 2013). Waste management can be done in an environment-friendly manner by employing the waste to energy approach (Kalia and Joshi 1995; Brunner and Rechberger 2015). Technologies based on different bioprocesses can help in the generation of either energy or value-added products from solid waste without the associated pollution problems (Fig. 4.1) and can thus provide sustainable solution for coping up with the huge amount of MSW generation (Khardenavis et al. 2013).

This chapter focuses on the pre-treatment methods for different types of solid wastes, process networks as well as methanogenic and thermophilic processes. A

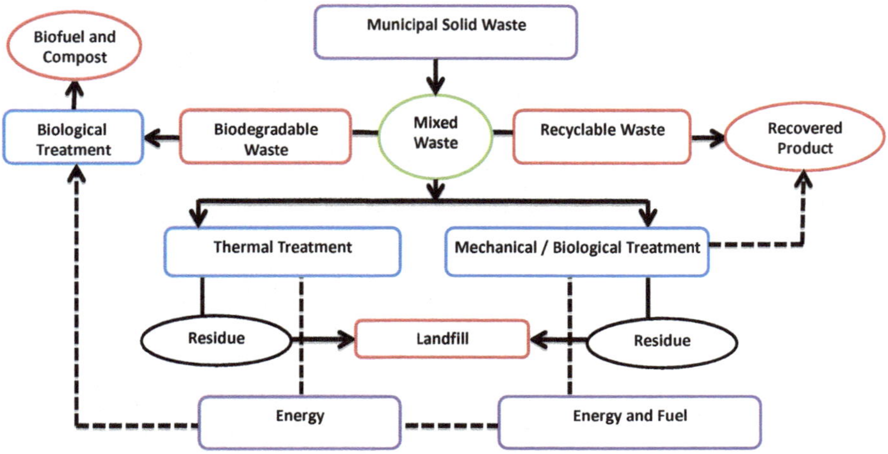

Fig. 4.1 Role of bioprocesses in solid waste management

large amount of hydrogen is required in many industries like oil processing and refineries. Considering the significance of biohydrogen in catering to the above industrial demand, along with an additional benefit of waste management, the biohydrogen production from MSW will also be discussed.

4.2 Pre-treatment Process for MSW in Bioprocesses

For bioprocessing of any waste, it is desirable that its quantity and impact on the environment have to be minimized first. In general, the pre-treatment can be a physical, chemical, thermal, or a biological process which alters the nature of the waste in terms of quality, hazardous nature, ease of handling and recovery (Torres and Lloréns 2008; Rani et al. 2012; Ariunbaatar et al. 2014; Cesaro and Belgiorno 2014). Sustainable waste management is possible by using the above techniques provided the maximum output in terms of a value-added product can be achieved with minimum inputs for which the operating parameters and pre-treatment technologies available at a particular waste management site have to be thoroughly studied and optimized for maximum efficiency (Jain et al. 2015; Kalia et al. 2016). The classification of wastes should be revised, and multiple measures of waste management should be applied simultaneously so that the available knowledge can be applied effectively for waste management (Pandit et al. 2015). Pre-treatment methods are often chosen based upon the type of waste, energy and cost efficiency, and environmental safety (Table 4.1). Saritha et al. (2012) have extensively reviewed the significance of various pre-treatment methods and their role in making the lignocellulosic biomass amenable to various bioprocesses. Biological and mechanical treatment (BMT) has been suggested to be an effective ISWM pre-treatment method for municipal waste (Hong et al. 2006). Recently, the application of termites called 'termiactors' has been reported for successful degradation of cotton remnants, coconut shells, jute, etc. which otherwise resist composting or anaerobic digestion (Abbasi and Gajalakshmi 2015). Bioprocesses based on the use of different microbes have also been applied for pre-treatment of solid waste with an aim for enhanced degradation and efficient management (Table 4.2). Significance of pre-treatment was demonstrated during solid state fermentation of agricultural waste, wheat straw, which resulted in optimum xylanase activity of *A. brasiliensis* (Ho and Lau 2014). Even hazardous matter like ash from MSW incinerators could be used as a source of rare earth elements after proper pre-treatment and purification (Funari et al. 2016). Substrate-specific application of different pre-treatment methods currently in use for management of diverse range of solid wastes is discussed.

4.2.1 Organic Solid Waste

Organic fraction of MSW refers to any waste of plant or animal origin which can be degraded easily by microorganisms. Lignocellulosic biomass is the most abundant organic waste with approximately 50 million tons of annual production worldwide

Table 4.1 Methods for management of different types of solid waste

S.N	Method	Type of waste	Product	Significance	References
Methods involving bioprocesses					
1.	Composting	Organic matter like food waste, leaves	Compost, biogas	Eco-friendly, cost-effective	Haug (2015)
2.	Windrow composting	Farm waste like animal manure and crop residues	Compost, biogas	Eco-friendly, cost-effective	Hellmann et al. (1997)
3.	Biofuel	Urban wastes, organic waste, sewage sludge, agricultural waste	Carboxylic acid, alcohols, ketones	Economical	Crutzen et al. (2016)
4.	Anaerobic digestion	Organic waste, sewage sludge	Biogas, fertilizer	Eco-friendly, cost-effective	Kamali et al. (2016)
5.	Mechanical biological treatment	Recyclable waste and municipal and industrial waste	Fuels, fertilizers and recycled material	Combines multiple methods, used for a variety of wastess, less residue	Sadhukhan et al. (2016)
6.	UASB	Organic waste	Biogas, heat	Solids can be digested for a long period	Pat-Espadas et al. (2016)
7.	Landfill	All solid waste	Landfill gas (methane)	Cost-effective	Powell et al. (2015)
Mechanical and chemical methods					
8.	Biodrying	Organic waste	–	Reduces the overall weight of the waste, cost-effective	Tuorto et al. (2015)
9.	Gasification	Organic waste	Syngas	Efficient than burning	Parihar et al. (2016)
10.	Recycling	Metals, paper, plastic, glass, textiles, e-waste	New product from same material	Reduces energy usage	He et al. (2015)
11.	Mechanical heat treatment	All wastes	–	Does not degrade the waste	Oluleye et al. (2016)
12.	Plasma arc waste disposal	Organic waste	Synthetic gas, electricity	Eco-friendly	Buyantuev et al. (2016)
13.	Pyrolysis	Organic waste, plastic waste	Charcoal, methanol, syngas, biochar, recycled plastic	Useful for industrial purposes	Zhou et al. (2015)

(continued)

Table 4.1 (continued)

S.N	Method	Type of waste	Product	Significance	References
14.	Incineration	Organic waste	Heat	Cheap but produces pollutants like flue gas, ash, etc.	Liu et al. (2015b)
15.	Hydrolysis	MSW, sewage, sludge	Ethanol	Enhances the digestion rate of waste	Biffa (2003)
16.	Hydropulping	Paper waste	Paper recovery	Exclusive for paper waste	Zaman (2013)
17.	Solid waste to protein	Cellulosic waste	Protein	Recovery of nutrient from waste	GOI (2001)

Table 4.2 Application of different microorganisms in pre-treatment of various types of solid wastes for enhanced degradation

S.N.	Waste	Microbe	Enzyme involved	References
1.	Plastic	*Pseudomonas* sp., *Clostridium* sp., *Aspergillus* sp.	2,4-Pentanidione esterase, laccase	Caruso (2015)
2.	Xenobiotics	*Alcaligenes* (2,4-diclorophenoxyacetic acid), *Acinetobacter* (4-chlorobenzene)	Intradiol dioxygenase	Guzik et al. (2013)
3.	Nylon	White-rot fungi	Manganese peroxidase	Deguchi et al. (1997)
4.	Cellulose	*Penicillium* sp.	Cellulase	Gao et al. (2012)
5.	Other carbohydrates	*Bacillus licheniformis* (starch)	Amylase	Siddique et al. (2014)
6.	Protein	*Peptococcus* sp.	Protease	Suthanthararajan et al. (2006)
7.	Fatty acids	*Syntrophomonas* sp.	Oxidase	McInerney et al. (1981)
8.	Chitin	*Metarhizium* sp.	Chitinase	Pinto et al. (1997)
9.	Keratin	*Serratia* sp. HPC1383	Protease	Khardenavis et al. (2009)
10.	Sewage sludge	Sulphate-reducing bacteria	Protease, phosphatase	Van den Brand et al. (2015)

(Xu et al. 2015). Though similar statistics are not accurately available for the Indian scenario, a recent survey has reported some information on the quantum of agro-residues generated which have been shown to be dominated by rice straw and husk, wheat straw, sugarcane tops and bagasse (Saritha et al. 2012). Its complex structure makes it difficult to break the polymer and release the sugar. The lignocellulosic material has to be made more susceptible to hydrolysis by pre-treatment of the waste by different methods to make it more prone towards the breakdown. Though many different methods are reported for the pre-treatment, viz. enzyme treatment, milling, treatment with acids, ammonia fibre explosion, etc. (Agbor et al. 2014), greater use of eco-friendly pre-treatment methods is being stressed recently. Comparative degradation of oil palm mesocarp fibre by superheated steam and hot compressed water was reported in a study with highest hemicellulose removal (94.8%) and acetic acid production being achieved by pre-treatment with hot compressed water. The corresponding glucose generation was observed to be 98.1% (g/g-substrate) on hot water treatment followed by milling which used 9.6 MJ/kg of palm fibre (Zakaria et al. 2015). Novel reactor systems can also have tremendous potential for managing organic fraction of MSW along with recovery of energy in the form of biomethane (Kalia et al. 2000). This was demonstrated efficiently in a case of methane generation from vegetable waste in a four-chambered anaerobic baffled reactor wherein the efficiency of the process was depicted by the carbon recovery factor which was found to be as high as 96% (Gulhane et al. 2016). Application of upstream pre-treatment methods has also been shown to result in improved efficiency of anaerobic digestion. Of the various mechanical pre-treatment methods, viz. milling, maceration, and heat treatment tested in olive oil production process, pre-treatment by milling resulted in 71% increase in methane production followed by 70% increase during thermal treatment, while 32% increase in methane generation was obtained by pre-treatment with enzymatic maceration (Donoso-Bravo et al. 2016). Among the various agricultural residues, rice straw is a relatively recalcitrant waste obtained in huge amounts which can serve as a potent substrate for biofuel production after pre-treatment. Thermal pre-treatment of rice straw with sulphuric acid, followed by consolidated bioprocessing with *Clostridium thermocellum* at 55 °C, was capable of achieving 90% hemicellulose removal. Higher H_2 production of 35.7 mMol L^{-1} was observed in a case of pre-treated sample in comparison to untreated sample (33.3 mMol L^{-1}), and ethanol production was also increased after pre-treatment (Zhang et al. 2015a). Methane production was also demonstrated to improve during anaerobic digestion of rice straw pre-treated with alkali followed by biological processing (Shu et al. 2015). Introduction of aerobic partial composting step was shown to result in improved efficiency of waste digestion by dry method which was attributed to the removal of easily degradable compounds (Brummeler and Koster 1990). In contrast to the beneficial effects of pre-treatment of organic waste in improvement of bioprocess efficiency, few reports have indicated that this may not be so in all the cases. A combination of microwave heating and hydrogen peroxide treatment for the organic fraction of MSW was shown to have negative effect on efficiency of anaerobic digestion due to a long lag phase (Shahriari et al. 2012). It has also been suggested in a study that acclimatization of microbial population could also be effectively used for enhancing the efficiency in management of organic wastes as an alternative to pre-treatment. This was effectively demonstrated in a study by Pandit et al. (2016) who enriched and selected the ligno-hemicellulose-degrading organisms in anaerobic bioreactor with rice straw, vegetable waste, and food waste degradation. Biogas yield was observed to be 1080 mL/g VS, 1160 mL/g VS, and 1410 mL of biogas/g VS added, respectively, for the above substrates with microbial communities having ligno-hemicellulose digestion abilities being abundant in case of rice straw. Lipids can also be a good source of biomethane production; however the presence of long-chain fatty acids (LCFAs) could result in inhibition of methanogenesis. Sharma et al. (2015) have pointed out that this drawback can be overcome by managing the long-chain fatty acid production through any of the pre-treatment methods and/or acclimatization strategies discussed above. Apart from various forms of biofuels produced from organic waste, such waste can have potential for production of other value-added products such as polyhydroxybutyrate (PHB). PHB is a biopolymer with properties similar to synthetic plastics, which can be produced by different bacteria (Kalia et al. 2003a). *Bacillus cereus* was shown to produce 1685 mg PHB from 1 L of pre-hydrolysed pea shell slurry (Patel et al. 2012). Controlled generation of polyhydroxyalkanoate has been demonstrated in the presence of bacterial cultures like *B. cereus* and *B. thuringiensis* by addition of volatile fatty acids generated from solid wastes like vegetable peels thereby adding value to the waste management process (Kumar et al. 2013, 2016).

4.2.2 Recalcitrant and Toxic Waste

Of the total MSW collected globally, approximately 95% goes to landfills (Kurniawan et al. 2006). For the material which are not easily biodegradable, landfill disposal is the only feasible option. Even such chemicals may require some degree of pre-treatment for reducing their hazardous effect, one of them being pre-oxidation which can enhance the biodegradation of such compounds in soil. Two ways were suggested for landfill management: a storage grade depository with minimum outside emissions and a controlled bioreactor for higher methanogenesis (Komilis et al. 1999). Batch type MSW reactors spiked with nano-TiO_2 at various pH and ionic strengths showed reduction in Ti concentrations which became stable in 12–24 h. This study was helpful as engineered nanomaterials were conventionally disposed off in landfills, and a pre-treatment method provided the benefit of safe disposal of nanomaterial (Dulger et al. 2016). In yet another study, pre-treatment of toluene, trichloroethylene, and pentachlorophenol by pre-oxidation using hydrogen peroxide followed by microbial degradation showed that pre-oxidation was found to be effective for microbial degradation of trichloroethylene and pentachlorophenol (Carberry and Benzing 1991). Fenton's process was another very cheap and simple process which had been reported as an after-treatment for landfill leachate for proper subsequent biological treatment (Lee et al. 1996; Bae et al. 1997; Tang and Huang 1997; Lopez et al. 2004; Deng and Englehardt 2006). Improvement in bioremediation of toxic waste metals including lead, zinc, manganese, and copper was shown to be achieved when such waste was pre-treated with *Metaphire posthuma*, which was a vermicomposting agent with urease activity (Sahariah et al. 2015). Bioremediation by a novel approach involving biosurfactant produced by *Pseudomonas aeruginosa* strain was also demonstrated to have tremendous scope for application in enhanced removal of multiple metal contaminants from contaminated soils (Juwarkar et al. 2008).

4.2.3 Recyclable Waste

It is best to sort the recyclable wastes out at the first place before the MSW is sent for processing so that they can be used for respective product recovery. A resource recovery scavenging approach has been proposed for recyclable wastes in which it is suggested to incorporate the services of scavengers for manually separating the recyclable wastes from the garbage (Forbid and Busch 2015; Tong and Tao 2016). Plastic comprises a recyclable part of MSW, and various separation and treatment techniques are being explored to get recycled plastic with good quality and properties similar to virgin plastics (Singh et al. 2016). Also, PHA is a good alternative to plastic. Various avenues for PHA production from wastes can prove useful in the problems faced with plastic waste management as well as reuse of waste (Singh et al. 2009, 2015; Kumar et al. 2015; Patel et al. 2016; Ray and Kalia 2017). Incineration ash, which is another major component of MSW, can find application in production of cement, but the presence of chlorine in such ash needs to be addressed. Pre-treatment of ash for fixing of chlorine has been suggested through formation of chloro-aluminate (Kikuchi 2001). The beneficial effect of pre-treatment was also shown in the medical cotton industry waste treatment where the alkaline pre-treatment of recycled medical cotton industry waste was found to result in 20% improvement in biogas production in comparison to non-treated cotton (Ismail and Talib 2016). Even toxic wastes such as tannery wastes containing chromium possessed the potential for being recycled for production of green and pink $Cr–Al_2O_3$ nanoparticles through application of altered sol–gel method (da Costa et al. 2016).

4.3 Process Network for MSW

Improper waste management leads to misuse of land resources, increased prices of land as well as health issues (Tan et al. 2013). Thus, the major dependency on landfills has to be minimized. Moreover, efficient alternatives are to be explored for ideal waste management. To set up solid waste facilities, one should thoroughly evaluate the sites considered based on landfill termination indicators like total organic carbon, fluorescence spectra, methanogenesis, lignocellulose, humus, etc. (Zheng et al. 2015). For the selection of sites for waste treatment, a tool is proposed which functions using analytical network process (ANP) as a first step and data envelopment analysis (DEA) is applied as second step. Thus, the selection of sites is handled as a multiple attribute decision-making method (Khadivi and Ghomi 2012). Another example is that of hierarchical analytical network process (HANP) which provides a framework suggesting the technology best suited for waste management in a particular country (Khan and Faisal 2008; Nixon et al. 2013). Such tools are of a great help for investors as well as researchers. Flow rate for the leachate at a landfill site can be modelled using a neural network-based leachate prediction method (NN-LEAP), and accordingly proper control methods can be implemented (Karacaa and Özkayab 2006). Neural networks could be applied as smart sensors for real-time process monitoring in waste management systems for which prior estimate of the quantity of waste to be managed in future is very essential. This prediction of generation of waste in the future can be facilitated by data-driven models as has been effectively demonstrated for assessing the amount of waste generated in Tehran through application of combination of wavelet transform–adaptive neuro-fuzzy inference system and wavelet transform–artificial neural network (Noori et al. 2009).

Process variable monitoring has been a consistent problem (Hamalainen and Seppalainen 1986; Prakasham et al. 2011) which can be accomplished by applying appropriate sensor technologies. Soft sensors provide a good alternative for measuring the secondary variables from which the values of unmeasured variables are deduced. Bolf et al. (2007) have demonstrated the use of two such sensors for composting of tobacco solid waste with the help of neural network-based models. Soft sensors can be more accurate and reliable than hardware sensors. Wireless sensor network (WSN) is used in various fields including solid waste management wherein the architecture of the tool provides data from the garbage bin filling area to the remote server relying upon sensor nodes and data transfer nodes (DTN) which in turn provides the solution (Longhi et al. 2012). Such tools are useful for the organization of resources for solid waste management. Genetic algorithms mimic mutation and selection which are the basis of evolution and work on the principle of survival of the fittest. A population of individuals (considered solutions) undergoes a sequence of transformations, and the next generation of solutions is selected. So the best 'individual' which comes up is the best solution. There is a report where a combination of ANN and a genetic algorithm has been used. Neural network has been used there for modelling biogas production, whereas the use of genetic algorithm has been reported for optimizing the production (Haider et al. 2008; Kana et al. 2012). By-product exchange networks among several companies and increasing scavengers with the use of decomposers will facilitate waste management (Geng et al. 2007). Geographic information system (GIS) and geographical information technology (GIT) can together serve decision-making systems for helping in the selection of landfill sites (Javaheri et al. 2006). A vision system trained in neural network and integrated to a robotic line with tracking algorithm has been developed for identifying motion images and recognizing solid urban waste (García-Hernández et al. 2016). The supermatrix aspect of analytical hierarchy technique is useful to get alternatives for waste management through value assessments (Shamshiry et al. 2015).The ultimate goal of all the neural networks should be the generation of as much energy as possible.

4.4 Methanogenesis of Leachate from MSW

Fresh leachate from MSW is organic wastewater which otherwise has nuisance value due to its negative impact on the environment and can be used as a potential biomethane generator (Luo et al. 2015). It can be more suitable than any other waste leachates with the potential to generate up to 0.4 m^3/kg methane provided various types of reactors are used in the presence of concentrated biomass.

Simulated landfill reactors are being used to increase the biomethane production from MSW. Acidogenesis suppresses hydrolysis and methanogenesis, and it could be overcome by altering chemical environment or by the complete oxidation of inhibitors. Vavilin et al. (2006) demonstrated that, stable methanogenic conditions could be achieved in the simulated reactors where leachate exchange was done with methanogenic reactors or which were subjected to initial aeration. To analyse the effect of silver ions and silver nanoparticles on methanogenesis, the process was monitored up to about 200 days in the presence of these two entities (Yang et al. 2013). Methanogenesis was found to be significantly inhibited by nanosilver owing to its high toxicity and easy bioavailability in landfill leachates. This emphasized the need for careful designing of operations to be carried out depending upon the location of waste treatment and the components being added from surrounding industries. Calcium was another species frequently encountered in MSW leachate which demonstrated major inhibition of methanogenesis under mesophilic condition when present together with ammonia nitrogen and aquatic humic substances in the ratio 15:1:10 (Lei et al. 2016). At higher calcium concentrations, inhibition of methanogenesis was attributed to the lethal effect of calcium on sludge and inhibition of mass transfer (Xia et al. 2016). Also, it has been observed that, from among the components of landfill leachates, ammonia was likely to be more problematic in the near future in comparison to heavy metals and xenobiotics and hence needed to be taken care of for preventing process failure (Kjeldsen et al. 2002). Leachate recirculation enhanced the methanogenesis as mass transfer was promoted along with efficient production and utilization of volatile fatty acids (He et al. 2005; Nair et al. 2014; Huang et al. 2016; Ko et al. 2016). A similar report showed the advantage of effluent recirculation in anaerobic reactor during biomethanation of vegetable waste in an ABR which resulted in higher biogas production. This was attributed to the effective management of pH during different reactions of methanogenic pathway where acidic pH was considered to be inhibitory to the methanogenic process. The study demonstrated the pH-dependent segregation of microbial community in different chambers of ABR and subsequent role of effluent recirculation in maintaining the microbial plasticity without compromising the biogas generation (Gulhane et al. 2016, 2017). The provision of equalization tanks before digesters helped in maintaining the acetate and propionate levels within proper range thereby avoiding clogging by facilitating removal of dissolved Ca^{2+} resulting in optimum methane production (Lozecznik et al. 2012). In Poland, hydromechanically separated waste was anaerobically digested to give 309–361 $dm^3/kg\ VS_{fed}$ biogas under thermophilic environment which increased to 494 $dm^3/kg\ VS_{fed}$ under mesophilic conditions on addition of 50% sewage sludge (Borowski 2015a).

To assess the microbes present in a solid waste treatment plant, cultivation-independent sequencing techniques can prove helpful (Thakur et al. 2015). Monitoring the microbes together with the composition of the leachate in simulated studies could help in understanding the biodegradation process in a full-scale landfill. There is also a need to study the community dynamics for waste management systems in detail since the exact role of many microbes was still not understood. In a closed municipal landfill area at Gouzikeng in China, only 2% methanogens were detected in the leachate with the dominance of Methanomicrobiales and Methanosarcinales in total methanogenic population (Huang et al. 2003). In a lab-level simulation study, solid waste showed higher percentage of cellulose degraders like *Firmicutes* along with methanogenic bacteria belonging to Methanobacteriales, Methanomicrobiales, and Methanosarcinaceae. It was reported that the presence of higher initial cellulose and hemicellulose contents favoured methanogens and gave high methane yield, while organisms from *Bacteroidetes* and *Thermotogae* were also found to be abundant during this period (Bareither et al. 2013). In yet another simulation with MSW taken out from a landfill in Austin, Texas, the study indicated the dominance of Methanobacteriaceae during the entire testing period (Fei et al. 2015). High-throughput 454 pyrosequencing was done for a MSW anaerobic reactor in Madrid wherein the study pointed out that though proteolytic bacteria were dominant throughout the study period, increased biodiversity was observed over the time period along with the presence of some rare organisms whose role was unknown in the process (Cardinali-Rezendea et al. 2016).

In addition to the above applications, the use of leachate in other areas will certainly prove helpful. Co-digesting food waste with a fresh leachate from MSW could be applied for improving the efficiency of anaerobic digestion process and subsequent enhancement in biogas owing to the metal elements required for digestion of food waste which were supplied by the fresh leachate (Zhang et al. 2015b). Such type of synergistic combinations of wastes could enable optimum degradation of waste. Leachate from organic fractions of MSW could also be used as a source of other value-added products such as polyhydroxyalkanoates (Korkakaki et al. 2016; Patel et al. 2016).

4.5 Thermophilic Process in MSW Treatment

Anaerobic digestion of organic wastes can help in efficiently managing such wastes through bioenergy generation (Pantaleo et al. 2013) whose efficiency is affected by the system temperature, availability of buffering capacity, pH, etc. in the system. Thus, controlling ammonia concentrations, pH and process temperature will ensure safe digestion of MSW.

For anaerobic digestion, temperature is not only important for the metabolism of microbes but also for the optimum rate of the chemical reactions (Palatsi et al. 2010). Anaerobic digestion has been divided broadly into two types based on the temperature at which it is carried out: mesophilic (carried out at 30–42 °C) and thermophilic (carried out at 43–55 °C). Thermophilic digestion processes have high efficiency, whereas mesophilic processes are more stable ones. Two-phase anaerobic digestion systems are being used for thermophilic as well as mesophilic systems for achieving higher methane yields (Leite et al. 2016; Meng et al. 2016). Thermophilic processes show high destruction of pathogens and better dewaterability of the digested biosolids. The growth rate of microorganisms has been reported to be 27–60% higher for thermophilic processes, and thus, same extent of biodegradation and methanogenesis occurs in almost half of the time in thermophilic process

(Fernández-Rodríguez et al. 2013). Evaluation of appropriate hydraulic retention times (HRT) for efficient biomethane production from solid waste under thermophilic conditions demonstrated that of the eight HRTs tested from 25 to 3.5 days, maximum gas production (methane, about 5.4 L/L/days) was seen at 4.5 days. This was attributed to higher microbial activities at low HRT (Zahedi et al. 2013a). Similar experiments on hydrogen production under similar conditions revealed that highest hydrogen production of 57% (v/v) was obtained at an organic loading rate of 110 g TVS/L/d and HRT of 0.5 day (Zahedi et al. 2013b). In a study, simultaneous digestion of biowaste and sewage sludge in an anaerobic digester was compared under both mesophilic and thermophilic conditions of which thermophilic process proved more stable and efficient (Yu et al. 2014). Food industry wastewater was subjected to joint mesophilic anaerobic–thermophilic aerobic treatment process such that the residue from the mesophilic process was added to thermophilic reactor and thermophilic reactor sludge was again added to mesophilic reactor. The combined process resulted in 90% removal of organic matter, and the presence of *Prevotella*, *Aminobacterium*, *Geobacillus* and unclassified *Actinobacteria* in the biomass was found to be common in these two reactors (Jang et al. 2015).

Optimal ammonia concentration helps in proper buffering of methanogenic medium and increases stability; however, high levels of ammonia have been reported to inhibit the microbial activity and hinder the digestion process (Hejnfelt and Angelidaki 2009). Anaerobic digestion study for cattle wastes at thermophilic conditions showed that at greater loading rates and pH >7.8, inhibition occurred by total ammonia higher than 1700 mg/L (Varel et al. 1977). An increase in pH from 7 to 8 increased the free ammonia levels at which the unhindered operation of digesters could be ensured by proper selection of parameters like C/N ratio, operation temperature, and the use of microbes adapted to higher ammonia concentration (Rajagopal et al. 2013). The continuous functioning of anaerobic digesters in the presence of high ammonia can be ensured provided the microbes are first acclimatized to the higher concentration of ammonia (Yenigün and Demirel 2013). Pretreatment of substrate could also help in reducing the toxic effect of ammonia as was observed in a study by Madrini et al. (2016) wherein addition of 31.5–47.2% natural zeolite during food waste composting helped in decreasing the ammonia emission at thermophilic stage and it also acted as an adsorbent. However, precaution needs to be taken during operation of such temperature-phased anaerobic digestion systems to avoid extra exposure of waste to thermophilic temperatures in order to reduce the ammonia released during protein degradation which has been reported to affect the methanogenesis in the subsequent mesophilic step (Borowski 2015b).

Other pre-treatment strategies such as thermo-chemical pre-treatments have tremendous potential to be exploited for better management of MSW. The composting time of rice straw was shown to be reduced to below 10 days by pre-treating the substrate with alkali followed by continuous thermophilic degradation (Hosseini and Aziz 2013). In another study, municipal biowaste was pressed to separate the liquid fraction from solid fraction of which the liquid was subjected to anaerobic digestion, while the solid fraction was sent for composting. Of the two treatments, anaerobic digestion was observed to be most effective with highest specific gas production of 0.92 m^3/kg at an organic loading of 4.7 kg VS/m^3 under thermophilic conditions (Micolucci et al. 2016).

4.6 Biohydrogen Production from MSW

Hydrogen is the fuel for the coming years since it is completely renewable, recyclable, and eco-friendly. Biohydrogen is the hydrogen which is produced biologically, most commonly by microbes like archaea, bacteria, and algae (Kumar et al. 1995; Porwal et al. 2008; Patel and Kalia 2013). Through biological hydrogen production, eco-friendly energy carrier can be generated using renewable energy sources like biomass (Kumar et al. 2014). For efficient biohydrogen production, efforts are being continually focussed on research on gene manipulation of microorganism, metabolic engineering, photo-fermentation, advanced reactors, matrices for whole cell immobilization, etc. (Sonakya et al. 2001; Das and Veziroglu 2008; Patel et al. 2014). The characterization of MSW for its physicochemical properties can provide insight into the composition of MSW components which might prove helpful for deciding upon the amenability of the substrate for H_2 production. In a study, Cheng et al. (2016) characterized MSW by multiple techniques, viz. scanning microscopy, Fourier transform infrared spectroscopy, etc., followed by fermentative hydrogen production along with rapid production of fatty acids which were then utilized as substrates for improved methane production. The methane production rates in the two-stage process were observed to be 14.8% and 15.1% higher for food waste and sewage sludge, respectively, in comparison to one-stage process.

Biohydrogen production can be achieved either by photo-fermentation or dark fermentation in the presence of two key enzymes, nitrogenase and hydrogenase (Patel et al. 2010, 2017). Figure 4.2 describes in brief the production of biohydrogen through microbial route. Purple non-sulphur bacteria (PNS) are mainly researched for photo-fermentation (Basak and Das 2007) with very few reports about continuous fermentation of complex substrates and consortia of microbes for biohydrogen production (Hawkes et al. 2002). Thermal processes have also been applied for producing hydrogen as demonstrated by Agarwal et al. (2013) during pyrolysis of kitchen vegetable waste in a packed bed reactor. Hydrogen has also been reported to be produced during the anaerobic digestion of organic waste; however, methanogens are known to consume H_2 which can result in lower hydrogen yields, and this can be overcome by reducing the HRTs of anaerobic digestion.

The rates of hydrogen and specific hydrogen production were increased when 1% v/v crude glycerol was added to industrial municipal solid waste. For this, thermophilic, dry-dark batch-mode fermentation was operated (Zahedi et al. 2016). In a study, it was found that mesophilic incubation showed high hydrogen production from organic solid wastes when the anaerobic treatment was followed by acetylene treatment and headspace of batch reactors was being vented and flushed with nitrogen (Valdez-Vazquez et al. 2006). Pre-treatment for complex food waste by ultrasonication followed by anaerobic digestion was found to result in increase in H_2

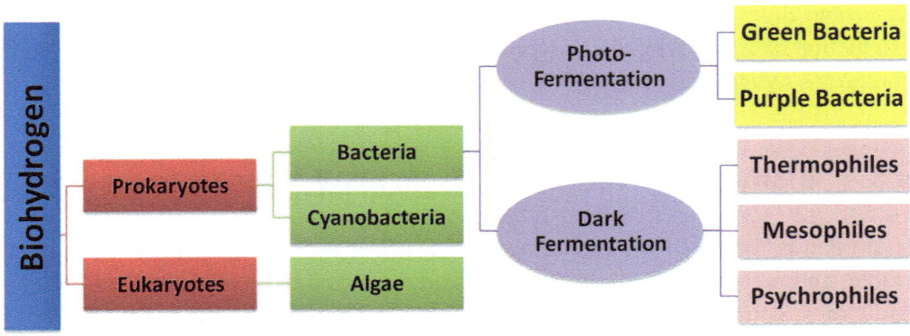

Fig. 4.2 Biohydrogen-producing microorganisms

production when higher total solids were used and pre-treatment time for ultrasonication was increased (Gadhe et al. 2014). Biohydrogen production from kitchen waste was improved when it was pre-treated with heat ranging from 90 to 200 °C with maximum production of 81.27 mL/g VS at 200 °C during the anaerobic fermentation (Li et al. 2014). Biohydrogen was also shown to be produced by dark fermentation of hydrolysate of food waste prepared by the action of glucoamylase and protease enzymes produced by *Aspergillus awamori* and *Aspergillus oryzae* (Han et al. 2016). H_2 production has been shown to be improved when organic fraction of MSW and mixed sewage sludge were co-digested in ratio of 5:1 at total solid concentration of 20% (Tyagi et al. 2014). Combination of paper waste and food waste as a feed was also reported to have desirable effect on enhanced H_2 generation (Pendyala et al. 2013).

The impact of surfactant on organic fraction of MSW was tested under thermophilic conditions in batch cultures with improved hydrogen production being observed in presence of T80 and PEG 6000. The analysis of microbial consortium showed the presence of *Enterobacter*, *Escherichia*, *Buttiauxella* and *Pantoea* as the hydrogen-producing bacteria (Elsamadony et al. 2015). Calcined red mud (CRM) was used for the pre-treatment of brewers' spent grain. The highest specific hydrogen production (198.62 mL/g-VS) was for 10 g/L CRM which provided required pH for hydrogen production by cellulose hydrolysis (Zhang and Zang 2016). Microbial treatment can also be employed for treatment of industrial wastewater as reported by Hensley et al. (2016) during the treatment by *Thermococcus* when biohydrogen production was explored from protein and α-glucosides in industrial wastewater. Supplementation of nutrients such as orthophosphate and pyrophosphate on H_2 production from municipal solid waste composting leachate was reported with enhanced H_2 production being observed in presence of orthophosphate which in addition to serving as a nutrient and energy source was also effective in relieving the inhibition by excess Ca^{2+} (Liu et al. 2015a).

In addition to hydrogen production by microbial consortia, studies have also been conducted on efficient hydrogen production by pure bacterial isolates. The first report of efficient biohydrogen production from MSW by pure microorganism was that of *Bacillus* sp. where two *Bacillus* strains were used to convert organic fraction of MSW into hydrogen resulting in generation of 61 mL of H_2/g volatile solids (Shah et al. 2016). Biohydrogen production by algae has also been reported to be beneficial owing to the ease and low cost of the process (Kothari et al. 2015). Genome mining can help to find the hydrogen-producing organisms and can help to exploit them for waste management (Kalia et al. 2003b). Desired biohydrogen production genes can give better production through transgenic expression with the help of genomic tools (Kalia and Purohit 2008).

4.7 Fungi in MSW Management

Fungi have been widely reported for their bioremediation potential and hence could be exploited to the fullest for waste management (Deshmukh et al. 2016). More than half the amount of MSW consists of organic matter of which cellulose is a main component due to which there is tremendous scope for managing such waste through activity of cellulolytic fungi. A study showed potential use of cellulolytic *Trichoderma*, *Chaetomium* and *Aspergillus* sp. for MSW management (Gautam et al. 2010; Jose et al. 2014). *Trichoderma reesei* and *Aspergillus niger* have also been exploited for solid waste degradation and cellulase production (Abdullah et al. 2016). *Moniliophthora roreri* cellulase enzyme has been used for agricultural waste degradation like rice husks (Solís et al. 2016). Controlling the community dynamics of fungi during the process of composting can increase the rate of humification (Zhao et al. 2016). In a study, *Aspergillus niger* was used for bioleaching as a substitute process for MSW fly ash management. It was found effective for leaching of manganese, zinc as well as aluminium (Wu and Ting 2006). Vermicompost has been found to be beneficial for the increase of fungi in composting process (Paul et al. 2011). A study was done in Spain to analyse the effect of addition of MSW to an arid soil. It was observed that the occurrence of both bacteria and fungi was increased and the soil community was altered in 17 years (Bastida et al. 2008). *R. oryzae* has been used for organic acid formation in waste foundry sand (Asokan et al. 2016). Soybean waste could be fermented by fungus *Fusarium* to obtain anticancer compound camptothecin (Bhalkar et al. 2016). Lignocellulose waste could be pretreated with white-rot fungi to help its further degradation (Rouches et al. 2016). Similarly, when willow sawdust was treated with fungus *Abortiporus biennis*, there was about 40% increase in biomethane potential. This effect was further enhanced on combination with alkali treatment (Alexandropoulou et al. 2017).

Molecular biology techniques can prove to be useful to study fungal communities in waste management. The population of fungi was studied in a composting process by cloning the ITS regions at various time points and phospholipid fatty acid analysis for fungal biomass assay in a composting plant (Hultman et al. 2010). Fungi can be considered as beneficial organisms for improving soil quality at metal-contaminated landfill sites and open waste dumps (Breza-Boruta et al. 2016; Chandran 2016; Long et al. 2016). *Phanerochaete chrysosporium* was reportedly subjected to nanoparticle-aided immobilization and applied for the remediation of leachate at landfill sites (Hu et al. 2016). Impact of fungi was tested to obtain compost from MSW with greater agronomic potential. Windrow composting was performed at different turning frequencies in the presence of fungal consortia comprising of *Trichoderma viride*, *Aspergillus niger* and *Aspergillus flavus*. Turning frequency of 1 week was found to be more economical and beneficial with respect to composting and safe end products (Awasthi et al. 2014). The fungal consortia made the process more effective as fungi were able to quickly decompose organic material at high mineralization rates thereby converting the waste into humus-like substances (Said-Pullicino et al. 2007; Lin et al. 2011). Simultaneous pre-treatment and saccharification by fungi has also helped in increasing the saccharification yields from agricultural wastes like rice straw (Dhiman et al. 2015).

4.8 Opinion

An important aspect of waste management is proper pre-treatment of different wastes so that further it can be disposed in the environment in a safe manner. Bioprocessing seems to be a significant strategy for providing eco-friendly solutions to waste management. There is ample scope for further optimizing the bioprocesses through application of genomic tools including next-generation sequencing techniques for identifying the microbial community responsible for efficient bioprocessing. Diverse range of pre-treatment methods should be explored to facilitate waste digestion. Co-digestion of different types of waste and combining bioprocessing with different waste treatments will help in better waste management with good energy recovery. The end products of waste treatment processes can have applications in novel unexplored areas.

Acknowledgements The authors are thankful to the Director of CSIR-National Environmental Engineering Research Institute (CSIR-NEERI) for providing necessary facilities for this work [KRC Manuscript No.: CSIR-NEERI/KRC/2017/JULY/EBGD/15]. Radhika Deshmukh is grateful to the Council of Scientific and Industrial Research (CSIR), India, for the award of Senior Research Fellowship. Funds from 12th Five-Year Plan Project ESC0108 are also gratefully acknowledged.

References

Abbasi SA, Gajalakshmi S (2015) Disposal of municipal solid waste with in situ termireactors: proof-of-concept. Biores Bioproc 2:1–5. https://doi.org/10.1186/s40643-015-0050-z

Abdullah JJ, Greetham D, Pensupa N, Tucker GA, Du C (2016) Optimizing cellulase production from municipal solid waste (MSW) using solid state fermentation (SSF). J Fundam Renewable Energy Appl 6. https://doi.org/10.4172/2090-4541.1000206

Agarwal M, Tardio J, Mohan SV (2013) Biohydrogen production from kitchen based vegetable waste: effect of pyrolysis temperature and time on catalysed and non-catalysed operation. Bioresour Technol 130:502–509. https://doi.org/10.1016/j.biortech.2012.11.094

Agbor V, Carere C, Cicek N, Sparling R, Levin D (2014) Biomass pretreatment for consolidated bioprocessing (CBP). Advances in biorefineries. Woodhead Publishing, Sawston, pp 234–258. https://doi.org/10.1533/9780857097385.1.234

Alexandropoulou M, Antonopoulou G, Fragkou E, Ntaikou I, Lyberatos G (2017) Fungal pretreatment of willow sawdust and its combination with alkaline treatment for enhancing biogas production. J Environ Manag 203:704–713. https://doi.org/10.1016/j.jenvman.2016.04.006

Ariunbaatar J, Panico A, Esposito G, Pirozzi F, Lens PN (2014) Pretreatment methods to enhance anaerobic digestion of organic solid waste. Appl Energy 123:143–156. https://doi.org/10.1016/j.apenergy.2014.02.035

Asokan P, Suji D, Rajesh R, Elayarajah B (2016) Treatment of waste foundry sand using fungi: an approach for the application in manufacture of concrete. Int J Appl Eng Res 11:1849–1853

Astrup TF, Tonini D, Turconi R, Boldrin A (2015) Life cycle assessment of thermal waste-to-energy technologies: review and recommendations. Waste Manag 37:104–115. https://doi.org/10.1016/j.wasman.2014.06.011

Awasthi MK, Pandey AK, Khan J, Bundela PS, Wong JW, Selvam A (2014) Evaluation of thermophilic fungal consortium for organic municipal solid waste composting. Bioresour Technol 168:214–221. https://doi.org/10.1016/j.biortech.2014.01.048

Bae J, Kin S, Chang H (1997) Treatment of landfill leachates: ammonia removal via nitrification and denitrification and further COD reduction via Fenton's treatment followed by activated sludge. Water Sci Technol 36:341–348. https://doi.org/10.1016/S0273-1223(97)00736-1

Bareither C, Wolfe GL, McMahon KD, Benson CH (2013) Microbial diversity and dynamics during methane production from municipal solid waste. Waste Manag 33:1982–1992. https://doi.org/10.1016/j.wasman.2012.12.013

Basak N, Das D (2007) The prospect of purple non-sulfur (PNS) photosynthetic bacteria for hydrogen production: the present state of the art. World J Microbiol Biotechnol 23:31–42. https://doi.org/10.1007/s11274-006-9190-9

Bastida F, Kandeler E, Hernández T, García C (2008) Long-term effect of municipal solid waste amendment on microbial abundance and humus-associated enzyme activities under semiarid conditions. Microb Ecol 55:651–661. https://doi.org/10.1007/s00248-007-9308-0

Bhalkar BN, Bedekar PA, Kshirsagar SD, Govindwar SP (2016) Solid state fermentation of soybean waste and an up-flow column bioreactor for continuous production of camptothecine by an endophytic fungus *Fusarium oxysporum*. RSC Adv 6:56527–56536. https://doi.org/10.1039/C6RA08956A

Biffward Program on Sustainable Resource Use (2003) Thermal methods of municipal waste treatment. C-Tech Innovation Ltd, Chester

Bolf N, Kopci CN, Briski F, Gomzi Z (2007) Software sensors for monitoring of a solid waste composting process. Chem Pap 61:98–102. https://doi.org/10.2478/s11696-007-0005-8

Borowski S (2015a) Co-digestion of the hydromechanically separated organic fraction of municipal solid waste with sewage sludge. J Environ Manag 147:87–94. https://doi.org/10.1016/j.jenvman.2014.09.013

Borowski S (2015b) Temperature-phased anaerobic digestion of the hydromechanically separated organic fraction of municipal solid waste with sewage sludge. Int Biodeterior Biodegrad 105:106–113. https://doi.org/10.1016/j.ibiod.2015.08.022

Brandelli A, Sala L, Kalil SJ (2015) Microbial enzymes for bioconversion of poultry waste into added-value products. Food Res Int 73:3–12. https://doi.org/10.1016/j.foodres.2015.01.015

Breza-Boruta B, Lemanowicz J, Bartkowiak A (2016) Variation in biological and physicochemical parameters of the soil affected by uncontrolled landfill sites. Environ Earth Sci 75:1–13. https://doi.org/10.1007/s12665-015-4955-9

Brummeler TE, Koster IW (1990) Enhancement of dry anaerobic batch digestion of the organic fraction of municipal solid waste by an aerobic pretreatment step. Biol Wastes 31:199–210. https://doi.org/10.1016/0269-7483(90)90159-P

Brunner PH, Rechberger H (2015) Waste to energy–key element for sustainable waste management. Waste Manag 37:3–12. https://doi.org/10.1016/j.wasman.2014.02.003

Buttol P, Masoni P, Bonoli A, Goldoni S, Belladonna V, Cavazzuti C (2007) LCA of integrated MSW management systems: case study of the Bologna District. Waste Manag 27:1059–1070. https://doi.org/10.1016/j.wasman.2007.02.010

Buyantuev SL, Guiling N, Kondratenko AS, Ye J, Bazarsadaev ET, Khmelev AB, Guo S (2016) Waste industrial processing of boron-treated by plasma arc to produce the melt and fiber materials. In: Proceedings of second international conference on electrical systems, technology and information 2015 (ICESTI 2015), Springer, Singapore, pp 353–361

Carberry JB, Benzing TM (1991) Peroxide pre-oxidation of recalcitrant toxic waste to enhance biodegradation. Water Sci Technol 23:367–376

Cardinali-Rezendea J, Rojas-Ojeda P, Nascimento AMA, Sanza JL (2016) Proteolytic bacterial dominance in a full-scale municipal solid waste anaerobic reactor assessed by 454 pyrosequencing technology. Chemosphere 146:519–525. https://doi.org/10.1016/j.chemosphere.2015.12.003

Caruso G (2015) Plastic degrading microorganisms as a tool for bioremediation of plastic contamination in aquatic environments. J Pollut Eff Cont 3:e112. https://doi.org/10.4172/2375-4397.1000e112

Cesaro A, Belgiorno V (2014) Pretreatment methods to improve anaerobic biodegradability of organic municipal solid waste fractions. Chem Eng J 240:24–37. https://doi.org/10.1016/j.cej.2013.11.055

Chandran S (2016) Managing former landfill sites –a case study of eco-restoration from Kochi, Kerala. Green Chem Technol Lett 1:82–85. https://doi.org/10.18510/gctl.2015.1113

Cheng J, Ding L, Lin R, Liu M, Zhou J, Cen K (2016) Physicochemical characterization of typical municipal solid wastes for fermentative hydrogen and methane co-production. Energ Convers Manage 117:297–304. https://doi.org/10.1016/j.enconman.2016.03.016

Crutzen PJ, Mosier AR, Smith KA, Winiwarter W (2016) N O release from agro-biofuel production negates global warming reduction by replacing fossil fuels. In: Crutzen PJ (ed) A pioneer on atmospheric chemistry and climate change in the anthropocene. Springer International Publishing, Cham, pp 227–238

da Costa CG, Peixoto JA, de Souza DR, Romão LP, Macedo ZS (2016) Recycling of chromium wastes from the tanning industry to produce ceramic nanopigments. Green Chem 18:5342–5356. https://doi.org/10.1039/C6GC01562J

Das D, Veziroglu TN (2008) Advances in biological hydrogen production processes. Int J Hydrogen Energy 33:6046–6057. https://doi.org/10.1016/j.ijhydene.2008.07.098

Deguchi T, Kakezawa M, Nishida T (1997) Nylon biodegradation by lignin-degrading fungi. Appl Environ Microbiol 63:329–331

Deng Y, Englehardt JD (2006) Treatment of landfill leachate by the Fenton process. Water Res 40:3683–3694. https://doi.org/10.1016/j.watres.2006.08.009

Deshmukh R, Khardenavis AA, Purohit HJ (2016) Diverse metabolic capacities of fungi for bioremediation. Indian J Microbiol 56:247–264. https://doi.org/10.1007/s12088-016-0584-6

Dhiman SS, Haw JR, Kalyani D, Kalia VC, Kang YC, Lee JK (2015) Simultaneous pretreatment and saccharification: green technology for enhanced sugar yields from biomass using a fungal consortium. Bioresour Technol 179:50–57. https://doi.org/10.1016/j.biortech.2014.11.059

Donoso-Bravo A, Ortega-Martinez E, Ruiz-Filippi G (2016) Impact of milling, enzyme addition, and steam explosion on the solid waste biomethanation of an olive oil production plant. Bioprocess Biosyst Eng 39:331–340. https://doi.org/10.1007/s00449-015-1519-z

Dulger M, Sakallioglu T, Temizel I, Demirel B, Copty NK, Onay TT, Uyguner-Demirel CS, Karanfil T (2016) Leaching potential of nano-scale titanium dioxide in fresh municipal solid waste. Chemosphere 144:1567–1572. https://doi.org/10.1016/j.chemosphere.2015.10.037

Elsamadony M, Tawfik A, Suzuki M (2015) Surfactant-enhanced biohydrogen production from organic fraction of municipal solid waste (OFMSW) via dry anaerobic digestion. Appl Energy 149:272–282. https://doi.org/10.1016/j.apenergy.2015.03.127

Fei X, Zekkos D, Raskin L (2015) Archaeal community structure in leachate and solid waste is correlated to methane generation and volume reduction during biodegradation of municipal solid waste. Waste Manag 36:184–190. https://doi.org/10.1016/j.wasman.2014.10.027

Fernández-Rodríguez J, Pérez M, Romero LI (2013) Comparison of mesophilic and thermophilic dry anaerobic digestion of OFMSW: kinetic analysis. Chem Eng J 232:59–64. https://doi.org/10.1016/j.cej.2013.07.066

Forbid GT, Busch G (2015) Effective integration of scavengers in solid waste recycling and material recovery. Future Dir Munic Solid Waste Manag Afr 88
Funari V, Bokhari SN, Vigliotti L, Meisel T, Braga R (2016) The rare earth elements in municipal solid waste incinerators ash and promising tools for their prospecting. J Hazard Mater 301:471–479. https://doi.org/10.1016/j.jhazmat.2015.09.015
Gadhe A, Sonawane SS, Varma MN (2014) Ultrasonic pretreatment for an enhancement of biohydrogen production from complex food waste. Int J Hydrogen Energy 39:7721–7729. https://doi.org/10.1016/j.ijhydene.2014.03.105
Gao L, Gao F, Wang L, Geng C, Chi L, Zhao J, Qu Y (2012) N-glycoform diversity of cellobiohydrolase I from *Penicillium decumbens* and synergism of nonhydrolytic glycoform in cellulose degradation. J Biol Chem 287:15906–15915. https://doi.org/10.1074/jbc.M111.332890
García-Hernández M, Flores A, Elizalde E, García-Arredondo AJ, Gutiérrez-Muro MA, Hevia-Montiel N (2016) Recognizing motion images solid waste jumbled on a neural network with a simple tracking performed in an automatic and robotic recycling line. Intelligent Computing Systems. Springer International Publishing, pp 116–124
Gautam SP, Bundela PS, Pandey AK, Awasthi MK, Sarsaiya S (2010) Screening of cellulolytic fungi for management of municipal solid waste. J Appl Sci Environ Sanitation 5:367–371
Geng Y, Zhu Q, Haight M (2007) Planning for integrated solid waste management at the industrial park level: a case of Tianjin, China. Waste Manag 27:141–150. https://doi.org/10.1016/j.wasman.2006.07.013
GOI (2001) Solid waste management manuals from Government of India. Ministry of Urban Development Government of India
Gulhane M, Khardenavis AA, Karia S, Pandit P, Kanade GS, Lokhande S, Vaidya AN, Purohit HJ (2016) Biomethanation of vegetable market waste in an anaerobic baffled reactor: effect of effluent recirculation and carbon mass balance analysis. Bioresour Technol 215:100–109. https://doi.org/10.1016/j.biortech.2016.04.039
Gulhane M, Pandit P, Khardenavis A, Singh D, Purohit H (2017) Study of microbial community plasticity for anaerobic digestion of vegetable waste in anaerobic baffled reactor. Renew Energy 101:59–66. https://doi.org/10.1016/j.renene.2016.08.021
Guzik U, Hupert-Kocurek K, Wojcieszyńska D (2013) Intradiol dioxygenases—the key enzymes in xenobiotics degradation. In: Chamy R, Rosenkranz F (eds) Biodegradation of hazardous and special products. InTech Rijeka, Croatia, pp 129–153. https://doi.org/10.5772/56205
Haider MA, Pakshirajan K, Singh A, Chaudhry S (2008) Artificial neural network-genetic algorithm approach to optimize media constituents for enhancing lipase production by a soil microorganism. Appl Biochem Biotechnol 144:225–235. https://doi.org/10.5772/5620510.1007/s12010-007-8017-y
Hamalainen RP, Seppalainen TO (1986) The analytic network process in energy policy planning. Socio Econ Plan Sci 20:399–405. https://doi.org/10.5772/5620510.1016/0038-0121(86)90054-6
Han W, Ye M, Zhu AJ, Huang JG, Zhao HT, Li YF (2016) A combined bioprocess based on solid-state fermentation for dark fermentative hydrogen production from food waste. J Clean Prod 112:3744–3749. https://doi.org/10.5772/5620510.1016/j.jclepro.2015.08.072
Harun R, Singh M, Forde GM, Danquah MK (2010) Bioprocess engineering of microalgae to produce a variety of consumer products. Renew Sust Energ Rev 14:1037–1047. https://doi.org/10.5772/5620510.1016/j.rser.2009.11.004
Haug R (2015) The practical handbook of compost engineering. CRC Press, Boca Raton
Hawkes FR, Dinsdale R, Hawkes DL, Hussy I (2002) Sustainable fermentative hydrogen production: challenges for process optimisation. Int J Hydrogen Energy 27:1339–1347. https://doi.org/10.5772/5620510.1016/S0360-3199(02)00090-3
He R, Shen D, Wang J, He Y, Zhu Y (2005) Biological degradation of MSW in a methanogenic reactor using treated leachate recirculation. Process Biochem 40:3660–3666. https://doi.org/10.5772/5620510.1016/j.procbio.2005.02.022

He Z, Li G, Chen J, Huang Y, An T, Zhang C (2015) Pollution characteristics and health risk assessment of volatile organic compounds emitted from different plastic solid waste recycling workshops. Environ Int 77:85–94. https://doi.org/10.5772/5620510.1016/j.envint.2015.01.004

Hejnfelt A, Angelidaki I (2009) Anaerobic digestion of slaughterhouse by-products. Biomass Bioenergy 33:1046–1054. https://doi.org/10.5772/5620510.1016/j.biombioe.2009.03.004

Hellmann B, Zelles L, Palojarvi A, Bai Q (1997) Emission of climate-relevant trace gases and succession of microbial communities during openwindrow composting. Appl Environ Biotechnol 63:1011–1018

Hensley SA, Moreira E, Holden JF (2016) Hydrogen production and enzyme activities in the hyperthermophile *Thermococcus paralvinellae* grown on maltose, tryptone, and agricultural waste. Front Microbiol 7:167. https://doi.org/10.5772/5620510.3389/fmicb.2016.00167

Ho HL, Lau LY (2014) Bioprocessing of agricultural wastes as optimised carbon source and optimisation of growth conditions for xylanase production by *Aspergillus brasiliensis* in agitated solid state fermentation (SsF). J Biodivers Biopros Dev 1:125. https://doi.org/10.5772/5620510.4172/2376-0214.1000125

Hong RJ, Wang GF, Guo RZ, Cheng X, Liu Q, Zhang PJ, Qian GR (2006) Life cycle assessment of BMT-based integrated municipal solid waste management: case study in Pudong, China. Resour Conserv Recycl 49:129–146. https://doi.org/10.5772/5620510.1016/j.resconrec.2006.03.007

Hosseini SM, Aziz HA (2013) Evaluation of thermochemical pretreatment and continuous thermophilic condition in rice straw composting process enhancement. Bioresour Technol 133:240–247. https://doi.org/10.5772/5620510.1016/j.biortech.2013.01.098

Hu L, Zeng G, Chen G, Dong H, Liu Y, Wan J, Chen A, Guo Z, Yan M, Wu H, Yu Z (2016) Treatment of landfill leachate using immobilized *Phanerochaete chrysosporium* loaded with nitrogen-doped TiO nanoparticles. J Hazard Mater 301:106–118. https://doi.org/10.5772/5620510.1016/j.jhazmat.2015.08.060

Huang L, Chen Y, Zhou H, Luo S, Lan C, Qu L (2003) Characterization of methanogenic archaea in the leachate of a closed municipal solid waste land. FEMS Microbiol Ecol 46:171–177. https://doi.org/10.5772/5620510.1016/S0168-6496(03)00218-6

Huang W, Wang Z, Guo Q, Wang H, Zhou Y, Nga WJ (2016) Pilot-scale landfill with leachate recirculation for enhanced stabilization. Biochem Eng J 105:437–445. https://doi.org/10.5772/5620510.1016/j.bej.2015.10.013

Hultman J, Vasara T, Partanen P, Kurola J, Kontro MH, Paulin L, Auvinen P, Romantschuk M (2010) Determination of fungal succession during municipal solid waste composting using a cloning-based analysis. J Appl Microbiol 108:472–487. https://doi.org/10.5772/5620510.1111/j.1365-2672.2009.04439.x

Ismail ZZ, Talib AR (2016) Recycled medical cotton industry waste as a source of biogas recovery. J Clean Prod 112:4413–4418. https://doi.org/10.5772/5620510.1016/j.jclepro.2015.06.069

Jain S, Jain S, Wolf IT, Lee J, Tong YW (2015) A comprehensive review on operating parameters and different pretreatment methodologies for anaerobic digestion of municipal solid waste. Renew Sust Energ Rev 52:142–154. https://doi.org/10.5772/5620510.1016/j.rser.2015.07.091

Jang HM, Ha JH, Park JM, Kim MS, Sommer SG (2015) Comprehensive microbial analysis of combined mesophilic anaerobic thermophilic aerobic process treating high-strength food wastewater. Water Res 73:291–303. https://doi.org/10.5772/5620510.1016/j.watres.2015.01.038

Javaheri H, Nasrabadi T, Jafarian MH, Rowshan GR, Khoshnam H (2006) Site selection of municipal solid waste landfills using analytical hierarchy process method in a geographical information technology environment in Giroft. J Environ Health Sci Eng 3:177–184

Jose AG, Shukla S, Shukla H, Sandhu SS (2014) Isolation and screening of potential cellulolytic fungi for the management of municipal solid waste. Int J Recent Sci Res 5:2298–2300

Juwarkar AA, Dubey KV, Nair A, Singh SK (2008) Bioremediation of multi-metal contaminated soil using biosurfactant-a novel approach. Indian J Microbiol 48:142–146. https://doi.org/10.5772/5620510.1007/s12088-008-0014-5

Kalia VC (2016) Microbial factories: biofuels, waste treatment. Springer, New Delhi

Kalia VC, Joshi AP (1995) Conversion of waste biomass (pea-shells) into hydrogen and methane through anaerobic digestion. Bioresour Technol 53:165–168. https://doi.org/10.5772/5620510.1016/0960-8524(95)00077-R
Kalia VC, Purohit HJ (2008) Microbial diversity and genomics in aid of bioenergy. J Ind Microbiol Biotechnol 35:403–419. https://doi.org/10.5772/5620510.1007/s10295-007-0300-y
Kalia VC, Sonakya V, Raizada N (2000) Anaerobic digestion of banana stem waste. Bioresour Technol 73:191–193. https://doi.org/10.5772/5620510.1016/S0960-8524(99)00172-8
Kalia VC, Chauhan A, Bhattacharyya G (2003a) Genomic databases yield novel bioplastic producers. Nat Biotechnol 21:845–846. https://doi.org/10.5772/5620510.1038/nbt0803-845
Kalia VC, Lal S, Ghai R, Mandal M, Chauhan A (2003b) Mining genomic databases to identify novel hydrogen producers. Trends Biotechnol 21:152–156. https://doi.org/10.5772/5620510.1016/S0167-7799(03)00028-3
Kalia VC, Prakash J, Koul S (2016) Biorefinery for glycerol rich biodiesel industry waste. Indian J Microbiol 56:113–125. https://doi.org/10.5772/5620510.1007/s12088-016-0583-7
Kamali M, Gameiro T, Costa MEV, Capela I (2016) Anaerobic digestion of pulp and paper mill wastes–an overview of the developments and improvement opportunities. Chem Eng J 298:162–182. https://doi.org/10.5772/5620510.1016/j.cej.2016.03.119
Kana EBG, Oloke JK, Lateef A, Adesiyan MO (2012) Modeling and optimization of biogas production on saw dust and other co-substrates using artificial neural network and genetic algorithm. Renew Energy 46:276–281. https://doi.org/10.5772/5620510.1016/j.renene.2012.03.027
Karacaa F, Özkayab B (2006) NN-LEAP: a neural network-based model for controlling leachate flow-rate in a municipal solid waste landfill site. Environ Model Softw 21:1190–1197. https://doi.org/10.5772/5620510.1016/j.envsoft.2005.06.006
Kawai K, Tasaki T (2016) Revisiting estimates of municipal solid waste generation per capita and their reliability. J Mater Cycles Waste Manag 18:1–13. https://doi.org/10.5772/562050.1007/s10163-015-0355-1
Khadivi MR, Ghomi SMT (2012) Solid waste facilities location using of analytical network process and data envelopment analysis approaches. Waste Manag 32:1258–1265. https://doi.org/10.5772/5620510.1016/j.wasman.2012.02.002
Khan S, Faisal MN (2008) An analytic network process model for municipal solid waste disposal options. Waste Manag 28:1500–1508. https://doi.org/10.5772/5620510.1016/j.wasman.2007.06.015
Khardenavis AA, Kapley A, Purohit HJ (2009) Processing of poultry feathers by alkaline keratin hydrolyzing enzyme from *Serratia* sp. HPC 1383. Waste Manag 29:1405–1415. https://doi.org/10.5772/5620510.1016/j.wasman.2008.10.009
Khardenavis AA, Wang JY, Ng WJ, Purohit HJ (2013) Management of various organic fractions of municipal solid waste via recourse to VFA and biogas generation. Environ Technol 34:2085–2097. https://doi.org/10.5772/5620510.1080/09593330.2013.817446
Kikuchi R (2001) Recycling of municipal solid waste for cement production: pilot-scale test for transforming incineration ash of solid waste into cement clinker. Resour Conserv Recycl 31:137–147. https://doi.org/10.5772/5620510.1016/S0921-3449(00)00077-X
Kim SK, Ravichandran YD, Khan SB, Kim YT (2008) Prospective of the cosmeceuticals derived from marine organisms. Biotechnol Bioprocess Eng 13:11–523. https://doi.org/10.5772/5620510.1007/s12257-008-0113-5
Kjeldsen P, Barlaz MA, Rooker AP, Baun A, Ledin A, Christensen TH (2002) Present and long-term composition of MSW landfill leachate: a review. Crit Rev Environ Sci Technol 32:297–336. https://doi.org/10.5772/5620510.1080/10643380290813462
Ko JH, Yang F, Xu Q (2016) The impact of compaction and leachate recirculation on waste degradation in simulated landfills. Bioresour Technol 211:72–79. https://doi.org/10.5772/5620510.1016/j.biortech.2016.03.070
Komilis DP, Ham RK, Stegmann R (1999) The effect of municipal solid waste pretreatment on landfill behavior: a literature review. Waste Manag Res 17:10–19. https://doi.org/10.5772/5620510.1034/j.1399-3070.1999.00005.x

Korkakaki E, Mulders M, Veeken A, Rozendal R, van Loosdrecht MC, Kleerebezem R (2016) PHA production from the organic fraction of municipal solid waste (OFMSW): overcoming the inhibitory matrix. Water Res 96:74–83. https://doi.org/10.5772/5620510.1016/j.watres.2016.03.033

Kothari R, Pandey A, Kumar V, Tyagi VV (2015) Algae-based biohydrogen: current status of bioprocess routes, economical assessment, and major bottlenecks. Algae and environmental sustainability. Springer, New Delhi, pp 77–86

Kumar A, Jain SR, Sharma CB, Joshi AP, Kalia VC (1995) Increased H2 production by immobilized microorganisms. World J Microbiol Biotechnol 11:156–159. https://doi.org/10.5772/5620510.1007/BF00704638

Kumar P, Patel SK, Lee JK, Kalia VC (2013) Extending the limits of *Bacillus* for novel biotechnological applications. Biotechnol Adv 31:1543–1561. https://doi.org/10.5772/5620510.1016/j.biotechadv.2013.08.007

Kumar P, Pant DC, Mehariya S, Sharma R, Kansal A, Kalia VC (2014) Ecobiotechnological strategy to enhance efficiency of bioconversion of wastes into hydrogen and methane. Indian J Microbiol 54:262–267. https://doi.org/10.5772/5620510.1007/s12088-014-0467-7

Kumar P, Mehariya S, Ray S, Mishra A, Kalia VC (2015) Biotechnology in aid of biodiesel industry effluent (glycerol): biofuels and bioplastics. In: Kalia VC (ed) Microbial factories. Springer, New Delhi, pp 105–119. https://doi.org/10.5772/5620510.1007/978-81-322-2598-0

Kumar P, Ray S, Kalia VC (2016) Production of co-polymers of polyhydroxyalkanoates by regulating the hydrolysis of biowastes. Bioresour Technol 200:413–419. https://doi.org/10.5772/5620510.1016/j.biortech.2015.10.045

Kurniawan TA, Lo WH, Chan GY (2006) Degradation of recalcitrant compounds from stabilized landfill leachate using a combination of ozone-GAC adsorption treatment. J Hazard Mater 137:443–455. https://doi.org/10.5772/5620510.1016/j.jhazmat.2006.02.020

Lee DS, Park SJ, Yoon JY (1996) Molecular weight distribution of humic substances as changed by Fenton's reagent and determination of rate constant of OH radical humic substances in metropolitan landfill leachate. J Korean Soc Environ Eng 18:999–1008

Lei Y, Dang Y, Lan Z, Sun D (2016) Effects of multiple inhibitory components on anaerobic treatment processes in municipal solid waste incineration leachate. Appl Microbiol Biotechnol 100:5123–5130. https://doi.org/10.1007/s00253-016-7341-y

Leite WR, Gottardo M, Pavan P, Belli Filho P, Bolzonella D (2016) Performance and energy aspects of single and two phase thermophilic anaerobic digestion of waste activated sludge. Renew Energy 86:1324–1331. https://doi.org/10.1016/j.renene.2015.09.069

Li M, Xia T, Zhu C, Xi B, Jia X, Wei Z, Zhu J (2014) Effect of short-time hydrothermal pretreatment of kitchen waste on biohydrogen production: fluorescence spectroscopy coupled with parallel factor analysis. Bioresour Technol 172:382–390. https://doi.org/10.1016/j.biortech.2014.09.040

Lin H, Wang B, Zhunag R, Zhou Q, Zhao Y (2011) Artificial construction and characterization of a fungal consortium that produces cellulolytic enzyme system with strong wheat straw saccharification. Bioresour Technol 102:10569–10576. https://doi.org/10.1016/j.biortech.2011.08.095

Liu Q, Chen W, Zhang X, Yu L, Zhou J, Xu Y, Qian G (2015a) Phosphate enhancing fermentative hydrogen production from substrate with municipal solid waste composting leachate as a nutrient. Bioresour Technol 190:431–437. https://doi.org/10.1016/j.biortech.2015.01.139

Liu Z, Dang Y, Li C, Sun D (2015b) Inhibitory effect of high NH –N concentration on anaerobic biotreatment of fresh leachate from a municipal solid waste incineration plant. Waste Manag 43:188–195. https://doi.org/10.1016/j.biortech.2011.11.114

Long XE, Wang J, Huang Y, Yao H (2016) Microbial community structures and metabolic profiles response differently to physiochemical properties between three landfill cover soils. Environ Sci Pollut Res 15:15483–15494. https://doi.org/10.1007/s11356-016-6681-6

Longhi S, dell'Inf DI, delle Marche, Italy A, Marzioni D, Alidori E, Di Buo G (2012) Solid waste management architecture using wireless sensor network technology. In: 5th international con-

ference on New Technologies, Mobility and Security (NTMS). doi:https://doi.org/10.1109/NTMS.2012.6208764
Lopez A, Pagano M, Volpe A, Di Pinto AC (2004) Fenton's pre-treatment of mature landfill leachate. Chemosphere 54:1005–1010. https://doi.org/10.1016/j.chemosphere.2003.09.015
Lozecznik S, Sparling R, Clark SP, VanGulck JF, Oleszkiewicz JA (2012) Acetate and propionate impact on the methanogenesis of landfill leachate and the reduction of clogging components. Bioresour Technol 104:37–43. https://doi.org/10.1016/j.biortech.2011.09.123
Luo J, Qiana G, Liu J, ZP X (2015) Anaerobic methanogenesis of fresh leachate from municipal solid waste: a brief review on current progress. Renew Sust Energ Rev 49:21–28. https://doi.org/10.1016/j.rser.2015.04.053
Madrini B, Shibusawa S, Kojima Y, Hosaka S (2016) Effect of natural zeolite (clinoptilolite) on ammonia emissions of leftover food-rice hulls composting at the initial stage of the thermophilic process. J Agric Meteorol 72:12–19. https://doi.org/10.2480/agrmet.D-15-00012
McInerney MJ, Bryant MP, Hespell RB, Costerton JW (1981) *Syntrophomonas wolfei* gen. nov. sp. nov., an anaerobic, syntrophic, fatty acidoxidizing bacterium. Appl Environ Microbiol 41:1029–1039
Meng Y, Jost C, Mumme J, Wang K, Linke B (2016) An analysis of single and two stage, mesophilic and thermophilic high rate systems for anaerobic digestion of corn stalk. Chem Eng J 288:79–86. https://doi.org/10.1016/j.cej.2015.11.072
Micolucci F, Gottardo M, Cavinato C, Pavan P, Bolzonella D (2016) Mesophilic and thermophilic anaerobic digestion of the liquid fraction of pressed biowaste for high energy yields recovery. Waste Manag 48:227–235. https://doi.org/10.1016/j.wasman.2015.09.031
Moser A (2012) Bioprocess technology: kinetics and reactors. Springer Science & Business Media, New York
Mussatto SI, Machado EM, Martins S, Teixeira JA (2011) Production, composition, and application of coffee and its industrial residues. Food Bioprocess Technol 4:661–672. https://doi.org/10.1007/s11947-011-0565-z
Nair A, Sartaj M, Kennedy K, Coelho NM (2014) Enhancing biogas production from anaerobic biodegradation of the organic fraction of municipal solid waste through leachate blending and recirculation. Waste Manag Res 32:939–946. https://doi.org/10.1177/0734242X14546036
Nixon JD, Dey PK, Ghosh SK, Davies PA (2013) Evaluation of options for energy recovery from municipal solid waste in India using the hierarchical analytical network process. Energy 59:215–223. https://doi.org/10.1016/j.energy.2013.06.052
Noori R, Abdoli MA, Farokhnia A, Abbasi M (2009) Results uncertainty of solid waste generation forecasting by hybrid of wavelet transform-ANFIS and wavelet transform-neural network. Expert Syst Appl 36:9991–9999. https://doi.org/10.1016/j.eswa.2008.12.035
Oluleye G, Smith R, Jobson M (2016) Modelling and screening heat pump options for the exploitation of low grade waste heat in process sites. Appl Energy 169:267–286. https://doi.org/10.1016/j.apenergy.2016.02.015
Palatsi J, Illa J, Prenafeta-Boldú FX, Laureni M, Fernandez B, Angelidaki I, Flotats X (2010) Long-chain fatty acids inhibition and adaptation process in anaerobic thermophilic digestion: batch tests, microbial community structure and mathematical modelling. Bioresour Technol 101:2243–2251. https://doi.org/10.1016/j.biortech.2009.11.069
Paleologos EK, Caratelli P, El Amrousi M (2016) Waste-to-energy: an opportunity for a new industrial typology in Abu Dhabi. Renew Sust Energ Rev 55:1260–1266. https://doi.org/10.1016/j.rser.2015.07.098
Pan I, Bomba D, Sen SK (2012) Composting of common organic wastes using microbial inoculants. Biotech 2:127–134. https://doi.org/10.1007/s13205-011-0033-5
Pandey BK, Vyas S, Pandey M, Gaur A (2016) Municipal solid waste to energy conversion methodology as physical, thermal, and biological methods. Curr Sci Perspect 2:39–46
Pandit PD, Gulhane MK, Khardenavis AA, Vaidya AN (2015) Technological advances for treating municipal waste. In: Kalia VC (ed) Microbial factories, vol 1, (4.1). Springer, New York, pp 217–229

Pandit PD, Gulhane MK, Khardenavis AA, Purohit HJ (2016) Mining of hemicellulose and lignin degrading genes from differentially enriched methane producing microbial community. Bioresour Technol 216:923–930. https://doi.org/10.1016/j.biortech.2016.06.021

Pantaleo A, Gennaro BD, Shah N (2013) Assessment of optimal size of anaerobic co-digestion plants: an application to cattle farms in the province of Bari (Italy). Renew Sust Energ Rev 20:57–70. https://doi.org/10.1016/j.rser.2012.11.068

Parihar AKS, Sethi V, Banerjee R (2016) Sustainability assessment of biomass gasification based distributed power generation in India. In: Proceedings of the first international conference on recent advances in bioenergy research. Springer, New Delhi, pp 213–225

Patel SKS, Kalia VC (2013) Integrative biological hydrogen production: an overview. Indian J Microbiol 53:3–10. https://doi.org/10.1007/s12088-012-0287-6

Patel SKS, Purohit HJ, Kalia VC (2010) Dark fermentative hydrogen production by defined mixed microbial cultures immobilized on ligno-cellulosic waste materials. Int J Hydrogen Energy 35:10674–10681. https://doi.org/10.1016/j.ijhydene.2010.03.025

Patel SKS, Singh M, Kumar P, Purohit HJ, Kalia VC (2012) Exploitation of defined bacterial cultures for production of hydrogen and polyhydroxybutyrate from pea-shells. Biomass Bioenergy 36:218–225. https://doi.org/10.1016/j.biombioe.2011.10.027

Patel SKS, Kumar P, Mehariya S, Purohit HJ, Lee JK, Kalia VC (2014) Enhancement in hydrogen production by co-cultures of *Bacillus* and *Enterobacter*. Int J Hydrogen Energy 39:14663–14668. https://doi.org/10.1016/j.ijhydene.2014.07.084

Patel SKS, Lee JK, Kalia VC (2016) Integrative approach for producing hydrogen and polyhydroxyalkanoate from mixed wastes of biological origin. Indian J Microbiol 56:293–300. https://doi.org/10.1007/s12088-016-0595-3

Patel SKS, Lee JK, Kalia VC (2017) Dark-fermentative biological hydrogen production from mixed biowastes using defined mixed cultures. Indian J Microbiol 57:171–176. https://doi.org/10.1007/s12088-017-0643-7

Pat-Espadas AM, Razo-Flores E, Rangel-Mendez JR, Ascacio-Valdes JA, Aguilar CN, Cervantes FJ (2016) Immobilization of biogenic Pd (0) in anaerobic granular sludge for the biotransformation of recalcitrant halogenated pollutants in UASB reactors. Appl Microbiol Biotechnol 100:1427–1436. https://doi.org/10.1007/s00253-015-7055-6

Paul JJ, Karmegam N, Daniel T (2011) Municipal solid waste (MSW) vermicomposting with an epigeic earthworm, *Perionyx ceylanensis Mich*. Bioresour Technol 102:6769–6773. https://doi.org/10.1016/j.biortech.2011.03.089

Pendyala B, Chaganti SR, Lalman JA, Heath DD, Shanmugam SR, Veeravalli SS (2013) Using a food and paper-cardboard waste blend as a novel feedstock for hydrogen production: influence of key process parameters on microbial diversity. Int J Hydrogen Energy 38:6357–6367. https://doi.org/10.1016/j.ijhydene.2013.03.003

Pinto ADS, Barreto CC, Vainstein MH, Schrank A, Ulhoa CJ (1997) Purification and characterization of an extracellular chitinase from the entomopathogen *Metarhizium anisopliae*. Can J Microbiol 43:322–327

Porwal S, Kumar T, Lal S, Rani A, Kumar S, Cheema S, Purohit HJ, Sharma R, Patel SK, Kalia VC (2008) Hydrogen and polyhydroxybutyrate producing abilities of microbes from diverse habitats by dark fermentative process. Bioresour Technol 99:5444–5451. https://doi.org/10.1016/j.biortech.2007.11.011

Powell JT, Townsend TG, Zimmerman JB (2015) Estimates of solid waste disposal rates and reduction targets for landfill gas emissions. Nat Clim Chang 6:162–165. https://doi.org/10.1038/nclimate2804

Prakasham RS, Sathish T, Brahmaiah P (2011) Imperative role of neural networks coupled genetic algorithm on optimization of biohydrogen yield. Int J Hydrogen Energy 36:4332–4339. https://doi.org/10.1016/j.ijhydene.2011.01.031

Quirós R, Gabarrell X, Villalba G, Barrena R, García A, Torrente J, Font X (2015) The application of LCA to alternative methods for treating the organic fiber produced from autoclaving unsorted municipal solid waste: case study of Catalonia. J Clean Prod 107:516–528. https://doi.org/10.1016/j.jclepro.2014.04.018

Rajagopal R, Massé DI, Singh G (2013) A critical review on inhibition of anaerobic digestion process by excess ammonia. Bioresour Technol 143:632–641. https://doi.org/10.1016/j.biortech.2013.06.030

Rani RU, Kumar SA, Kaliappan S, Yeom IT, Banu JR (2012) Low temperature thermo-chemical pretreatment of dairy waste activated sludge for anaerobic digestion process. Bioresour Technol 103:415–424. https://doi.org/10.1016/j.biortech.2011.09.124

Ray S, Kalia VC (2017) Microbial co-metabolism and polyhydroxyalkanoate co-polymers. Indian J Microbiol 57:39–47. https://doi.org/10.1007/s12088-016-0622-4

Rouches E, Herpoël-Gimbert I, Steyer JP, Carrere H (2016) Improvement of anaerobic degradation by white-rot fungi pretreatment of lignocellulosic biomass: a review. Renew Sust Energ Rev 59:179–198. https://doi.org/10.1016/j.rser.2015.12.317

Sadhukhan J, Ng KS, Martinez-Hernandez E (2016) Novel integrated mechanical biological chemical treatment (MBCT) systems for the production of levulinic acid from fraction of municipal solid waste: a comprehensive techno-economic analysis. Bioresour Technol 215:131–143. https://doi.org/10.1016/j.biortech.2016.04.030

Sahariah B, Goswami L, Kim KH, Bhattacharyya P, Bhattacharya SS (2015) Metal remediation and biodegradation potential of earthworm species on municipal solid waste: a parallel analysis between *Metaphire posthuma* and *Eisenia fetida*. Bioresour Technol 180:230–236. https://doi.org/10.1016/j.biortech.2014.12.062

Said-Pullicino D, Erriquens FG, Gigliotti G (2007) Changes in the chemical characteristics of water-extractable organic matter during composting and their influence on compost stability and maturity. Bioresour Technol 98:1822–1831. https://doi.org/10.1016/j.biortech.2006.06.018

Saritha M, Arora A, Lata (2012) Biological pretreatment of lignocellulosic substrates for enhanced delignification and enzymatic digestibility. Indian J Microbiol 52:122–130. https://doi.org/10.1007/s12088-011-0199-x

Shah AT, Favaro L, Alibardi L, Cagnin L, Sandon A, Cossu R, Casella S, Basaglia M (2016) *Bacillus* sp. strains to produce bio-hydrogen from the organic fraction of municipal solid waste. Appl Energy 176:116–124. https://doi.org/10.1016/j.apenergy.2016.05.054

Shahriari H, Warith M, Hamoda M, Kennedy KJ (2012) Anaerobic digestion of organic fraction of municipal solid waste combining two pretreatment modalities, high temperature microwave and hydrogen peroxide. Waste Manag 32:41–52. https://doi.org/10.1016/j.wasman.2011.08.012

Shamshiry E, Mokhtar MB, Abdulai AM, Komoo I (2015) Using the analytic hierarchy process to enhance sustainable solid waste management: case study of Langkawi Island, Malaysia. Environ Qual Manag 24:51–64. https://doi.org/10.1002/tqem.21404

Sharholy M, Ahmad K, Mahmood G, Trivedi RC (2008) Municipal solid waste management in Indian cities–a review. Waste Manag 28:459–467. https://doi.org/10.1016/j.wasman.2007.02.008

Sharma P, Khardenavis AA, Purohit HJ (2015) Metabolism of long chain fatty acids (LCFAs) in methanogenesis. In: Kalia VC (ed) Microbial factories, vol 1. (4.4). Springer, New York, pp 279–291

Shu CH, Jaiswal R, Shih JS (2015) Improving biodegradation of rice straw using alkaline and *Aspergillus niger* pretreatment for methane production by anaerobic co-digestion. J Bioprocess Biotech 5:256. https://doi.org/10.4172/2155-9821.1000256

Siddique F, Hussain I, Mahmood MS, Ahmed SI, Iqbal A (2014) Isolation and characterization of a highly thermostable alpha-amylase enzyme produced by *Bacillus licheniformis*. Pak J Agric Sci 51:299–304

Singh M, Patel SK, Kalia VC (2009) *Bacillus subtilis* as potential producer for polyhydroxyalkanoates. Microb Cell Factories. https://doi.org/10.1186/1475-2859-8-38

Singh M, Kumar P, Ray S, Kalia VC (2015) Challenges and opportunities for customizing polyhydroxyalkanoates. Indian J Microbiol 55:235–249. https://doi.org/10.1007/s12088-015-0528-6

Singh N, Hui D, Singh R, Ahuja IPS, Feo L, Fraternali F (2016) Recycling of plastic solid waste: a state of art review and future applications. Compos B Eng. https://doi.org/10.1016/j.compositesb.2016.09.013

Solís N, Chiriboga C, Ávalos R, Rueda D, Albán C, Manjunatha B, Rueda B, Selvanayagam M (2016) Use of Cellulase enzyme obtained from *Monilia (Moniliophthora roreri)* for treatment of solid waste of cob, rice husks and cocoa shell. J Appl Pharm Sci 6:66–70. https://doi.org/10.7324/JAPS.2016.60311

Sonakya V, Raizada N, Kalia VC (2001) Microbial and enzymatic improvement of anaerobic digestion of waste biomass. Biotechnol Lett 23:1463–1466. https://doi.org/10.1023/A:1011664912970

Subramanian PM (2000) Plastics recycling and waste management in the US. Resour Conserv Recycl 28:253–263. https://doi.org/10.1016/S0921-3449(99)00049-X

Suthanthararajan R, Chitra K, Umamaheswari B, Ravindranath E, Rajamani S (2006) U.S. Patent Application No. 11/368,347

Tan S, Hashima H, Ho WS, Lim JS, Lee CT (2013) Optimal process network of municipal solid waste management in Iskandar Malaysia. J Clean Prod 71:48–58. https://doi.org/10.1016/j.jclepro.2013.12.005

Tang WZ, Huang CP (1997) Stoichiometry of Fenton's reagent in the oxidation of chlorinated aliphatic organic pollutants. Environ Technol 18:13–23. https://doi.org/10.1080/09593330.1997.9618467

Thakur N, Khardenavis A, Purohit HJ (2015) Recent advances in factors and methods for stimulation of biomethane production. Recent Adv DNA Gene Sequences 9:3–13. https://doi.org/10.2174/235209220901160126152924

Tong X, Tao D (2016) The rise and fall of a "waste city" in the construction of an "urban circular economic system": the changing landscape of waste in Beijing. Resour Conserv Recycl 107:10–17. https://doi.org/10.1016/j.resconrec.2015.12.003

Torres ML, Lloréns MD (2008) Effect of alkaline pretreatment on anaerobic digestion of solid wastes. Waste Manag 28(11):2229–2234. https://doi.org/10.1016/j.wasman.2007.10.006

Tuorto SJ, Brown CM, Bidle KD, McGuinness LR, Kerkhof LJ (2015) BioDry: an inexpensive, low-power method to preserve aquatic microbial biomass at room temperature. PloS One. https://doi.org/10.1371/journal.pone.0144686

Tyagi VK, Campoy RA, Álvarez-Gallego CJ, García LR (2014) Enhancement in hydrogen production by thermophilic anaerobic co-digestion of organic fraction of municipal solid waste and sewage sludge–optimization of treatment conditions. Bioresour Technol 164:408–415. https://doi.org/10.1016/j.biortech.2014.05.013

Valdez-Vazquez I, Ríos-Leal E, Carmona-Martínez A, Muñoz-Páez KM, Poggi-Varaldo HM (2006) Improvement of biohydrogen production from solid wastes by intermittent venting and gas flushing of batch reactors headspace. Environ Sci Technol 40:3409–3415. https://doi.org/10.1021/es052119j

Van den Brand TPH, Roest K, Chen GH, Brdjanovic D, van Loosdrecht MCM (2015) Occurrence and activity of sulphate reducing bacteria in aerobic activated sludge systems. World J Microbiol Biotechnol 31:507–516. https://doi.org/10.1007/s11274-015-1807-4

Varel VH, Isaacson HR, Bryant M (1977) Thermophilic methane production from cattle waste. Appl Environ Microbiol 33:298–307

Vavilin VA, Jonsson S, Ejlertsson J, Svensson BH (2006) Modelling MSW decomposition under landfill conditions considering hydrolytic and methanogenic inhibition. Biodegradation 17:389–402. https://doi.org/10.1007/s10532-005-9009-7

HY W, Ting YP (2006) Metal extraction from municipal solid waste (MSW) incinerator fly ash—chemical leaching and fungal bioleaching. Enzym Microb Technol 38:839–847. https://doi.org/10.1016/j.enzmictec.2005.08.012

Xia Y, He PJ, HX P, Lu F, Shao LM, Zhang H (2016) Inhibition effects of high calcium concentration on anaerobic biological treatment of MSW leachate. Environ Sci Pollut Res 23:7942–7948. https://doi.org/10.1007/s11356-016-6052-3

Xu J, Wang X, Hu L, Xia J, Wu Z, Xu N, Dai B, Wu B (2015) A novel ionic liquid-tolerant *Fusarium oxysporum* BN secreting ionic liquid-stable cellulase: consolidated bioprocessing of pretreated lignocellulose containing residual ionic liquid. Bioresour Technol 181:18–25. https://doi.org/10.1016/j.biortech.2014.12.080

Yang Y, Gajaraj S, Wall JD, Hu Z (2013) A comparison of nanosilver and silver ion effects on bioreactor landfill operations and methanogenic population dynamics. Water Res 47:3422–3430. https://doi.org/10.1016/j.watres.2013.03.040

Yenigün O, Demirel B (2013) Ammonia inhibition in anaerobic digestion: a review. Process Biochem 48:901–911. https://doi.org/10.1016/j.procbio.2013.04.012

Yu D, Kurola JM, Lahde K, Kyamalainen M, Sinkkonen A, Romantschuk M (2014) Biogas production and methanogenic archaeal community in mesophilic and thermophilic anaerobic co-digestion processes. J Environ Manag 143:54–60. https://doi.org/10.1016/j.jenvman.2014.04.025

Zahedi S, Sales D, Romero LI, Solera R (2013a) Optimisation of the two-phase dry-thermophilic anaerobic digestion process of sulphatecontaining municipal solid waste: population dynamics. Bioresour Technol 148:443–452. https://doi.org/10.1016/j.biortech.2013.09.002

Zahedi S, Sales D, Romero LI, Solera R (2013b) Hydrogen production from the organic fraction of municipal solid waste in anaerobic thermophilic acidogenesis: influence of organic loading rate and microbial content of the solid waste. Bioresour Technol 129:85–91. https://doi.org/10.1016/j.biortech.2012.11.003

Zahedi S, Solera R, García-Morales JL, Sales D (2016) Effect of the addition of glycerol on hydrogen production from industrial municipal solid waste. Fuel 180:343–347. https://doi.org/10.1016/j.fuel.2016.04.063

Zakaria MR, Norrrahim MNF, Hirata S, Hassan MA (2015) Hydrothermal and wet disk milling pretreatment for high conversion of biosugars from oil palm mesocarp fiber. Bioresour Technol 181:263–269. https://doi.org/10.1016/j.biortech.2015.01.072

Zaman AU (2013) Identification of waste management development drivers and potential emerging waste treatment technologies. Int J Environ Sci Technol 10:455–464. https://doi.org/10.1007/s13762-013-0187-2

Zhang J, Zang L (2016) Enhancement of biohydrogen production from brewers spent grain by calcined-red mud pretreatment. Bioresour Technol 209:73–79. https://doi.org/10.1016/j.biortech.2016.02.110

Zhang W, Lei Z, Li A (2015a) Anaerobic co-digestion of food waste with MSW incineration plant fresh leachate: process performance and synergistic effects. Chem Eng J 259:795–805. https://doi.org/10.1016/j.cej.2014.08.039

Zhang H, Wang YJ, Lü F, Chai LN, Shao LM, He PJ (2015b) Effects of dilute acid pretreatment on physicochemical characteristics and consolidated bioprocessing of rice straw. Waste Biomass Valor 6:217–223. https://doi.org/10.1007/s12649-014-9341-4

Zhao X, He X, Xi B, Gao R, Tan W, Zhang H, Li D (2016) The evolution of water extractable organic matter and its association with microbial community dynamics during municipal solid waste composting. doi:https://doi.org/10.1016/j.wasman.2016.07.018

Zheng W, Lü F, Bolyard SC, Shao L, Reinhart DR, He P (2015) Evaluation of monitoring indicators for the post-closure care of a landfill for MSW characterized with low lignin content. Waste Manag 36:222–229. https://doi.org/10.1016/j.wasman.2014.10.031

Zhou H, Wu C, Onwudili JA, Meng A, Zhang Y, Williams PT (2015) Polycyclic aromatic hydrocarbons (PAH) formation from the pyrolysis of different municipal solid waste fractions. Waste Manag 36:136–146. https://doi.org/10.1016/j.wasman.2014.09.014

Processes of Microbial Transformation and Physical Removal of Polychlorinated Biphenyls (PCBs) in Wastewater Treatment

5

Soliver C. Fusi, Alisha Y. Chan, and Birthe V. Kjellerup

Abstract

Polychlorinated biphenyls (PCBs) are persistent organic pollutants that can cause detrimental effects when humans and animals are exposed to them in the environment. The chemical persistence of these molecules has rendered it difficult to remove from the environment. The objective of this chapter is to discuss the sources of PCBs, factors influencing their transformation, and overall processes of removal from the environment. In addition, microbial transformation of PCBs and physical removal in wastewater treatment plants will be discussed. Finally, future means of PCB elimination, based on methods currently in use in similar systems, will be proposed.

Keywords

Biotransformation • Dechlorination • Environment • Wastewater • Pollutants

---

Soliver C. Fusi and Alisha Y. Chan contributed evenly as first authors.

S.C. Fusi
Department of Chemical and Biomolecular Engineering, University of Maryland, College Park, MD 20742, USA

A.Y. Chan • B.V. Kjellerup (✉)
A. James Clark School of Engineering, Department of Civil and Environmental Engineering, University of Maryland, 1147 Glenn L. Martin Hall, College Park, MD 20742, USA
e-mail: bvk@umd.edu

H.J. Purohit et al. (eds.), *Optimization and Applicability of Bioprocesses*,
https://doi.org/10.1007/978-981-10-6863-8_5

5.1 Introduction

Polychlorinated biphenyls (PCBS) are persistent organic pollutants created by heating together two benzene rings in the presence of chlorine gas (State of Oregon, Department of Environmental Quality 2003). They are characterized by their phenyl rings and their ability to bond with one to ten chlorines, as depicted in Fig. 5.1. Each degree of chlorination is called a homologue. There are 209 different PCB configurations, called congeners. Though PCBs were anthropogenically produced for over 40 years, they were first discovered in 1865 as a by-product of coal tar (America Unites 2016). Though some PCBs were found naturally occurring in the environment, most PCBs were synthesized (US EPA 2016b). PCBs were first produced in the 1920s for use in nonflammable oils, hydraulic fluids, circuits, and the like (Cairns and Siegmund 1981). Their resistance to both acid and alkali, nonflammability, electrical insulating properties, and overall chemical stability led to their mass production (US EPA 2016b). Since they have a high octanol-water partition coefficient (−LogKow) shown in Table 5.1, PCBs preferentially sorb to particulate matter rather than remain in solute form (Erickson 1997). Insolubility and toxicity in water generally increase with additional chlorines (Abramowicz 1995; Erickson 1997). However, 12 congeners are particularly toxic due to their structure being similar to dioxins (White and Birnbaum 2009).

PCBs are commonly referred to by the trade name Aroclor, given to them by their major producer, Monsanto Corporation. One of the first reported health effects

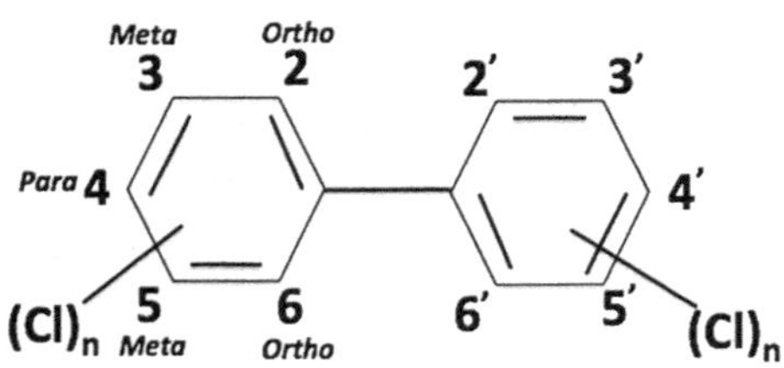

Fig. 5.1 The structure of a PCB showing the ten possible chlorine locations on the phenyl ring

Table 5.1 Octanol-water partition coefficients across the homologue groups

Homologue group	Molecular weight	$-\text{LogK}_{ow}$
Mono-CB	188.7	4.5–4.7
Di-CB	223.1	5.0–5.6
Tri-CB	257.6	5.6–6.1
Tetra-CB	292	5.9–6.7
Penta-CB	326.4	6.4–7.5
Hexa-CB	360.9	7.1–8.3
Hepta-CB	395.3	7.9
Octa-CB	429.8	8.4–8.6
Nona-CB	464.2	9.1
Deca-CB	498.7	9.6

$-\text{LogK}_{ow}$ = partition coefficient water/octanol. Values in table were acquired from Pereira (2004)

of PCBs was by a US Public Health Service senior official in 1936 (America Unites 2016). Several other health effects such as chloracne and cancer and harmful effects to the immune, reproductive, nervous, and endocrine systems were later linked to PCBs (Clapp et al. 2008; ATSDR 2014; US EPA 2016b). Discovery of these health effects let the EPA to ban the production of PCBs in 1979 and set forth rules that slowly prevented several industrial applications of PCBs (America Unites 2016). However, PCBs were not banned worldwide until 2001, and these regulations still allowed for the continued use of PCB-containing materials, some of which are still in use today (Sowers and May 2013). Currently, an estimated 150 million pounds of PCBs are dispersed throughout the environment, and 290 million pounds remain in landfills (Bans and Manufacture 1979). An estimated total of 3.4 billion pounds of PCBs were produced in the 1900s, and these chemicals are still present in the natural environment today, bioaccumulating in living systems and matriculating through wastewater (Illinois Department of Public Health 2009; US EPA 2016b).

5.2 Sources of PCBs in Wastewater

Major sources of PCBs in wastewater are from runoff, leaching from landfills, and improper disposal of chemical waste (US EPA 2016b). Previously disposed capacitors, transformers, inks, lubricants, adhesives, and other PCB-containing products are present in unsecured landfills and currently leach toxic PCB waste into water systems during rain events (State of Oregon, Department of Environmental Quality 2003). One notable example of PCBs' journey into our environment is through the improper disposal of PCB-laden waste into the Hudson River by General Electric (GE) between 1947 and 1977. The Hudson River was listed as a Superfund site, deeming GE responsible for riding the river of PCBs. Millions of cubic yards of PCB-contaminated sediment were dredged out. Such efforts have proven unsuccessful and PCBs are still present in soil and water (EPA 2016a).

The low vapor pressure and hydrophobicity of PCBs allow partitioning between water and the atmosphere (A risk-Management 2001). Also due to the partitioning of PCBs between water and other medium, PCBs have been present in biosolids after the wastewater treatment process and used as fertilizer. They have found their way into groundwater and dispersed into the atmosphere, enabling them to travel to different areas of our ecosystem (State of Oregon Department of Environmental Quality 2003). Cold temperatures cause atmospheric PCBs to condense back to the ground and into surface water. PCBs partition and sorb to organic components and therefore can be found in nearly every compartment of the environment (A risk-Management 2001). Although PCBs are hydrophobic and tend to sorb to particulate, organic matter, desorption, and rain events typically place PCBs back into wastewater (Montague 1993; Sowers and May 2013). A diagram summarizing the sources of PCBs is shown in Fig. 5.2. Major sources include landfill runoff, PCB-contaminated biosolids used in fertilizer, atmospheric deposition, and improper waste disposal. Other sources of PCB in the environment include the release of effluent water that contains PCBs even after the wastewater treatment process.

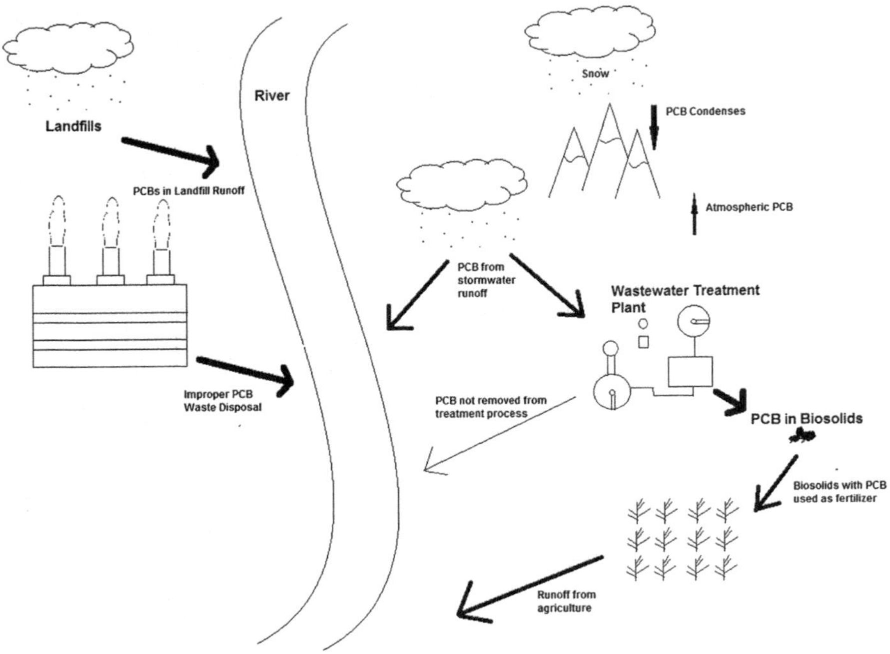

Fig. 5.2 Sources of PCB in the environment

5.3 Factors in PCB Transformation: Pre-wastewater Treatment

Several modeling techniques have been applied in natural systems for quantification of PCBs present. An example is that the identification of the percent of each PCB congener is used to define characteristic properties of PCB-contaminated water sources (Rodenburg et al. 2010). These PCB quantifications have been used to determine the processes by which PCBs are transformed. This method is validated by its ability to determine the Aroclors present in a given environmental sample. The prevalence of specific PCB congeners is theoretically used to classify a specific sample as the product of dechlorination, wastewater treatment, an Aroclor, etc. (Rodenburg et al. 2010).

PCB dechlorination preferentially occurs at the *meta* and *para* positions of the biphenyl rings due to steric hindrance (Su et al. 2016). For instance, PCB 4, having a chlorine attached to one *ortho* position, is often used as the indicator of the end-point of PCB dechlorination. PCB 4, shown in Fig. 5.3, and PCBs 1, 19, 25, 26, 32, and 49 as well as the lack of PCB 11 are identified as potential indicator of advanced dechlorination (Rodenburg et al. 2010). That is, the prevalence of *ortho*-substituted congeners, with low molecular weight, in a sample, is representative of microbial dechlorination. Dechlorinating activities have been observed in combined wastewater treatment plant (WWTP) influents that were not present in stormwater runoff

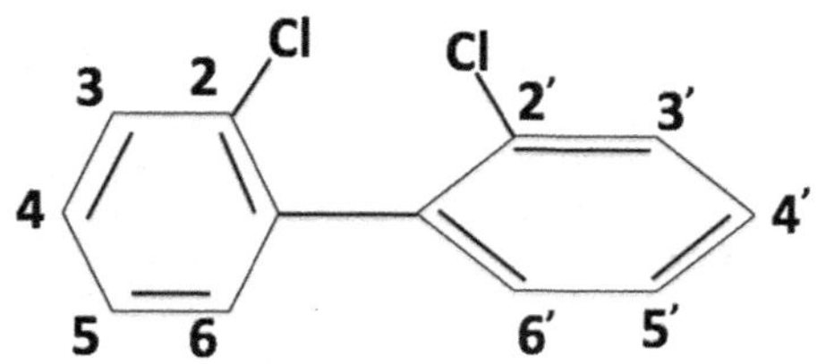

Fig. 5.3 Depiction of PCB 4 showing steric hindrance. PCB 4, having chlorines on both ortho positions, if used as an indicator of dechlorination. The different structure of the right biphenyl ring symbolizes the rotation of the rings

alone (Rodenburg et al. 2010). This shows that there is microbial anaerobic dechlorination activity taking place prior to the wastewater treatment process, likely in sewers and contaminated groundwater. Nonbiological transformation processes do not result in *ortho*-chlorinated compounds. They can cleave the rings or chemically dechlorinate, but not with preference for specific chlorine.

5.4 Microbial PCB Transformation

PCB transformation and complete mineralization require a combination of anaerobic dechlorinating and aerobic degradation activity (Abramowicz 1995). Identification of the involved bacteria as well as the conditions in which they thrive is key to continued work on PCB transformation and removal. Organohalide respiration, the process by which microorganisms derive energy by using PCBs as an electron acceptor, was noted downstream of the Hudson River. This serves as evidence of anaerobic dechlorination of PCBs in freshwater, estuary, and marine systems. Anaerobic dechlorination is often followed by aerobic degradation, performed by *Burkholderia xenovorans* strain LB400 or other aerobic bacteria ubiquitously present in the environment (Patureau and Trably 2006; Payne et al. 2013). The diversity of these microorganisms may result in their potential presence in a variety of sources such as direct waste inputs, soil particles from stormwater runoff, and sediments from groundwater and aquifers.

PCBs with less than four chlorines can be broken down by the activation of the catabolic biphenyl pathway present in certain aerobic bacteria (Bedard 2004). The catabolic biphenyl pathway is a four-step enzymatic process that turns biphenyl into benzoic acid and two hydroxy-penta-2,3-dienoic acid. Pentatonic acid can then be converted to acetyl-CoA and used in the tricarboxylic acid cycle (Bedard 2004). Unfortunately it becomes more difficult to degrade PCBs as the number of chlorines increases; thus reductive dechlorination is required and is most often the limiting step.

5.4.1 Anaerobic Dechlorination

Naturally occurring anaerobic microbial biodegradation of PCBs has been observed in several environments such as soil, sediment, and wastewater biosolids (Flanagan and May 1993). The main limiting factor in this anaerobic dechlorination process remains the bioavailability of PCBs, which is reduced due to their hydrophobicity and thus partitioning between the aqueous and particulate phases (Patureau and Trably 2006). Highly chlorinated PCBs can be reductive dechlorinated resulting in PCBs with a degree of chlorination of five and often less (Bertin et al. 2011). Lightly chlorinated congener is susceptible to complete degradation by aerobic microorganisms (Pieper 2005). Anaerobic dechlorination of higher chlorinated PCBs seemed to stagnate after about 40% dechlorination was obtained (Patureau and Trably 2006). These results were attributed to inadequate incubation periods of 4 months which reduced the adaptation time needed for the bacteria. This is consistent with increased dechlorination rates with increased acclimation time (Patureau and Trably 2006).

Anaerobic dechlorination rates are influenced by the PCB source, degree of chlorination and other chemical characteristics, inoculum incubation time, thermophilic versus mesophilic conditions, total solids thus bioavailability, pH, and likely other, yet to be distinguished factors (Chang et al. 1999; Bertin et al. 2011). Since such conditions can greatly influence PCB transformation rates, various models have been developed to aid in predicting dechlorinating patterns (Hughes et al. 2010). A few distinct trends include the affinity for highly chlorinated PCBs to be preferentially dechlorinated under anaerobic conditions (Patureau and Trably 2006). Parallely, lightly chlorinated PCBs are preferentially degraded under aerobic conditions and not reductively dechlorinated. More degradation in aerobic environments versus dechlorination during anaerobic conditions has been attributed to enhanced bioavailability in aerobic processes since the lowly chlorinated congeners have a higher solubility than highly chlorinated congeners (Patureau and Trably 2006). Stalling in anaerobic activity is noticed when congeners are no longer flanked (Payne et al. 2013). Dechlorination resulting in up to 80% by mass loss of overall PCBs (compared to 25% in the control) was observed in mesocosms augmented with both anaerobic and aerobic microorganisms. No increase in lightly chlorinated congeners was noted in this system. This supports sequential anaerobic dechlorination proceeded by aerobic cleavage and degradation of PCBs (Payne et al. 2013).

The bacteria capable of anaerobic dechlorination are as examples: *Dehalobium chlorocoercia* DF-1, *Dehalococcoides* CBDB1, *Dehalococcoides mccartyii*195, *Desulfitobacterium*, and *Dehalobacter*, along with phylotypes DEH-10 and SF-2 (Maphosa et al. 2012; Payne et al. 2013; Sowers and May 2013). In addition, *Desulfuromonas* and *Dehalospirillum multivorans* are also capable of dechlorinating dioxins and chlorinated solvents in groundwater (Leeson et al. 2004; Pieper 2005). The benefits of anaerobic dechlorination lie in its reduction of dioxin-like activity of PCBs (Tiedje et al. 1993). The products of anaerobic degradation, typically lightly chlorinated PCBs, are left readily available for aerobic degradation (Abramowicz 1995).

5.4.2 Aerobic Degradation

PCB transformation preferentially occurs on *meta-*, *para-*, and then *ortho*-substituted sites (Tiedje et al. 1993). The presence of *ortho*-substituted PCBs after a year incubation under anaerobic conditions and lack of *ortho* dechlorination showed the need for further work on aerobic PCB transforming microbes (Tiedje et al. 1993; Wu et al. 2000). Aerobic PCB transformation occurs through the 2,3-dioxygenase pathway (Abramowicz 1995; Haddock et al. 1995). The resulting compound, chlorobenzoic acids, can be further reduced to CO_2, H_2O, Cl, and biomass (Abramowicz 1995). Aerobic bacteria capable of PCB transformation include the genera *Pseudomonas*, *Burkholderia*, *Achromobacter*, *Comamonas*, *Ralstonia*, *Acinetobacter*, *Rhodococcus*, *and Bacillus* (Maltseva et al. 1999; Pieper 2005; Kjellerup et al. 2012; Zhang et al. 2015).

B. xenovorans strain LB400 is one of the most capable PCB degraders that is nonpathogenic (Mikszewski 2004). LB400 can be naturally found in soil rhizospheres of certain species of grass plants. Both *Alcaligenes eutrophus* H850 and *Pseudomonas* spp. have been seen to degrade highly chlorinated PCBs through the 3,4-dioxygenase pathway, as opposed to the well-known 2,3-dioxygenase pathway (Furukawa 1994). There is evidence of LB400's capability to metabolize representatives of all major structural classes of PCBs (Bopp 1986). LB400 has been shown to have the capability to aerobically degrade over 20 different PCB congeners (Chain et al. 2006).

Ralstonia eutropha strain H850 has also been shown to degrade a large number of chlorinated aromatic compounds and chemically related pollutants including PCBs (Abbey et al. 2003). H850 is commonly found in soil and water and is considered nonpathogenic. They have the ability to live in both aerobic and anaerobic environments. Both *B. xenovorans* and *R. eutropha* have the ability to degrade PCBs with chlorinated 2,3-positions. The biphenyl oxygenase has an acute affinity for 2- and 2,4-chlorophenyl rings at the *ortho* position. As a result, they can use oxygenolytic dehalogenation to spontaneously produce dihydroxybiphenyl (Mikszewski 2004). *Rhodococcus* sp. strain RHA1 is another promising aerobic PCB-degrading bacterium. It has been successfully used in post anaerobic (Jorge et al. 2006). RHA1 has been shown to degrade both *ortho-* and *meta*-substituted compounds and has been shown to degrade highly chlorinated biphenyls up to heptachlorobiphenyls (Seto et al. 1995). *Bacillus* strain IITR-54^T was isolated from aged PCB-contaminated soil (Manickam et al. 2014). *Ralstonia* strains have been isolated from PCB-contaminated dredged soils and have been attributed with up to 91% of degradation of Aroclor 1242 (Adebusoye et al. 2008). *Comamonas*, among others, has been shown to preferentially attack para-chlorinated congeners, which leads to problematic meta-cleavage products (Maltseva et al. 1999). *Achromobacter*, isolated from sewage, has been shown to degrade mono- and dichlorinated PCBs but becomes limited with an increased degree of chlorination (Ahmed and Focht 1973).

These strains have been successfully identified for their PCB transformation capabilities. The use of preferential dechlorination (*meta*, *para*, *ortho*, respectively), microbial capabilities, and their limitations can be used in conjunction with physical methods of PCB transformation in wastewater treatment plants.

5.5 Physical Removal of PCB

5.5.1 Sand Filtration

Sand filtration is often used as a final clarifying treatment step at wastewater treatment plants (Huisman and Wood 1974). These filters are the final step in the treatment process before chlorination to remove pathogens (bacteria and viruses) and are intended to remove particulate matter that might still be present in the effluent. As PCBs are hydrophobic, they have a tendency to bond to particles in water thus also to particles in sand filters. In these filters, biofilms are present and are involved in both physical filtration process and microbial processes (United States 1999). Biofilm can be defined as an "assemblage of surface-associated microbial cells that is enclosed in an extracellular polymeric substance matrix" (Donlan 2002). They are characterized by their utility of surface area, such as those provided in sand filters, to establish large colonies consisting of complex microbial communities. One successful use of sand filters is through bioaugmentation with estrogen-degrading bacteria, which have been shown to remove endocrine disruptors in wastewater (Haig et al. 2016). Overall, sand filtration in wastewater treatment has the benefits of physical filtration, chemical sorption, aerobic assimilation, and anaerobic dechlorination by embedded bacteria (Lesikar and Texas Agricultural Extension Service 1999; Kjellerup and Edwards 2013). Prior use of sand filtration, as aforementioned, for chemical removal shows promise for utility in PCB removal.

5.5.2 Activated Carbon Filtration

Activated carbon (AC), produced from organic material, including coconut, wood, and coal, is used in wastewater treatment due to its sorption capabilities (United States 2000). It is widely used to adsorb many different types of organic contaminants (Marmarelli and Sherbondy 2010). One of the most significant benefits of AC is its large surface area and porosity. One pound of activated carbon can have the surface area equivalence of six football fields (about 30,000m^2), making it an ideal adsorbent group for organic chemicals (Marmarelli and Sherbondy 2010). AC mainly tends to sorb organic compounds with molecular weights above 50 g/mol and boiling points above 50 °C (Carbtrol Corporation 1990). It is also limited by the innate variations in pH, temperature, and flowrate of WWTP (United States 2000).

Factors that influence AC's sorption characteristics include molecular weight, hydrocarbon saturation, index of refraction, and solubility. It also tends to be more effective at sorbing compounds with high refractive indices, molecular weights and boiling points, and low solubility (Wastewater Engineering 1991). PCBs fit this criteria in that they have high-molecular-weight (188–498 g/mol) indexes of refraction, high boiling points (256–456 °C), and low solubility (7.6E-4-9.3 g/m^3, with most congeners falling at the lower end) (Erickson 1997). PCB desorption rates decrease with increasing chlorination and hydrophobicity (Zimmerman et al. 2004). Also noted is the decrease in PCB concentration in the aqueous phase with increased contact time (Zimmerman et al. 2004).

AC has proven effective in PCB sequestration in sediment (Choi et al. 2014). AC decreases the bioavailability of PCBs to benthic organisms but also does not inhibit microbial dechlorination of PCBs when used as a carrier of dechlorinating microorganisms (Werner et al. 2005; Kjellerup et al. 2014). Studies showed that 2% by mass of dry sediment had reduction of aqueous PCB concentrations of over 98% after a 6-month treatment period (Werner et al. 2005). Additionally, significantly lower desorption was witnessed in AC-amended sediment (7%) compared to the untreated counterpart (74%). Recent field studies have demonstrated the reduction of PCB uptake by benthic organisms by up to 99% as well as the reduction of PCB concentration at sediment equilibrium by 93% (Beckingham and Ghosh 2011). Additionally, the efficacy of AC treatment increased with increasing dose and decreasing particle size (McLeod et al. 2007). The activated carbon-water distribution coefficient was notably similar in two different sediments hinting that similar such results may occur in different PCB-contaminated sources (Werner et al. 2005). The addition of AC into the sand filters to aid in PCB sorption should be evaluated due to activated carbon increasing the overall surface area and the high adsorption ability allowing more PCBs to potentially adsorb (Margot et al. 2013).

5.6 Future Technologies and Perspectives

As PCBs have been circulating in the environment for more than 40 years, they have found their way to various compartments of the environment. PCBs initially released directly into bodies of water may likely be adsorbed onto organic matter in soils (A Risk-Management 2001). Temporal and spatial deposition patterns as well as the various characteristics of the 209 congeners decrease the feasibility of PCB source tracking. For that reason, setting out to identify specific PCB discharge sources and types of PCBs coming from each source will not be an easy feat. Fortunately, various research groups have already begun defining specific congeners as indicators of PCB sources. Similarly, other groups have begun linking certain congeners and groups of congeners to specific transformation processes. Though this has mainly been done in pre-wastewater treatment modeling, similar techniques can be employed to model PCB transformation throughout the wastewater treatment process. Identification of the microbial characteristics of the various steps in the treatment process, in conjunction with the congeners used as indicators, can map out process specific microbial transformations of PCBs in WWTP.

High PCB dechlorination and degradation have been witnessed in studies using subsequent anaerobic and aerobic treatment. Thus, establishing biofilm using aerobes such as *B. xenovorans* strain LB400 could be done in aerated processes in WWTP, focusing on overcoming the limitation of its ability to grow on PCBs. Similarly, application of DF-1 and other anaerobic organohalide-respiring organisms can take place in anaerobic processes in WWTP. Highly chlorinated congeners could be dechlorinated in sedimentation basins that had been bioaugmented with anaerobic bacteria discussed here. Subsequently, lightly chlorinated congeners, both naturally occurring and as a result of anaerobic dechlorination, could be biodegraded and mineralized in aerated bioreactors bioaugmented with the aforementioned aerobic bacteria. The correlation between incubation and acclimation period of these microorganisms and removal efficiency in various WWTPs should be evaluated. Site characteristics that affect PCB transformation such as pH, temperature, and oxygen concentration should be included. This compilation will be used to define a feasible medium between desired removal and achievable incubation periods for specific WWTPs.

Establishing biofilm using dechlorinating bacteria in sand filters will allow for the removal of particulate matter as well as the dechlorination of PCBs, thus optimizing the final stage of the wastewater treatment process. In addition, the sand filtration process may also be involved in enhancement of microbial PCB dechlorination if used as a delivery vessel for organohalide-respiring microorganisms. Thus, AC-amended sand filters seem promising for long-term PCB sequestration while aiding in biological PCB transformation. The benefit of such an approach lies in the small amount of unwanted PCB desorption witness with the use of only a small amount of AC. In addition, AC reduces PCB bioavailability to benthic organisms, but not to the microorganisms needed in PCB transformation. This is beneficial since following sand filtration, the wastewater is discharged directly into natural bodies of water.

A comparison of influent and effluent wastewater, where PCBs characteristic of dechlorination are more present in effluent than influent, can be an indication of dechlorination within the wastewater treatment process. Research geared toward determining PCB transformation activity prior to wastewater treatment can form the basis for transformation within the treatment process. That is, a collective representation of PCB congeners that is characteristic of individual water sources can serve in generating an influent PCB profile for various WWTPs, which is a more feasible feat than source tracking. Such a collection will allow for more conclusive descriptions of specific PCB transformations that occur downstream from the influent into the WWTP. A similar approach can be used for defining a working profile for each process step in the WWTP. Here, the congeners' characteristic of each step, along with the microbial diversity will be the bases of the mapped profile. Prior research on PCB transformation has provided the necessary foundation for defining processes of microbial transformation of PCBs in wastewater treatment.

Acknowledgments The authors would like to acknowledge SERDP project no. ER-2135 and the LSAMP-2016-2017 URP program for their financial support. Additionally, we would like to thank Mrs. Elaine Wilson, District of Columbia Water and Sewer Authority, Washington, DC, USA.

References

A risk-management strategy for PCB-contaminated sediments (2001) National Academy Press, Washington, DC. http://site.ebrary.com/id/10072156

Abbey A-MI, Beaudette LA, Lee H, Trevors JT (2003) Polychlorinated biphenyl (PCB) degradation and persistence of a gfp-marked *Ralstonia eutropha* H850 in PCB-contaminated soil. Appl Microbiol Biotechnol 63:222–230. https://doi.org/10.1007/s00253-003-1380-x

Abramowicz D (1995) Aerobic and anaerobic PCB biodegradation in the environment. Environ Health Perspect 103:97–99

Adebusoye SA, Ilori MO, Picardal FW, Amund OO (2008) Extensive biodegradation of polychlorinated biphenyls in Aroclor 1242 and electrical transformer fluid (Askarel) by natural strains of microorganisms indigenous to contaminated African systems. Chemosphere 73:126–132. Gerssen-Gondelach10.1016/j.chemosphere.2008.04.055

Ahmed M, Focht DD (1973) Oxidation of polychlorinated biphenyls by *Achromobacter* PCB. Bull Environ Contam Toxicol 10:70–72. https://doi.org/10.1007/BF01685874

America Unites (2016) The history of PCBs. http://malibuunites.com/the-history-of-pcbs/

ATSDR (2014) ATSDR case studies in environmental medicine Polychlorinated Biphenyls (PCBs) Toxicity. U.S. Department of Health and Human Services. http://www.atsdr.cdc.gov/csem/pcb/docs/pcb.pdf

Beckingham B, Ghosh U (2011) Field-scale reduction of PCB bioavailability with activated carbon amendment to river sediments. Environ Sci Technol 45:10567–10574. https://doi.org/10.1021/es202218p

Bedard DL (2004) Polychlorinated biphenyls in aquatic sediments: environmental fate and outlook for biological treatment. In: Häggblom MM, Bossert ID (eds) Dehalogenation. Springer, US, pp 443–465. doi:https://doi.org/10.1007/0-306-48011-5_17

Bertin L, Capodicasa S, Fedi S, Zannoni D, Marchetti L, Fava F (2011) Biotransformation of a highly chlorinated PCB mixture in an activated sludge collected from a membrane biological reactor (MBR) subjected to anaerobic digestion. J Hazard Mater 186:2060–2067. https://doi.org/10.1016/j.jhazmat.2010.12.131

Bopp LH (1986) Degradation of highly chlorinated PCBs by *Pseudomonas* strain LB400. J Ind Microbiol 1:23–29. https://doi.org/10.1007/BF01569413

Cairns T, Siegmund EG (1981) PCBs: regulatory history and analytical problems. Anal Chem 53:1183A–1193A. https://doi.org/10.1021/ac00234a002

Carbtrol Corporation (1990) Activated carbon systems for PCB removal. http://www.carbtrol.com/images/white-papers/pcb_removal.pdf

Chain PSG, Denef VJ, Konstantinidis KT, Vergez LM, Agulló L, Reyes VL, Tiedje JM (2006) *Burkholderia xenovorans* LB400 harbors a multireplicon, 9.73-Mbp genome shaped for versatility. Proc Natl Acad Sci 103:15280–15287. https://doi.org/10.1073/pnas.0606924103

Chang BV, Chou SW, Yuan SY (1999) Microbial dechlorination of polychlorinated biphenyls in anaerobic sewage sludge. Chemosphere 39:45–54. https://doi.org/10.1016/S0045-6535(98)00587-6

Choi Y, Cho YM, Luthy RG (2014) In situ sequestration of hydrophobic organic contaminants in sediments under stagnant contact with activated carbon. 1. Column studies. Environ Sci Technol 48:1835–1842. https://doi.org/10.1021/es403335g

Clapp RW, Jacobs MM, Loechler EL (2008) Environmental and occupational causes of cancer new evidence, 2005–2007. Rev Environ Health 23:1–37. https://doi.org/10.1515/REVEH.2008.23.1.1

Donlan RM (2002) Biofilms: microbial life on surfaces. Emerg Infect Dis 8:881–890. https://doi.org/10.3201/eid0809.020063

de Souza PM (2004) Polychlorinated dibenzo-p-dioxins (PCDD), dibenzofurans (PCDF) and polychlorinated biphenyls (PCB): main sources, environmental behaviour and risk to man and biota. QuÃ mica Nova 27(6)

EPA Bans PCB Manufacture (1979) Phases out uses. USEPA. https://www.epa.gov/aboutepa/epa-bans-pcb-manufacture-phases-out-uses

Erickson MD (1997) Analytical chemistry of PCBs, 2nd edn. CRC Press, New York. Print

Flanagan WP, May RJ (1993) Metabolite detection as evidence for naturally occurring aerobic PCB biodegradation in Hudson River sediments. Environ Sci Technol 27:2207–2212. https://doi.org/10.1021/es00047a030

Furukawa K (1994) Molecular genetics and evolutionary relationship of PCB-degrading bacteria. Biodegradation 5:289–300. https://doi.org/10.1007/BF00696466

Haddock JD, Horton JR, Gibson DT (1995) Dihydroxylation and dechlorination of chlorinated biphenyls by purified biphenyl 2,3-dioxygenase from *Pseudomonas* sp. strain LB400. J Bacteriol 177:20–26

Haig SJ, Gauchotte-Lindsay C, Collins G, Quince C (2016) Bioaugmentation mitigates the impact of estrogen on coliform-grazing protozoa in slow sand filters. Environ Sci Technol 50:3101–3110. https://doi.org/10.1021/acs.est.5b05027

Hughes AS, Van Briesen JM, Small MJ (2010) Identification of structural properties associated with polychlorinated biphenyl dechlorination processes. Environ Sci Technol 44:2842–2848. https://doi.org/10.1021/es902109w

Huisman L, Wood WE (1974) Slow sand filtration. World Health Organization, Geneva. http://www.who.int/water_sanitation_health/publications/ssf9241540370.pdf?ua=1

Illinois Department of Public Health (2009) Polychlorinated biphenyls (PCBs) fact sheet. http://www.idph.state.il.us/envhealth/factsheets/polychlorinatedbiphenyls.htm

Jorge L, Alan MC, Michael R, John F, Olga V, Tamara V, James M (2006) Degradation of Aroclor 1242 Dechlorination products in sediments by *Burkholderia xenovorans* LB400(ohb) and *Rhodococcus* sp. strain RHA1(fcb). Appl Environ Microbiol 72:2476–2482. https://doi.org/10.1128/AEM.72.4.2476-2482.2006

Kjellerup BV, Edwards S (2013) Application of biofilm covered activated carbon particles as a microbial inoculum delivery system for enhanced bioaugmentation of PCBs in contaminated sediment. SERDP Report ER-2135. Washington, DC. https://clu-in.org/download/contaminantfocus/pcb/PCBs-Seds-ER-2135-FR-PI.pdf

Kjellerup BV, Naff C, Edwards SJ, Ghosh U, Baker JE, Sowers KR (2014) Effects of activated carbon on reductive dechlorination of PCBs by organohalide respiring bacteria indigenous to sediments. Water Res 52:1–10

Kjellerup BV, Paul P, Ghosh U, May HD, Sowers KR, Kjellerup BV, Sowers KR (2012) Spatial distribution of PCB dechlorinating bacteria and activities in contaminated soil. Appl Environ Soil Sci:e584970. https://doi.org/10.1155/2012/584970

Leeson A, Beevar E, Henry B, Fortenberry J, Coyle C (2004) Principles and practices of enhanced anaerobic bioremediation of chlorinated solvents (No. NFESC-TR-2250-ENV). Naval Facilities Engineering Service Center Port Hueneme, CA. http://www.dtic.mil/dtic/tr/fulltext/u2/a511850.pdf

Lesikar BJ, Texas Agricultural Extension Service (1999) Sand filter. Texas Agricultural Extension Service, Texas A and M University System, College Station

Maltseva OV, Tsoi TV, Quensen JF, Fukuda M, Tiedje JM (1999) Degradation of anaerobic reductive dechlorinationproducts of Aroclor 1242 by four aerobic bacteria. Biodegradation 10:363–371. https://doi.org/10.1023/A:1008319306757

Manickam N, Singh NK, Bajaj A, Kumar RM, Kaur G, Kaur N, Mayilraj S (2014) *Bacillus mesophilum* sp. nov., strain IITR-54T, a novel 4-chlorobiphenyl dechlorinating bacterium. Arch Microbiol 196:517–523. https://doi.org/10.1007/s00203-014-0988-9

Maphosa F, Lieten SH, Dinkla I, Stams AJ, Smidt H, Fennell DE (2012) Ecogenomics of microbial communities in bioremediation of chlorinated contaminated sites. Front Microbiol 3:351. https://doi.org/10.3389/fmicb.2012.00351

Margot J, Kienle C, Magnet A, Weil M, Rossi L, de Alencastro LF, Barry DA (2013) Treatment of micropollutants in municipal wastewater: ozone or powdered activated carbon? Sci Total Environ 461–462:480–498. https://doi.org/10.1016/j.scitotenv.2013.05.034

Marmarelli J, Sherbondy J (2010) Removing PCBs from groundwater with activated carbon. Environmental Science and Engineering Magazine. http://ese.dgtlpub.com/2010/2010-01-31/pdf/removing_PCBs_from_groundwater.pdf

McLeod PB, van den Heuvel-Greve MJ, Luoma SN, Luthy RG (2007) Biological uptake of polychlorinated biphenyls by *Macoma balthica* from sediment amended with activated carbon. Environ Toxicol Chem/SETAC 26:980–987. https://doi.org/10.1897/06-278R1.1

Mikszewski A (2004) Emerging technologies for the in situ remediation of PCB-contaminated soils and sediments: bioremediation and nanoscalezero-valent iron. U.S. Environmental Protection Agency. https://clu-in.org/download/techdrct/tdbio_of_pcbs_paper.pdf

Montague P (1993) Environmental Research Foundation. How we got here – Part 1: The history of chlorinated diphenyl (PCBs)

Patureau D, Trably E (2006) Impact of anaerobic and aerobic processes on polychlorobiphenyl removal in contaminated sewage sludge. Biodegradation 17:9–17. https://doi.org/10.1007/s10532-005-1920-4

Payne RB, Fagervold SK, May HD, Sowers KR (2013) Remediation of polychlorinated biphenyl impacted sediment by concurrent bioaugmentation with anaerobic halo respiring and aerobic degrading bacteria. Environ Sci Technol 47:3807–3815. https://doi.org/10.1021/es304372t

Pieper DH (2005) Aerobic degradation of polychlorinated biphenyls. Appl Microbiol Biotechnol 67:170–191. https://doi.org/10.1007/s00253-004-1810-4

Rodenburg LA, Du S, Fennell DE, Cavallo GJ (2010) Evidence for widespread dechlorination of polychlorinated biphenyls in groundwater, landfills, and wastewater collection systems. Environ Sci Technol 44:7534–7540. https://doi.org/10.1021/es1019564

Seto M, Kimbara K, Shimura M, Hatta T, Fukuda M, Yano K (1995) A novel transformation of polychlorinated biphenyls by *Rhodococcus* sp. strain RHA1. Appl Environ Microbiol 61:3353–3358

Sowers KR, May HD (2013) In situ treatment of PCBs by anaerobic microbial dechlorination in aquatic sediment: are we there yet? Curr Opin Biotechnol 24:482–488. https://doi.org/10.1016/j.copbio.2012.10.004

State of Oregon, Department of Environmental Quality (2003) Fact sheet: sources of polychlorinated biphenyls. http://www.deq.state.or.us/lq/cu/nwr/PortlandHarbor/docs/SourcePCBs.pdf

Su G, Huang L, Shi R, Liu Y, Lu H, Zhao Y, Zheng M (2016) Thermal dechlorination of PCB-209 over Ca species-doped Fe_2O_3. Chemosphere 144:81–90. https://doi.org/10.1016/j.chemosphere.2015.08.022

Tiedje JM, Quensen JF, Chee-Sanford J, Schimel JP, Boyd SA (1993) Microbial reductive dechlorination of PCBs. Biodegradation 4:231–240. https://doi.org/10.1007/BF00695971

United States (1999) Wastewater technology fact sheet: intermittent sand filters. United States Environmental Protection Agency, Washington, DC. http://purl.access.gpo.gov/GPO/LPS45368

United States (2000) Wastewater technology fact sheet: granular activated carbon adsorption and regeneration. United States Environmental Protection Agency, Office of Water, Washington, DC. http://nepis.epa.gov/Exe/ZyPDF.cgi?Dockey=P1001QTK.PDF

US EPA (2016a) Hudson river cleanup. Retrieved April 12, 2016. https://www3.epa.gov/hudson/cleanup.html#quest1

US EPA (2016b) Learn about polychlorinated biphenyls (PCBs) [Policies and guidance]. Retrieved April 12, 2016. https://www.epa.gov/pcbs/learn-about-polychlorinated-biphenyls-pcbs

Wastewater engineering (1991) 3rd edn. Metcalf and Eddy, Inc. Revised by George Tchobanoglous and Franklin L. Burton. McGraw-Hill, Inc., New York, p 1334. ISBN 0-07-041690-7

Werner D, Higgins CP, Luthy RG (2005) The sequestration of PCBs in Lake Hartwell sediment with activated carbon. Water Res 39:2105–2113. https://doi.org/10.1016/j.watres.2005.03.019

White SS, Birnbaum LS (2009) An overview of the effects of dioxins and dioxin-like compounds on vertebrates, as documented in human and ecological epidemiology. J Environ Sci Health Part C Environ Carcinog Ecotoxicol Rev 27:197–211. https://doi.org/10.1080/10590500903310047

Wu Q, Sowers KR, May HD (2000) Establishment of a polychlorinated biphenyl-dechlorinating microbial consortium, specific for doubly flanked chlorines in a defined, sediment-free medium. Appl Environ Microbiol 66:49–53. https://doi.org/10.1128/AEM.66.1.49-53.2000

Zhang H, Jiang X, Lu L, Xiao W (2015) Biodegradation of polychlorinated biphenyls (PCBs) by the novel identified cyanobacterium *Anabaena* PD-1. PloS One 10:e0131450. https://doi.org/10.1371/journal.pone.0131450

Zimmerman JR, Ghosh U, Millward RN, Bridges TS, Luthy RG (2004) Addition of carbon sorbents to reduce PCB and PAH bioavailability in marine sediments: physicochemical tests. Environ Sci Technol 38:5458–5464. https://doi.org/10.1021/es034992v

Seqestration Options for Phosphorus in Wastewater

6

Varsha Jha, Sampada Puranik (Chande), and Hemant J. Purohit

Abstract

Inefficient wastewater treatment introduces huge amount of nutrients mainly phosphorus and nitrogen to the natural waterbodies. Excessive phosphate in the water leads to the growth of algae or eutrophication. One-third of the aquatic ecology has been destroyed by eutrophication worldwide including China, Japan, Europe, South Asia and South Africa. Artificial eutrophication affects the water ecology around the world by decreasing the quality standards of water and alters the ecosystem structure and function. Phosphorus is known to be a limiting factor, and it is crucial to remove the phosphate from the effluent prior to exoneration into waterbodies.

Intracellular phosphate content of certain important species of bacteria influences phosphate removal in wastewater treatment. A variety of polyphosphate-accumulating organisms (PAOs) are involved. Under alternating anaerobic and aerobic conditions, these PAOs store phosphate in the form of polyphosphate. Among PAOs, *Accumulibacter* sp., *Pseudomonas* sp., *Aeromonas hydrophila*, *Tetrasphaera* sp. and gram-positives are the major role players as phosphate removers. As compared to chemical method, biological way of nutrient removal proved to be cost-effective, and it reduces the sludge production. An integrative approach towards phosphoregulation is a key aspect of dealing with the problem.

V. Jha • H.J. Purohit
Environmental Biotechnology and Genomics Division, CSIR – National Environmental Engineering and Research Institute (CSIR-NEERI), Nehru Marg, Nagpur 440020, Maharashtra, India

S. Puranik (Chande) (✉)
Environmental Biotechnology and Genomics Division, National Environmental Engineering Research Institute, CSIR-NEERI, Nagpur 440020, Maharashtra, India

Department of Internal Medicine, Yale University School of Medicine, New Haven, CT, USA
e-mail: sampada.chande@yale.edu

H.J. Purohit et al. (eds.), *Optimization and Applicability of Bioprocesses*,
https://doi.org/10.1007/978-981-10-6863-8_6

Keywords
Enhanced biological phosphorus removal (EBPR) • Polyphosphate (poly P) • Polyphosphate-accumulating organisms (PAOs)

6.1 Introduction

Wastewater management, eutrophication and phosphoregulation have an indispensable connection. Water scarcity is the immense problem faced worldwide due to increased growth of population, climate change and inefficient wastewater management. It was reported in the fourth World Water Development Report that only 20% of globally produced wastewater is currently receiving proper treatment (UN report 2012). There is an urgent need to conserve water and reuse the properly treated water for agricultural or other non-portable use.

To meet the needs other than drinking purposes, 1 billion gallons of treated wastewater have been used in the United States. An EPA estimate suggests that almost 91 billion is spent to assert and improve treatment systems all over the nation.

The discharge of untreated wastewater/improperly treated wastewater leads to the major calamity for water sources called "eutrophication" as this wastewater contains a huge amount of nutrients in it. According to UNEP newsletter and technical publication, the untreated wastewater or wastewater treated by conventional mechanical-biological techniques still contains 25–40 mg/l and 6–10 mg/l of nitrogen and phosphorus, respectively [1].

Eutrophication is the enrichment of waterbodies when a huge amount of nutrient-containing wastewater is dumped into it (Yewalkar-Kulkarni et al. 2016). It causes excessive growth of algae leading to the condition called algal bloom, causing the decrease in dissolved oxygen content and death of normal aquatic flora and fauna. Decomposition of these dead matters releases nutrients which amplify the process of eutrophication.

Dodds et al. (2008) showed that combined cost of $2.2 billion (approx.) was spent annually for recreational water usage, waterfront real estate, recovery of lost biodiversity and drinking water as a result of eutrophication in US freshwaters.

To check this deleterious effect, it is essential to control nutrient amount, and prime focus has been given to removal of phosphorus. Wastewater treatment plants or nutrient removal plants take advantage of polyphosphate-accumulating organisms (PAOs) which have the capability of accumulation of polyphosphate by removing phosphate from wastewater. Alternating anaerobic and aerobic condition is required for phosphate removal. Under anaerobic conditions PAOs uptake volatile fatty acid (VFA), e.g. acetate; polyhydroxybutyrate (PHB) is formed from this acetate and further used for cell growth and polyphosphate synthesis under aerobic conditions (Strom 2006). Based on the statistics for wastewater treatment, biological removal seems to be a simple and ecologically balanced way for phosphate removal and to curb eutrophication. Through this chapter we have tried to compile the issues regarding nutrients in wastewater; their ill effects, i.e. eutrophication; how this eutrophication and changing climate is related to each other; and the measures taken to get rid of eutrophication and reduce the nutrient loading into natural waterbodies by WWTPs. In this chapter, we have also illustrated the role of (meta)genomics approach to reveal the bacterial community structure so that the better modelling and designing of the WWTPs could be possible and best quality effluent would be generated to circumvent the harmful effects of effluent and sludge disposal to natural waterbodies. Not only the harmful effects of algal bloom but their use in WWTPs in nutrient sequestration is also discussed.

6.2 Nutrient Issues in Wastewater Management

The combined effluents that come from domestic use, industrial use, urban runoff and agricultural runoff are considered as wastewater. Wastewater management is a process of treatment of wastewater/sewage prior to introducing it to the waterbodies so that the ecology of the water should be maintained. Wastewater contains various pollutants such as:

- Plant nutrients (phosphorus and nitrogen mainly)
- Heavy metals (cadmium, mercury, nickel, lead and zinc)
- Pathogens (bacteria, viruses and other microorganisms)
- Other organic pollutants (UN report 2015)

All these pollutants have detrimental effects on both environment and human health; therefore, it is indispensable to treat the wastewater before disposing it to the natural water sources. If the wastewater management is neglected, then it will lead to two major impacts: one is chemical and nutrient contamination and the other one is microbial pollution.

6.2.1 Phosphate/Nutrient Induced in the Sewage

Wastewater comes from the industries, urban runoff and household; these are the major sources which introduce the nutrients into the sewage (Romero et al. 2013; UN report 2015). Household wastewater mainly consists grey water (kitchen and bathing wastewater) and black water (excreta, urine and faecal sludge) (UN report 2015). SeaWeb in their newsletter reported sewage and septic tanks as a major source of nutrient pollution as many soap and detergents contain phosphorus, whereas human excreta are known to be nitrogen rich [2]. In relation to microbiota, diversity and species richness are the factors to be considered for balancing the nutrient flow in any ecosystem. It has been found in many ecosystems that functional diversity dictates equilibrium in the ecosystem rather than species number (Mulder et al. 2012). This is an important factor as diversity is dictated by environment and nutrient cycling. Environmental sensing of phosphate has important roles in wastewater treatment (phosphate removal). High phosphate causes eutrophication in an aqueous environment. The intracellular phosphate content of certain important species influences phosphate removal in wastewater treatment (Ruiz-Martinez et al. 2015). Thus phosphate homeostasis could in part influence the rate of removal of phosphate in wastewater and reduce eutrophication.

6.2.2 Nutrient Management in Sewage

Various physical and chemical methods like filtration, membrane technologies and precipitation, respectively, are used to remove phosphate, but the biological method (EBPR) is known to remove phosphate in a cost-effective manner. It also reduces the phosphate level to acceptable standards (Strom 2006) and is also more environmentally friendly (Gunther et al. 2009). According to FAO Corporate Document Repository, after treatment of the sewage water by conventional methods, it still has the nutrients in concentration, phosphorus (P), 10 mg/l; potassium (K), 30 mg/l; and nitrogen (N), 50 mg/l [3]. EBPR is known to achieve around <0.1 mg/L effluent P levels (Barnard 2006). Phosphate removal by struvite precipitation was also studied recently (Lu et al. 2016).

6.2.3 Consequences of Untreated/Partially Treated Sewage Disposal to the Different Waterbodies

The biggest drawback of sewage disposal is that it demolishes the aquatic biodiversity by harmful effect of eutrophication. It also causes dissolved oxygen depletion to satisfy the BOD of organic matter present in sewage (Diaz and Rosenberg 2008). The introduction of sewage also leads to other harmful effects like the introduction of pathogenic organisms and production of unpleasant smelling gases, e.g. H_2S (Klein and Perera 2002; Topare et al. 2011). The sewage disposal intensifies the presence of faecal coliforms (Rim-Rukesh and Agbozu 2013); these pathogenic organisms are known to cause waterborne diseases. According to an article in GE step ahead 2015, annually around 1 lakh causalities occur in India due to these waterborne diseases like cholera, jaundice, diarrhoea and typhoid [4]. The partially treated sewage when introduced into fresh waterbodies greatly reduces biological and physicochemical qualities of receiving waterbodies. Rim-Rukesh and Agbozu (2013) studied the impacts of partially treated wastewater on Epie Creek (Nigeria). In their study, they use Malaysian Water Quality Index (WQI) to assay the water quality. They report that Epie Creek is equitably polluted on the basis of their findings: dissolved oxygen (DO) 3.73–5.20 mg/l, chemical oxygen demand (COD) 17.3–53.2 mg/l, biochemical oxygen demand (BOD) 12.4–36.7 mg/l, total faecal coliforms 2,120–20,800 cfu/ml, total phosphorus (TP) 0.73–1.73 mg/l and ammoniacal nitrogen 4.10–5.0 mg/l. With the obvious effect of destroying the waterbody composition, untreated wastewater has long-time consequences including the destruction of waterbody ecosystem by altering the species distribution.

6.3 Eutrophication

The natural ageing process of waterbodies is called eutrophication. In this process of natural ageing, a large, profound, nutrient-poor lake successively turns in to be nutrient rich, and with the course of time, it becomes a pond and then converts to a marsh. The anthropogenic activities have increased the rate of this process, and it became so common that the term eutrophication itself sounds like a terrible condition of waterbodies. In the United States alone, eutrophication accounts for almost one-half of the impaired lake area and 60% of damaged river (Smith 2003). The United Nations Environment Programme (UNEP) in their newsletter and technical publication, volume 3, states that the eutrophic waterbodies are classified into four classes based on their nutrient concentrations (mainly nitrogen and phosphorus): these are oligotrophic, mesotrophic, eutrophic and hypereutrophic containing phosphorus and nitrogen (Table 6.1) [1].

6.3.1 Source of Nutrient to the Waterbodies

Nutrients can be deposited to the aquatic systems by two means either naturally or by human activities (anthropogenic). It will take centuries for a lake to become eutrophic by natural means (Gao 2015), but the anthropogenic activities speed up the process by heavy deposition of nutrients (Chislock et al. 2013; Erisman et al. 2013). Natural rock weathering (Carpenter 2008) and atmospheric depositions (Anderson 2002) are the examples of natural ways, while erosion and leaching from fertilized agricultural areas, development of aquaculture and sewage from cities, urban runoff and industrial wastewater are the main source of anthropogenic activities. Expansion of aquaculture plays a role in eutrophication by discharging the unused animal food and excreta of fish into the water (Klein and Perera 2002). It is quite easy to control the point source of nutrient pollution, whereas the non-point sources are difficult to control.

6.3.2 Consequences of Eutrophication

Ultimately whatever may be the source of nutrient to the waterbodies, the effect will almost always be the same. Availability of the high concentration of nutrients enhances the primary productivity of the waterbodies by increasing the metabolic

Table 6.1 Classification of eutrophic waterbodies based on nitrogen and phosphorus concentration

Sr. No.	Status of waterbody	Average total phosphorus (μg/l)	Average total nitrogen (μg/l)
1.	Oligotrophic	8.0	661
2.	Mesotrophic	26.7	753
3.	Eutrophic	84.4	1875
4.	Hypereutrophic	>200	High

rates which result in the increased production of phytoplankton and *Cyanobacteria* out of which some are poisonous and some are non-poisonous. Apart from nutrient availability, physical factors like temperature, renewal of water and light also play an essential role (Klein and Perera 2002). It was observed that in the Chesapeake Bay during the spring season owning to nutrient-rich environment, phytoplankton biomass increases (Anderson 2002). Blooms of *Cyanobacteria* result into foul-smelling scum and cause problems in drinking water by reducing its taste (Carpenter 2008); it also prevents the penetration of light to the bottom (Lehtiniemi et al. 2005). The major effect of eutrophication is depletion of oxygen and formation of dead zone near the bottom. After the eventual death of algal blooms during the process of decomposition, bacteria consume oxygen, and even some bacteria use sulphates (SO_4^{2-}), and as a result, free S^{2-} take up available dissolved O_2 and make it unavailable for organisms (Klein and Perera 2002); this creates hypoxic condition and results into loss of actual biodiversity by killing normal aquatic flora and fauna. In 2005, 146 coastal marine dead zones had been documented around the world, out of which 43 were in the United States (Dybas 2005). The extensive dead zone was created during summer due to nutrient discharge from the Mississippi and Atchafalaya rivers in the northern Gulf of Mexico (Rabalais et al. 2002). Diaz in 2008 observed that dead zones have been reported to affect more than 245,000 km^2, which accounts for more than 400 marine ecosystems. Smith (2003) mentioned there were 3,164 reported events of poisoning and 148 deaths of humans in the Asia-Pacific region alone. Economic losses may exceed US $1 million per event, and monitoring efforts may cost up to US $50,000 for each affected area. According to NOAA forecast, the prediction of dead zone ranges from 5204 to 6823 mi^2 in the Gulf of Mexico in the year 2016. This forecast is based on the nutrient runoff, river and stream data from the United States Geological Survey (USGS). According to USGS around 20,800 metric tons of phosphorus and 146,000 metric tons of nitrate are received by the Gulf of Mexico from Mississippi and Atchafalaya rivers in May 2016. This makes 25% and 12% above the long term (1980–2015) of nitrate and phosphorus, respectively [5].

6.3.3 Correlation of Eutrophication and Climate Change

Not only by the excessive loading of nutrient by the anthropogenic activities but the climate change also affected the eutrophication. Climate change is mainly the consequence of pollution (especially introduction of greenhouse gases through the burning of fuels) caused by anthropogenic activities (Vijayavenkataraman et al. 2012). Basically, we look at eutrophication as the excessive growth of cyanobacteria (harmful algal blooms). The climate change (altered rainfall patterns, increased storms, melting glaciers and warming soil) increases the signs of eutrophication (Jeppesen et al. 2010; Jeppesen et al. 2011). Rising temperature correlates with eutrophication by various ways like increasing the nutrient loading due to increased mineralizing rate and reducing the water level which causes the concentration of existing nutrients; this is accompanied by low intense storm which causes nutrient loading due to soil erosion (Rustad et al. 2001; Brookshire et al. 2011). Paerl and Paul (2012) studied the anthropogenic and climatic control on a harmful algal bloom. They state that growth of cyanobacteria and the bloom-forming ability are highly influenced by the nutrient enrichment as well as climatic changes like hydrologic changes, global warming and increased frequencies and intensities of tropical storms and droughts. Temperature rise has a positive impact on the cyanobacterial growth (Watkinson et al. 2005). The cyanobacterial photoprotective and photosynthetic pigments absorb light and add on to the increased water temperature (Paerl and Paul 2012).

So far we have discussed the direct effect of climate change on eutrophication and growth of (cyano) harmful algal blooms, but climate change is also influenced indirectly by the consequence of eutrophication, i.e. dead zones. Recently, Altieri and Gedan (2015) reviewed the impact of climate change on dead zone. They also examined various climatic parameters like temperature, precipitation, wind, storm, ocean acidification and sea-level rise and concluded that it can affect the O_2 availability and response to hypoxia. The main reason of dead zone is the appearance of a hypoxic condition in the waterbodies and death of aquatic fauna due to low oxygen availability. The temperature rise plays a key role in promoting the hypoxic condition. Warm water has low capacity to hold oxygen; thus the rise in temperature causes low availability to aquatic animals (Altieri and Gedan 2015). Another effect of the rise in temperature is that it causes stratification of the surface water and prevents mixing of oxygenated water (Cloern 2001). The climate change is also known to affect the time and rate of production of phytoplanktons (Winder and Sommer 2012). Similar kind of observation was made by Alheit et al. (2005); they observed early bloom formation in the Baltic Sea for the warmer period. As compared to phytoplanktons, their grazers are more sensitive towards temperature change; in many examples, it is seen that with the increase in temperature, grazing ability also increases. Climate warming also results in widening of existing dead zones and change in duration by the following ways: sea-level rise, season stretching and hypoxic thermal kill zones (Altieri and Gedan 2015).

Thus, climate change has an impact on eutrophication. It seems subtle if we take into consideration a short time period. But, over the course of time, this pools together to create more impact on enhancing rates of eutrophication.

6.3.4 Members of Algal Bloom

Among the bloom-forming organism, blue-green algae (cyanobacteria) usually override other algal species (Smith 2001; Wang and Lei 2016). Paerl and Paul (2012) reported about the algal bloom-forming toxigenic cyanobacteria, and these are *Anabaena*, *Cylindrospermopsis*, *Microcystis* and *Oscillatoria* (*Planktothrix*), while others are *Phaeocystis* and several dinoflagellates (*Prorocentrum*, *Gymnodinium*, *Dinophysis*) (Klein and Perera 2002). The composition slightly differs for different waterbodies, but the effects remain the same.

6.3.5 Remedy of Eutrophication

In order to mitigate the harmful effects of eutrophication, various approaches suggested by many people are compiled by Chislock et al. (2013). These approaches are:

- Diversion of excess nutrients
- Altering nutrient ratios
- Application of potent algaecides and herbicides

But it seems to be more important to check the primary causative agent of eutrophication, i.e. nutrient loading. To circumvent eutrophication, only the control of reactive nitrogen is not enough. Measures to control phosphorus are essential and must be included in management programmes (Carpenter 2008). In the early 1990s, the deposition of about 50% of total P to Tar-Pamlico watershed of the Albemarle-Pamlico estuarine systems in North California by the largest phosphorus mine reduced to more than 90%, placing the example of point source control (Anderson 2002). This is the first line as the deposition of excessive phosphorus in any form is controlled beforehand. To eradicate phosphate from the point sources, various strategies are applied by the wastewater treatment plants; these are discussed in details in the other section of this chapter.

6.4 Phosphoregulation

So far we have discussed the phosphate content in wastewater, phosphate pollution in the natural waterbodies and its harmful effects. But phosphate itself does not harm directly; it's the excessive growth of algal bloom which is responsible for the detrimental effects on water ecosystem. In this section, we discuss the importance of phosphate in life, phosphate regulation in wastewater and other things.

6.4.1 Phosphate Flow in Ecosystems

Phosphorus is considered as the fifth most essential element for growth. Not only bacteria but plants as well as humans all require phosphorus in their nutrition. It plays a role in various biological processes like synthesis and maintenance of membrane and is involved in cell signalling as a second messenger and component of genetic material and energy metabolism (Santos-Beneit 2015). Bergwitz and Juppnner (2011) have reported that, if not properly regulated, phosphate can cause many diseases in humans like muscle myopathy, tumour-induced osteomalacia, cardiomyopathy, neuropathy and haemolysis. Misregulation of phosphate in the environment causes an environmental problem (Santos-Beneit 2015).

In general, bacteria utilizes the inorganic form of phosphorus (phosphate ion, i.e. Pi) present in its environment. In the case of unavailability of inorganic phosphate,

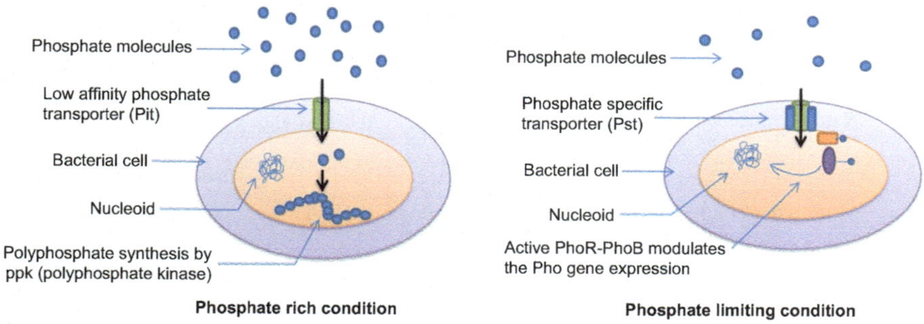

Fig. 6.1 The phosphorus utilization by bacteria and cellular components involved in the utilization

the bacteria use organic compounds containing phosphate like phosphonates and glycerol-3-phosphate. They have two types of transporters for phosphate ion depending upon the availability of Pi; these are Pi-specific transporter (Pst) and low-affinity phosphate transporter (Pit) expressed under low and high availability of Pi, respectively. In the case of organic compounds containing phosphate, related transporters and enzymes are involved in their metabolism. Under the Pi-rich condition, the bacteria take up an excess amount of inorganic phosphate and with the help of enzyme polyphosphate kinase (ppk) forms a long polymer called polyphosphate (poly P) which severs as a source of energy under Pi-starving conditions (Fig. 6.1) (Santos-Beneit 2015).

6.4.2 Phosphorus Sequestration in Natural Waterbodies

We know that natural waterbodies receive phosphorus from both natural and anthropogenic activities (which may include the point source or non-point source) such as agricultural runoff and municipal and industrial sewage effluents (Howarth et al. 2000). As per United Nations Environment Programme (UNEP), the excessive nutrient loading can be prevented by wetlands; this ascertains a natural way to eradicate the problem of eutrophication. The basic concept is that wetland soil adsorbed the phosphorus present in the effluent/polluted water, whereas the nitrate is released as nitrogen in the atmosphere, after its conversion. Díaz et al. (2012) report the effectiveness of constructed wetland to reduce the agricultural runoff pollutants before discharging it in Sacramento-San Joaquin river system (California).

Many studies reported the phosphorus removal by adsorption methods, e.g. on activated aluminium oxide (Genz et al. 2004), oxide tailings (Zeng et al. 2004), zeolite (Karapinar 2009), ferrihydrite (Carabante et al. 2010) and titanium dioxide (Delaney et al. 2011). Zamparas et al. (2012) studied the effectiveness of modified bentonite (Zenith/Fe) in phosphorus sequestration in natural waterbodies and report 80% removal of phosphorus. Earlier used sand, gravels and soil do not remove nutrient with great efficiencies (Park 2009). Recent researches suggest that the industrial wastes and by-products would aid to improve the nitrogen and

Table 6.2 List of agricultural by-products (ABPs) used for phosphorus removal

Sr. No.	Form	Agricultural by-products (biosorbents)	References
1.	Natural	Sawdust of Aleppo pine	Benyoucef and Amrani (2011)
		Palm surface fibre	Ismail (2012)
2.	Modified	Diethylenetriamine – cross-linked cotton stalk and wheat stalk	Xu et al. (2011)
		2-hidroxypropyltrimethyl ammonium chloride modified coconut shell fibre	De Lima et al. (2012)

phosphorus removal if used as the substrate (Ahmad et al. 2016). Ahmad et al. (2016) reviewed the use of water treatment sludge as a substrate in constructed wetlands with improved efficiencies in nutrient removal.

Some materials can form hazardous species; e.g. it has been viewed in many cases that aluminium causes toxicity to living organisms (Haghseresht 2004). The agricultural by-products (ABPs) can be successfully used as biosorbent as these can be a prevalent source because of its low cost, eco-friendliness, and utilization of agricultural wastes (Nguyen (TAH) et al. 2012). The agricultural by-products are either used in its natural form or can be modified in order to increase their effectiveness (Table 6.2). The efficiency of ABPs is also governed by parameters like pH, temperature, adsorbent dosage, interfering ions and contact time (Nguyen (TAH) et al. 2012).

6.4.3 Phosphorus Sequestration Through Designed Bioreactors

Strom (2006) reported that phosphorous can be removed by various methods like filtration of particulate phosphate, membrane technologies, precipitation, crystallization, adsorption, constructed wetlands and enhanced biological phosphorous removal – EBPR. Various reactors and treatment techniques have been designed and practised with the aim of improving the effluent quality (to reduce the nutrient loading into natural water system). In this section, we discuss the bioreactors used in nutrient removal except for EBPR system; because of its huge serving in the nutrient removal, it is discussed separately. Seow et al. (2016) reviewed wastewater treatment technologies and discussed aerobic granulation, biofilm technology and microbial fuel cell techniques and their merits and demerits. All three techniques are used to treat various kinds of wastewater, but aerobic granulation shows effective results in phosphate/nutrient removal. The aerobic granulation technique is known to remove nutrients (nitrogen and phosphorus) from the slaughterhouse wastewater, livestock wastewater and domestic sewage with the values 99.3% and 83.5%, 73% and 70%, and maximum volumetric conversion rates for nitrogen and phosphorus were 0.17 and 0.24 kg/m^3 in respective wastewater (Othman et al. 2013; Liu et al. 2015; Pronk et al. 2015). In the Netherlands under the trade name of Nereda®, a full-scale domestic sewage treatment plant is operated using aerobic granulation technique (Pronk et al. 2015).

Electrocoagulation is a technique used with increasing rate to treat not only the industrial wastewater but also raw domestic sewage as well as tertiary treated to reduce the microbes, nutrients and other personal care pollutants. Energy efficiency and less expensive nature make it popular to use in the wastewater treatment process. Percent mean reduction of phosphorus in both raw and tertiary-treated wastewater sample was found to be 95.79% and 96.33%, respectively (Symonds et al. 2015). Ozyonar and Karagozoglu (2011) perceived 98% phosphorus removal from domestic sewage by using electrocoagulation technique and concluded that aluminium electrode effectively removes phosphorus.

The microalgal culture is used to remove phosphorus because of its ability to assimilate this nutrient for its growth in photobioreactor which employed light source for the cultivation of microalgae. Abdel-Raouf et al. (2012) reviewed the use of microalgae in wastewater treatment. He states different treatment systems, like immobilized cell system, dialysis culture, tubular photobioreactor, stabilization pond and algal mat, and also the use of harvested biomass for energy production. The microalgal membrane technique is highly popular because of its ease in harvesting the biomass. Singh and Thomas (2012) worked on a unique microalgae membrane photoreactor (mMR) with the aim of furnishing the effluent obtained from domestic wastewater treating aerobic membrane bioreactor. They reported good rate of removal of NO_3, NO_2 and PO_4 under both batch and continuous modes of operation. The algal-bacterial biofilm reactor reported to remove phosphorus, nitrogen and carbon from domestic wastewater at 10 days HRT (hydraulic retention time), $85 \pm 9\%$, $70 \pm 8\%$ and $91 \pm 3\%$, respectively (Posadas et al. 2013). Sukacova et al. (2015) studied the phosphorus removal by microalgal biofilm under different light conditions and reported $97 \pm 1\%$ of phosphorus under nonstop synthetic illumination.

6.4.4 Mechanism of Phosphate Sequestration

Wastewater treatment plants use enhanced biological phosphorus removal (EBPR) process to remove phosphate from sewage. The EBPR process is basically an activated sludge process functioning under anaerobic and aerobic conditions and introduces the influent wastewater in anaerobic phase (Kristiansen et al. 2013); this enriches the polyphosphate-accumulating organisms (PAOs). Under the anaerobic condition where the influent wastewater is mixed, the bacteria are able to uptake the carbon source/volatile fatty acid (e.g. acetate) and convert it into polyhydroxybutyrate (PHB), deriving the energy from stored poly P, and release orthophosphate in medium (Martin et al. 2006; Wong and Beiko 2015). The reducing equivalent required is obtained from either conversion of stored glycogen to PHB and CO_2 or by partial oxidation of acetyl CoA through the TCA cycle (Skennerton et al. 2015). Under aerobic conditions poly P is synthesized by accumulating the phosphate present outside the cell; it also synthesizes glycogen by using the energy from anaerobically stored PHB. Thus the rate of P accumulation is much higher than the rate it is released during the anaerobic phase; this is how the PAOs are able to remove the

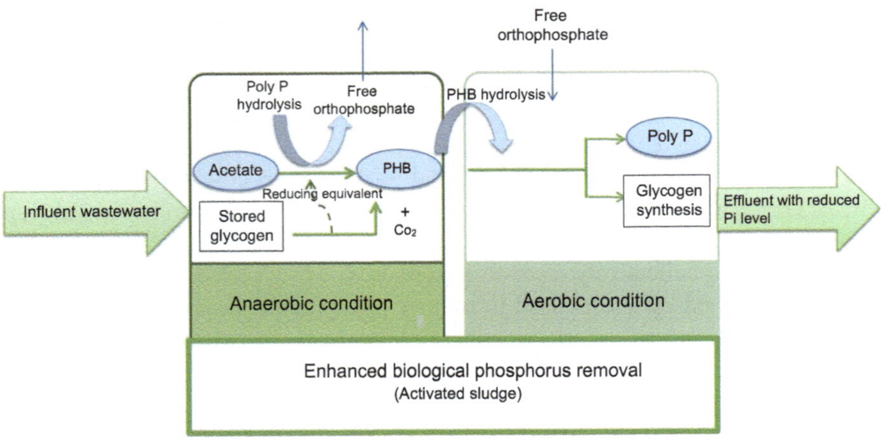

Fig. 6.2 The mechanism involved in EBPR process in wastewater treatment plant

higher amount of phosphate from the wastewater (Lu et al. 2016). This mechanism is diagrammatically represented (Fig. 6.2).

6.4.5 Microbial Community Structure: Its Relation to Phosphorus

Enhanced biological phosphorus removal (EBPR) process is not performed by a single dominant bacterial species, but a mixed population of bacteria is able to utilize carbon source/VFA, store PHB, accumulate poly P, store and maintain glycogen, etc. Some heterotrophic bacteria are known to accumulate polyphosphate and use it as the source of phosphate and energy under phosphate-starving condition (Santos-Beneit 2015). Many researchers have found pure cultures, e.g. members of genera *Acinetobacter*, *Tetrasphaera*, *Microlunatus* and *Lampropedia*, with the traits of PAOs, but they were not found to be significant for wastewater treatment plants (Seviour et al. 2003). In 2009, Gunther et al. developed a dual polyphosphate/DNA fluorescent staining approach which explores the knowledge of noncultivable PAOs. Fluorescent staining determines the number of poly P granules, whereas DNA contents and cell size explain the active PAOs. For isolation and identification of the representative PAOs, cell sorting and terminal restriction fragment length polymorphism (T-RFLP) profiling of 16S rRNA gene were used. The dual staining divides the complex community into subcommunities on the basis of their growth rate and poly P content. 16s rRNA gene sequencing verified the staining technique specificity and generated clone library. This library showed a low diversity composed of phylotypes of *Candidatus Accumulibacter* and members of *Pseudomonas* and *Tetrasphaera* genera. The tetracycline (TC) and 4′-6′-diamidino-2-phenylindole (DAPI) are considered as a reliable method for the detection of PAO activity. In other studies, it was reported that the *Accumulibacter* sp. and *Tetrasphaera* sp. are the abundant poly P accumulators (Oehmen et al. 2007; Nielsen et al. 2010; Nguyen et al. 2011). The *Tetrasphaera* sp. comprises 30–35% of the microbial community, whereas *Accumulibacter* sp. forms only 3–10% (Gu et al. 2008; He et al. 2008; Nielsen et al. 2010; Nguyen et al. 2011). Sidat et al. (1999) carried out a study using a culture-dependent method and isolated 39 monocultures from the sludge procured from Johannesburg full-scale BNR system, out of which 24 isolates showed the phosphorus-accumulating ability. Gram negative forms 58% and gram positive forms 42% of total phosphorus-accumulating population isolated from sludge. Their study showed that gram positive and *Pseudomonas* sp. form the 50% of the population.

For successful operation of EBPR and its further optimization, it is essential to have the knowledge of PAO biodiversity (Blackall et al. 2002). To explore the microbial community dynamics, Ju and Zhang (2015) revealed a statistical method based on the correlation between bacterial community networks and the associated taxonomic affiliations through this predict co-occurrence and co-exclusion patterns of the community. Through this work they studied the 16s rRNA gene sequencing data, and species-species association (SSA) network was constructed which comprised of 3899 pairwise significant SSA correlations which connect 170 species-level OTUs. During their work, they found that apart from the closely related bacteria (taxonomically) which share the similar niche taxonomically, less related species were also observed and posed competition to community assembly.

6.5 GAOs as Competitor of PAOs

Glycogen-accumulating organisms (GAOs) are the potential competitor of PAOs for volatile fatty acids, and they do not accumulate phosphorus (Oehmen et al. 2006; Gu et al. 2008). *Candidatus Competibacter phosphatis* (B12%) and *Defluviicoccus vanus* (9%) were found to be the abundant GAOs in EBPR system (Saunders et al. 2003; Burow et al. 2007). If *Competibacter* sp. is present along with *Accumulibacter* sp. in WWTPs, then it will hamper the EBPR process by not achieving the target level of phosphorus removal (Zhang et al. 2011). Metabolically the GAOs are the same as that of PAOs except they do not accumulate poly P and use it as a source of energy under anaerobic phase (Cydzik-Kwiatkowska and Zielińska 2016).

6.6 Carbon and Phosphorus Metabolism in EBPR

EBPR system places a good example where one can study the correlated carbon and phosphorus metabolism. All PAOs have the same or related genes and their function for phosphate and central carbon metabolism. Here we take *Accumulibacter* as an example organism to discuss the metabolic process involved in EBPR.

The phosphate metabolism of PAO involves the gene for low-affinity phosphate transporters (*pitA*) and phosphate-specific transporters (*pstABC*) and genes responsible for the formation (*ppk1*) and hydrolysis (*ppx*) of polyphosphate. The metabolic pathways are almost similar for PHA production from acetate, glycolysis and the

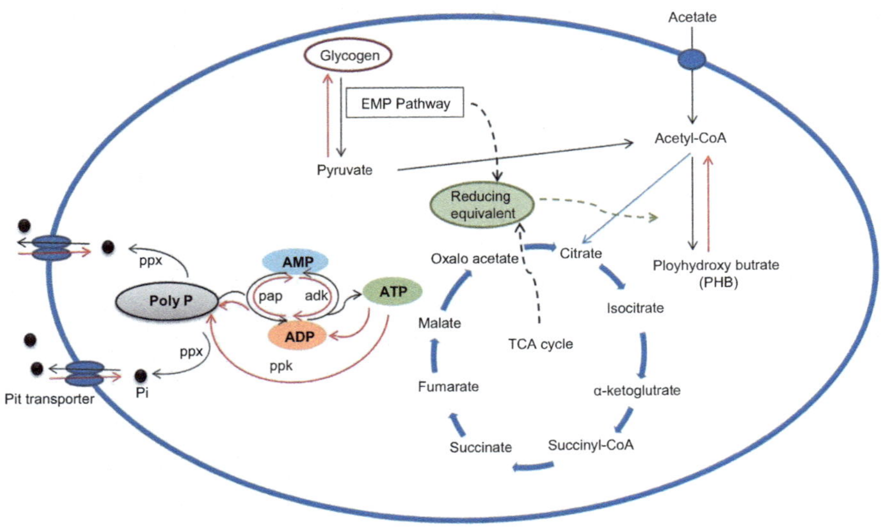

Fig. 6.3 Metabolic event that occur in the *Accumulibacter* EBPR system. Events that occur in anaerobic conditions are depicted with black arrows and aerobic events with red arrows; black dot (•) represents the inorganic phosphate, (⬭) represents the reducing equivalent, (⬭) represents AMP, (⬭) represents ADP and (⬭) represents ATP

TCA cycle (Skennerton et al. 2015). There is a mystery regarding *Accumulibacter* carbon metabolism, either the EMP or ED pathway is followed during anaerobic glycolysis (Oehmen et al. 2007). Recently sequencing of all the *Accumulibacter* genomes showed the presence of the EMP pathway (Hesselmann et al. 2000). The source of reducing the power required during the formation of PHB presented another topic of conflict. Reducing power was either provided by degradation of store glycogen or by anaerobic operation of the citric acid cycle, i.e. TCA cycle (Skennerton et al. 2015). For anaerobic TCA to occur, it is mandatory to reoxidize the reduced quinones; quinol reductase helps in this process which is found in the genome of *Accumulibacter* UW-1 (Martin et al. 2006). A pictorial representation of this metabolic event that occur in *Accumulibacter* is given (Fig. 6.3). The biochemical network of EBPR is thus much more complex than just utilization and removal of phosphate as it has inputs from central energy metabolism. But, studying this in detail will help engineer bacteria through potentially suggesting a carbon source which ultimately assists phosphate removal. Such conditioned consortia would most likely have increased EBPR.

6.7 Metagenomic Approach

Due to the immense contribution of (meta)genomic approaches in exploring the phylogenetic composition and functional prospective of a complex community in the EBPR process, this section is purely devoted to metagenomics and its application in WWT plants. Initially, the metagenomic approaches were used to identify the uncultivable microbes; now these approaches are used to study the dynamics of bacterial communities (Meena et al. 2015; Ambardar et al. 2016; Sharma and Lal 2017). The first metagenomic study was applied to obtain a draught genome of *Accumulibacter phosphatis* UW-1 from lab-scale EBPR reactors (Martin et al. 2006). Phylogenetic studies are not only limited to 16s rRNA gene but also carried out by using *ppk1* gene. Study of these genes categorized the *Accumulibacter* into two clades, i.e. type I and type II, which is again subdivided into IA-IE and IIA-IIF, respectively (Flowers et al. 2013; Skennerton et al. 2015). Deep knowledge was gathered from the independent work of many researchers who use next-generation sequencing methods and expand our knowledge regarding the genetic variety of *Accumulibacter*; their work produces additional draught genomes, and these are as follows: one of *A. phosphatis* UW-2 from clade IA (Flowers et al. 2013), one of *Accumulibacter* sp. strain HKU-1 from Class IB (Mao et al. 2014) and one from clade IA, one from clade IC, three from clade IIC and three from clade IIF (Skennerton et al. 2015). Thus as a number of draught genomes for these bacteria increase, our knowledge and ability to build an effective as well as safe consortia for EPBR increase. It is quite difficult to compile the metagenomic studies; we have tried to gather the information and compile in a tabular form for easy understanding (Table 6.3).

6.8 Algae: Problem as Well as Boon

So far we have seen the algal community as the culprit of the worst situation of the aquatic systems called eutrophication, and we have also discussed its harmful effects. The ability of the algae to show robust growth in the presence of nutrient-rich conditions is only the dark shade of algae. This nature of algae can be exploited by using algae in the nutrient removal systems in WWT plants. This idea has been used in the early 1950s, and now this has been used in many lab scales and pilot-scale plants with more innovative techniques and improved efficiencies. Many researchers have reported the use of microalgal culture in the wastewater treatment with the ability of microalgae in nutrient stripping as well as removal of heavy metals and toxic compounds, and the biomass can be successfully used for other purposes (Kumar and Goyal 2010; Pittman et al. 2011; Abdel-Raouf et al. 2012; Ruiz-Martinez et al. 2012; Yewalkar-Kulkarni et al. 2016). The obtained biomass can be further used for the production of biofuels like biodiesel (Schenk et al. 2008), fertilizers and animal feed and pharmaceutical industry (Singh and Thomas 2012). The use of microalgae in wastewater treatment is popular because of its skill to use solar energy for biomass production which to some extent shorts down the cost of the process and can be grown in the outdoor solar bioreactors (photobioreactor) (Abdel-Raouf et al. 2012). For the enhanced removal of nutrient, either the monoculture or the polyculture is being used. Ruiz-Martinez et al. (2012) in their work used a polyculture with the idea that the strain will get selected and evolve according to the need and condition.

Table 6.3 The metagenomic approaches used for defining the microbial community structure and bacterial abundance

Sr. No.	Sample collection site	Wastewater type	Metagenomic approach	Bacterial abundance	References
1.	13 Danish full-scale wastewater treatment plants		Illumina sequencing of 16s rRNA amplicon (V_4 region)	*Nitrotoga*	Saunders et al. (2016)
2.	Aalborg east wastewater treatment plant, Denmark, (57.044565°N) (10.047598°E)	Domestic wastewater	Illumina sequencing and q-FISH	*Accumulibacter* clade IIA strain UW-1	Albertsen et al. (2012)
3.	30 Danish full-scale wastewater treatment plants		MAR-FISH	*Candidatus halomonas phosphatis*	Nguyen (HTT) et al. (2012)
4.	Municipal wastewater treatment plants, China		PCR-DGGE and FISH	*Pseudomonas* and *Acinetobacter* (in suspended sludge) *Nitrosomonas*, *Nitrospira* and *Nitrobacter* sp.(in biofilm)	Bai et al. (2016)
5.	Municipal wastewater treatment plants, China		PCR-DGGE	*β-Proteobacteria* (*Candidatus Accumulibacter*, *Rhodocyclus* or *Dechloromonas*), *Bacteroidetes* and *γ-Proteobacteria* (*Pseudomonas*, *Alcaligenes* and *Acinetobacter*)	Lv et al. (2014)
6.	9 lab- and full-scale EBPR and 4 non-EBPR systems		16s rRNA sequence analysis and FISH	Novel cohesive clusters with 7 subgroup in *γ-Proteobacteria*	Kong et al. (2002)
7.	8 activated sludge samples Municipal wastewater treatment plants		Illumina HTS-based metagenomics	Proteobacteria followed by *Actinobacteria*, *Chloroflexi*, *Bacteroidetes*	Ju et al. (2014)

(continued)

Table 6.3 (continued)

Sr. No.	Sample collection site	Wastewater type	Metagenomic approach	Bacterial abundance	References
8.	Skagen wastewater treatment plants, Skagen, Denmark	Industrial wastewater (fish)	MAR-FISH	PAOs related to *Rhodocyclus*, 2 morphotypes related to *Tetrasphera*	Kong et al. (2005)
9.	25 Danish full-scale wastewater treatment plants		FISH	*Accumulibacter* clades IA, II A, II C, II D	Nielsen et al. (2012)
				Tetrasphera clades I, II, III	

q-FISH quantitative fluorescent in situ hybridization, *MAR-FISH* micro-autoradiography fluorescent in situ hybridization, *PCR-DGGE* polymerase chain reaction denaturing gradient gel electrophoresis, *HTS* high-throughput sequencing

Like bacteria microalgae are also known to store phosphorus in the form of polyphosphate (poly P) and will use it later on under the phosphate starvation conditions (Eixler et al. 2006; Larsdotter 2006; Powell et al. 2008, 2009, 2011; Rao et al. 2011). Apart from this biological way of phosphorus removal, chemical reactions are also known to remove the phosphate in the culture. As the photosynthesis occurs, it will reduce the CO_2 which results in the increase in the pH. This increased pH causes the precipitation of phosphate after complex formation with metal ions (Pires et al. 2013). There are multiple safe ways to remove phosphorus, and it depends upon environmental conditions prevalent in waterbodies and level of eutrophication whether to use a single or a combination of ecologically safe methods.

6.8.1 Microalga Used in WWT

Cai et al. (2013) discussed the efficiency of microalgal species in nutrient removal from various wastewater sources. They account the efficacy of *Chlorella* sp., *C. pyrenoidosa*, *C. sorokiniana*, *C. vulgaris*, *Scenedesmus* sp., *S. obliquus*, *Oscillatoria* sp. and *Arthrospira* sp. in the removal of nitrogen and phosphorus. Likewise various others have reported different microalgae in wastewater treatment, *Cyanobacteria*, *Phormidium* sp. (Abdel-Raouf et al. 2012), *Chlorella* (*Chlorella* sp.), *Scenedesmus dimorphus* (*S. dimorphus*), *Chlorella vulgaris* (*C. vulgaris*), *Scenedesmus quadricauda* (*S. quadricauda*) (Singh and Thomas 2012), *Spirulina platensis* (Lodi et al. 2003), *Chlamydomonas globosa*, *Chlorella minutissima* and *Scenedesmus bijuga* (Bhatnagar et al. 2011). Many people have reported the effective removal of nutrient in wastewater with the aid of microalgae culture (Table 6.4).

Table 6.4 List of microalgae culture used in nutrient removal

Sr. No.	Microalga culture	Efficiency of nutrient removal		References
		Nitrogen (%)	Phosphorus (%)	
1.	*Chlorella vulgaris*	86	78	Pires et al. (2013)
2.	*Microalgae* (order Chlorococcales) and *Cyanobacteria*	67.2	97.8	Ruiz-Martinez et al. (2012)
3.	*Chlorella* sp., *Chlorella vulgaris, Scenedesmus quadricauda* and *Scenedesmus dimorphus*	72 and 92	82 and 92	Singh and Thomas (2012)
4.	*Spirulina*	84–96	72–87	Olguin et al. (2003)
5	*Botryococcus braunii*	79.63	100	Sydney et al. (2011)
6.	*Neochloris oleoabundans*	99	100	Wang and Lan (2011)
7.	*Phormidium bohneri*	82	85	Pires et al. (2013)
8.	Microalgal consortium	70 ± 8	85 ± 9	Posadas et al. (2013)
9.	*Algal biomass*	83 ± 25	91 ± 12	Mulbry et al. (2008)

6.8.2 Current Advances in the Phosphorus Sequestration by Algae

Pires et al. (2013) have reviewed the recent improvement in the use of microalgae culture in the wastewater treatment with the three different culturing techniques, i.e. suspended cells, immobilized cells and microalgae consortia (microalgae and bacteria). In suspended cell culture, the microalgae were found to remove average nutrient concentrations, 35.5 mg/l of NH_4^+, 0.40 mg/l of NO_3^- and 3.89 mg/l of PO_4^{3-}, and it was observed that initial higher cell density gives better efficiencies of removal (Abdel-Raouf et al. 2012). For the immobilization of cells, alginate and carrageenan polymers were used (Pires et al. 2013), *Chlorella vulgaris* (Tam and Wong 2000), *Dunaliella salina* (Pires et al. 2013), *Scenedesmus obliquus* (Ruiz-Marin et al. 2010), *Scenedesmus rubescens* (Shi et al. 2007) and *Scenedesmus* sp. (Fierro et al. 2008; Zhang et al. 2008; He and Xue 2010). Various studies have been carried out using the microalgae and bacteria consortia. This proves to be an economical way to remove nutrients as microalgae provide O_2 and bacteria provide CO_2 by its action of degradation of organic compounds (Munoz and Guieysse 2006; Park et al. 2008; Subashchandrabose et al. 2011). De-bashan and Bashan (2004) reported 100% ammonium, 15% nitrate and 36% phosphorus removal from municipal wastewater by immobilizing the microalgae-bacterium consortium (*Chlorella vulgaris/Azospirillum brasilense* and *Chlorella sorokiniana/Azospirillum brasilense*) in alginate beads.

The expensive immobilizing material and its fragility during long-term operation use lead to the innovative idea of using biofilm photobioreactor (Guzzon et al. 2008; Boelee et al. 2011; Zamalloa et al. 2013). Posadas et al. (2013) studied the carbon, nitrogen and phosphorus removal with the efficiencies 91 ± 3%, 70 ± 8% and 85 ±

9%, respectively. A study of phosphorus removal under different light regime was carried out by using microalgal biofilm (biofilm photobioreactor), and it was reported that under continuous artificial illumination, this system can remove 97 ± 1% of total phosphorus from wastewater, while only 36–41% is removed when illuminated for 12 h. The species found to be present in biofilm were *Phormidium autumnale*, *Pseudanabaena* sp., *Chroococcus* sp., coccal green alga, *Monoraphidium contortum*, Diatoms, *Scenedesmus acutus* and *Cymbella minuta*; they were subject to seasonal variation in abundance, but some remains constant throughout the same like *Pseudanabaena* sp. remains always dominant and *Cymbella minuta* remains always rear (Sukacova et al. 2015).

Perspective

Huge amount of money is spent by the government for the construction and upgradation of wastewater treatment plants, but it did not achieve the appropriate goal of making wastewater fit for discharge. Phosphorus (nutrient pollution) in the waterbodies causes eutrophication which alters the water ecology. Wastewater treatment plants exploit the ability of PAOs for the phosphorus removal from wastewater. This proves to be effective, but still there is a scope of research for making these bacteria more efficient and achieve above the discharge limit. Metagenomics and genomic techniques revealed a lot of information about the community dynamics and give the insight about the molecular level which gives the idea of metabolic fluxes. For better understanding of this process, system biology approaches must be applied along with other fruitful strategies like metatranscriptomics, meta-metabolomics or community metabolomics, proteomics, etc.

Acknowledgment Authors highly acknowledge Director, CSIR-NEERI for providing facilities for this work [KRC manuscript no. CSIR-NEERI/KRC/2017/July/EBGD/16]. Varsha Jha and Sampada (Puranik) Chande is supported by UGC Junior Research Fellowship and a postdoctoral fellowship from NIH-Training grant T32 at Yale School of Medicine respectively.

References

Abdel-Raouf N, Al-Homaidan AA, Ibraheem IBM (2012) Microalgae and wastewater treatment. Saudi J Biol Sci 19:257–275. https://doi.org/10.1016/j.sjbs.2012.04.005

Ahmad T, Ahmad K, Alam M (2016) Sustainable management of water treatment sludge through 3 'R'concept. J Clean Prod 124:1–13. https://doi.org/10.1016/j.jclepro.2016.02.073

Albertsen M, Hansen LBS, Saunders AM, Nielsen PH, Nielsen KL (2012) A metagenome of a full-scale microbial community carrying out enhanced biological phosphorus removal. ISME J 6:1094–1106. https://doi.org/10.1038/ismej.2011.176

Alheit J, Möllmann C, Dutz J, Kornilovs G, Loewe P, Mohrholz V, Wasmund N (2005) Synchronous ecological regime shifts in the central Baltic and the North Sea in the late 1980s. ICES J Mar Sci J Conseil 62:1205–1215. https://doi.org/10.1016/j.icesjms.2005.04.024

Altieri AH, Gedan KB (2015) Climate change and dead zones. Glob Chang Biol 21:1395–1406. https://doi.org/10.1111/gcb.12754

Ambardar S, Gupta R, Trakroo D, Lal R (2016) High throughput sequencing: an overview of sequencing chemistry. Indian J Microbiol 56:394–404. https://doi.org/10.1007/s12088-016-0606-4

Anderson DM, Glibert PM, Burkholder JM (2002) Harmful algal blooms and eutrophication: nutrient sources, composition, and consequences. Estuaries 25:704–726. https://doi.org/10.1007/BF02804901

Bai Y, Zhang Y, Quan X, Chen S (2016) Nutrient removal performance and microbial characteristics of a full-scale IFAS-EBPR process treating municipal wastewater. Water Sci Technol 73:1261–1268. https://doi.org/10.2166/wst.2015.604

Barnard J (2006) Requirements for achieving effluent phosphorus of less than 0.1 mg/L. WERF Workshop: 9–11

Benyoucef S, Amrani M (2011) Removal of phosphorus from aqueous solutions using chemically modified sawdust of Aleppo pine (*Pinus halepensis Miller*): kinetics and isotherm studies. Environmentalist 31:200–207. https://doi.org/10.1007/s10669-011-9313-1

Bergwitz C, Jüppner H (2011) Phosphate sensing. Adv Chronic Kidney Dis 18:132–144. https://doi.org/10.1053/j.ackd.2011.01.004

Bhatnagar A, Chinnasamy S, Singh M, Das KC (2011) Renewable biomass production by mixotrophic algae in the presence of various carbon sources and wastewaters. Appl Energ 88:3425–3431. https://doi.org/10.1016/j.apenergy.2010.12.064

Blackall LL, Crocetti GR, Saunders AM, Bond PL (2002) A review and update of the microbiology of enhanced biological phosphorus removal in wastewater treatment plants. A Van Leeuw J Microb 81:681–691. https://doi.org/10.1023/A:1020538429009

Boelee NC, Temmink H, Janssen M, Buisman CJN, Wijffels RH (2011) Nitrogen and phosphorus removal from municipal wastewater effluent using microalgal biofilms. Water Res 45:5925–5933. https://doi.org/10.1016/j.watres.2011.08.044

Brookshire ENJ, Gerber S, Webster JR, Vose JM, Swank WT (2011) Direct effects of temperature on forest nitrogen cycling revealed through analysis of long term watershed records. Glob Chang Biol 17:297–308. https://doi.org/10.1111/j.1365-2486.2010.02245.x

Burow LC, Kong Y, Nielsen JL, Blackall LL, Nielsen PH (2007) Abundance and ecophysiology of *Defluviicoccus* spp., glycogen-accumulating organisms in full-scale wastewater treatment processes. Microbiology 153:178–185. https://doi.org/10.1099/mic.0.2006/001032-0

Cai T, Park SY, Li Y (2013) Nutrient recovery from wastewater streams by microalgae: status and prospects. Renew Sustain Energy Rev 19:360–369. https://doi.org/10.1016/j.rser.2012.11.030

Carabante I, Grahn M, Holmgren A, Hedlund J (2010) In situ ATR–FTIR studies on the competitive adsorption of arsenate and phosphate on ferrihydrite. J Colloid Interface Sci 351:523–531. https://doi.org/10.1016/j.jcis.2010.07.064

Carpenter SR (2008) Phosphorus control is critical to mitigating eutrophication. Proc Natl Acad Sci U S A 105:11039–11040. https://doi.org/10.1073/pnas.0806112105

Chislock MF, Doster E, Zitomer RA, Wilson AE (2013) Eutrophication: causes, consequences, and controls in aquatic ecosystems. Nat Educ Knowl 4:10

Cloern JE (2001) Our evolving conceptual model of the coastal eutrophication problem. Mar Ecol Prog Ser 210:223–253. https://doi.org/10.3354/meps210223

Cydzik-Kwiatkowska A, Zielińska M (2016) Bacterial communities in full-scale wastewater treatment systems. World J Microbiol Biotechnol 32:1–8. https://doi.org/10.1007/s11274-016-2012-9

de Lima ACA, Nascimento RF, de Sousa FF, Josue Filho M, Oliveira AC (2012) Modified coconut shell fibers: a green and economical sorbent for the removal of anions from aqueous solutions. Chem Eng J 185:274–284. https://doi.org/10.1016/j.cej.2012.01.037

De-Bashan LE, Bashan Y (2004) Recent advances in removing phosphorus from wastewater and its future use as fertilizer (1997–2003). Water Res 38:4222–4246. https://doi.org/10.1016/j.watres.2004.07.014

Delaney P, McManamon C, Hanrahan JP, Copley MP, Holmes JD, Morris MA (2011) Development of chemically engineered porous metal oxides for phosphate removal. J Hazard Mater 185:382–391. https://doi.org/10.1016/j.jhazmat.2010.08.128

Diaz RJ, Rosenberg R (2008) Spreading dead zones and consequences for marine ecosystems. Science 321:926–929. https://doi.org/10.1126/science.1156401

Díaz FJ, Anthony TO, Dahlgren RA (2012) Agricultural pollutant removal by constructed wetlands: implications for water management and design. Agric Water Manag 104:171–183. https://doi.org/10.1016/j.agwat.2011.12.012

Dodds WK, Bouska WW, Eitzmann JL, Pilger TJ, Pitts KL, Riley AJ, Thornbrugh DJ (2008) Eutrophication of US freshwaters: analysis of potential economic damages. Environ Sci Technol 43:12–19. https://doi.org/10.1021/es801217q

Dybas CL (2005) Dead zones spreading in world oceans. Bioscience 55:552–557. https://doi.org/10.1641/0006-3568(2005)055[0552

Eixler S, Karsten U, Selig U (2006) Phosphorus storage in Chlorella vulgaris (Trebouxiophyceae, Chlorophyta) cells and its dependence on phosphate supply. Phycologia 45:53–60. https://doi.org/10.2216/04-79.1

Erisman JW, Galloway JN, Seitzinger S, Bleeker A, Dise NB, Petrescu AMR, Leach AM, de Vries W (2013) Consequences of human modification of the global nitrogen cycle. Philos Trans R Soc Lond Ser B Biol Sci 368:20130116. https://doi.org/10.1098/rstb.2013.0116

Fierro S, del Pilar S-SM, Copalcua C (2008) Nitrate and phosphate removal by chitosan immobilized Scenedesmus. Bioresour Technol 99:1274–1279. https://doi.org/10.1016/j.biortech.2007.02.043

Flowers JJ, He S, Malfatti S, del Rio TG, Tringe SG, Hugenholtz P, McMahon KD (2013) Comparative genomics of two '*Candidatus Accumulibacter*' clades performing biological phosphorus removal. ISME J 7:2301–2314. https://doi.org/10.1038/ismej.2013.117

Gao W, Chen Y, Liu Y, Guo HC (2015) Scientometric analysis of phosphorus research in eutrophic lakes. Scientometrics 102:1951–1964. https://doi.org/10.1007/s11192-014-1500-7

Genz A, Kornmüller A, Jekel M (2004) Advanced phosphorus removal from membrane filtrates by adsorption on activated aluminium oxide and granulated ferric hydroxide. Water Res 38:3523–3530. https://doi.org/10.1016/j.watres.2004.06.006

AZ G, Saunders A, Neethling JB, Stensel HD, Blackall LL (2008) Functionally relevant microorganisms to enhanced biological phosphorus removal performance at full-scale wastewater treatment plants in the United States. Water Environ Res 80:688–698. https://doi.org/10.2175/106143008X276741

Günther S, Trutnau M, Kleinsteuber S, Hause G, Bley T, Röske Müller S (2009) Dynamics of polyphosphate-accumulating bacteria in wastewater treatment plant microbial communities detected via DAPI (4′, 6′-diamidino-2-phenylindole) and tetracycline labeling. Appl Environ Microb 75:2111–2121. https://doi.org/10.1128/AEM.01540-08

Guzzon A, Bohn A, Diociaiuti M, Albertano P (2008) Cultured phototrophic biofilms for phosphorus removal in wastewater treatment. Water Res 42:4357–4367. https://doi.org/10.1016/j.watres.2008.07.029

Haghseresht F (2004) Comparison of the factors that affect performances of Phoslock and Alum. Phoslock Water Solutions Ltd. Internal Report. IR 002/04

He S, Xue G (2010) Algal-based immobilization process to treat the effluent from a secondary wastewater treatment plant (WWTP). J Hazard Mater 178:895–899. https://doi.org/10.1016/j.jhazmat.2010.02.022

He S, AZ G, McMahon KD (2008) Progress toward understanding the distribution of Accumulibacter among full-scale enhanced biological phosphorus removal systems. Microb Ecol 55:229–236. https://doi.org/10.1007/s00248-007-9270-x

Hesselmann RPX, Von Rummell R, Resnick SM, Hany R, Zehnder AJB (2000) Anaerobic metabolism of bacteria performing enhanced biological phosphate removal. Water Res 34:3487–3494. https://doi.org/10.1016/S0043-1354(00)00092-0

Howarth RW, Anderson DM, Church TM, Greening H, Hopkinson CS, Huber WC, Wiseman WJ (2000) Clean coastal waters: understanding and reducing the effects of nutrient pollution. National Academy Press, Washington, DC. ISBN: 0-309-06948-3, 405 p

Ismail ZZ (2012) Kinetic study for phosphate removal from water by recycled date-palm wastes as agricultural by-products. Int J Environ Stud 69:135–149. https://doi.org/10.1080/00207233.2012.656975

Jeppesen E, Moss B, Bennion H, Carvalho L, De Meester L, Feuchtmayr H, Liboriussen L (2010) Interaction of climate change and eutrophication. In: Kernan M, Battarbee RW, Moss B (eds) Climate change impacts on freshwater ecosystems. Blackwell Publishing Ltd, Somerset, pp 119–151. ISBN: 978-1-4051-7913-3

Jeppesen E, Kronvang B, Olesen JE, Audet J, Søndergaard M, Hoffmann CC, Beklioglu M (2011) Climate change effects on nitrogen loading from cultivated catchments in Europe: implications for nitrogen retention, ecological state of lakes and adaptation. Hydrobiologia 663:1–21. https://doi.org/10.1007/s10750-010-0547-6

Ju F, Zhang T (2015) Bacterial assembly and temporal dynamics in activated sludge of a full-scale municipal wastewater treatment plant. ISME J 9:683–695. https://doi.org/10.1038/ismej.2014.162

Ju F, Guo F, Ye L, Xia Y, Zhang T (2014) Metagenomic analysis on seasonal microbial variations of activated sludge from a full-scale wastewater treatment plant over 4 years. Environ Microbiol Rep 6:80–89. https://doi.org/10.1111/1758-2229.12110

Karapınar N (2009) Application of natural zeolite for phosphorus and ammonium removal from aqueous solutions. J Hazard Mater 170:1186–1191. https://doi.org/10.1016/j.jhazmat.2009.05.094

Klein G, Perera P (2002) Eutrophication and health. Office for Official Publications of the European Commission, Luxembourg. World Health Organization, Geneva

Kong Y, Ong SL, Ng WJ, Liu WT (2002) Diversity and distribution of a deeply branched novel proteobacterial group found in anaerobic–aerobic activated sludge processes. Environ Microbiol 4:753–757. https://doi.org/10.1046/j.1462-2920.2002.00357.x

Kong Y, Nielsen JL, Nielsen PH (2005) Identity and ecophysiology of uncultured actinobacterial polyphosphate-accumulating organisms in fullscale enhanced biological phosphorus removal plants. Appl Environ Microb 71:4076–4085. https://doi.org/10.1128/AEM.71.7.4076-4085.2005

Kristiansen R, Nguyen HTT, Saunders AM, Nielsen JL, Wimmer R, Le VQ, Nielsen KL (2013) A metabolic model for members of the genus Tetrasphaera involved in enhanced biological phosphorus removal. ISME J 7:543–554. https://doi.org/10.1038/ismej.2012.136

Kumar R, Goyal D (2010) Waste water treatment and metal (Pb2+, Zn2+) removal by microalgal based stabilization pond system. Indian J Microbiol 50:34. https://doi.org/10.1007/s12088-010-0063-4

Larsdotter K (2006) Microalgae for phosphorus removal from wastewater in a Nordic climate, p 11. PhD thesis, School of Biotechnology, Royal Institute of Technology, Stockholm, Sweden. ISBN 91-7178-288-5

Lehtiniemi M, Engström-Öst J, Viitasalo M (2005) Turbidity decreases anti-predator behaviour in pike larvae, Esox lucius. Environ Biol Fish 73:1–8. https://doi.org/10.1007/s10641-004-5568-4

Liu Y, Kang X, Li X, Yuan Y (2015) Performance of aerobic granular sludge in a sequencing batch bioreactor for slaughterhouse wastewater treatment. Bioresour Technol 190:487–491. https://doi.org/10.1016/j.biortech.2015.03.008

Lodi A, Binaghi L, Solisio C, Converti A, Del Borghi M (2003) Nitrate and phosphate removal by Spirulina platensis. J Ind Microbiol Biotechnol 30:656–660. https://doi.org/10.1007/s10295-003-0094-5

Lu YZ, Wang HF, Kotsopoulos TA, Zeng RJ (2016) Advanced phosphorus recovery using a novel SBR system with granular sludge in simultaneous nitrification, denitrification and phosphorus removal process. Appl Microbiol Biotechnol:1–8. https://doi.org/10.1007/s00253-015-7249-y

Lv JH, Yuan LJ, Chen X, Liu L, Luo DC (2014) Phosphorus metabolism and population dynamics in a biological phosphate-removal system with simultaneous anaerobic phosphate stripping. Chemosphere 117:715–721.0020doi:https://doi.org/10.1016/j.chemosphere.2014.10.018

Mao Y, Yu K, Xia Y, Chao Y, Zhang T (2014) Genome reconstruction and gene expression of "Candidatus Accumulibacter phosphatis" Clade IB performing biological phosphorus removal. Environ Sci Technol 48:10363–10371. https://doi.org/10.1021/es502642b

Martín HG, Ivanova N, Kunin V, Warnecke F, Barry KW, McHardy AC, Dalin E (2006) Metagenomic analysis of two enhanced biological phosphorus removal (EBPR) sludge communities. Nat Biotechnol 24:1263–1269. https://doi.org/10.1038/nbt1247

Meena KK, Kumar M, Mishra S, Ojha SK, Wakchaure GC, Sarkar B (2015) Phylogenetic study of methanol oxidizers from Chilika-Lake sediments using genomic and metagenomic approaches. Indian J Microbiol 55:151–162. https://doi.org/10.1007/s12088-015-0510-3

Mulbry W, Kondrad S, Pizarro C, Kebede-Westhead E (2008) Treatment of dairy manure effluent using freshwater algae: algal productivity and recovery of manure nutrients using pilot-scale algal turf scrubbers. Bioresour Technol 99:8137–8142. https://doi.org/10.1016/j.biortech.2008.03.073

Mulder C, Boit A, Mori S, Arie Vonk J, Dyer SD, Faggiano L, Marquet PA (2012) 1 Distributional (In) congruence of biodiversity-ecosystem functioning. Adv Ecol Res 46:1

Munoz R, Guieysse B (2006) Algal–bacterial processes for the treatment of hazardous contaminants: a review. Water Res 40:2799–2815. https://doi.org/10.1016/j.watres.2006.06.011

Nguyen HTT, Le VQ, Hansen AA, Nielsen JL, Nielsen PH (2011) High diversity and abundance of putative polyphosphate-accumulating Tetrasphaera-related bacteria in activated sludge systems. FEMS Microbiol Ecol 76:256–267. https://doi.org/10.1111/j.1574-6941.2011.01049.x

Nguyen TAH, Ngo HH, Guo W, Nguyen TV (2012a) Phosphorous removal from aqueous solutions by agricultural by-products: a critical review. J Water Sustain 2:193–207

Nguyen HTT, Nielsen JL, Nielsen PH (2012b) 'Candidatus Halomonas phosphatis', a novel polyphosphate-accumulating organism in full-scale enhanced biological phosphorus removal plants. Environ Microb 14:2826–2837. https://doi.org/10.1111/j.1462-2920.2012.02826.x

Nielsen PH, Mielczarek AT, Kragelund C, Nielsen JL, Saunders AM, Kong Y, Vollertsen J (2010) A conceptual ecosystem model of microbial communities in enhanced biological phosphorus removal plants. Water Res 44:5070–5088. https://doi.org/10.1016/j.watres.2010.07.036

Nielsen PH, Saunders AM, Hansen AA, Larsen P, Nielsen JL (2012) Microbial communities involved in enhanced biological phosphorus removal from wastewater—a model system in environmental biotechnology. Curr Opin Biotechnol 23:452–459. https://doi.org/10.1016/j.copbio.2011.11.027

Oehmen A, Saunders AM, Vives MT, Yuan Z, Keller J (2006) Competition between polyphosphate and glycogen accumulating organisms in enhanced biological phosphorus removal systems with acetate and propionate as carbon sources. J Biotechnol 123:22–32. https://doi.org/10.1016/j.jbiotec.2005.10.009

Oehmen A, Lemos PC, Carvalho G, Yuan Z, Keller J, Blackall LL, Reis MA (2007) Advances in enhanced biological phosphorus removal: from micro to macro scale. Water Res 41:2271–2300. https://doi.org/10.1016/j.watres.2007.02.030

Olguín EJ, Galicia S, Mercado G, Pérez T (2003) Annual productivity of *Spirulina* (Arthrospira) and nutrient removal in a pig wastewater recycling process under tropical conditions. J Appl Phycol 15:249–257. https://doi.org/10.1023/A:1023856702544

Othman I, Anuar AN, Ujang Z, Rosman NH, Harun H, Chelliapan S (2013) Livestock wastewater treatment using aerobic granular sludge. Bioresour Technol 133:630–634. https://doi.org/10.1016/j.biortech.2013.01.149

Ozyonar F, Karagozoglu B (2011) Operating cost analysis and treatment of domestic wastewater by electrocoagulation using aluminium electrodes. Pol J Environ Stud 20:173–179

Paerl HW, Paul VJ (2012) Climate change: links to global expansion of harmful cyanobacteria. Water Res 46:1349–1363. https://doi.org/10.1016/j.watres.2011.08.002

Park WH (2009) Integrated constructed wetland systems employing alum sludge and oyster shells as filter media for P removal. Ecol Eng 35:1275–1282. https://doi.org/10.1016/j.ecoleng.2009.05.015

Park Y, Je KW, Lee K, Jung SE, Choi TJ (2008) Growth promotion of *Chlorella ellipsoidea* by co-inoculation with Brevundimonas sp. isolated from the microalga. Hydrobiologia 598:219–228. https://doi.org/10.1007/s10750-007-9152-8

Pires JCM, Alvim-Ferraz MCM, Martins FG, Simões M (2013) Wastewater treatment to enhance the economic viability of microalgae culture. Environ Sci Pollut Res Int 20:5096–5105. https://doi.org/10.1007/s11356-013-1791-x

Pittman JK, Dean AP, Osundeko O (2011) The potential of sustainable algal biofuel production using wastewater resources. Bioresour Technol 102:17–25. https://doi.org/10.1016/j.biortech.2010.06.035

Posadas E, García-Encina PA, Soltau A, Domínguez A, Díaz I, Muñoz R (2013) Carbon and nutrient removal from centrates and domestic wastewater using algal–bacterial biofilm bioreactors. Bioresour Technol 139:50–58. https://doi.org/10.1016/j.biortech.2013.04.008

Powell N, Shilton AN, Pratt S, Chisti Y (2008) Factors influencing luxury uptake of phosphorus by microalgae in waste stabilization ponds. Environ Sci Technol 42:5958–5962. https://doi.org/10.1021/es703118s

Powell N, Shilton A, Chisti Y, Pratt S (2009) Towards a luxury uptake process via microalgae–defining the polyphosphate dynamics. Water Res 43:4207–4213. https://doi.org/10.1016/j.watres.2009.06.011

Powell N, Shilton A, Pratt S, Chisti Y (2011) Phosphate release from waste stabilisation pond sludge: significance and fate of polyphosphate. Water Sci Technol 63:1689–1694. https://doi.org/10.2166/wst.2011.336

Pronk M, de Kreuk MK, de Bruin B, Kamminga P, Kleerebezem R, van Loosdrecht MCM (2015) Full scale performance of the aerobic granular sludge process for sewage treatment. Water Res 84:207–217. https://doi.org/10.1016/j.watres.2015.07.011

Rabalais NN, Turner RE, Wiseman WJ Jr (2002) Gulf of Mexico hypoxia, AKA "the dead zone". Annu Rev Ecol Syst 33:235–263. https://doi.org/10.1146/annurev.ecolsys.33.010802.150513

Rao PH, Kumar RR, Raghavan BG, Subramanian VV, Sivasubramanian V (2011) Application of phycoremediation technology in the treatment of wastewater from a leather-processing chemical manufacturing facility. Water SA 37:07–14

Rim-Rukeh A, Agbozu E (2013) Impact of partially treated sewage effluent on the water quality of recipient Epie Creek Niger Delta, Nigeria using Malaysian Water Quality Index (WQI). J Appl Sci Environ Manag 17:5–12

Romero E, Garnier J, Lassaletta L, Billen G, Le Gendre R, Riou P, Cugier P (2013) Large-scale patterns of river inputs in southwestern Europe: seasonal and interannual variations and potential eutrophication effects at the coastal zone. Biogeochemistry 113:481–505. https://doi.org/10.1007/s10533-012-9778-0

Ruiz-Marin A, Mendoza-Espinosa LG, Stephenson T (2010) Growth and nutrient removal in free and immobilized green algae in batch and semicontinuous cultures treating real wastewater. Bioresour Technol 101:58–64. https://doi.org/10.1016/j.biortech.2009.02.076

Ruiz-Martinez A, Garcia NM, Romero I, Seco A, Ferrer J (2012) Microalgae cultivation in wastewater: nutrient removal from anaerobic membrane bioreactor effluent. Bioresour Technol 126:247–253. https://doi.org/10.1016/j.biortech.2012.09.022

Ruiz-Martínez A, Serralta J, Romero I, Seco A, Ferrer J (2015) Effect of intracellular P content on phosphate removal in *Scenedesmus* sp. experimental study and kinetic expression. Bioresour Technol 175:325–332. https://doi.org/10.1016/j.biortech.2014.10.081

Rustad LEJL, Campbell J, Marion G, Norby R, Mitchell M, Hartley A, Gurevitch J (2001) A meta-analysis of the response of soil respiration, net nitrogen mineralization, and aboveground plant growth to experimental ecosystem warming. Oecologia 126:543–562. https://doi.org/10.1007/s004420000544

Santos-Beneit F (2015) The Pho regulon: a huge regulatory network in bacteria. Front Microbiol 6:402. https://doi.org/10.3389/fmicb.2015.00402

Saunders AM, Oehmen A, Blackall LL, Yuan Z, Keller J (2003) The effect of GAOs (glycogen accumulating organisms) on anaerobic carbon requirements in full-scale Australian EBPR (enhanced biological phosphorus removal) plants. Water Sci Technol 47:37–43

Saunders AM, Albertsen M, Vollertsen J, Nielsen PH (2016) The activated sludge ecosystem contains a core community of abundant organisms. ISME J 10:11–20. https://doi.org/10.1038/ismej.2015.117

Schenk PM, Thomas-Hall SR, Stephens E, Marx UC, Mussgnug JH, Posten C, Hankamer B (2008) Second generation biofuels: high-efficiency microalgae for biodiesel production. Bioenergy Res 1:20–43. https://doi.org/10.1007/s12155-008-9008-8

Seow TW, Lim CK, Norb MHM, Mubarakb MFM, Lam CY, Yahya A, Ibrahim Z (2016) Review on wastewater treatment technologies. Int J Appl Environ Sci 11:111–126

Seviour RJ, Mino T, Onuki M (2003) The microbiology of biological phosphorus removal in activated sludge systems. FEMS Microbiol Rev 27:99–127. https://doi.org/10.1016/S0168-6445(03)00021-4

Sharma A, Lal R (2017) Survey of (meta) genomic approaches for understanding microbial community dynamics. Indian J Microbiol 57:23–58. https://doi.org/10.1007/s12088-016-0629-x

Shi J, Podola B, Melkonian M (2007) Removal of nitrogen and phosphorus from wastewater using microalgae immobilized on twin layers: an experimental study. J Appl Phycol 19:417–423. https://doi.org/10.1007/s10811-006-9148-1

Sidat M, Bux F, Kasan HC (1999) Polyphosphate accumulation by bacteria isolated from activated sludge. Water SA 25:175–179

Singh G, Thomas PB (2012) Nutrient removal from membrane bioreactor permeate using microalgae and in a microalgae membrane photoreactor. Bioresour Technol 117:80–85. https://doi.org/10.1016/j.biortech.2012.03.125

Skennerton CT, Barr JJ, Slater FR, Bond PL, Tyson GW (2015) Expanding our view of genomic diversity in *Candidatus Accumulibacter clades*. Environ Microbiol 17:1574–1585. https://doi.org/10.1111/1462-2920.12582

Smith VH (2001) Blue-green algae in eutrophic fresh waters. Lake Line 21:34–36

Smith VH (2003) Eutrophication of freshwater and coastal marine ecosystems a global problem. Environ Sci Pollut Res Int 10:126–139. https://doi.org/10.1065/espr2002.12.142

Strom PF (2006) Technologies to remove phosphorus from wastewater. Rutgers University, New Brunswick, p 18

Subashchandrabose SR, Ramakrishnan B, Megharaj M, Venkateswarlu K, Naidu R (2011) Consortia of cyanobacteria/microalgae and bacteria: biotechnological potential. Biotechnol Adv 29:896–907. https://doi.org/10.1016/j.biotechadv.2011.07.009

Sukačová K, Trtílek M, Rataj T (2015) Phosphorus removal using a microalgal biofilm in a new biofilm photobioreactor for tertiary wastewater treatment. Water Res 71:55–63. https://doi.org/10.1016/j.watres.2014.12.049

Sydney EB, Da Silva TE, Tokarski A, Novak AC, De Carvalho JC, Woiciecohwski AL, Soccol CR (2011) Screening of microalgae with potential for biodiesel production and nutrient removal from treated domestic sewage. Appl Energy 88:3291–3294. https://doi.org/10.1016/j.apenergy.2010.11.024

Symonds EM, Cook MM, McQuaig SM, Ulrich RM, Schenck RO, Lukasik JO, Breitbart M (2015) Reduction of nutrients, microbes, and personal care products in domestic wastewater by a benchtop electrocoagulation unit. Sci Rep 5:9380. https://doi.org/10.1038/srep09380

Tam NFY, Wong YS (2000) Effect of immobilized microalgal bead concentrations on wastewater nutrient removal. Environ Pollut 107:145–151. https://doi.org/10.1016/S0269-7491(99)00118-9

Topare NS, Attar SJ, Manfe MM (2011) Sewage/wastewater treatment technologies: a review. Chem Commun 1:18–24

UN report (2012) Managing water under uncertainty and risk, The United Nations world water development report 4, UN Water Reports, World Water Assessment Programme

UN report (2015) Wastewater management-A UN-Water analytical brief, New York

Vijayavenkataraman S, Iniyan S, Goic R (2012) A review of climate change, mitigation and adaptation. Renew Sustain Energy Rev 16:878–897. https://doi.org/10.1016/j.rser.2011.09.009

Wang B, Lan CQ (2011) Biomass production and nitrogen and phosphorus removal by the green alga Neochloris oleoabundans in simulated wastewater and secondary municipal wastewater effluent. Bioresour Technol 102:5639–5644. https://doi.org/10.1016/j.biortech.2011.02.054

Wang LP, Lei K (2016) Rapid identification and quantification of *Aureococcus anophagefferens* by qPCR method (Taqman) in the Qinhuangdao coastal area: a region for recurrent Brown tide breakout in China. Indian J Microbiol 56:491–497. https://doi.org/10.1007/s12088-016-0619-z

Watkinson AJ, O'Neil JM, Dennison WC (2005) Ecophysiology of the marine cyanobacterium, *Lyngbya majuscula* (Oscillatoriaceae) in MoretonBay, Australia. Harmful Algae 4:697–715. https://doi.org/10.1016/j.hal.2004.09.001

Winder M, Sommer U (2012) Phytoplankton response to a changing climate. Hydrobiologia 698:5–16. https://doi.org/10.1007/s10750-012-1149-2

Wong DHJ, Beiko RG (2015) Transfer of energy pathway genes in microbial enhanced biological phosphorus removal communities. BMC Genomics 16:526. https://doi.org/10.1186/s12864-015-1752-5

Xu X, Gao Y, Gao B, Tan X, Zhao YQ, Yue Q, Wang Y (2011) Characteristics of diethylenetriamine-crosslinked cotton stalk/wheat stalk and their biosorption capacities for phosphate. J Hazard Mater 192:1690–1696. https://doi.org/10.1016/j.jhazmat.2011.07.009

Yewalkar-Kulkarni S, Gera G, Nene S, Pandare K, Kulkarni B, Kamble S (2016) Exploiting phosphate-starved cells of *Scenedesmus* sp. for the treatment of raw sewage. Indian J Microbiol:1–9. https://doi.org/10.1007/s12088-016-0626-0

Zamalloa C, Boon N, Verstraete W (2013) Decentralized two-stage sewage treatment by chemical–biological flocculation combined with microalgae biofilm for nutrient immobilization in a roof installed parallel plate reactor. Bioresour Technol 130:152–160. https://doi.org/10.1016/j.biortech.2012.11.128

Zamparas M, Gianni A, Stathi P, Deligiannakis Y, Zacharias I (2012) Removal of phosphate from natural waters using innovative modified bentonites. Appl Clay Sci 62:101–106. https://doi.org/10.1016/j.clay.2012.04.020

Zeng L, Li X, Liu J (2004) Adsorptive removal of phosphate from aqueous solutions using iron oxide tailings. Water Res 38:1318–1326. https://doi.org/10.1016/j.watres.2003.12.009

Zhang E, Wang B, Wang Q, Zhang S, Zhao B (2008) Ammonia–nitrogen and orthophosphate removal by immobilized *Scenedesmus* sp. isolated from municipal wastewater for potential use in tertiary treatment. Bioresour Technol 99:3787–3793. https://doi.org/10.1016/j.biortech.2007.07.011

Zhang Z, Li H, Zhu J, Weiping L, Xin X (2011) Improvement strategy on enhanced biological phosphorus removal for municipal wastewater treatment plants: full-scale operating parameters, sludge activities, and microbial features. Bioresour Technol 102:4646–4653. https://doi.org/10.1016/j.biortech.2011.01.017

Web Links

Halls S, Yamazaki K, Water quality: the impact of Eutrophication.http://www.unep.or.jp/ietc/publications/short_series/lakereservoirs-3/fwd.asp

SeaWeb, Ocean issue briefs: nutrients. http://www.seaweb.org/resources/briefings/nutrient.php

FAO corporate document repository, Wastewater treatment and use in agriculture. http://www.fao.org/docrep/T0551E/t0551e03.htm#1.%20wastewater%20characteristics%20and%20effluent%20quality%20parameters

GE Step Ahead, Why should we reuse wastewater. http://www.moneycontrol.com/gestepahead/article.php?id=954013&cid=2

NOAA Headquarters, Average 'dead zone' predicted for Gulf of Mexico in 2016: outlook incorporates multiple hypoxia models for the second year. https://www.sciencedaily.com/releases/2016/06/160610094739.htm

Bioremediation of Terrestrial Oil Spills: Feasibility Assessment

7

Pradnya Nagkirti, Azhar Shaikh, Gowdaman Vasudevan, Vasundhara Paliwal, and Prashant Dhakephalkar

Abstract

Extensive industrialization and urbanization have led to an escalating global demand for petroleum and its products, thereby causing an increasing probability of marine as well as terrestrial oil spills. The need for efficient, sustainable as well as cost-effective oil spill clean-up technology therefore becomes inevitable. Over the years majority of research has focused on marine oil spills with far lesser reports on remediating terrestrial oil spills, which are far more complex and challenging for remediation. This study aims to provide brief information on the current technologies available for remediation of terrestrial oil spills. A comparative analysis of physical, chemical, thermal and bioremediation techniques has been reviewed in terms of their advantages, disadvantages and efficiency and also their feasibility assessment.

Keywords

Bioremediation • Oil spills • Marine • Hazard • Emulsion

7.1 Introduction to Terrestrial Oil Spills

Probability of accidental spillage of petroleum products especially during transportation and storage has increased with the escalating global demand for petroleum and has thus created a need for efficient clean-up technologies (Das and Chandran 2011). During the transport and exploitation of crude oil, 1.3 million litres of oil enter the environment each year (Brooijmans et al. 2009). The volume of crude oil

P. Nagkirti • A. Shaikh • G. Vasudevan • V. Paliwal (✉) • P. Dhakephalkar (✉)
Department of Microbiology, Bioenergy Group, MACS-Agharkar Research Institute, G.G. Agarkar Road, Pune 411004, Maharashtra, India
e-mail: vapaliwal@gmail.com; pkdhakephalkar@aripune.org

H.J. Purohit et al. (eds.), *Optimization and Applicability of Bioprocesses*,
https://doi.org/10.1007/978-981-10-6863-8_7

Table 7.1 Some of the major oil spills around the world are enlisted below

Date	Location	Spill type	Tons of spilled crude oil
2010	USA, Gulf of Mexico	Deepwater Horizon	560,000–585,000
1991	Iraq and Kuwait, Persian Gulf	Gulf War oil spill	270,000–820,000
1980	Mexico, Gulf of Mexico	Ixtoc	454,000–480,000
1979	Trinidad and Tobago	Atlantic Empress/Aegean Captain	287,000
1992	Uzbekistan	Fergana Valley	285,000
1983	Iran, Persian Gulf	Nowruz field platform	260,000
1991	Angola	ABT Summer Offshore	260,000
1991	Italy, Mediterranean Sea near Genoa	MT Haven	144,000
1989	USA, Prince William Sound, Alaska	Exxon Valdez oil tanker	35,065–103,896

Modified table of Wood (2002)

discharge happening naturally has been projected to be 600,000 mt/year (Kvenvolden and Cooper 2003). Accidental or anthropogenic release of hydrocarbons has become a major source of aquatic and terrestrial contamination (Das and Chandran 2011). Terrestrial oil spills receive less attention than the reports of marine/aquatic oil spills (Fingas 2012). Table 7.1 provides several reported major oil spills around the world. In terrestrial oil spill, there is a vertical diffusion of the oil, thereby causing entrapment of toxic contaminants deep into the subsurface of the soil. Also the spreading and movement of oil across the surface are more complicated on land as compared to water (Fingas 2012). The accumulation of these hydrocarbons causes extensive damage to animal and plant tissues leading to death or mutations in long term (Khashayar and Mahsa 2010).

7.2 Properties of Crude Oil

Crude oil (petroleum) is a combination of organic compounds, where every compound has its unique chemical and physical properties. Crude oil extracted from different reservoirs exhibits a wide range of physical and chemical properties. These properties should be considered instantaneously during crude oil spillage, as it may assist in selecting the technique for removal of oil from the contaminated sites.

7.2.1 Density

Density of oil is measured according to the standard method D 5002 of ASTM (American Society for Testing and Materials) and expressed in g/ml (Fingas 2014). The density of spilled oil increases overtime, where most of the volatile components are lost due to evaporation. Two density-related properties of oils can be used.

7.2.1.1 Specific Gravity

It is the ratio of densities of oil to pure water, at a specified temperature.

7.2.1.2 American Petroleum Institute (API) Gravity

Depending on the API gravity, the petroleum liquid can be measured as heavy or light compared to water. If the API gravity of petroleum liquid is greater than 10, then it is lighter and floats on water, while if it is less than 10, then it is heavier and sinks.

Oils having high densities tend to have low API gravities and vice versa (Fingas 2014).

7.2.2 Pour Point

The lowest temperature, which allows the oil to flow, is pour point of the oil (Speight 2015). For example, highly viscous oils, due to their viscosity, may not flow at its pour point. During oil spillage, it is essential to study the pour point of oil, which provides a rough idea about the temperature at which flow ceases (www.oilproduction.net/files/Introduction).

7.2.3 Flash Point

It is the temperature required for the fuel to form an ignitable vapour once exposed to flame (www.oilproduction.net/files/Introduction). With respect to safety of spill clean-up operations, flash point becomes an important factor to consider. Under ambient conditions, gasoline and light fuels may get ignited causing serious hazards when spilled (Fingas 2014). Crude oil spillages that are fresh will have lower flash points, until the evaporation or dispersion of lighter components occur (www.oilproduction.net/files/Introduction).

7.2.4 Viscosity

Viscosity is defined in terms of resistance of the fluid to flow. Fluids having low viscosity flow more easily. Temperature also affects viscosity of a fluid; the lower the temperature, the higher is the viscosity. In case of an oil spillage, it is viscosity that usually determines the rate at which spreading of the oil will take place as well as the degree to which it will penetrate the soil (www.oilproduction.net/files/Introduction). The lesser the content of asphaltenes, the lower will be the viscosity of crude oil and vice versa (Fingas 2014).

7.2.5 Emulsion Formation

Water-in-oil emulsion is the formation of small globules of water in oil. It is important to know the possibility of oil to form an emulsion, what are the physical characteristics and whether the emulsion is stable. If spillage occurs on water surface, then characteristics of emulsions are different than that of parent crude oil. The water-in-oil emulsion has imperative significance towards the removal of crude oil in spillage (www.oilproduction.net/files/Introduction).

7.2.6 Sulphur Content

Along with hydrocarbons, crude oil also consists of sulphur, oxygen, nitrogen, organometallic complexes such as nickel and vanadium and dissolved gases, such as hydrogen sulphide. At an oil spillage site, sulphur content becomes a safety concern for the clean-up worker. Also, the burning of crude oil with high sulphur content leads to production of sulphur dioxide which is toxic to the environment (www.oilproduction.net/files/Introduction).

7.2.7 Volatile Organic Compounds (VOCs)

Volatile organic compounds (especially aromatic hydrocarbons) commonly found in petroleum are benzene, toluene, ethylbenzene and xylenes (BTEX). During accidental spills, these compounds frequently enter soil, groundwater, etc. due to their high solubility and mobility. BTEX are known to be hazardous carcinogenic and neurotoxic compounds; hence, removal of these components from spill sites becomes necessary (www.oilproduction.net/files/Introduction).

7.2.8 Composition of Crude Oil

The main constituents of oil can be broadly grouped into saturates (branched, unbranched and cyclic alkanes), aromatics (monocyclic and polycyclic), resins and asphaltenes (Fingas 2014). The outermost layer of oil is constituted by saturates, while the innermost is composed of asphaltenes due to their large molar masses. The microbial degradation of different fractions of crude oil is of the following order: alkanes>light aromatics (monocyclic)>cycloalkanes>heavy aromatics (polyaromatics)>asphaltenes. Resins are the most easily degraded fraction of crude oil as they are light polar molecules. Polyaromatic hydrocarbons (PAHs) such as naphthalene, phenanthrene, etc. are referred to as low molecular weight or light PAHs as they are composed of two or three cyclic rings, while those made up of four or more cyclic rings (e.g. pyrene, chrysenes, etc.) have high molar masses and are referred to as heavy PAHs and are known to be potential carcinogenic compounds.

Asphaltenes are highly complex and biodegradation resistant due to their viscous nature and high molar mass (Fingas 2014).

7.3 Hazardous Effects

Soil contaminated with crude oil poses a major global problem. Various hazardous effects are associated with oil spills. Air, ground and water pollution due to various petroleum products result in extinction of numerous species of plants and animals. Once they reach the environment, it leads to damage primarily by its blocking effect of oil layer to water, nutrients, oxygen and access to light.

In terrestrial habitats, crude oil gets absorbed by plant and soil particles thereby affecting the soil fertility leading to poor growth of plants. It also adversely affects the germination of seeds resulting in a decrease of agricultural productivity of the soil (Wang et al. 2008). It also has serious effects to human health, either from internal exposure to oil (injection or inhalation) or from external exposure through skin and eye irritation. Health risk is associated with direct contact to vapours of contaminants of soil (Thapa et al. 2012). Exposure to petroleum hydrocarbon products may cause acute toxic effects such as neurological problems and ocular (eye) and respiratory system issues. Along with migraines and headaches, people living in affected areas showed nausea, throat infections and nose and eye irritations.

7.4 Different Methods to Clean Up Terrestrial Oil Spills

To avoid problems associated with oil spills, it is necessary to remove a maximum portion of the crude oil from the spillage area and decontaminate the site. For determining the efficacy of a remediation technique, cost is an important factor. The cost analysis for a remediation technique would require considering the implementation cost as well as the cost of averting the oil from harming the environment. An optimal spill response would be the one which minimizes cost of operation as well as environmental deterioration (Prendergast and Gschwend 2014). Some of the factors influencing the cost associated with cleaning up oil spills are the type of product spilled, location and time of spillage, local and national laws and clean-up strategy. Location, oil type, total spill amount and depth of contamination are most important factors determining the per-unit amount cost (Khan et al. 2004). Conventional physiochemical methods are available to clean up oil from the contaminated area.

7.4.1 Physical Methods

The primary response option for the clean-up of oil-contaminated sites is the physical restraint and recovery of free oil. Commonly used physical methods include the following.

7.4.1.1 Soil Excavation

This is a mechanical removal of contaminated soil from the area either for burying or burning. It is suggested that soil excavation is an efficient way to remove contaminated soil. Disadvantage associated with this technique is that it is expensive, and the area, from which soil is removed, becomes more prone to the erosion (Araruna et al. 2004).

7.4.1.2 Soil Washing

An ex situ method that uses a combination of water and chemical additives to remove the contaminant. In this process, fine soil such as silt and clay are separated from coarse particles like gravel. Crude oil contaminants incline towards fine soil particles rather than coarse soil. At the time of washing of soil, silt and clay particles can be removed mechanically and separated from uncontaminated coarse soil (Wood 2002; Arwidsson et al. 2010). This uncontaminated coarse soil is then retained as backfill, while contaminated soil is further treated or disposed. Soil washing technique particularly targets contaminants such as semi-volatile organic compounds (SVOCs) and petroleum residuals. Soil washing can be a cost-effective technique for soil sites with gravels. Based on the contaminant concentration, quantity and site-specific conditions, the average cost for soil washing technique can be around $170 US/t (Khan et al. 2004).

7.4.1.3 Soil Vapour Extraction (SVE)

It is a cost-effective, in situ method for removal of contaminants in soil that are volatile. With the help of specially designed system, the volatile components of crude oil are removed from soil in the form of vapour. Vacuum is applied through underground wells which pulls up the contaminants to the surface in the form of vapour or gas. Sometimes air is also used to enhance the process. SVE is mainly used to remove chlorinated hydrocarbons (e.g. trichloroethylene) from contaminated soil (Albergaria et al. 2012) and at sites with comparatively porous and homogeneous contaminated unsaturated zone. The process suffers from disadvantages like large energy requirement causing economical constraints. Also, lighter fuels such as gasoline can be successfully removed by SVE, while the heavier fuels (e.g. diesel, kerosene and heating oils) are not readily removed by this method. Sites with high moisture levels and those having groundwater table less than a metre underneath the land surface are not appropriate for SVE technique. The operational costs of SVE are between $20 and $50 US/t of contaminated soil (Environmental Protection Agency report) (Khan et al. 2004).

7.5 Thermal Clean-Up Methods

Thermal methods have been shown to be more effective in some soils, e.g. clays, where majority of other clean-up methods fail (Araruna et al. 2004).

7.5.1 Thermal Desorption

In this method, crude oil-contaminated soil is heated at 200–1,000°F; this causes the contaminants with low boiling point to get vaporize or desorb (physically separate) and release from soil (Ukpaka and Amadi 2014). Thermal desorber causes either complete or partial decomposition of organic substituent, depending upon the temperature used or type of organic compound present in the contaminated soil for VOCs; low-temperature thermal desorption is used, while higher temperature is mainly used for PAHs. Thermal desorption changes the form of contaminant into a further treatable form, whereas incineration intends to destroy it completely (Singh et al. 2009). Thermal desorber has been shown to decontaminate crude oil-polluted soil with 90% efficiency (Ezeji et al. 2005).The major pitfalls of this technique is that it is expensive, time-consuming and hazardous. Using thermal desorption technique, the overall remediation cost of petroleum-contaminated soils varies from $50 to $330 US/mt that include cost of initial setup, transportation charges, excavation and replacement of treated soil (Khan et al. 2004).

7.5.2 Incineration

High temperatures are used to burn the contaminants from crude oil-contaminated soil. According to US EPA, incineration occurs at high temperature ranging from 1,600 to 2,500 °F. Hazardous wastes including crude oil get destroyed at this high temperature, while other toxic elements get reduced to basic elements such as hydrogen, carbon, chlorine, nitrogen, etc. These elements then combine with oxygen to form nontoxic compounds like water, carbon dioxide and nitrogen oxide. Disadvantages associated with this technique are requirement of high energy and operational cost, large space involvement in the clean-up process and dangers of environmental pollution (Fericelli 2011).

7.6 Chemical Methods

Chemical methods, mostly dispersants, have consistently been used in many countries as a response alternative. Dispersants are surfactants, which are sprayed on the oil slick using aircraft for reducing the water/oil interfacial tension thereby promoting dissolution of oil (Prendergast and Gschwend 2014). However, heavier crude oil fractions tend to be difficult to remove using dispersants resulting in escalation of clean-up costs (Etkin 1999). The cost of removing oil from offshore using dispersants has been reported to be around $7350/tonne (Etkin 1999).

7.7 Biological Methods

Remediation by physical methods tends to be costly and thus not applicable on large scale. Biological treatments offer a cost-effective and sustainable alternative for hydrocarbon-contaminated sites. For contaminant removal, bioremediation is most often the primary method of choice. Different microorganisms or plants can be used to decontaminate polluted sites (Chandra et al. 2013).

7.7.1 Phytoremediation

Phytoremediation involves the use of plants for in situ treatment of polluted sites to remove, immobilize and/or degrade contaminated soil. Phytoremediation has been shown as a promising technology for in situ removal of petroleum hydrocarbons. During phytoremediation for degradation of petroleum hydrocarbons, synergy between microbes and plants present in the rhizosphere is employed (Fingas 2014; Hou et al. 2015). Different phytoremediation techniques can be used such as phyto-extraction, phytodegradation, phytostabilization and phytovolatilization based on their applicability and the nature of contaminant (Dhankher et al. 2012). Phytoremediation provides numerous advantages over other remedial approaches like cost-effectiveness, applicable at any geographical site that can sustain plant growth, and improves soil quality and structure. As compared to other remediation strategies that cost around \$75–100 per m^3, the cost of plant growth-promoting rhizobacteria (PGPR)-enhanced phytoremediation is \$25–50 per m^3(Greenberg and Gerwing 2009). Despite all the advantages, phytoremediation also suffers from certain challenges such as the requirement of large amount of biomass that can grow quickly (especially in poor-quality soils), selecting the right combination of plant-bacterial species and presence of high concentration of petroleum hydrocarbons (Fingas 2014). Table 7.2 gives the feasibility of physical methods and the cost required to carry out each technique.

7.7.2 Bioremediation

Bioremediation is an unconventional method for treating contaminated sites (Liu et al. 2010). It is a biological process, wherein microbial capacities are exploited in favour of removal of toxic and hazardous chemicals from basic resources of environment.

This technology uses naturally occurring organisms to degrade perilous compounds into less or nontoxic compounds. The first successful application of bioremediation for petroleum-contaminated soil came from Exxon Valdez oil spill of 1989 (Chandra et al. 2013). This created a remarkable interest in the potential use of bioremediation technology as a primary clean-up strategy. There are numerous advantages of using bioremediation over other prevailing methods such as it is less expensive and eliminates the waste permanently (Perelo 2010). The residues left

Table 7.2 Feasibility assessment of physical and chemical technology for remediation

Technology	Process efficiency	Ease of operation	Capital expenses	Applicability	References
Physical methods					
Soil vapour extraction (SVE)	The volatile components of crude oil are removed from soil in the form of vapour	In situ technology for remediating contaminated soils with unsaturated volatile and semi-volatile organic compounds	\$20–\$50 per tonne	Mainly used to remove chlorinated hydrocarbons (e.g. trichloroethylene) from contaminated soil	Imamura et al. (1997)
Thermal desorption	Thermal desorption changes the form of contaminant into a more treatable form, causes either complete or partial decomposition of organic compounds	Contaminated soil is heated to temp of 200–1,000 °F. The contaminants with low boiling point get vaporize or desorb	\$50–\$330 per tonne	For VOCs, low-temperature thermal desorption is used while higher temperature is mainly used for PAHs. Thermal desorber has been shown to decontaminate crude oil-polluted soil with 90% efficiency	Ezeji et al. (2005)
Chemical methods	Reducing the water/oil interfacial tension thereby promoting dissolution of oil	Dispersants have been routinely used as a response option	\$7,350 per tonne	Light oil removed efficiently, however heavier crude oil fractions tend to be difficult to remove using dispersants resulting in escalation of clean-up costs	Etkin (1999)
Phytoremediation	Involves the use of plants for in situ treatment of polluted sites to extract, immobilize, contain and/or degrade contaminated soil promising technology for removal of petroleum hydrocarbons in situ	Degradation of petroleum hydrocarbons, synergy between microbes and plants present in the rhizosphere is employed	\$75–100 per m^3	Different phytoremediation techniques can be used, such as phytoextraction, phytodegradation, phytostabilization and phytovolatilization	Dhankher et al. (2012)

after bioremediation treatment are generally nontoxic which consist of carbon dioxide, water and cell biomass, whereas physical and chemical methods usually transfer the contaminant from one environmental area to another. Since it is a natural process, it has greater public acceptance (Shukla et al. 2010). In nature, bioremediation can occur either aerobically or anaerobically.

7.8 Mechanism of Bioremediation

7.8.1 Aerobic Process

Aerobic biodegradation involves the breakdown of hydrocarbon by microorganism using oxygen as terminal electron acceptor. Biodegradation of petroleum hydrocarbon takes place at a faster rate under aerobic condition (Fritsche and Hofrichter 2008). For example, enzymes like monooxygenases and dioxygenases catalyse the degradation of aromatic and aliphatic hydrocarbons by the addition of oxygen molecule (Karigar and Rao 2011). Some of the microorganisms that carry out aerobic biodegradation include *Bacillus*, *Corynebacterium*, *Staphylococcus*, *Streptococcus*, *Shigella*, *Alcaligenes*, *Acinetobacter*, *Escherichia*, *Klebsiella* and *Enterobacter* (Kafilzadeh et al. 2011).

The four fractions of crude oil are saturates, aromatics, resins and asphaltene compounds, and among them saturates and aromatic fractions are biodegraded in the marine and in terrestrial environments, while resins and asphaltene are highly resistant to microbial degradation (Bergauer et al. 2005; Yemashova et al. 2007). Degradation of alkanes can occur by terminal oxidation or subterminal oxidation pathway and is classified as terminal or subterminal (van Beilen et al. 2006). In terminal oxidation pathway, the terminal methyl group of alkane gets oxidized to form primary alcohol by enzyme monooxygenase, which is further oxidized to aldehyde and finally into fatty acid by alcohol dehydrogenase and aldehyde dehydrogenase. Fatty acid can further be converted into acetyl CoA by β-oxidation (Rojo 2009), or it can undergo ω-hydroxylation at the terminal methyl group (ω position) producing ω-hydroxy fatty acid by the enzyme ω-fatty acid monooxygenase, which is transformed into dicarboxylic acid and treated by β-oxidation (Coon 2005).This acetyl CoA enters into TCA cycle, which liberates CO_2 and generates energy in the form of ATP (Gandy and Gandy 1980). In subterminal oxidation pathway, alkane is oxidized at terminal position to form secondary alcohol and converted into corresponding ketone (Okoh 2006), which is further oxidize to ester by enzyme Baeyer-Villiger monooxygenase. The ester is hydroxylated to form alcohol and fatty acid (Kotani et al. 2007).

Aromatic degradation: For aromatics the rate of degradation mainly depends on the number of rings present in the molecule. The more the number of aromatic rings, the more resistant the molecule is to microbial degradation (Mrozik et al. 2003). Biodegradation of aromatic hydrocarbon takes place by the cleavage of the aromatic ring into straight-chain acid (Fig. 7.1). These reactions involve the inclusion of molecular oxygen into the ring structure, which is catalysed by the enzyme

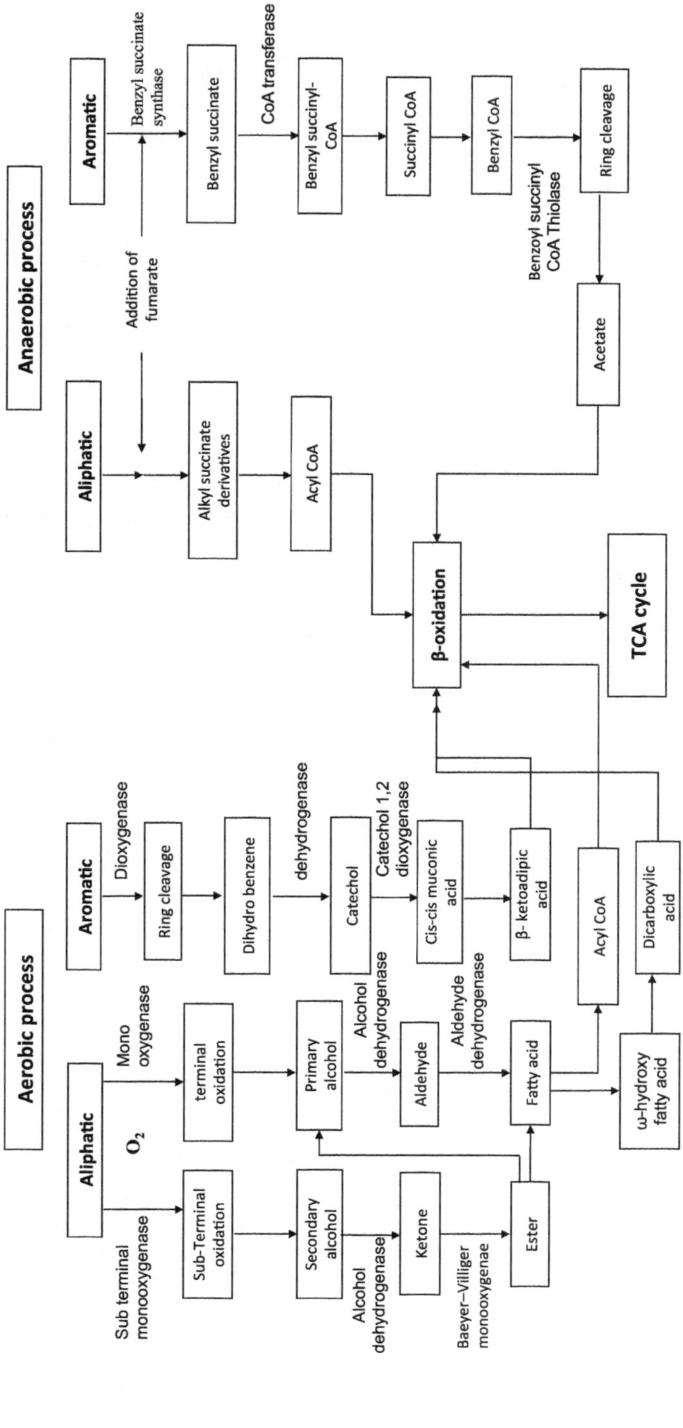

Fig. 7.1 Aerobic and anaerobic hydrocarbon degradation pathway

dioxygenases. The resulting compound dihydrobenzene is further converted to catechol by the action of the enzyme dehydrogenase and then cleaved between the closed hydroxylated carbon atoms to form cis-cis-muconic acid. The muconic acid is further metabolized into β-ketoadipic acid, which undergoes series of reaction and finally converted into succinic acid and acetyl CoA (Fuchs et al. 2011).

7.8.2 Anaerobic Process

Anaerobic hydrocarbon degradation usually occurs in deep and anoxic environments such as submerged soil, aquatic sediments and soil contaminated with natural oil and its residues. In such environments, biodegradation of hydrocarbons is carried out by facultative and obligatory anaerobic microorganisms (archaea and bacteria), which are identified as *Geobacter metallireducens* (Zang et al. 2013), *Desulfobaculatoluolica* Tol2 (Chakraborty and Coates 2004), *Aromatoleum aromaticum* (Wöhlbrand et al. 2007) and *Magnetospirillum magneticum* AMB-1 (Shinoda et al. 2005). These microorganisms use alternate electron acceptor such as Fe (III) (ferric ion reducers), sulphate (sulphate reducers), nitrate (denitrifying organisms) and CO_2 (methanogens), such as chlorate, manganese and chromate (Weelink et al. 2009, 2010). The use of electron acceptors other than oxygen is based on its availability and the competition among different types of microorganisms for electron donors (Lovley 2003). Microbial degradation, especially of recalcitrant hydrocarbons, is limited without the presence of oxygen, as most of the microbes require oxygen as their terminal electron acceptor for degradation. Anaerobic degradation of toluene has been studied in depth; the initial step is the incorporation of fumarate molecule into the methyl group of toluene to form benzyl succinate, which is catalysed by enzyme benzyl succinate synthase (Fig. 7.1). Subsequent reactions result in the oxidation of benzoyl succinate and eventually formation of benzoyl-CoA and acetyl CoA by enzyme CoA-transferase. Benzoyl-CoA, last intermediate of aromatic hydrocarbon metabolism, is then further broken down by β-oxidation (Biegert et al. 1996).

A pathway for hydrocarbon degradation and the nature of bacteria involved in degradation can be either aerobic or anaerobic depending upon the type of terminal electron acceptor available (Olajire and Essien 2014). Aerobic microbes grow quickly with high cell densities, while anaerobic culture requires long time to grow. Under aerobic process, more number of ATP and biomass is generated per molecule of hydrocarbon degraded. Thus, more significant degradation can occur in a short span of time in aerobic condition as compared to anaerobic process.

7.9 Physiochemical Factors Affecting Bioremediation

Although microbial community plays an important role in biodegradation process of petroleum hydrocarbons, the biodegradation rate principally depends on environmental conditions. An effective degradation of crude oil requires simultaneous action of several metabolically active microorganisms with favourable environmental conditions. A number of environmental factors play a significant role in effective hydrocarbon degradation.

7.9.1 Temperature

Temperature plays an imperative role in the petroleum hydrocarbon biodegradation, as it affects hydrocarbon solubility and soil moisture content, invariably, disturbing the microbial flora. Bioavailability of several saturates and aromatics to the microbes for degradation is dependent on temperature (Coulon et al. 2005). The biodegradation rate usually increases with increase in temperature; at low temperatures, viscosity of oil increases and also affects the volatility of highly toxic low-molecular-weight compounds, thus delaying biodegradation (Rahman et al. 2002).Temperatures between 30 and 40 °C produce maximum increase in the hydrocarbon metabolism rate, and beyond that there is an increase in toxicity of hydrocarbons to microbial membrane, leading to microbial cell death and reduction in biodegradation (Coulon et al. 2005). Optimum temperature differs for terrestrial and aquatic sample. In case of soil environments, the highest degradation rate occurs around 30–40 °C, while in aquatic biota like fresh water ecosystem, we witnessed a maximum hydrocarbon degradation at 20–30 °C, while for marine environment it was at 15–20 °C (Das and Chandran 2011).

7.9.2 pH

pH has a significant impact on bioremediation, controlling enzyme activities, transport processes and nutrient solubility (Cerqueira et al. 2011). pH is also one of the most important factors for effective degradation; it influences the metabolic activity and growth of microorganism, since most of the microbes require specific pH to carry out their metabolic activity; identification of optimum pH for microbial growth and activity would help in designing efficient bioremediation strategy (Rousk et al. 2009). The choice of the pH depends on the microorganism to be used for the degradation, for example, fungal strains are found to carry biodegradation at lower pH, while bacteria show higher degradation at pH values between 6.5 and 8.5 (Dibble and Bartha 1979). Maximum crude oil degradation occurs at pH 8.0; beyond it a decreasing efficiency has been observed. This indicates that extreme pH has inhibitory action on majority of the microbial degradation processes.

7.9.3 Salinity

Many microorganisms are capable of growing at high salinity like *Pseudomonas aeruginosa* which showed the highest growth and biodegradation activity at 3.5% salinity (Thavasi et al. 2007). At a NaCl concentration of 2.5%, *Rhodococcus erythropolis* showed better growth and hydrophobicity (de Carvalho and ds Fonseca 2005).

Rate of hydrocarbon degradation decreased if soil salinity increased beyond the optimum level. High salt concentration may decrease the water activity of the surrounding environments, which can in turn retard the cellular activities and growth of bacteria.

7.9.4 Nutrients

Nutrients like nitrogen, potassium, phosphorus and also iron are very important for successful bioremediation of hydrocarbon contaminants. The growth of microorganisms and hydrocarbon degradation can be greatly enhanced by supplementation with inorganic N and P. The optimum nutrient balance ratio of carbon/nitrogen/phosphorus equals 100:10:4 is required for hydrocarbon remediation (Thapa et al. 2012), though it may vary as per the need of the microorganism. Other than these nutrients, some micro- and macronutrients are also required for successful bioremediation (Leys et al. 2005).

7.9.5 Moisture

Soil moisture content can influence the biodegradation rate due to its effects on hydrocarbon bioavailability, diffusion processes, transfer of produced gases, oxygen availability in the soil and soil toxicity level (Borresen and Rike 2007; Yang et al. 2009). Low moisture content inhibits the growth of microorganism, while excess moisture affects the soil aeration decreasing the rate of hydrocarbon degradation. The water activity or water potential (aw) of aquatic environment is stable near 0.98; therefore, available water for microbial growth and metabolism is limited in turn affecting hydrocarbon degradation (Dibble and Bartha 1979).

7.9.6 Soil Type

Most of the contaminants like hydrocarbons tend to adsorb on soil making them unavailable to microorganism for degradation; therefore, in some condition, bioavailability of hydrocarbon also depends on soil type. Oxygen availability for bioremediation processes is affected by permeability, bulk density of the soil and moisture content. Soils like clay, which are fine textured and has low permeability, prevent diffusion of air and nutrients in soil that can affect bioremediation process (Sihag et al. 2014). Soil with granular or porous texture with high permeability enhances the biodegradation rate of hydrocarbon (Bali et al. 2002).

7.9.7 Redox Potential

Petroleum hydrocarbon degradation can occur at both aerobic and anaerobic condition. Redox potential and oxygen content differentiate between oxidizing and reducing conditions (Leahy and Colwell 1990). Based on the dominant electron acceptor, the redox environment condition affects the presence of bacteria in soil. Depending upon the redox potential, a desired reaction that is oxidation or reduction during the course of remediation can be detected. Hydrocarbon degradation under aerobic conditions is significantly faster, in which oxygen acts as electron acceptor; only when oxygen availability in the soil is depleted only then bacteria will utilize nitrate, sulphate or iron as alternate electron acceptor. When a particular electron acceptor has been consumed, the next redox process will be initiated; this sequence will be accompanied by a decline in the redox potentials (Leahy and Colwell 1990).

7.10 Bioremediation of Terrestrial Oil Spill

Bioremediation has proved to be one of the most sustainable oil spill clean-up strategy, but its full potential is yet to be tapped. The efficiency of soil remediation largely depends on the characteristics of soil matrix like porosity, density and air permeability (Tomei and Daugulis 2013). One of the major challenges of bioremediation is the slow outcome particularly the bioremediation of heavy fractions of crude oil. Bioremediation techniques are mainly of two types, in situ and ex situ. In situ bioremediation of contaminated soil is treatment on site, whereas in ex situ bioremediation, contaminated soil is unearthed and is later on transported to a treatment facility (Speight and Arjoon 2012; Fingas 2014).

Various reports are available where process of bioremediation has been studied at different oil-contaminated sites. Effect of different process parameters has been observed by using seven bacterial strains in the north region of the Shaanxi Province in China (Xu et al. 2009). Ex situ bioremediation techniques such as the use of bio-slurry reactor and land farming cells have been used for the management of heavily oil-contaminated soil. The use of slurry-phase biotreatment shows excellent reduction in oil concentration of up to 88% after just 2 months (Kuyukina et al. 2003). Effect of mesophilic (25 °C) and thermophilic temperatures (45 °C) on the process of bioremediation has been studied by Hesnawi and Mogadami (2013). Mesophilic temperature shows degradation up to 79.8%, while 62.4% degradation was observed at thermophilic temperature (Hesnawi and Mogadami 2013). Some of the bacteria reported for petroleum hydrocarbons are reported in Table 7.3.

7.10.1 In Situ Bioremediation

In situ techniques are more cost-effective and prove to be more useful when there is problem of deeper oil contamination (Speight and Arjoon 2012; Fingas 2014). The key initial step for in situ bioremediation is the assessment of site condition by identifying the presence of oleophilic bacteria among the native microbial population and the limiting conditions that may affect the remediation process (Fingas 2014). The execution of in situ bioremediation depends mainly on the subsurface hydrology, the nature of contamination and the extent of contaminated area. In general the productivity of in situ bioremediation is higher when soil subsurface permeability is

Table 7.3 Bacteria reported for petroleum hydrocarbon degradation

S. No.	Bacteria	Compound metabolized	References
1.	*Mycobacterium vanbaalenii* PYR-1	Benzo[a]pyrene	Moody et al. (2004)
2.	*Rhodococcus erythropolis*	Long chain c-alkanes	Adhikari et al. (2015)
	Gordonia terrae	Long chain c-alkanes	Adhikari et al. (2015)
3.	*Pseudomonas putida*	Toluene, naphthalene	Petersen (1996)
4.	*Dechloromonas* strain RCB	Benzene	Chakraborty and Coates (2005)
5.	*Geobacillus thermoleovorans*	n-hexadecane	Marchant et al. (2006)
6.	*Bacillus subtilis* DM-04 and *Pseudomonas aeruginosa* M and NM	Crude oil	Das and Mukherjee (2007)
7.	*Stenotrophomonas maltophilia* strain VUN 10,010	Polycyclic aromatic hydrocarbons	Dandie et al. (2004)
8.	*Methylococcus, Methylosinus, Methylocystis*	Aliphatic and chlorinated alkanes, alkenes and aromatic and chlorinated aromatics	McDonald et al. (2006)

more, the depth of contamination is 8–10 m, and shallow groundwater depth of 10 m is present below the ground surface. Depending on the level of contamination and its depth in the soil, the time required for in situ bioremediation may vary from 12 to 24 months. The advantages of this method are the complete removal of contaminant(s), lower health hazard to site workers and low operational cost, while the disadvantages include slow kinetics extending the degradation process and depth of soil (Speight and Arjoon 2012; Tomei and Daugulis 2013). Some of the in situ technologies are discussed below.

7.10.2 Bioventing

Bioventing is an in situ bioremediation technique where oxygen or air (at low airflow rates) is supplied into the unsaturated zone through gas injection. It is primarily used for removal of mid-weight petroleum compounds like diesel, jet fuel and gasoline (Garima and Singh 2014; Macaulay and Rees 2014). Soil permeability of the contaminated site plays an important role for the successful implementation of this technique (Garima and Singh 2014; Fingas 2014). Bioventing technique is of two types: active and passive. In passive bioventing, oxygen is supplied through atmospheric pressure into the injection wells, whereas in active bioventing, a blower or pump is used for forcing oxygen into the wells.

Figure 7.2 provides a diagrammatic view of bioventing technique. Bulman et al. (1993) reported the use of bioventing technology at a diesel spill site in Australia.

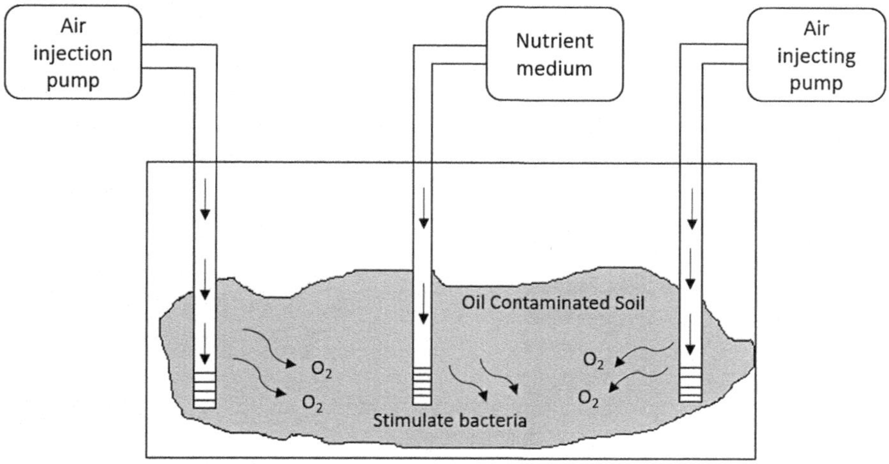

Fig. 7.2 A diagrammatic representation of in situ bioventing technique

After treatment in a 3-m-deep soil, the total hydrocarbon concentration reduced from 10% to 30% after 6 months. After another 6 months, the addition of nutrients further reduced total hydrocarbon concentration to 30% at a depth of 3.5 m (www.hawaii.edu/abrp/Technologies/biovent.html).

Bioventing is limited when the soil is of low permeability, low moisture content and low microbial concentration. Cost of bioventing mainly depends upon the concentration and type of contaminants, soil permeability, pumping rate and off-gas treatment. The approximate cost for bioventing might be around $10–$60/cubic yard; as the area increases for remediation, the cost can be decreased (Aggarwal et al. 1991). This technology does not demand for expensive equipment and has relatively low maintenance.

7.10.3 Biosparging

When there is some risk involved in remediation methods available, biosparging is the preferred option. It removes contaminants at the saturated zones (region below the water table); it also creates a blockade against migration of contaminants off-site (http://horizontaldrill.com). To facilitate aerobic degradation, the existing oleophilic microflora is provided with nutrients and oxygen (under high pressure). Figure 7.3 provides a diagrammatic view of biosparging. Kao et al. (2008) reported the effectiveness of biosparging at a petroleum hydrocarbon spill site contaminated with BTEX. In 10 months, 70% of BTEX was biodegraded using biosparging technology. Biosparging offers a cost-effective and easy to set up technique for removal of mid-weight hydrocarbons, while for polyaromatic hydrocarbons, the process has

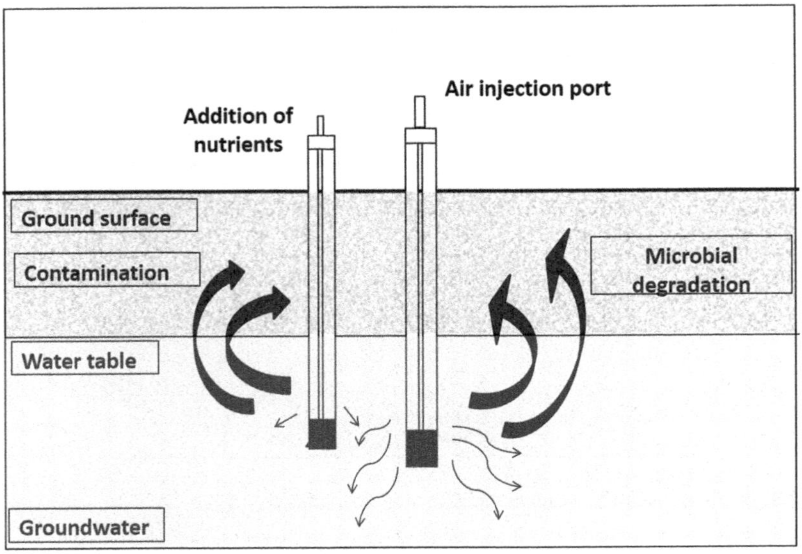

Fig. 7.3 A diagrammatic representation of in situ biosparging technique

been shown to be too slow (Garima and Singh 2014; Macaulay and Rees 2014). Heavier products (e.g. lubricating oils) usually take longer time to biodegrade, but still biosparging can be an option at such sites.

Biosparging in the vadose zone is an efficient method that is cost-effective than SVE or conventional air sparging. For biosparging the flow rates are low, and the costs for blowers, operation and maintenance are less (www.horizontaldrill.com). The costs of implementing biosparging using compressed air injection would be around 12.5–25 euro per m^3.

7.10.4 Bioslurping

It is relatively a recent in situ bioremediation method, combining three different techniques: vacuum-enhanced pumping, soil vapour extraction and bioventing (Macaulay and Rees 2014). Bioslurping simultaneously achieves aerobic biodegradation of the vadose zone by microbes as well as expulsion of liquid saturates (light nonaqueous phase) that are present in the capillary fringe and water table. Figure 7.4 provides a diagrammatic view of bioslurping technique. Petroleum hydrocarbons, like benzene, toluene, xylene, paraffins, naphthalenes, diesel oil and gasoline, were remediated using bioslurping technology (Lee and Lee 2010). Bioslurping is infeasible in fine-textured soils and soil with high moisture content. High moisture content causes reduction in air permeability, thereby decreasing oxygen transfer capability and finally causing inhibition of microbial activity in the contaminated soil. Bioslurping was carried out for the recovery of LNAPL by the US army, which was estimated at around \$56/gal LNAPL recovered. It has successfully been used

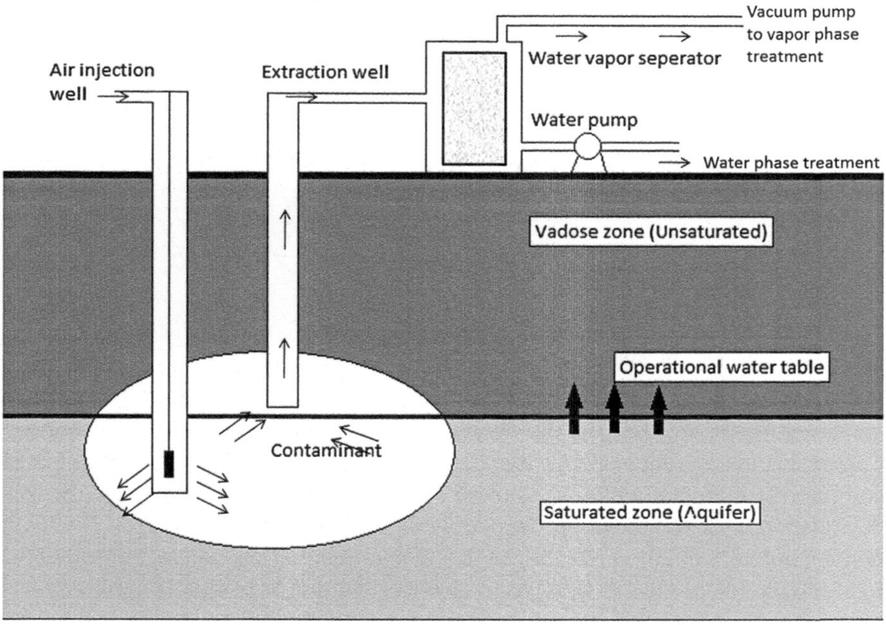

Fig. 7.4 A diagrammatic representation of in situ bioslurping technique

for clean-up of petroleum hydrocarbons and remediation of light nonaqueous phase liquid (LNAPL) pollution (Parker 1995). Previous results from many contaminated sites pointed that LNAPL recovery was better with bioslurping than with any other evaluated techniques.

7.11 Ex Situ Bioremediation

Ex situ bioremediation technologies offer advantages like reaction containment and process control. Therefore, ex situ bioremediation offers a possibility for enhancement of biodegradation kinetics and mass transfer, optimizing operating parameters and choice of additional pretreatments. On the other hand, the disadvantages of ex situ bioremediation are the risk factors related with the unearthing and relocation of the contaminated soil and the overall cost of the treatment (Tomei and Daugulis 2013). Depending on the different degrees of complexity of treatment, ex situ bioremediation can be classified as the following.

7.11.1 Land Farming

One of the most ubiquitously used technologies for petroleum hydrocarbon remediation is land farming (Speight and Arjoon 2012; Tomei and Daugulis 2013;

Fig. 7.5 A diagrammatic representation of land farming technique. Addition of nutrients and tilling will be carried out for efficient bioremediation

Camenzuli and Freidman 2015). Initially, nutrients are applied on the contaminated soil (up to 1 m thick), which is followed by tilling of the soil to accelerate the biodegradation and volatilization of petroleum hydrocarbons (Thapa et al. 2012; Nikolopoulou et al. 2013). Volatilization and microbial activity aid in removal of the lighter fractions of the oil, whereas the heavier fraction is predominantly removed by microbial degradation. Land farming can be customized according to the site-specific conditions like climate, type of soil and temperature. The supplement of nitrogen and phosphorous as well as contaminant degrading allochthonous prokaryotes has been shown to aid in the successful application of land farming (Thapa et al. 2012; Nikolopoulou et al. 2013). Figure 7.5 provides a diagrammatic view of land farming technique. Land farming is especially implied for the disposal of oily sludge and other petroleum refinery wastes. Land farming is inefficient in remediating certain hydrocarbon compounds like creosote, pentachlorophenol (PCP) and bunker C oil (Huang et al. 2004). Adequate monitoring and environmental safeguards are required. The limitation of land farming is that the degradation process is very slow and ineffective in degrading high molar mass PAHs (Macaulay and Rees 2014). Using land farming technique, Marin et al. (2005) reported an 80% biodegradation of total hydrocarbons in 11 months of refinery sludge located in semiarid region. The utilization of simple equipment and operability makes land farming a cost-competent technology (Tomei and Daugulis 2013). Typical costs for land farming range from $30 to $60 per tonne (Cookson 1995); as compared to other remediation technologies, costs of installation, operation and maintenance are relatively inexpensive. For remediation, it might take a period of 6 months up to 2 years, and the duration is more for contaminants with denser hydrocarbons (Hejazi 2002).

7.11.2 Biopiles

Biopile is a hybrid technique for treatment of petroleum hydrocarbons offering a combination of land farming and composting (Speight and Arjoon 2012). In this method, piles containing the contaminated soil are stimulated by addition of

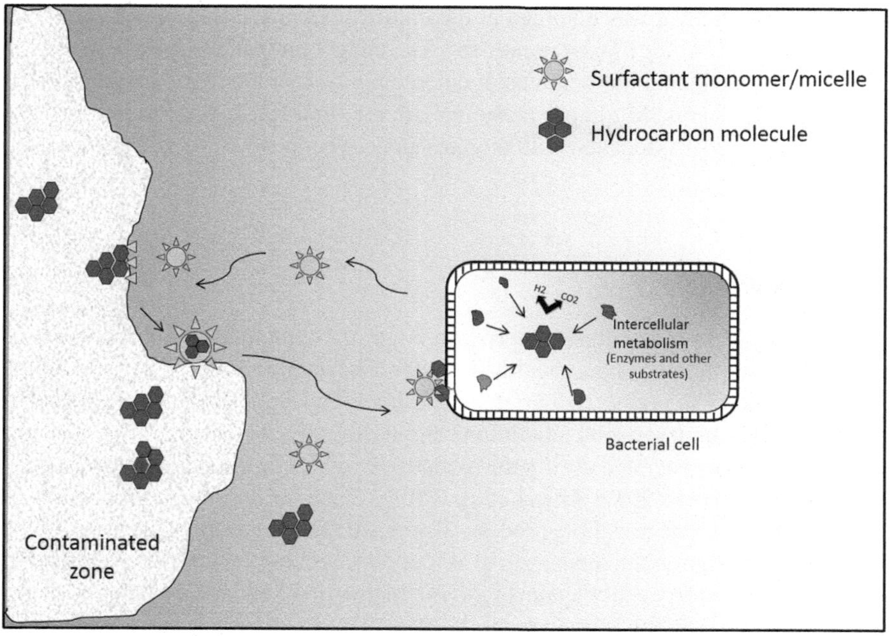

Fig. 7.6 An illustration on mechanism of action of biosurfactants

microbes and/or by addition of nutrients and water (Camenzuli and Freidman 2015; Iturbe and Lopez 2015). Petroleum-contaminated soils having sandy granulometry have especially been treated using biopile technique. The biodegrading activity of microbes aids in reduction of heavy fractions of oil (Iturbe and Lopez 2015). Figure 7.6 provides a diagrammatic view of biopile technique. The efficiency of biopile depends upon parameters such as soil and contaminant characteristics as well as weather conditions. Biopiling is effectively carried out in cold regions like Arctic grounds, where temperature is increased and prolonged over a period, thus optimizing microbial activity (Filler et al. 2008). The nature of soil plays an important role as it influences migration of microbes, as well as permeability of air and nutrients into the soil (Iturbe and Lopez 2015).

Biopiles are efficient in remediating a majority of the petroleum compound contaminants like nonhalogenated and halogenated volatile organic compounds (VOCs), semi-VOCs and pesticides (Singh et al. 2009). Further, biopile can also prevent off-site movement of contaminants and facilitate monitoring of bioremediation process (Filler et al. 2008; Sanscartier et al. 2009; Singh et al. 2009). Major disadvantage of biopiling technique is excess air injection can cause soil drying, thus affecting the microbial activity (Sanscartier et al. 2009). Iturbe and Lopez (2015) reported the application of biopile technology for bioremediation of a hydrocarbon-contaminated site located in Mexico. A 77.7% heavy fraction hydrocarbon removal was obtained at the end of treatment using both biopile and land farming technology (as pretreatment).The budget for using biopiles depends on the nature of pollutant, the treatment method employed or the requirement of emission control systems. Minimum personnel are required for its process and maintenance, and the cost associated with biopile can be around US $130–$260/cubic yard (FRTR 1999).

7.11.3 Composting

Composting is a process that comprises mixing of contaminated soil with organic substances like manure and agricultural wastes. The organic material helps in reducing biodegradation period by enriching the microbial population in the soil. The process occurs under aerobic conditions and utilizes the advantage of the elevated pile temperature generated during the oxidative exothermic biodegradation reaction (Speight and Arjoon 2012; Thapa et al. 2012). This rise in temperature results in more efficient biodegradation process. Composting has several advantages over other bioremediation methods, such as ease of process, cost-effectiveness and higher efficiency especially against bioremediation of PAHs, chlorophenols, petroleum hydrocarbons, etc. (Prakash et al. 2015). Composting can be carried out to contaminated soil with pentachlorophenol (PCP), refinery sludge, insecticides, ethylene glycol in landfill sludges, explosive and polycyclic aromatic (PAHs) (Norris 1994; Cookson 1995). Good-quality compost can be obtained through process optimization, which can be further used as manure or as inoculant for treating other organic contaminated soil. Composting requires large area and excavation of contaminated soils, which may further lead to unrestrained release of volatile compounds. Using fresh biowaste, Van Gestel et al. (2003) reported an 85% degradation of diesel fuel under thermophilic composting, whereas, when mature compost was used, only 35% degradation was recorded. Costs might vary depending upon the amount of soil to be remediated, nature of the contaminant, design to be employed and availability of amendments. Composting systems are of three types: (1) windrow, (2) static pile and (3) in-vessel composting. Windrow composting of explosive-contaminated soils costs approximately $190 per cubic yard (Norris 1994; Cookson 1995), while estimated costs for the other two systems are much higher.

7.12 Enhanced Bioremediation

Bioremediation can also be enhanced by either bioaugmentation, biostimulation and other techniques. While bioaugmentation utilizes the catabolic potential of specialized microbes, biostimulation relies on enhancing the growth of indigenous microbes by supplementation of organic materials to the contaminated site. The deciding factor for implementation of either of these techniques largely depends on the degradation efficiency of the native microbial population and the degree of contamination of the site to be treated.

7.12.1 Bioaugmentation

The indigenous or native microbes may not always be able to restrict or degrade the contaminants as they may be lacking the required gene pool. Bioaugmentation is the introduction of a designed microbial population (or consortia) harbouring the necessary catabolic gene pool to effectively degrade pollutants present on the site (Paliwal et al. 2012; Macaulay and Rees 2014). Bioaugmentation has been shown to be successful in restoring sites contaminated with aromatic pollutants; however, it has its own limitations too. One such problem is the right strain selection. It is important to use strains having the potential to degrade the contaminant and survive the competition with autochthonous bacteria and predation by other organisms. Thus, a priori knowledge of the indigenous microflora inhabiting the site is important. Fast growth, ability to withstand high concentration of contaminant and capability of proliferating wide environmental conditions are also some of the important factors that should be considered for designing microbial bioaugmentation strategies (Mrozik and Piotrowska 2010). Sites that lack sufficient microbial population or the necessary catabolic gene pool are best suited for bioaugmentation. Bioaugmentation not only enhances the metabolic activity of the microbes but also influences the bioavailability of contaminants. This technique can be efficiently used on (1) contaminated sites that possess inadequate microbial biomass and (2) the indigenous microbial population incompetent to completely mineralize the contaminants. Several studies have revealed that bioaugmentation method using specific microbes has been successfully employed in soil contaminated with PAH and diesel oil and oily sludge (Mrozik and Piotrowska 2010; Tahhan et al. 2011).

7.12.2 Biostimulation

Biostimulation is the enhancement of the local microflora by supplementing it with nutrients or electron acceptors/donors and gaseous formulations. Biostimulation technique involves identification and adjustment of factors affecting the biodegradation rate by the autochthonous microorganisms (Tyagi et al. 2011; Macaulay and Rees 2014). The efficiency of biostimulation for degradation of oil depends on the properties of oil, the nutrients and the contaminated site. At a site contaminated with petroleum hydrocarbons, nitrogen and phosphorus are usually limiting, whereas carbon is present in a substantial amount. Thus, at such sites, addition of the rate-limiting nutrient may help in acceleration of decontamination rate by the native microbial population. Environmental factors such as pH and moisture content can also be enhanced to attain optimum microbial degradation (Tyagi et al. 2011). Sarkar et al. (2005) reported that biostimulation with inorganic fertilizers (N and P) and biosolids (organic matter from the treatment of waste) enhanced biodegradation up to 96% of soil contaminated with diesel and petroleum hydrocarbons.

7.12.3 Combined Bioaugmentation and Biostimulation

Bioaugmentation and biostimulation have been shown to enhance biodegradation of hydrocarbons in oil spillage sites. However, both the techniques when applied individually suffer from limitations (Wu et al. 2016). In bioaugmentation the survival and degradation ability of the introduced bacteria/consortia is highly dependent on environmental factors. Biostimulation may help in hastening the process if applied to sites that are oleophilic and microbe rich (Macaulay and Rees 2014). These indigenous microbes can be harvested by culturing and reintroducing them to the contaminated site. Thus, the problem of intolerance by the exogenous microbes to the introduced site may be solved by combining both bioaugmentation and biostimulation strategies. Hamdi et al. (2007) reported better dissipation rates for PAH (anthracene and pyrene) when PAH-contaminated soil was treated simultaneously with biostimulation and bioaugmentation. Wu et al. (2016) reported a microcosm study on an oil-contaminated soil bioaugmented with *Acinetobacter* SZ-1 strain KF453955 and biostimulated with nitrogen and phosphorous. After 6 weeks of incubation, efficiency of removal of total petroleum hydrocarbons (TPH) was 60% for bioaugmentation and 34% for biostimulation.

7.12.4 Biosurfactants

Bioavailability of oil/hydrocarbons to microorganism is one of the rate-limiting factors in bioremediation, thus delaying the process (Fingas 2014; Macaulay and Rees 2014). The use of biosurfactant offers a solution to this issue, as biosurfactant provides an increased surface area of the oil available for microbial degradation. Biosurfactants such as rhamnolipid (produced by *Pseudomonas aeruginosa*) are amphiphilic compounds of microbial origin. The water-retention capacity of oil-polluted sandy soil has been shown to be increased by addition of biosurfactants, thereby enhancing the bioavailability of hydrocarbons to the native oleophilic soil microbes. Figure 7.7 provides an illustration on the mechanism of action of biosurfactants. Biosurfactants have also been reported to reduce the adaptation time of microbes at the contaminated site (Macaulay and Rees 2014).

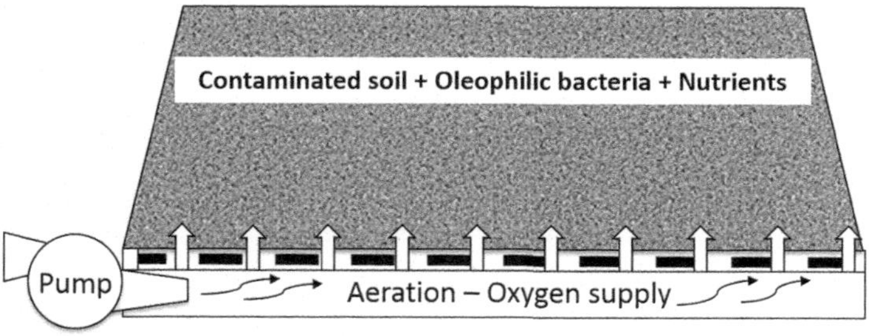

Fig. 7.7 A diagrammatic representation of biopile technique

It has been suggested that a site-specific combination of biosurfactant addition, bioaugmentation and biostimulation may provide an enhanced strategy for speeding up of bioremediation. Therefore, there is a need for more number of field trials as well as cost-efficient industrial scale up for biosurfactant production (Macaulay and Rees 2014).

Surfactants are known to enhance release of contaminants from soil, where a concentration of 1–2% (w/w) is used for washing contaminated soil. Biosurfactants were used to study its enhancement on bioremediation with the control on soil contaminated with 700 mg/kg of octane. Soil amended with biosurfactant was able to reduce 58% of octane concentration compared to the control. Bezza and Chirwa (2015) reported biosurfactant-producing microbes employed in petroleum hydrocarbon-contaminated soil, efficiently degrading creosote. Several studies have reported the effective removal of aromatic hydrocarbons such as PAHs and pentachlorophenol by rhamnolipids with 60–80% efficiency, depending on biosurfactant concentration and its contact time (Reis et al. 2013). Unconventional technologies such as solvents/surfactants or thermal methods for remediation of dense nonaqueous phase liquid (DNAPL) are highly site specific and cost around \$300 – \$600/cubic yard (cfpub.epa.gov). Table 7.4 gives the feasibility of bioremediation techniques and the cost required to carry out each technique.

7.13 To Conclude

Bioremediation is one of the foremost approaches for remediation of petroleum hydrocarbon contaminants in soil, as it utilizes the natural biological mechanisms of microbes to achieve the maximum possible removal of organic contaminants in soil (Das and Chandran 2011). Feasibility studies help us in identifying, screening and selecting remedial technologies for effective site clean-up and also to identify and evaluate alternative technologies for remediation of contaminant, such that it does not pose any risk to the environment and human health. The remediation technology

Table 7.4 Feasibility assessment of in situ and ex situ bioremediation methods

Technology	Process efficiency	Ease of operation	Capital expenses	Applicability	References
In situ					
Bioventing	That can be treated aerobically	Providing air/oxygen to the indigenous microorganisms	\$10–\$60 per cubic yard	Petroleum hydrocarbons, pesticides, solvents and preservatives	Aggarwal et al. (1991)
Biosparging	Creates blockade against migration of contaminants off-site	Enriches with oxygen, which stimulates the microbial break down of contaminants	12.5–25 euro per m^3	Mid-weight petroleum products like diesel fuel and jet fuel, lighter petroleum products like gasoline	Muehlberger et al. (1997) and Brown et al. (1996)
Bioslurping	Efficiently removes free-floating petroleum hydrocarbons (free product)	Adapts vacuum-enhanced dewatering method to remediate hydrocarbon-contaminated sites	\$56/gal LNAPL	Remediation of light nonaqueous phase liquid (LNAPL) pollution	Parker (1995)
Ex situ					
Land farming	Treating petroleum hydrocarbons, especially the lighter hydrocarbons	Contaminated soils, sediments or sludge are incorporated into the soil surface and aerated periodically by tilling to enhance microbial activity	\$30–\$60 per ton	Land farming is especially implied for the disposal of oily sludge and other petroleum refinery wastes	Hejazi (2002)
Biopiling	Effectively carried out in cold regions, where temperature is increased thus optimizing microbial activity	Enhancing the soil thermally and therefore increasing the microbiological activity of the contaminated soil	\$130–\$260 per cubic yard	Efficient in remediating most petroleum compounds as well as pesticides, nonhalogenated VOCs, halogenated VOCs and SVOCs	USEPA (1998) and FRTR (1999)

Composting	Nutrients supplied as readily degradable organic wastes like manure, yard wastes and food processing wastes that enhance microbial activity	Low capital and operating costs, ease of operation and also with relatively high treatment efficiency	190 per cubic yard	Applied to soils and sediments contaminated with pentachlorophenol (PCP), refinery sludge, insecticides, explosive-contaminated soil, ethylene glycol and polycyclic aromatic (PAHs)	Cookson (1995) and Norris (1994)
Bioaugmentation	Not only enhance the metabolic activity of the microbes but also influence the bioavailability of contaminants	Addition of microbial communities that can effectively degrade contaminant is introduced to the cleaning site	NA	Successfully employed in soil contaminated with PAH and diesel oil	Tahhan et al. (2011), Paliwal et al. (2012), and Macaulay and Rees (2014)
Biostimulation	Biostimulation with the addition of fertilizers has been shown to increase the number and activity of microbial populations, thus enhancing degradation in soils	Addition of nutrients, electron acceptor and oxygen can enhance the indigenous bacterial efficiency in bioremediation	NA	Biodegradation of petroleum hydrocarbons enhanced by up to 96% after the addition of biosolids and inorganic fertilizers	Sarkar et al. (2005)
Biosurfactants	Surfactants have the potential to enhance desorption of contaminants from soil; usually, 1–2% (w/w) of surfactant is used for washing contaminant soil	Aid biodegradation of hydrocarbons by increasing its bioavailability	$300–$600 per cubic yard	Biosurfactant-producing microbes employed in petroleum hydrocarbon-contaminated soil efficiently degraded creosote	Bezza and Chirwa (2015)
					Garcia-junco et al. (2003) and Mulligan and Eftekhari (2003)
				Polycyclic aromatic hydrocarbons (PAHs) and pentachlorophenol	

NA not available

to be employed ultimately depends upon its cost-effectiveness, the type of contaminant and also the time taken to achieve complete clean-up of the contaminants. Based on these criteria, the remediation technologies may vary from site to site; however, bioremediation technique is perhaps one of the most environmental friendly approaches especially for remediation of terrestrial oil spills.

Acknowledgements Pradnya Nagkirti is thankful to UGC for providing Junior Research Fellowship.

References

Adhikari D, Araki KS, Mukai M, Kai T, Kubota K, Kawagoe T, Kubo M (2015) Development of efficient bioremediation system for petroleum hydrocarbon contaminated soil based on hydrocarbon degrading bacteria and organic matter control. Austin J Biotechnol Bioeng 2:1048

Aggarwal PK, Means JL, Hinchee RE (1991) Formulation of nutrient solutions for in situ bioremediation. In: Hinchee RE, Olfenbuttel RF (eds) In situ bioreclamation: applications and investigations for hydrocarbon and contaminated site remediation. Butterworth-Heinemann, Stoneham, pp 51–66

Albergaria JT, Maria da Conceição M, Delerue-Matos C (2012) Remediation of sandy soils contaminated with hydrocarbons and halogenated hydrocarbons by soil vapour extraction. J Environ Manag 104:195–201. https://doi.org/10.1016/j.jenvman.2012.03.033

Araruna JT, Portes VLO, Soares APL, Silva MG, Sthel MS, Schramm DU, Vargas H (2004) Oil spills debris clean up by thermal desorption. J Hazard Mater 110:161–171. https://doi.org/10.1016/j.jhazmat.2004.02.054

Arwidsson Z, Elgh-Dalgren K, von Kronhelm T, Sjöberg R, Allard B, van Hees P (2010) Remediation of heavy metal contaminated soil washing residues with amino polycarboxylic acids. J Hazard Mater 173:697–704. https://doi.org/10.1016/j.jhazmat.2009.08.141

Bali G, Rallapalli R, Sullia SB, Shiralipour A, Kasturi S (2002) Environmental biotechnology: concepts, definitions and criteria Environmental biotechnology. APH Publishing Corporation, New Delhi, pp 1–29

Bergauer P, Fonteyne PA, Nolard N, Schinner F, Margesin R (2005) Biodegradation of phenol and phenol-related compounds by psychrophilic and cold-tolerant alpine yeasts. Chemosphere 59:909–918. https://doi.org/10.1016/j.chemosphere.2004.11.011

Bezza FA, Chirwa EMN (2015) Production and applications of lipopeptide biosurfactant for bioremediation and oil recovery by *Bacillus subtilis* CN2. Biochem Eng J 101:168–178. https://doi.org/10.1016/j.bej.2015.05.007

Biegert T, Fuchs G, Heider J (1996) Evidence that anaerobic oxidation of toluene in the denitrifying bacterium *Thauera aromatica* is initiated by formation of benzyl succinate from toluene and fumarate. Eur J Biochem 238:661–668. https://doi.org/10.1111/j.1432-1033.1996.0661w.x

Borresen MH, Rike AG (2007) Effects of nutrient content, moisture content and salinity on mineralization of hexadecane in an Arctic soil. Cold Reg Sci Technol 48:129–138. https://doi.org/10.1016/j.coldregions.2006.10.006

Brooijmans RJ, Pastink MI, Siezen RJ (2009) Hydrocarbon-degrading bacteria: the oil-spill clean-up crew. Microb Biotechnol 2:587–594. https://doi.org/10.1111/j.1751-7915.2009.00151.x

Brown RA, Hinchee RE, Norris RD, Wilson JT (1996) Bioremediation of petroleum hydrocarbons: a flexible, variable speed technology. Remed J 6:95–109. https://doi.org/10.1002/rem.3440060308

Bulman TL, Newland M, Wester A (1993) In situ bioventing of a diesel fuel spill. Hydrol Sci J 38:297–308. https://doi.org/10.1080/02626669309492676

Camenzuli D, Freidman BL (2015) On-site and in situ remediation technologies applicable to petroleum hydrocarbon contaminated sites in the Antarctic and Arctic. Polar Res 34:24492. https://doi.org/10.3402/polar.v34.24492

Cerqueira VS, Hollenbach EB, Maboni F, Vainstein MH, Camargo FA, Maria do Carmo RP, Bento FM (2011) Biodegradation potential of oily sludge by pure and mixed bacterial cultures. Bioresour Technol 102:11003–11010. https://doi.org/10.1016/j.biortech.2011.09.074

Chakraborty R, Coates JD (2004) Anaerobic degradation of monoaromatic hydrocarbons. Appl Microbiol Biotechnol 64:437–446. https://doi.org/10.1007/s00253-003-1526-x

Chakraborty R, Coates JD (2005) Hydroxylation and carboxylation—two crucial steps of anaerobic benzene degradation by *Dechloromonas* strain RCB. Appl Environ Microbiol 71:5427–5432. https://doi.org/10.1128/AEM.71.9.5427-5432.2005

Chandra S, Sharma R, Singh K, Sharma A (2013) Application of bioremediation technology in the environment contaminated with petroleum hydrocarbon. Ann Microbiol 63:417–431. https://doi.org/10.1016/j.wneu.2012.12.019

Cookson JT Jr (1995) Bioremediation engineering: design and application. McGraw-Hill, New York

Coon MJ (2005) Omega oxygenases: nonheme-iron enzymes and P450 cytochromes. Biochem Biophys Res Commun 338:378–385. https://doi.org/10.1016/j.bbrc.2005.08.169

Coulon F, Pelletier E, Gourhant L, Delille D (2005) Effects of nutrient and temperature on degradation of petroleum hydrocarbons in contaminated sub-Antarctic soil. Chemosphere 58:1439–1448. https://doi.org/10.1016/j.chemosphere.2004.10.007

Dandie CE, Thomas SM, Bentham RH, McClure NC (2004) Physiological characterization of Mycobacterium sp. strain 1B isolated from a bacterial culture able to degrade highmolecular-weight polycyclic aromatic hydrocarbons. J Appl Microbiol 97:246–255. https://doi.org/10.1111/j.1365-2672.2004.02087.x

Das N, Chandran P (2011) Microbial degradation of petroleum hydrocarbon contaminants: an overview. Biotechnol Res Int 2011:941810. https://doi.org/10.4061/2011/941810

Das K, Mukherjee AK (2007) Crude petroleum-oil biodegradation efficiency of Bacillus subtilis and Pseudomonas aeruginosa strains isolated from a petroleum-oil contaminated soil from North-East India. Bioresour Technol 98:1339–1345. https://doi.org/10.1016/j.biortech.2006.05.032

de Carvalho CC, ds Fonseca MMR (2005) Degradation of hydrocarbons and alcohols at different temperatures and salinities by *Rhodococcus erythropolis* DCL14. FEMS Microbiol Ecol 51:389–399. https://doi.org/10.1016/j.femsec.2004.09.010

Dhankher OP, Pilon-Smits EA, Meagher RB, Doty S (2012) Biotechnological approaches for phytoremediation. Plant biotechnology and agriculture. Elsevier, Amsterdam, pp 309–328

Dibble JT, Bartha R (1979) Effect of environmental parameters on the biodegradation of oil sludge. Appl Environ Microbiol 37:729–739

Etkin DS (1999) Estimating cleanup costs for oil spills. In: International oil spill conference. American Petroleum Institute, pp 35–39

Ezeji EU, Anyanwu BN, Ibekwe VI, Onyeze GOC (2005) Studies on the utilization of petroleum hydrocarbon by microorganisms isolated from oil-contaminated soil. Int J Nat Appl Sci 1:122–128. https://doi.org/10.4314/ijonas.v1i2.36073

Fericelli PD (2011) Comparison of sludge treatment by gasification vs. incineration. In: Ninth LACCEI Latin American and Caribbean Conference (LACCEI'2011), engineering fora smart planet, innovation, information technology and computational tools for sustainable development

Filler DM, Barnes DL, Johnson RA, Snape I (2008) Thermally enhanced bioremediation and integrated systems. In: Filler DM, Snape I, Barnes DL (eds) Bioremediation of petroleum hydrocarbons in cold regions. Cambridge University Press, Cambridge

Fingas M (2012) The basics of oil spill cleanup. CRC Press, Boca Raton

Fingas M (ed) (2014) Handbook of oil spill science and technology. Wiley, Hoboken

Fritsche W, Hofrichter M (2008) Aerobic degradation by microorganisms. In: Biotechnology set, 2nd edn. Wiley-VCH Verlag GmbH, Weinheim, pp 144–167

FRTR (1999) Biopiles. Federal Remediation Technologies Roundtable,USEPA, Washington, DC. http://www.frtr.gov/matrix2/section4/4_12.html

Fuchs G, Boll M, Heider J (2011) Microbial degradation of aromatic compounds—from one strategy to four. Nat Rev Microbiol 9:803–816. https://doi.org/10.1038/nrmicro2652

Gandy AF, Gandy ET (1980) Microbiology for environmental scientists and engineers. McGraw-Hill, New York

Garcia-junco M, Gomez-lahoz C, Niqui-arroyo JL, Ortega-calvo JJ (2003) Biosurfactant-and biodegradation-enhanced partitioning of polycyclic aromatic hydrocarbons from nonaqueous-phase liquids. Environ Sci Technol 37:2988–2996. https://doi.org/10.1021/es020197q

Garima T, Singh SP (2014) Application of bioremediation on solid waste management: a review. J Bioremed Biodegr 5:248. https://doi.org/10.4172/2155-6199.1000248

Greenberg B, Gerwing P (2009) Phytoremediation. Hazard Mater Manag:27–31

Hamdi H, Benzarti S, Manusadžianas L, Aoyama I, Jedidi N (2007) Bioaugmentation and biostimulation effects on PAH dissipation and soil ecotoxicity under controlled conditions. Soil Biol Biochem 39:1926–1935. https://doi.org/10.1016/j.soilbio.2007.02.008

Hejazi RF (2002) Oily sludge degradation study under arid conditions using a combination of Landfarm and Bioreactor Technologies. PhD thesis, Faculty of Engineering and Applied Science, Memorial University of Newfoundland, St John's, Canada

Hesnawi RM, Mogadami FS (2013) Bioremediation of Libyan crude oil-contaminated soil under mesophilic and thermophilic conditions. APCBEE Procedia 5:82–87. https://doi.org/10.1016/j.apcbee.2013.05.015

Hou J, Liu W, Wang B, Wang Q, Luo Y, Franks AE (2015) PGPR enhanced phytoremediation of petroleum contaminated soil and rhizosphere microbial community response. Chemosphere 138:592–598. https://doi.org/10.1016/j.chemosphere.2015.07.025

Huang XD, El-Alawi Y, Penrose DM, Glick BR, Greenberg BM (2004) A multi-process phytoremediation system for removal of polycyclic aromatic hydrocarbons from contaminated soils. Environ Pollut 130:465–476. https://doi.org/10.1016/j.envpol.2003.09.031

Imamura T, Kozaki S, Kuriyama A, Kawaguchi M, Touge Y, Yano T, Kawabata Y (1997) Bioaugmentation of TCE-contaminated soil with inducer-free microbes. In: Biotechnology in the sustainable environment. Springer US, Boston, pp 97–106. https://doi.org/10.1007/978-1-4615-5395-3_10

Iturbe R, López J (2015) Bioremediation for a soil contaminated with hydrocarbons. J Pet Environ Biotechnol 6:208. https://doi.org/10.4172/2157-7463.1000208

Kafilzadeh F, Sahragard P, Jamali H, Tahery Y (2011) Isolation and identification of hydrocarbons degrading bacteria in soil around Shiraz refinery. Afr J Microbiol Res 5:3084–3089

Kao CM, Chen CY, Chen SC, Chien HY, Chen YL (2008) Application of in situ biosparging to remediate a petroleum-hydrocarbon spill site: field and microbial evaluation. Chemosphere 70:1492–1499. https://doi.org/10.1016/j.chemosphere.2007.08.029

Karigar CS, Rao SS (2011) Role of microbial enzymes in the bioremediation of pollutants: a review. Enzyme Res 2011:805187. https://doi.org/10.4061/2011/805187

Khan FI, Husain T, Hejazi R (2004) An overview and analysis of site remediation technologies. J Environ Manag 71:95–122. https://doi.org/10.1016/j.jenvman.2004.02.003

Khashayar T, Mahsa T (2010) Biodegradation potential of petroleum hydrocarbons by bacterial diversity in soil. World Appl Sci J 8:750–755

Kotani T, Yurimoto H, Kato N, Sakai Y (2007) Novel acetone metabolism in a propane-utilizing bacterium, *Gordonia* sp. strain TY-5. J Bacteriol 189:886–893. https://doi.org/10.1128/JB.01054-06

Kuyukina MS, Ivshina IB, Ritchkova MI, Philp JC, Cunningham CJ, Christofi N (2003) Bioremediation of crude oil-contaminated soil using slurry-phase biological treatment and land farming techniques. Soil Sediment Contam 12:85–99. https://doi.org/10.1080/713610962

Kvenvolden KA, Cooper CK (2003) Natural seepage of crude oil into the marine environment. Geo-Mar Lett 23:140–146. https://doi.org/10.1007/s00367-003-0135-0

Leahy JG, Colwell RR (1990) Microbial degradation of hydrocarbons in the environment. Microbiol Rev 54:305–315

Lee H, Lee YJ (2010) A field scale study of serial treatment using slurping and in situ soil flushing to remediate an oil contaminated site. Asian J Chem 22:1535–1549

Leys NM, Bastiaens L, Verstraete W, Springael D (2005) Influence of the carbon/nitrogen/phosphorus ratio on polycyclic aromatic hydrocarbon degradation by *Mycobacterium* and *Sphingomonas* in soil. Appl Microbiol Biotechnol 66:726–736. https://doi.org/10.1007/s00253-004-1766-4

Liu W, Luo Y, Teng Y, Li Z, Ma LQ (2010) Bioremediation of oily sludge-contaminated soil by stimulating indigenous microbes. Environ Geochem Health 32:23–29. https://doi.org/10.1007/s10653-009-9262-5

Lovley DR (2003) Cleaning up with genomics: applying molecular biology to bioremediation. Nat Rev Microbiol 1:35–44. https://doi.org/10.1038/nrmicro731

Macaulay BM, Rees D (2014) Bioremediation of oil spills: a review of challenges for research advancement. Ann Environ Sci 8:9–37

Marchant R, Sharkey FH, Banat IM, Rahman TJ, Perfumo A (2006) The degradation of n-hexadecane in soil by thermophilic *Geobacilli*. FEMS Microbiol Ecol 56:44–54. https://doi.org/10.1111/j.1574-6941.2006.00061.x

Marin JA, Hernandez T, Garcia C (2005) Bioremediation of oil refinery sludge by landfarming in semiarid conditions: influence on soil microbial activity. Environ Res 98:185–195. https://doi.org/10.1016/j.envres.2004.06.005

McDonald IR, Miguez CB, Rogge G, Bourque D, Wendlandt KD, Groleau D, Murrell JC (2006) Diversity of soluble methane monooxygenase-containing methanotrophs isolated from polluted environments. FEMS MicrobiolLett 255:225–232. https://doi.org/10.1111/j.1574-6968.2005.00090.x

Moody JD, Freeman JP, PP F, Cerniglia CE (2004) Degradation of benzo [a] pyrene by *Mycobacterium vanbaalenii* PYR-1. Appl Environ Microbiol 70:340–345. https://doi.org/10.1128/AEM.70.1.340-345.2004

Mrozik A, Piotrowska-Seget Z (2010) Bioaugmentation as a strategy for cleaning up of soils contaminated with aromatic compounds. Microbiol Res 165(5):363–375. https://doi.org/10.1016/j.micres.2009.08.001

Mrozik A, Piotrowska-Seget Z, Labuzek S (2003) Bacterial degradation and bioremediation of polycyclic aromatic hydrocarbons. Pol J Environ Stud 12:1

Muehlberger EW, Harris K, Hicks P (1997) In situ biosparging of a large scale dissolved petroleum hydrocarbon plume at a Southwest Lumber mill. In: TAPPI proceedings of environmental conference and exhibition, vol 1

Mulligan CN, Eftekhari F (2003) Remediation with surfactant foam of PCP-contaminated soil. Eng Geol 70:269–279. https://doi.org/10.1016/S0013-7952(03)00095-4

Nikolopoulou M, Pasadakis N, Norf H, Kalogerakis N (2013) Enhanced ex situ bioremediation of crude oil contaminated beach sand by supplementation with nutrients and rhamnolipids. Mar Pollut Bull 77:37–44. https://doi.org/10.1016/j.marpolbul.2013.10.038

Norris RD (1994) Handbook of bioremediation. CRC Press, Boca Raton

Okoh AI (2006) Biodegradation alternative in the cleanup of petroleum hydrocarbon pollutants. Biotechnol Mol Biol Rev 1:38–50

Olajire AA, Essien JP (2014) Aerobic degradation of petroleum components by microbial consortia. J Pet Environ Biotechnol 5:195. https://doi.org/10.4172/2157-7463.1000195

Paliwal V, Puranik S, Purohit HJ (2012) Integrated perspective for effective bioremediation. Appl Biochem Biotechnol 166:903–924. https://doi.org/10.1007/s12010-011-9479-5

Parker J (1995) Bioslurping enhances free product recovery. Soil Groundw Cleanup, October

Perelo LW (2010) Review: in situ and bioremediation of organic pollutants in aquatic sediments. J Hazard Mater 177:81–89. https://doi.org/10.1016/j.jhazmat.2009.12.090

Prakash V, Saxena S, Sharma A, Singh S, Singh SK (2015) Treatment of oil sludge contamination by composting. J Bioremed Biodegr 6:284. https://doi.org/10.4172/2155-6199.1000284

Prendergast DP, Gschwend PM (2014) Assessing the performance and cost of oil spill remediation technologies. J Clean Prod 78:233–242. https://doi.org/10.1016/j.jclepro.2014.04.054

Rahman KSM, Thahira-Rahman J, Lakshmanaperumalsamy P, Banat IM (2002) Towards efficient crude oil degradation by a mixed bacterial consortium. Bioresour Technol 85:257–261. https://doi.org/10.1016/S0960-8524(02)00119-0

Reis RS, Pacheco GJ, Pereira AG, Freire DM (2013) Biosurfactants: production and applications. Biodegradation: life of science, ISBN 978–53

Rojo F (2009) Degradation of alkanes by bacteria. Environ Microbiol 11:2477–2490. https://doi.org/10.1111/j.1462-2920.2009.01948.x

Rousk J, Brookes PC, Bååth E (2009) Contrasting soil pH effects on fungal and bacterial growth suggest functional redundancy in carbon mineralization. Appl Environ Microbiol 75:1589–1596. https://doi.org/10.1128/AEM.02775-08

Sanscartier D, Zeeb B, Koch I, Reimer K (2009) Bioremediation of diesel-contaminated soil by heated and humidified biopile system in cold climates. Cold Reg Sci Technol 55:167–173. https://doi.org/10.1016/j.coldregions.2008.07.004

Sarkar D, Ferguson M, Datta R, Birnbaum S (2005) Bioremediation of petroleum hydrocarbons in contaminated soils: comparison of biosolids addition, carbon supplementation, and monitored natural attenuation. Environ Pollut 136:187–195. https://doi.org/10.1016/j.envpol.2004.09.025

Shinoda Y, Akagi J, Uchihashi Y, Hiraishi A, Yukawa H, Yurimoto H, Kato N (2005) Anaerobic degradation of aromatic compounds by Magnetospirillum strains: isolation and degradation genes. Biosci Biotechnol Biochem 69:1483–1491. https://doi.org/10.1271/bbb.69.1483

Shukla KP, Singh NK, Sharma S (2010) Bioremediation: developments, current practices and perspectives. Genet Eng Biotechnol J 3:1–20

Sihag S, Pathak H, Jaroli DP (2014) Factors affecting the rate of biodegradation of polyaromatic hydrocarbons. Int J Pure App Biosci 2:185–202

Singh A, Kuhad RC, Ward OP (2009) Biological remediation of soil: an overview of global market and available technologies. In: Advances in applied bioremediation. Springer, Berlin, pp 1–19. https://doi.org/10.1007/978-3-540-89621-0_1

Speight JG (2015) Handbook of petroleum product analysis. Wiley, New York

Speight JG, Arjoon KK (2012) Bioremediation of petroleum and petroleum products. Wiley, New York

Tahhan RA, Ammari TG, Goussous SJ, Al-Shdaifat HI (2011) Enhancing the biodegradation of total petroleum hydrocarbons in oily sludge by a modified bioaugmentation strategy. Int Biodeterior Biodegrad 65:130–134. https://doi.org/10.1016/j.ibiod.2010.09.007

Thapa B, Kc AK, Ghimire A (2012) A review on bioremediation of petroleum hydrocarbon contaminants in soil. Kathmandu Univ J Sci Eng Technol 8:164–170. https://doi.org/10.3126/kuset.v8i1.6056

Thavasi R, Jayalakshmi S, Balasubramanian T, Banat IM (2007) Effects of salinity, temperature, Ph and crude oil concentration on biodegradation of crude oil by *Pseudomonas aeruginosa*. J Biol Environ Sci 1:51–57

Tomei MC, Daugulis AJ (2013) *Ex situ* bioremediation of contaminated soils: an overview of conventional and innovative technologies. Crit Rev Environ Sci Technol 43:2107–2139. https://doi.org/10.1080/10643389.2012.672056

Tyagi M, da Fonseca MMR, de Carvalho CC (2011) Bioaugmentation and biostimulation strategies to improve the effectiveness of bioremediation processes. Biodegradation 22:231–241. https://doi.org/10.1007/s10532-010-9394-4

Ukpaka CP, Amadi SA (2014) Evaluation of low temperature thermal desertion (LTTD) method of treating cuttings. Am Multidiscip Int Res J 2(1):1–9

USEPA (1998) Biopiles Office of the Underground Storage Tank, US Environmental Protection Agency. Publication # EPA 510-B-95-007. http://www.epa.gov/swerust1/cat/biopiles.htm

van Beilen JB, Funhoff EG, van Loon A, Just A, Kaysser L, Bouza M, Witholt B (2006) Cytochrome P450 alkane hydroxylases of the CYP153 family are common in alkanedegrading eubacteria lacking integral membrane alkane hydroxylases. Appl Environ Microbiol 72:59–65. https://doi.org/10.1128/AEM.72.1.59-65.2006

Van Gestel K, Mergaert J, Swings J, Coosemans J, Ryckeboer J (2003) Bioremediation of diesel oil-contaminated soil by composting with biowaste. Environ Pollut 125:361–368. https://doi.org/10.1016/S0269-7491(03)00109-X

Wang J, Zhang Z, Su Y, He W, He F, Song H (2008) Phytoremediation of petroleum polluted soil. Pet Sci 5:167–171. https://doi.org/10.1007/s12182-008-0026-0

Weelink SA, Van Doesburg W, Saia FT, Rijpstra WI, Röling WF, Smidt H, Stams AJ (2009) A strictly anaerobic beta proteo bacterium Georgfuchsiatoluolica gen. nov., sp. nov. degrades aromatic compounds with Fe (III), Mn (IV) or nitrate as an electron acceptor. FEMS Microbiol Ecol 70:575–585

Weelink SA, van Eekert MH, Stams AJ (2010) Degradation of BTEX by anaerobic bacteria: physiology and application. Rev Environ Sci Biotechnol 9:359–385. https://doi.org/10.1007/s11157-010-9219-2

Wöhlbrand L, Kallerhoff B, Lange D, Hufnagel P, Thiermann J, Reinhardt R, Rabus R (2007) Functional proteomic view of metabolic regulation in "*Aromatoleum aromaticum*"strain EbN1. Proteomics 7:2222–2239. https://doi.org/10.1002/pmic.200600987

Wood LA (2002) Overview of remediation technologies. Terra Resources Ltd, Wolverine. 6 pp. Available online: www.terrawash.com/twp2.htm

Wu M, Dick WA, Li W, Wang X, Yang Q, Wang T, Chen L (2016) Bioaugmentation and biostimulation of hydrocarbon degradation and the microbial community in a petroleumcontaminated soil. Int Biodeterior Biodegrad 107:158–164. https://doi.org/10.1016/j.ibiod.2015.11.019

www.hawaii.edu/abrp/Technologies/biovent.html

www.horizontaldrill.com

www.oilproduction.net/files/Introduction

Xu JL, Huang TL, Tang ZX, Xiao ZQ (2009) Characteristics of highly efficient petroleum degrading bacteria in remediation of oil contaminated soil. Acta Petrolei Sin (Pet Process Sect) 25:570–576

Yang SZ, Jin HJ, Wei Z, He RX, Ji YJ, Li XM, SP Y (2009) Bioremediation of oil spills in cold environments: a review. Pedosphere 19:371–381. https://doi.org/10.1016/S1002-0160(09)60128-4

Yemashova NA, Murygina VP, Zhukov DV, Zakharyantz AA, Gladchenko MA, Appanna V, Kalyuzhnyi SV (2007) Biodeterioration of crude oil and oil derived products: a review. Rev Environ Sci Biotechnol 6:315–337. https://doi.org/10.1007/s11157-006-91188

Zhang T, Tremblay PL, Chaurasia AK, Smith JA, Bain TS, Lovley DR (2013) Anaerobic benzene oxidation via phenol in Geobacter metallireducens. Appl Environ Microbiol 79:7800–7806. https://doi.org/10.1128/AEM.03134-13

Role of Clostridial Nitroreductases in Bioremediation

8

Razia Kutty and George N. Bennett

Abstract

The use of clostridial enzymes in industry, environment, medical and economical aspects has been of immense significance and is well documented. Over decades the research has been focused on developing efficient strains and modified enzymes to be used in the bio-butanol production. The research interest in clostridial nitroreductase enzyme has increased recently due to its application in cancer treatment and bioremediation. To date, the report on the nitroreductases from *Clostridium* species is scarce. The function of these nitroreductases in bioremediation has been demonstrated. The structure of nitroreductases from *Escherichia coli* and *Bacillus subtilis* has been determined; however, very little documentation is available on clostridial nitroreductases. The information on the nitroreductase genes and proteins has become available from genomic and protein databases. However the significance of clostridial nitroreductases in its metabolism is not well understood. In this chapter, we discuss the findings on clostridial nitroreductases on the avenue for bioremediation and its applications in medical and industrial areas. The literature also aims to have comprehensive analysis of clostridial physiology based on the existing information.

Keywords

Bacillus • *Clostridium* • Biogas • Bioremediation • Digestion • Environment

---

R. Kutty (✉)
Department of Microbiology, Savitribai Phule Pune University, Ganeshkhind, Pune, Maharashtra, India
e-mail: razkutty@yahoo.co.in

G.N. Bennett
Department of Biochemistry & Cell Biology MS-140, Rice University, Houston, TX 77005-1892, USA

H.J. Purohit et al. (eds.), *Optimization and Applicability of Bioprocesses*,
https://doi.org/10.1007/978-981-10-6863-8_8

8.1 Introduction

Anaerobic microbiology is the main principle on which the anaerobic digestion processes and/or energy generation in the form of biogas works. The biological agents involved in this process are microorganisms (Boopathy et al. 1993, 1994; Bradley et al. 1997; Batstone and Virdis 2014; Bohutskyi et al. 2015; FitzGerald et al. 2015; Fuess and Garcia 2015). The fermentation process involved in industry and domestic purposes is also based on anaerobic digestion. The development of "omics" approaches and its integration with existing information would aid in further widening the scope of anaerobic microorganisms.

The clostridia are strictly anaerobic gram-positive bacteria characterized by their ability to form spores. Most species of the genus *Clostridium* occur at many places in the environment, particularly the soil. This genus includes both pathogenic and non-pathogenic species. *Clostridium acetobutylicum* was first isolated from corn in 1912 by Weizmann and has immense commercial applications. Clostridia possess fermentive potential and have been known for decades for production of solvents (Kim and Zeikus 1985; Jones and Woods 1986; Xie et al. 2015; Reeve and Reid 2016). Clostridia are microorganisms of enormous importance in industrial, medical and environmental applications. The genus *Clostridium* is widespread particularly in the soil environment. Intensive research focus is laid on investigations on their physiology to gain advantage in its potential applications. Currently it is also the focus of metabolic engineering as biochemistry and genetics have been well studied.

Many anaerobic bacteria have been used for nitroaromatic bioremediation measures in the environment as the nitroaromatic compounds undergo complete reduction to less toxic compounds under the action of anaerobic microorganism, viz. clostridia sp. (Daun et al. 1998; Lenke et al. 1998; Kitts et al. 2000; Esteve-Núñez et al. 2001; Li et al. 2014). *Clostridium acetobutylicum* has been reported to biotransform nitroaromatic explosives such as trinitrotoluene (Padda et al. 2003; Watrous et al. 2003; Kutty and Bennett 2005). The nitroreductase activity of clostridia is attributed to various enzymes. The main enzymes catalysing this reaction are hydrogenase and other oxidoreductases like pyruvate-ferredoxin and NADH-ferredoxin oxidoreductases. These enzymes are classified as ferredoxin-reducing enzyme and differ from oxygen insensitive and oxygen sensitive of aerobic and facultative bacteria in structure and mechanism (Angermaier et al. 1981; Angermier and Simon1983).

Nitroreductases and clostridial spores possess wide applications in the field of cancer prodrug therapy. Clostridial promoters and engineered spores from non-pathogenic *Clostridium sporogenes* have also been used to overexpress the prodrug-activating genes leading to improved antitumour activity (Theys et al. 2006; Liu et al. 2008). The tumour regression by *Clostridium*-directed enzyme prodrug therapy (CDEPT) which has been demonstrated in mouse xenograft model of human colon carcinoma indicates the advances in clinical development of CDEPT (Heap et al. 2014). Prodrugs are also chemically designed to achieve a higher rate of activation and sensitivity by the nitroreductases.

This chapter elaborates on the significance of nitroreductases from clostridia species with emphasis on bioremediation and biotechnological application perspective in order to gain further insights into the physiology of anaerobic *Clostridium* sp.

8.2 Nitroaromatic Compounds Occurrence, Toxicity and Bioremediation

Nitroaromatic compounds are non-natural toxic pollutants with applications in manufacture of diverse compounds like explosives, pesticides, pharmaceuticals, etc. The nitro-group contributes to their recalcitrance and they are degraded slowly in the environment (Ju and Parales 2010). Nitroaromatic explosive compounds 2,4,6-trinitrotoluene (TNT) and 2,4-dinitrotoluene (DNT) occur at contaminated defence sites. 1,3,5-Trinitro-1,3,5-triazine (RDX), cyclotetramethylene tetramine (HMX), some pesticides (atrazine) and anilines also contribute to toxic environmental pollution. Most of these nitroaromatic compounds and their intermediates are demonstrated to be potentially carcinogenic and mutagenic (Padda et al. 2003). These nitroaromatics are mentioned as priority pollutants in the US Environmental Protection Agency records and considered to be fatal to human health and environment. The incineration of organic compounds and photochemical reaction in the atmosphere contribute to these nitroaromatics compounds into the environment where it undergoes complex physical, chemical and biological processes. There is increased research interest currently on eco-friendly and economically feasible strategies to decontaminate these substances.

The bioremediation of nitroaromatic compounds has been well documented over the last two decades by pure cultures as well as mixed cultures of bacteria (Kroger et al. 2004; Caballero et al. 2005; Leungsakul et al. 2006; Mercimek et al. 2013). These processes may occur under aerobic or anaerobic conditions leading to mineralization and reduced intermediate formation, respectively (Watrous et al. 2003; Kutty and Bennett 2006).

The soil contaminated with di- and poly-nitroaromatics requires strict anaerobic conditions for complete removal which necessitates studies towards anaerobic biotransformation of these compounds. Diverse bacterial species have been shown to possess potential to biodegrade these nitroaromatic compounds (Kivisaar 2011, Ryan et al. 2011). The reported strains include *Pseudomonas*, *E.coli* and *Arthrobacter* (Fernández et al. 2009; Mulla et al. 2011; Kumar et al. 2012; Mercimek 2015). The nitroreductases are the major enzymes involved in these processes. These processes are extrapolated to the level of bioreactors. The studies also have included in situ field-scale investigations.

8.3 Nitroreductases from *Clostridium*

Many aerobic bacteria have been reported to possess nitroreductase enzyme with the potential to detoxify nitroaromatics (Kimet al. 2015; Martinez 2015). The nitroreductase enzymes in anaerobic bacteria reduce TNT to the corresponding amine derivative via formation of nitroso- and hydroxylamine intermediates which consequently undergoes mineralization (Fig. 8.1).

8.3.1 Biochemical Characterization

However reports on nitroreductases from *Clostridium* sp. are scarcely available (Table 8.1). Two nitroreductases have been identified in *Clostridium acetobutylicum* genome, belonging to oxygen-insensitive nitroreductase group, which were cloned in *E.coli* (Kutty and Bennett 2005). The overexpressed and purified enzymes NitA and NitB are flavoproteins with the potential to reduce trinitrotoluene. Earlier the nitroreductase enzyme has been reported in *Clostridium kluyveri*, *Clostridium* spec La 1, *Clostridium sporogenes*, *Clostridium perfringens* and *Clostridium bifermentans*. The iron-only hydrogenase from *Clostridium acetobutylicum* also possesses potential to reduce TNT as that of oxygen-insensitive nitroreductases (Angermaier and Simon 1983; Rafii and Carneglia 1993; Regan and Crawford 1994; Rafii and Coleman 1999). Hydrogenase and carbon monoxide dehydrogenase have been demonstrated to possess nitroreductase activity under strictly anaerobic conditions (Huang et al. 2000; Watrous et al. 2003). Overexpressed and purified hydrogenase enzyme from *Clostridium acetobutylicum* has been demonstrated to be a potent

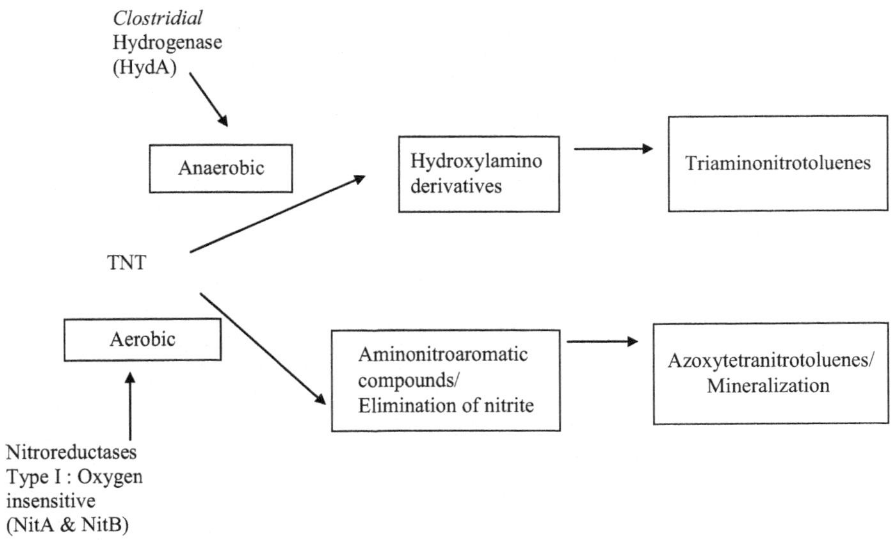

Fig. 8.1 Pathways of aerobic and anaerobic nitroreduction by clostridial nitroreductases

Table. 8.1 Relevant clostridial nitroreductases

Bacteria and nitroreductase names	Monomer sizes (kDa)	Main substrates	References
Clostridium acetobutylicum nitroreductase NitA	31	TNT, 2,4-DNT	Kutty and Bennett (2005)
Clostridium acetobutylicum nitroreductase NitB	23	TNT, 2,4-DNT	Kutty and Bennett (2005)
Clostridium perfringens	–	Azo dyes (Direct Blue 15), 4-nitrobenzoic acid, 1-amino-7-nitrofluorene, 1-nitropyrene	Rafii and Cerniglia (1993)
Clostridium acetobutylicum hydrogenase HydA	67	TNT	Watrous et al. (2003)
			Kutty and Bennett (2006)
Clostridium leptum	–	Nitrazepam	Rafii and Cerniglia (1997)
Carbon monoxide dehydrogenase from *Clostridium thermoaceticum*	–	TNT	Huang et al. (2000)

nitroreductase reductively degrading trinitrotoluene. The apparent Km of Fe hydrogenase for TNT was found to be 152 μM. The Fe hydrogenase was extremely sensitive to oxygen.

The purified carbon monoxide dehydrogenase (CODH) from *Clostridium thermoaceticum* demonstrated reductive degradation of trinitrotoluene with Michaelis constant (Km) and a turnover number (Kcat values) of 165 ± 43 μM and 400 ± 94 μM, respectively (Huang et al. 2000). NMR studies of carbon monoxide dehydrogenase using 13C nuclear magnetic resonance (NMR) demonstrated redox reaction catalysed by carbon monoxide dehydrogenase with formation of carbon dioxide (Seravalli and Ragdale 2008).

The nitroreductase enzymes, hydrogenase and carbon monoxide dehydrogenase (CODH) from clostridia are diverse from type I (oxygen insensitive) and type II (oxygen sensitive) nitroreductases of aerobic and facultative bacteria, respectively, in structure and mechanistic aspects. The metabolism of exogenous carbon source in these microorganisms generates reducing equivalents utilized by certain enzymes and/or cofactors leading to an unspecific detoxification reaction by reduction of nitrosubstituents (Peres and Agathos 2000; Rieger et al. 2002).

The oxidoreductases enzymes in clostridia contain low redox potential inorganic iron-sulphur centres, participating in electron transfer reactions, thereby contributing to the gratuitous activity by means of reduction of electron carriers. The characteristics of highly oxygen-sensitive clostridial hydrogenase are that they are bidirectional catalyzing the electron transfer to and from an electron carrier (Baffert

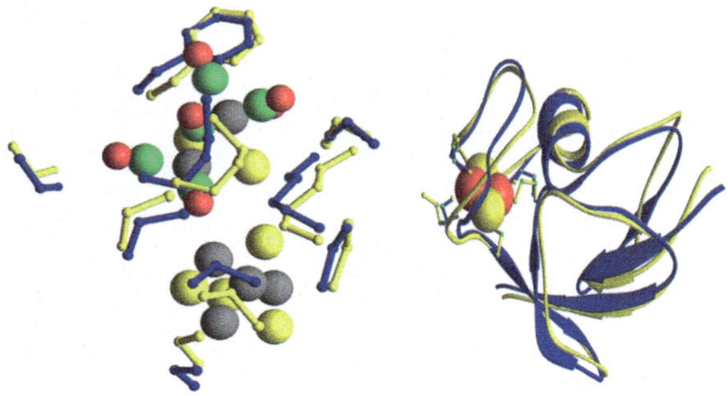

Fig. 8.2 The active site cluster (H cluster) with its polypeptide environment (**a**) and N-terminal end (**b**) of the *Clostridium acetobutylicum* hydrogenase shown as ribbon (yellow) generated with the known template of the iron hydrogenase from *C. pasteurianum* (blue) by using the geno3D protein modelling software. The cysteine sulphur atoms involved in the bonding of H cluster to the proteins are shown as yellow and iron atoms are rust. The [4Fe-4S] subclusters are shown in grey

et al. 2012, 2014). Clostridia being a fermentative anaerobe use this mechanism to produce hydrogen gas by disposing the reducing equivalents to reduce protons; alternatively, the synthesis of acetyl-CoA in *Clostridium thermoaceticum* occurs due to reduction of CO_2. Hydrogenase genes from the strains *C. acetobutylicum* P262 and ATCC 824 have been cloned and sequenced (Soucaille 2009). The availability of the sequence of *C. acetobutylicum* genome allows enzymes involved in TNT degradation to be studied. The availability of genome sequence of *C. acetobutylicum* allows identification of homologues of hydrogenases in *C. acetobutylicum* (Fig. 8.2).

The nitroreductases NitA and NitB which are demonstrated to be oxygen insensitive have been studied in *Clostridium acetobutylicum.* This enzyme possesses 89- and 17-fold greater specific activity, respectively, towards TNT as acceptor substrate and NADH as cofactor. This activity is with respect to the activity of the cell extracts from *E.coli* harbouring the control plasmid. 2,4-DNT served as better electron acceptor by these enzymes as observed by the greater rates of NADH oxidation. NitA enzyme showed 17-fold greater activity compared to that of NitB towards TNT as substrate. 2,6-DNT did not serve as substrate for NitB; however, NitA enzyme could reduce both 2,4-DNT and 2,6-dinitrotoluene (2,6-DNT). Nitrofurazone and the nitrosubstituted explosives, viz. hexahydro-1,3,5-trinitro-1,3,5-triazine (RDX) and octahydro-1,3,5,7-tetranitro-1,3,5,7-tetrazocine (HMX), were reduced by NitB, utilizing NADH. TNT reduction activity was 2.8-fold higher by NitB enzyme when NADH served as electron donor compared to NADPH. NitA enzyme showed K_m values of 8.51 μM and 555 μM for TNT and NADH, respectively.

NitA and NitB are flavoproteins utilizing FMN as a cofactor and consisting of a single nitroreductase domain. The FMN-binding domain in flavin reductase P of *Vibrio harveyi* consists of 11–15 amino acid segments (Tanner et al. 1996). NitA and NitB amino acid sequence also shows conservation of the corresponding residues. Arg^{203} and Arg^{208} residues in *E. coli* NfsA are shown to be involved in NADPH binding as demonstrated by site-directed mutagenesis of these residues which resulted in low catalytic activity; however, similar relative activities were observed with mutants when NADH was used as electron donor. NfsA and the flavin reductase P of *Vibrio harveyi* show conserved Arg^{203} residue; however, Arg^{208} in NfsA is substituted by Lys^{208} in flavin reductase P (Kobori et al. 2000). NitA does not utilize NADPH since the corresponding residues are substituted by Ile and Lys, respectively. NADH and NADPH both serve as electron donor in NitB which shows Arg^{203} and Arg^{208} replaced by Ala and His residues.

8.3.2 Characterization of Genes Encoding Nitroreductase

Novel genes encoding nitroreductase family proteins fused to a ferredoxin-like domain were found in the *Clostridium acetobutylicum* ATCC 824 genome sequence. Further characterization is required for understanding the physiological role of these genes. The association of ferredoxin with nitroreductase genes indicates the possible involvement of this Fe-S protein in nitroreductase function. The only report in our knowledge on unique ferredoxin domain containing N-terminal extension has been reported in the strain *Clostridium acetobutylicum* ATCC 824 (Kutty and Bennett 2007). The putative hydrogenase gene from *Clostridium acetobutylicum* ATCC 824 (*hydA*) was cloned and characterized. The *hydA* codes for a 64-KDa protein possessing strong identity with the proteins encoded by the [Fe] hydrogenase genes of *Desulfovibrio* and *Clostridium* species.

The transcriptional regulation has been demonstrated with the hydrogenase activity in strains producing acids, solvents and alcohols (Fontaine et al. 2002; Gorwa et al. 1996). It has been demonstrated that the *hydA* gene from *Clostridium acetobutylicum* P626 is expressed at the same level during both acidogenesis and solventogenesis. This HydA protein was 82% similar and 67% identical with *Clostridium pasteurianum* hydrogenase protein. The reports indicate that the HydA from *Clostridium acetobutylicum* ATCC 824 and *Clostridium acetobutylicum* P262 were encoded from monocistronic operon. Comparison of hydrogenase of *Clostridium acetobutylicum* ATCC 824 with Fe hydrogenase of *Clostridium acetobutylicum* P262 and *Clostridium pasteurianum* W5 indicated 70% identity and 80% similarity. The −10 (5′ TATAAT 3′) and −35 (5′TTTAGA) regions deduced from the major transcription start site in *Clostridium* are homologous to clostridial consensus regions that correspond to elements recognized by *B. subtilis* (σ43) and *E.coli* (σ70) RNA polymerases.

The *Clostridium acetobutylicum* sequence obtained can be implicated in the development of metabolic networks and flux models which link the genotype of the strain with its phenotype (Mcanulty et al. 2012). These genomic-scale models and networks can be used to predict the genetics and physiological processes involved in nitroreduction.

8.3.3 Structural Characterization of Clostridial Nitroreductases

There is limited information on the structure-function relation of nitroreductase from clostridial species. X-ray crystal structure of monomeric iron hydrogenase from *Clostridium pasteurianum* was reported to be 1.8 Å resolution. The H cluster represents the active site along with five distinct [Fe-S] clusters present in the enzyme. The H cluster contains six iron atoms in the form of [4Fe-4S] cubane which is covalently linked to [2Fe] subcluster via a non-protein ligand (Peters 1998). The structural details reported provide mechanistic details of hydrogen production and the physiological roles of Fe-S clusters (Fourmond et al. 2014).

In order to determine the 3D structure of *Clostridium acetobutylicum* ATCC 824 Fe-only hydrogenase, simulation modelling was carried out using the reported X-ray structure of *Clostridium pasteurianum* Fe hydrogenase as template. The RMS deviation of the generated model for its Cα atom was found to be 1.18 Å when compared to the X-ray crystal structure of *Clostridium pasteurianum* hydrogenase. This information was used in generation of superimposed stereoview of H cluster of *Clostridium acetobutylicum* ATCC 824 Fe-only hydrogenase and that of *Clostridium pasteurianum* Fe-only hydrogenase (Fig. 8.2). This information was used to synthesize antibodies against different portions of Fe hydrogenase, and the involvement of these portions in TNT reduction was confirmed (Kutty and Bennet 2006).

The crystal structure of nitroreductase family protein from *Clostridium difficile* (PDB/3H4O) and *Clostridium novyi* (PDB ID/3G14) has been determined at 1.50 and 1.75 Å, respectively. These are homodimeric flavoproteins.

8.3.4 Evolution and Diversity of Nitroreductases

Hydrogenase is the key enzyme involved in microbial energy metabolism. It functions in reversible oxidation of molecular hydrogen. Recently it has been reported that Fe hydrogenases are diverse, and each organism may harbour several homologue hydrogenases (Girbal et al. 2005). Significant difference was obtained between 16S rRNA gene and [Fe-Fe] hydrogenase-based phylogenic studies. The scarce information about the phylogeny of [Fe-Fe] hydrogenase requires extensive investigations on its diversity analysis as it had lower resolution compared to 16S rRNA tree (Schmidt et al. 2010).

8.3.5 Clostridial Nitroreductases in Post-genomic Era

The *Clostridium acetobutylicum* genome sequence was reported by Nolling et al. (2001). The genome consists of 3.94 Mb chromosome and 192 kb megaplasmid containing many genes involved in solvent production (Girbal and Soucaille 1998; Gonzalez-Pajuelo et al. 2005). The genome also indicates the presence of numerous genes encoding nitroreductase enzymes (CAC0718, CAC0748, CAC0850, CAC1418, CAC2311, CAC2508, CAC3359, CAC3555); however, the nitroreductase activity of these genes is yet to be investigated. This information has allowed comparison with genomes of microorganisms having closer phylogenetic relations, gene order and operons. The reports demonstrate that the ferredoxin is involved in initial reduction of nitro-groups (Angermaier et al. 1981). The hydrogenase functions by interacting with low midpoint redox potential electron carriers which can accept electrons from other oxidoreductases and get reduced. Ferredoxins provide this function in clostridia (Angermaier and Simon 1983). These genes encoding ferredoxins and polyferredoxins are found in *Clostridium acetobutylicum* ATCC 824 genome which may be involved in TNT transformation. Polyferredoxins are characterized by the presence of multiple iron-sulphur centres identical to that of bacterial ferredoxins. These domains are formed as a consequence of intrasequence gene duplication, transposition and fusion during evolution (Guerrini et al. 2008). The duplicated domains can also be formed without four conserved cysteine residues and changed binding characteristics. An operon encoding methyl viologen encoding hydrogenase in *Methanobacterium thermoautotrophicum* contains $12Fe_4S_4$ (Weiss and Thauer 1993; Reeve et al. 1989; Steigerwald et al. 1992). Polyferredoxin in *Methanosarcina barkeri* is encoded by *fmhF* gene which is associated as a component of *formylmethanofuran dehydrogenase* (Vorholt et al. 1996). The reported structural characteristics of polyferredoxin in both studies imply its role in the reaction mechanism of hydrogenase.

The transcriptomic and the proteomic data from the genomic studies will lead to new findings about the gene expression and regulation network. Various strategies can also be developed for metabolic engineering in the post-genomic era.

8.4 Conclusions

Nitroreductases from clostridia species are reported to be potent enzymes from bioremediation perspective. These enzymes can be engineered for applications in soil remediation.

References

Angermaier L, Simon H (1983) On the reduction of aliphatic and aromatic nitrocompounds by Clostridia, the role of ferredoxin and its stabilization. Hoppe Seylers Z Physiol Chem 364:961–975. https://doi.org/10.1515/bchm2.1983.364.2.961

Angermaier L, Hein F, Simon H (1981) Investigations on the reduction of aliphatic and aromatic nitrocompounds by Clostridia species and enzyme systems. In: Bothe H, Trebst A (eds) Biology of inorganic nitrogen and sulfur, proceedings in life sciences. Berlin, Springer, pp 266–275

Baffert C, Sybirna K, Ezanno P, Lautier T, Hajj V, Meynial-Salles I (2012) Covalent attachment of FeFe hydrogenases to carbon electrodes for direct electron transfer. Anal Chem 84:7999–8005. https://doi.org/10.1021/ac301812s

Baffert C, Fourmond V, Leger C, Meynial-Salles I, Soucaille P, Sybirna K (2014) Covalent attachment of FeFe hydrogenases to graphite electrode and inhibition studies. J Biol Chem 19:S280–S280

Batstone DJ, Virdis B (2014) The role of anaerobic digestion in the emerging energy economy. Curr Opin Biotechnol 27:142–149. https://doi.org/10.1016/j.copbio.2014.01.013

Bohutskyi P, Ketter B, Chow S, Adams KJ, Betenbaugh MJ, Allnutt FC, Bouwer EJ (2015) Anaerobic digestion of lipid-extracted *Auxenochlorella protothecoides* biomass for methane generation and nutrient recovery. Bioresour Technol 183:229–239. https://doi.org/10.1016/j.biortech.2015.02.012

Boopathy R, Wilson M, Kulpa CM (1993) Anaerobic removal of 2,4,6-trinitrotoluene (TNT) under different electron accepting conditions: laboratory study. Water Environ Res 65:271–275. https://doi.org/10.2175/WER.65.3.12

Boopathy R, Wilson M, Montemagno CD, Manning JF, Kulpa CF (1994) Biological transformation of 2,4,6-trinitrotoluene (TNT) by soil bacteria isolated from TNT-contaminated soil. Bioresour Technol 47:19–24. https://doi.org/10.1016/0960-8524(94)90023-X

Bradley PC, Chapelle FH, Landmeyer JE, Schumacher JG (1997) Potential for intrinsic bioremediation of a DNT-contaminated aquifer. Ground Water 35:12–17. https://doi.org/10.1111/j.1745-6584.1997.tb00054

Caballero A, Lazaro JJ, Ramos JL, Esteve-Nunez A (2005) PnrA, a new nitroreductase-family enzyme in the TNT-degrading strain *Pseudomonas putida* JLR11. Environ Microbiol (8):1211–1219. https://doi.org/10.1111/j.1462-2920.2005.00801.x

Daun GL, Reuss M, Knackmuss HJ (1998) Biological treatment of TNT-contaminated soil Anaerobic cometabolic reduction and interaction of TNT and metabolites with soil components. Environ Sci Technol 32:1956–1963. https://doi.org/10.1021/es970949u

Esteve-Nunez A, Caballero A, Ramos JL (2001) Biological degradation of 2,4,6-Trinitrotoluene. Microbiol Mol Biol Rev 3:335–352. https://doi.org/10.1128/mmbr.65.3.335-352.2001

Fernández M, Duque E, Pizarro-Tobías P, Van Dillewijn P, Wittich R, Ramos JL (2009) Microbial responses to xenobiotic compounds. Identification of genes that allow *Pseudomonas putida* KT2440 to cope with 2,4,6-trinitrotoluene. Microb Biotechnol 2:287–294. https://doi.org/10.1111/j.1751-7915.2009.00085.x

FitzGerald JA, Allen E, Wall DM, Jackson SA, Murphy JD, Dobson AD (2015) *Methanosarcina* play an important role in anaerobic co-digestion of the seaweed Ulva lactuca: taxonomy and predicted metabolism of functional microbial communities. PLoS One 10(11):e0142603. https://doi.org/10.1371/journal.pone.0142603. eCollection 2015

Fontaine L, Meynial-Salles I, Girbal L, Yang X, Croux C, Soucaille P (2002) Molecular characterization and transcriptional analysis of adhE2, the gene encoding the NADH-dependent aldehyde/alcohol dehydrogenase responsible for butanol production in alcohologenic cultures of *Clostridium acetobutylicum* ATCC 824. J Bacteriol 184:821–830. https://doi.org/10.1128/JB.184.3.821-830.2002

Fourmond V, Baffert C, Ezanno P, Leger C, Greco C, Bruschi M (2014) The oxidative inactivation of FeFe hydrogenase reveals the plasticity of the H-cluster. J Biol Inorg Chem 19:S232–S232. https://doi.org/10.1038/nchem.1892

Fuess LT, Garcia ML (2015) Bioenergy from stillage anaerobic digestion to enhance the energy balance ratio of ethanol production. J Environ Manag 162:102–114. https://doi.org/10.1016/j.jenvman.2015.07.046

Girbal L, Soucaille P (1998) Regulation of solvent production in *Clostridium acetobutylicum*. Trends Biotechnol 16:11–16. https://doi.org/10.1016/S0167-7799(97)01141-4

Girbal L, von Abendroth G, Winkler M, Benton PMC, Meynial-Salles I (2005) Homologous and heterologous overexpression in *Clostridium acetobutylicum* and characterization of purified clostridial and algal Fe-only hydrogenases with high specific activities. Appl Environ Microbiol 71:2777–2781. https://doi.org/10.1128/AEM.71.5.2777-2781.2005

González-Pajuelo M, Meynial-Salles I, Mendes F, Andrade JC (2005) Metabolic engineering of *Clostridium acetobutylicum* for the industrial production of 1, 3-propanediol from glycerol. Metab Eng 7:329–336. https://doi.org/10.1016/j.ymben.2005.06.001

Gorwa MF, Croux C, Soucaille P (1996) Molecular characterization and transcriptional analysis of the putative hydrogenase gene of *Clostridium acetobutylicum* ATCC 824. J Bacteriol 178:2668–2675. https://doi.org/10.1128/jb.178.9.2668-2675.1996

Guerrini O, Burlat B, Léger C, Guigliarelli B, Soucaille P, Girbal L (2008) Characterization of two 2 [4Fe4S] ferredoxins from *Clostridium acetobutylicum*. Curr Microbiol 56:261–267. https://doi.org/10.1007/s00284-007-9072

Heap JT, Theys J, Ehsaan M, Kubiak AM, Dubois L, Paesmans K, Mellaert LV, Knox R, Kuehne SA, Lambin P, Minton NP (2014) Spores of Clostridium engineered for clinical efficacy and safety cause regression and cure of tumors in vivo. Oncotarget 5:1761–1769. 10.18632/oncotarget.1761

Huang S, Lindahl PA, Wang C, Bennett GN, Rudolph FB, Hughes JB (2000) 2,4,6-Trinitrotoluene reduction by carbon monoxide dehydrogenase from *Clostridium thermoaceticum*. Appl Environ Microbiol 66:1474–1478. https://doi.org/10.1128/aem.66.4.1474-1478.2000

Jones DT, Woods DR (1986) Acetone-butanol fermentation revisited. Microbiol Rev 50:484–524

Ju K, Parales RE (2010) Nitroaromatic compounds, from synthesis to biodegradation. Microbiol Mol Biol Rev 74:250–272. https://doi.org/10.1128/mmbr.00006-10

Kivisaar M (2011) Evolution of catabolic pathways and their regulatory systems in synthetic nitroaromatic compounds degrading bacteria. Mol Microbiol 82:265–268. https://doi.org/10.1111/j.1365-2958.2011.07824.x

Kim BH, Zeikus JG (1985) Importance of hydrogen metabolism in regulation of solventogenesis by *Clostridium acetobutylicum*. Dev Ind Microbiol 26:1–14

Kitts CL, Green CE, Otley RA, Alvarez MA, Unkefer PJ (2000) Type I nitroreductases in soil enterobacteria reduce TNT (2,4,6,-trinitrotoluene) and RDX (hexahydro-1,3,5-trinitro-1,3,5 triazine). Can J Microbiol 46:278–282

Kobori T, Sasaki H, Lee WC, Zenno S, Saigo K, Murphy ME, Tanokura M (2000) Structure and site-directed mutagenesis of a flavoprotein from *Escherichia coli* that reduces nitrocompounds. J Biol Chem 276:2816–2823. https://doi.org/10.1074/jbc.m002617200

Kröger M, Schumacher ME, Risse H, Fels G (2004) Biological reduction of TNT as part of a combined biological–chemical procedure for mineralization. Biodegradation 15:241–248. https://doi.org/10.1023/b:biod.0000042914.53688.63

Kutty R, Bennett GN (2005) Biochemical characterization of trinitrotoluene transforming oxygen-insensitive nitroreductases from *Clostridium acetobutylicum* ATCC 824 over expressed in *E coli*. Arch Microbiol 184:158–167. https://doi.org/10.1007/s00203-005-0036-x

Kutty R, Bennett GN (2006) Studies on inhibition of transformation of 2,4,6-trinitrotoluene catalyzed by Fe-only hydrogenase from *Clostridium acetobutylicum*. J Ind Microbiol Biotechnol 33:368–376. https://doi.org/10.1007/s10295-005-0067-y

Kutty R, Bennett GN (2007) Characterization of a novel ferredoxin with N-terminal extension from *Clostridium acetobutylicum* ATCC 824. Arch Microbiol 187:161–169. https://doi.org/10.1007/s00203-006-0184-7

Lenke HW, Warrelmann J, Daun G, Hund K, Steglen U, Walter U, Knackmuss HJ (1998) minated soil. 2 Biologically induced immobilization of the contaminants and full-scale application. Environ Sci Technol 32:1964–1971. https://doi.org/10.1021/es970950t

Leungsakul T, Johnson GR, Wood TK (2006) Protein engineering of the 4-Methyl-5-nitrocatechol monooxygenase from *Burkholderia sp.* Strain DNT for enhanced degradation of nitroaromatics. Appl Environ Microbiol 72:3933–3939. https://doi.org/10.1128/aem.02966-05

Li RW, Giarrizzo JG, Wu S, Li W, Duringer JM, Craig AM (2014) Metagenomic insights into the RDX-degrading potential of the ovine rumen microbiome. PLoS One 9:e110505. https://doi.org/10.1371/journal.pone.0110505

Liu SC, Ahn GO, Kioi M, Dorie MJ, Patterson AV, Brown JM (2008) Optimized clostridium-directed enzyme prodrug therapy improves the antitumor activity of the novel DNA cross-linking agent PR-104. Cancer Res 68:7995–8003. https://doi.org/10.1158/0008-5472.CAN-08-1698

McAnulty MJ, Yen J, Freedman BG, Senger RS (2012) Genome-scale modeling using flux ratio constraints to enable metabolic engineering of clostridial metabolism in silico. BMC Syst Biol 6:42. https://doi.org/10.1186/1752-0509-6-42

Nölling J, Breton G, Omelchenko MV, Makarova KS, Zeng Q, Gibson R (2001) Genome sequence and comparative analysis of the solvent-producing bacterium *Clostridium acetobutylicum*. J Bacteriol 183:4823–4838. https://doi.org/10.1128/JB.183.16.4823-4838.2001

Padda RS, Wang C, Hughes JB, Kutty R, Bennett GN (2003) Mutagenicity of nitroaromatic degradation compounds. Environ Toxicol Chem 22:2293–2297. https://doi.org/10.1111/j.1462-2920.2005.00801.x

Peters JW (1998) X-ray crystal structure of the Fe-Only Hydrogenase (CpI) from *Clostridium pasteurianum* to 1.8 Angstrom Resolution. Science 282:1853–1858. https://doi.org/10.1126/science.282.5395.1853

Peres CM, Agathos SN (2000) Biodegradation of nitroaromatic pollutants: from pathways to remediation. Biotechnol Annu Rev 6:197–220. https://doi.org/10.1016/s1387-2656(00)06023-3

Rafii F, Cerniglia CE (1993) Comparison of the azoreductase and nitroreductase from Clostridium perfringens. Appl Environ Microbiol 59:1731–1734

Rafii F, Coleman T (1999) Cloning and expression in *Escherichia coli* of an azoreductase gene from *Clostridium perfringens* and comparison with azoreductase genes from other bacteria. J Basic Microbiol 39:29–35

Reeve BW, Reid SJ (2016) Glutamate and histidine improve both solvent yields and the acid tolerance response of *Clostridium beijerinckii* NCP 260. J Appl Microbiol 120:1271–1281. https://doi.org/10.1111/jam.13067

Reeve JN, Beckler GS, Cram DS, Hamilton PT, Brown JW, Krzycki JA, Kolodziej AF, Alex L, Orme-Johnson WH, Walsh CT (1989) A hydrogenase-linked gene in Methanobacterium thermoautotrophicum strain ΔH encodes a polyferredoxin. Proc Natl Acad Sci U S A 86:3031–3035

Regan KM, Crawford RL (1994) Characterization of *Clostridium bifermentans* and its biotransformation of 2,4,6-trinitrotoluene (TNT) and 1,3,5-triaza-1,3,5-trinitrocyclohexane (RDX). Biotechnol Lett 16:1081–1086. https://doi.org/10.1007/BF01022407

Rieger P, Meier H, Gerle M, Vogt U, Groth T, Knackmuss H (2002) Xenobiotics in the environment: present and future strategies to obviate the problem of biological persistence. J Biotechnol 94:101–123. https://doi.org/10.1016/s0168-1656(01)00422-9

Ryan A, Kaplan E, Laurieri N, Lowe E, Sim E (2011) Activation of nitrofurazone by azoreductases: multiple activities in one enzyme. Sc Rep 1:63. https://doi.org/10.1038/srep00063

Schmidt O, Drake HL, Horn MA (2010) Hitherto unknown [Fe-Fe]-Hydrogenase gene diversity in anaerobes and anoxic enrichments from a moderately acidic fen. Appl Environ Microbiol 76:2027–2031. https://doi.org/10.1128/aem.02895-09

Seravalli J, Ragsdale SW (2008) 13C NMR characterization of an exchange reaction between CO and CO2 catalyzed by carbon monoxide dehydrogenase. Biochemist 47:6770–6781. https://doi.org/10.1021/bi8004522

Soucaille P (2009) Metabolic engineering of *Clostridium acetobutylicum* for the production of butanol at high yield. New Biotechnol 25:S324

Steigerwald VJ, Pihl TD, Reeve JN (1992) Identification and isolation of the polyferredoxin from *Methanobacterium thermoautotrophicum* strain delta H. Proc Natl Acad Sci U S A 89:3031–3035

Tanner JJ, Lei B, Tu S, Krause KL (1996) Flavin reductase P: structure of a dimeric enzyme that reduces flavin. Biochemist 42:13531–13539. https://doi.org/10.1021/bi961400v

Theys J, Pennington O, Dubois L, Anlezark G, Vaughan T, Mengesha A, Landuyt W, Anné J, Burke PJ, Dûrre P, Wouters BG, Minton NP, Lambin P (2006) Repeated cycles of Clostridium-directed enzyme prodrug therapy result in sustained antitumour effects in vivo. Br J Cancer 95:1212–1219. https://doi.org/10.1038/sj.bjc.6603367

Vorholt JA, Vaupel M, Thauer RK (1996) A polyferredoxin with eight [4Fe–4S] clusters as a subunit of molybdenum formylmethanofuran dehydrogenase from *Methanosarcina barkeri*. Eur J Biochem 236:309–317. https://doi.org/10.1111/j.1432-1033.1996.t01-1-00309.x

Watrous MM, Clark S, Kutty R, Huang S, Rudolph FB, Hughes JB, Bennett GN (2003) 2,4,6 trinitrotoluene reduction by an Fe-only hydrogenase in *Clostridium acetobutylicum*. Appl Environ Microbiol 69:1542–1547. https://doi.org/10.1128/AEM.69.3.1542-1547.2003

Weiss DS, Thauer RK (1993) Methanogenesis and the unity of biochemistry. Cell 72:819–822

Xie B-T, Liu Z-Y, Tian L, Li F-L, Chen X-H (2015) Physiological response of Clostridium ljungdahlii DSM 13528 of ethanol production under different fermentation conditions. Bioresour Technol 177:302–307. https://doi.org/10.1016/j.biortech.2014.11.1

Activated Sludge Process and Energy

9

J. Rajesh Banu, U. Ushani, and R. Yukesh Kannah

Abstract

Sludge is the chief by-product of wastewater treatment plants, and further clearance of sludge is a major and complex environmental issue. Stabilization of sludge is done to diminish the pathogen content, eradicate unpleasant odors, and diminish the potential for decomposition. The cost to treat surplus amounts of waste activated sludge is very high, accounting for up to 60% of the entire operational costs of a treatment plant. Sludge degradation is carried out by many treatment methods, which are followed by energy production. Sludge degradation involves many pretreatment approaches: biological, chemical, thermal, mechanical, etc. Every pretreatment ends with anaerobic/aerobic digestion, which leads to biogas production. Among these methods, anaerobic digestion is very effective due to its eco-friendly impact and high potential for energy recovery. The rate-limiting step of anaerobic/aerobic digestion is hydrolysis of higher organic compounds to smaller molecules. These pretreatment methods play a key role in enhancing hydrolysis which, in return, increases biogas production with varying energy and cost consumption. Many research studies are ongoing to figure out the best pretreatment methods with high biogas production and low energy and cost consumption. To cut down energy and cost utilization, many pretreatments are combined together, such as thermochemical and chemobiological pretreatments, etc.

Keywords

Energy • Hydrolysis • Pretreatment • Sludge • Stabilization

J. Rajesh Banu (✉) • U. Ushani • R. Yukesh Kannah
Department of Civil Engineering, Regional Centre of Anna University, Tirunelveli, Tamil Nadu, India
e-mail: rajeshces@gmail.com

H.J. Purohit et al. (eds.), *Optimization and Applicability of Bioprocesses*,
https://doi.org/10.1007/978-981-10-6863-8_9

9.1 Introduction

The activated sludge process (ASP) is an extensively applicable process in treatment of wastewater at present and is also a well-known biological process. The ASP is commonly used for treating wastewater from both domestic and industrial sectors. It has countless benefits; for example, it can be operated at low cost with high treatment efficiency. However, it results in the production of huge quantities of waste activated sludge (WAS). The excess organic matter (sludge) produced by the abovementioned process has to be suitably subjected to a treatment process before its eventual clearance. Although a lot of approaches have been introduced to treat the sludge, among them, anaerobic digestion (AD) has been advised as an apt and probable one, as it employs a complex microbial community to treat the sludge and results in generation of energy-rich methane (Uma et al. 2012a, b, 2013a, b; Merrylin et al. 2013a; Ariunbaatar et al. 2014; Kavitha et al. 2015a; Ushani et al. 2016). Different pretreatment methods have emerged so far to progress degradation effectiveness (i.e., reduction of solids and killing of disease-causing organisms); however, the achievement of every method hitherto has involved sludge-based techniques and therefore this entails an assessment approach for further implementation (Poornima et al. 2014; Gayathri et al. 2015; Godvin Sharmila et al. 2015). Based on the demand for improved functioning of the ASP and reactor operation, attention has been focused on gaining further insights into flocculation, given the extreme importance of floc properties, which determine the behavior of activated sludge in full-scale processes such as settling, dewatering, and gravity drainage. The foremost problem associated with ASP is the generation of an enormous quantity of sludge as a derivative of conversion of solubilized materials into bioflocs and generated gases (carbon dioxide, methane, nitrogen, sulfur dioxide, etc.).

The current method of sludge clearance involves treatment, transfer, and removal. Thus appropriate clearance of waste activated sludge is an immense challenge for wastewater operatives. Over time, the treatment of activated sludge has changed from a purely anaerobic method to a simple aerobic method, and then to a combination of anaerobic and aerobic methods. In practical terms, the ideal method is to reduce surplus sludge generation at the time of treatment before its production. Hence the management and removal of the surplus sludge generated from biological wastewater plants are among the main imperative financial and ecological concerns (Rajesh Banu et al. 2006, 2007a, b, 2014; Rajesh Banu and Kaliappan 2007). The latest policies concerning sludge management and clearance that have been followed in numerous nations, plus community and ecological crises, have led to increased attention to emergent practices. There are new integrated strategies aimed at decreasing or minimizing surplus generation of sludge (Suresh Karthik Kumara et al. 2014).This section focuses on appraisal of various types of pretreatment processes for improving rates of bioenergy production.

9.2 Various Pretreatment Methods

Various types of pretreatment technologies are discussed in the following sections, with an outline of particular pretreatments implemented and a summary of their outcomes. This section mainly describes the particular pretreatments illustrated in Fig. 9.1.

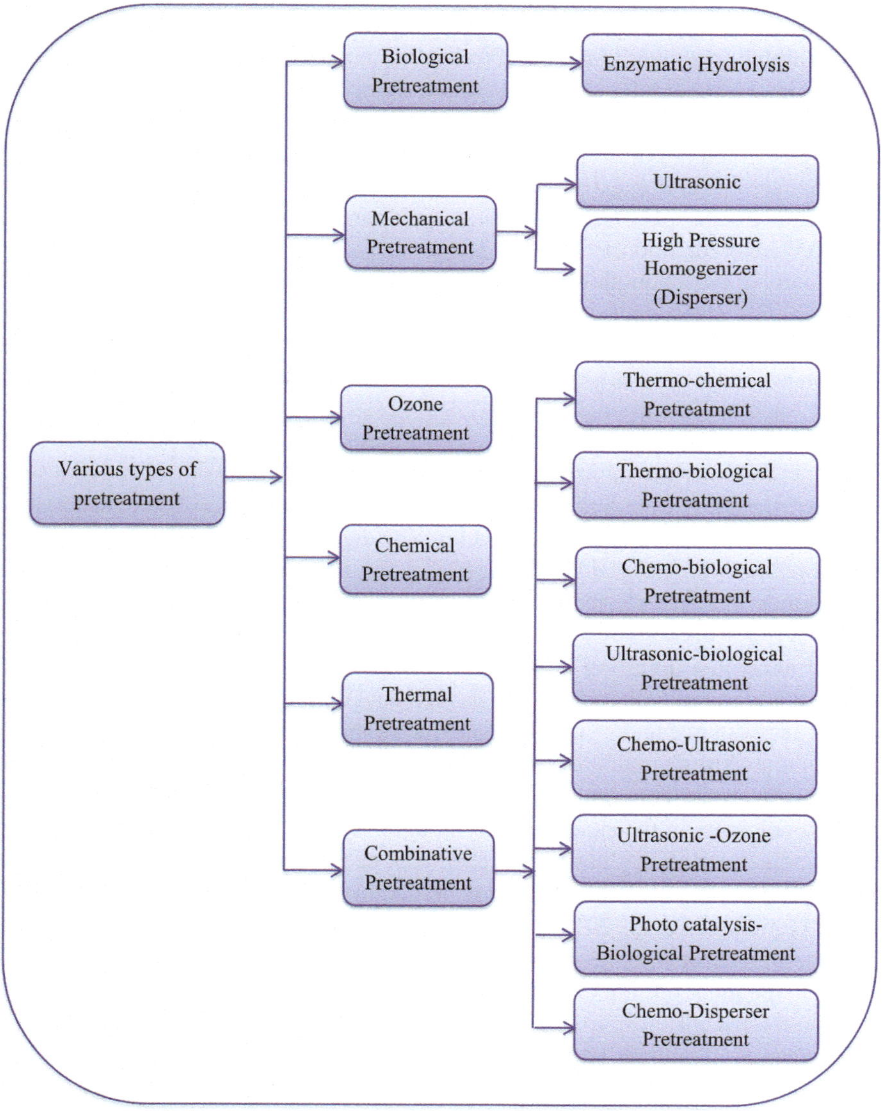

Fig. 9.1 Schematic flow diagram of pretreatments

9.2.1 Biological Pretreatment

Anaerobic and aerobic digestion are commonly used in most wastewater treatment plants (WWTPs) to stabilize unnecessary sludge. Aerobic digestion is a biological treatment process that uses long-term aeration to stabilize and decrease the overall mass of organic waste by biologically abolishing volatile solids (VS). Pollutants are oxidized into carbon dioxide (CO_2), water (H_2O), and biomass. By aerating oxygen through the aerators, the oxidation process can be significantly enhanced. Aerobic digestion is a more common procedure all over the world than other biological methods. During digestion, bacteria play a major role in fragmenting waste products. Aerobic treatment frequently produces a more enriched waste character than that attained by anaerobic treatment (Rajesh Banu et al. 2016). The aims of aerobic digestion are to stabilize sludge and decrease pathogens, smell, volume, and volatility. The key factors governing the action of aerobic digesters are the same as those for other aerobic biological processes and include waste sludge characteristics, oxygen requirement, pH, temperature, mixing, and solids retention time (Water Environment Federation 1992).

Anaerobic digestion plays a major role for its capability to promote conversion of organic substances into biogas. Thereby it can decrease the volume of the final sludge for clearance and destroy most pathogens while controlling odor problems. Greenhouse gases can also be reduced by this renewable energy (Jagadish et al. 2012). Anaerobic digestion is a profitable process and produces biogas from different substrates (Fantozzi and Buratti 2009; Rajesh Banu et al. 2014; Kavitha et al. 2014a, b, 2016a). Anaerobic digestion comprises four phases, which happen instantaneously in sludge digesters. The first phase involves hydrolysis of complex organic molecules into soluble organic molecules. This hydrolysis is the rate-limiting step, which restricts biogas yields. The second stage involves fermentation of organic compounds into volatile fatty acids. A further step includes production of acetic acid and CO_2, followed by methane and carbon dioxide production. The microbiology of anaerobic digestion is difficult to analyze and involves acidogenesis and numerous bacterial collections, such as methanogen bacteria. This biogas is valued as an energy source and is used for many domestic purposes. During the anaerobic digestion, a slow degradation rate of sludge occurs due to the rate-limiting step of sludge hydrolysis. This could be due to the lower degradability of cell membranes and exopolymers in the sludge (Merrylin et al. 2013a). To promote anaerobic and aerobic digestion ability, biological, chemical, ozone, thermal, and mechanical disintegration practices have functioned as pretreatment (Li et al. 2007; Poornima et al. 2014; Gopikumar et al. 2016; Kavitha et al. 2015a; Sowmya et al. 2015). These practices break down cell walls of microbes and discharge organic compounds found in the inner portion of cells (Sowmya et al. 2015). Biological pretreatment is a more eco-friendly way to effectively solubilize organic substances in the sludge than other pretreatment procedures. Use of hydrolytic bacteria, which produce enzymes, is a better biological pretreatment procedure with the advantage of high biogas yield, a mild working environment, and less energy requirements (Merrylin et al. 2013b). An overview of biological pretreatment is shown in Table 9.1.

Table 9.1 Overview of biological pretreatment

References	Type of sludge	Optimized conditions	Results
Kavitha et al. (2015c)	Municipal waste activated sludge	*Bacillus jerish 03 and Bacillus jerish* 04 (protease), 40 °C, pH 6.5, 42 h	COD solubilization 14.2%
			SS reduction 10%
			Biogas production (Rb) 1.21 L/(g VS d)
Kavitha et al. (2015d)	Municipal waste activated sludge	*Planococcus jake* 01 (biosurfactant), 40 °C, 42 h	COD solubilization 9.09%
			SS reduction 13.57%
			Biogas production (Rb) 0.225 (L/g VS)
Gopikumar et al. (2016)	Municipal waste activated sludge	*Bacillus licheniformis*, 15 days of hydraulic retention time	Anaerobic digestion, COD solubilization 40%
			SS reduction 43%
			Biogas yield 189.61 mL/g VS
Lakshmi et al. (2014)	Municipal waste activated sludge	*Bacillus jerish 03 and Bacillus jerish* 04 (lysozyme)	COD solubilization 16.2%
			SS reduction 17%
Merrylin et al. (2013b)	Municipal waste activated sludge	*Bacillus licheniformis* (protease), 55 °C, 24 h, pH 6.5	COD solubilization 7.2%
			SS reduction 10%
			Aerobic digestion, SS reduction 52.4%

COD chemical oxygen demand, *Rb* maximum biogas production rate, *SS* suspended solids, *VS* volatile solids

9.2.1.1 Enzymatic Hydrolysis

9.2.1.1.1 Treatment Implementation and Outcome

Activated sludge degradation by enzymes is more powerful. For anaerobic digestion, enzymatic treatment is the best practice. A lytic enzyme plays a key role in upgrading the hydrolysis step to acidogenesis, which is widely known for its rate-limiting step in anaerobic digestion, though enzymatic pretreatment of activated sludge is more expensive (Erikkson et al. 2002).Therefore, cultivating microorganisms that secrete a hydrolytic enzyme has become an operative approach. The hydrolytic enzyme lysozyme has the capability to degrade bacterial cell walls, which consist of polysaccharides as substrates (Xin et al. 2015). Lysozyme enhances cell lysis by breaking down cell walls and can also degrade organic molecules in sludge, which in turn enhances sludge stabilization (He et al. 2014). According to Lakshmi et al. (2014) *Bacillus jerish* 03 and *B. jerish* 04 secrete lysozyme enzymes, which have the ability to solubilize sludge. Protease has the individual ability to hydrolyze peptide bonds of proteins into peptide chains and amino acids, significantly increasing the effectiveness of sludge degradation. Protease enzyme has ability to procatalyze peptide bonds, which enhances sludge degradation (Luo et al. 2013). According to many researchers (Parmar et al. 2001; Ayol 2005; Ayol and

Dente 2005), hydrolytic enzymes such as protease, amylase, and cellulase have the capacity to improve the effectiveness of sludge degradation. Kavitha et al. (2014a) reported that *Bacillus jerish* 03 and *jerish* 04 secrete protease and amylase, which increases sludge degradation through a mixture of these enzymes. Watson et al. (2004) stated that the *Acinetobacter* family (*Raoultella* and *Pandoraea*) also exhibit protease-secreting activities. Lipases can hydrolyze triacylglycerols, which leads to sludge degradation. Dharmsthiti and Kuhasuntisuk (1998) treated restaurant wastewater by using *Pseudomonas aeruginosa*, which can secrete lipase, and achieved worthwhile results. Felice et al. (1997) treated olive oil factory wastewater using cultivated *Yarrowia lipolytica* ATCC 20255, and the chemical oxygen demand (COD) was substantially reduced.

Activated sludge flocs provide a frame for microorganisms, extracellular polymeric substances (EPSs), and other inorganic and organic compounds. EPSs are central to floc formation in the activated sludge (Nguyen et al. 2014). EPSs are the building material for bacterial clearance and remain attached to the cells' outer surfaces to form biofilms. EPSs mostly consist of distinct organic ingredients, such as polysaccharides, proteins, and some macromolecules (Monique et al. 2008; Wang et al. 2014). The main role of EPSs in sludge flocs is to provide protective action against a high-shear environment (Guo and Xu 2011). This protective property inhibits diffusion of enzymes and other lytic bacterial actions for sludge degradation (Zhang et al. 2015). Hence, the destruction of the EPS flocs probably tends to enhance sludge floc disintegration. Thus, sludge pretreatments using mechanical (ultrasonic and dispenser) methods are essential. This methodology should be applied to the sludge prior to the hydrolysis treatment to boost the degradation of sludge which, in turn, increases biogas yields.

9.2.2 Mechanical Pretreatment

9.2.2.1 Ultrasonic Disintegration

9.2.2.1.1 Treatment Implementation and Outcome

A mechanical pretreatment method such as ultrasonication is an evolving operative procedure to improve disintegration of the sludge (Pilli et al. 2011). Ultrasonication is a recognized procedure for the breakdown of sludge flocs and breaks down microbial cells, thus degradation of the organic compounds and reduction of particle size occur (Harrison 1991). The main principle behind ultrasonication is the construction of cavitation bubbles in the aqueous phase (Ngoc et al. 2013). In the aqueous phase, infrequent elevated-pressure (compression) and low-pressure (rarefaction) sequences are created. Small vacuum bubbles are produced during rarefaction. Aggressive shrinking of bubbles occurs when the bubbles stretch to a size at which they cannot withstand the equilibrium between the pressure and the viscosity forces (Kavitha et al. 2016b). An ultrasonic pretreatment setup is shown in Fig. 9.2. The hydrodynamic shear forces developed by acoustic cavitation act on the compounds responsible for the sludge disintegration (Zhang et al. 2007). Moreover, it has been

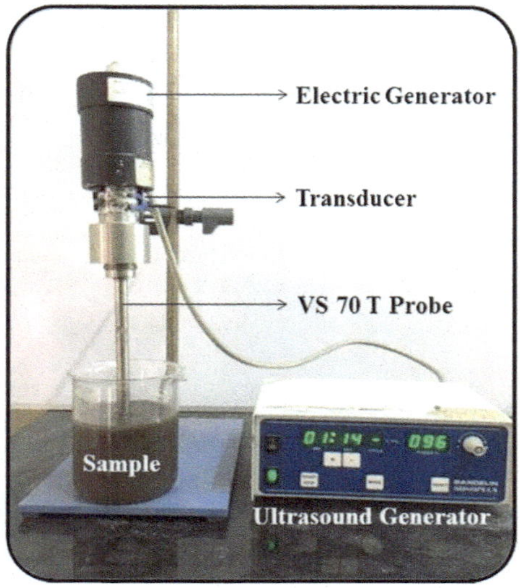

Fig. 9.2 Example of an ultrasonic pretreatment setup

shown that the cavitational bubble collapse creates forceful native heating (≈5000 K) and high pressure (≈100 atm) (Suslick 1998). Because of this, ultrasonication leads to cell lysis (Jianguo et al. 2009) and promotes biogas generation (Jianguo et al. 2014).

In ultrasonication, rapid disintegration is achieved at the beginning of the process, which makes EPSs more accessible for detachment and disintegration by the cavitational forces. According to many studies, fast disintegration of sludge is obtained at an early stage (Cho et al. 2012; Sahinkaya and Sevimli 2013). After a particular time, however, the cavitational effects plateau and the discharge of soluble chemical oxygen demand (SCOD) declines, possibly because of the drop in easily available degradable organic substances and the existence of additional unmanageable organic substances for the action of the cavitational force. More time and a large amount of energy are required for ultrasonic disintegration of recalcitrant organics (Zhang et al. 2008). In one study, a 1 min sonic treatment time with a working rate of 20 kHz and 200 W of supplied energy was found to optimize sludge degradation (Gayathri et al. 2015). In 1999, the German company Hielscher started equipping various wastewater treatment plants throughout the world with an ultrasonic disintegration setup with up to 48 kW of power. Some of these arrangements upgraded the biogas production by up to 25% (Braguglia et al. 2011). An overview of ultrasonic pretreatment is shown in Table 9.2.

9.2.2.2 High-Pressure Homogenizer

9.2.2.2.1 Treatment Implementation and Outcome

A high-pressure homogenizer (HPH) is one of the most frequently used approaches for a large-scale setup. An HPH uses a purely mechanical process, which is induced by forcing a fluidic product through a thin gap (the homogenizing nozzle) at high

Table 9.2 Overview of ultrasonic pretreatment

References	Type of sludge	Optimized conditions	Results
Sowmya et al. (2015)	Dairy waste activated sludge	pH 6.8, frequency 20 kHz	SCOD release 750 mg/L
		Power supply 200 W	Biogas production 4.0697 L/(g VS d)
Gayathri et al. (2015)	Dairy waste activated sludge	Frequency 20 kHz	COD solubilization 21.87%
		Power supply 200 W	SS reduction 13.57%
			Biogas production 0.9044 L/(g VS d)
Uma Rani et al. (2013b)	Dairy waste activated sludge	Frequency 20 kHz	COD solubilization 59%
		Power supply 200 W	SS reduction 46%
			Biogas production 80%

COD chemical oxygen demand, *SCOD* soluble chemical oxygen demand, *SS* suspended solids, *VS* volatile solids

pressure (150–200 MPa). The flocs are exposed to these narrow gaps to obtain cavitation and shear pressures, resulting in cell disintegration (Mata-Alvarez et al. 2000; Zhang et al. 2012). These pretreatment approaches are not common for municipal solid waste, but they are more generally used with other substrates such as lignocellulosic materials, manure, and wastewater treatment plant sludge. Use of bead mill, electroporation, and liquefaction pretreatments of municipal solid waste have been studied on a laboratory scale, whereas use of rotary drum, screw press, disc screen shredder, food waste disposer, and piston press treatments have been successfully applied at full scale. Both electroporation and liquefaction pretreatments cause destruction of cellular structures, with a softening effect similar to that of the aerobic digestion process (Parawira 2007; Kavitha et al. 2016c). The homogenization pressure and HPH cycle count are the working factors that generally influence the sludge disintegration percentage, which is related to the cell disruption efficacy (Kleinig and Middelberg 1998). It is usually believed that microbial cell disruption is more effective by means of higher homogenization pressure. However, mechanical approaches to cell destruction are energy demanding, requiring additional energy input for a setup with greater homogenization pressure with numerous cycles. Consequently, it is compulsory to optimize the functional settings to acquire the desired degradation effectiveness with low energy intake. According to Zhang et al. (2013), a single homogenization cycle with a pressure of 30 MPa is an energy-saving step for a sludge sample with total solids (TS) of 2.483%. The benefits of mechanical pretreatment comprise no odor generation, easy applicability, good dewaterability of the final anaerobic residue, and reasonable energy consumption. The disadvantages include no noteworthy effect on pathogen elimination and the probability of instrument blockage or scaling (Perez-Elvira et al. 2006; Toreci et al. 2009).

9.3 Ozone Pretreatment

Ozone (O_3) is a dominant oxidant and disinfectant. It has the capability to break down strong cell walls bounded by extracellular polysaccharides and transform the intracellular compounds (Sowmya et al. 2015). This process of conversion results in release of more soluble and refractory compounds, which can be oxidized easily, being more biodegradable and flexible to the anaerobic bacteria (Carballa et al. 2007), for an increased rate of biogas production. Ozonation pretreatment is an alternative process of chemical pretreatment (Carrere et al. 2010). Normally in chemical pretreatment there will be the result of accumulation of residues of externally added chemicals, increasing the salt concentration (Kianmehr et al. 2010). For sludge pretreatment, ozone is beneficial when compared with other methods of pretreatment (Sri Bala Kameswari et al. 2011). The advantage of this pretreatment, as stated above, is that it also acts as a disinfecting agent to reduce microorganism levels in waste activated sludge (Weemaes et al. 2000) and is also more efficient in reducing odors. An ozone pretreatment setup is shown in Fig. 9.3.

Fig. 9.3 Example of an ozone pretreatment setup

9.3.1 Treatment Implementation and Outcome

Organic compounds available in waste activated sludge can be effectively treated using an ozonation process. The process results in enhanced solubilization of organic compounds. The sludge floc chemical structure may be destroyed during ozonation pretreatment. Hence the outcomes are based on the rates of gas transfer and the reaction taking place (Yan et al. 2009). The influence of ozone pretreatment on waste activated sludge was studied in two optimized conditions: (1) an ozone gas flow rate of 0.075 L/min; and (2) an ozone dosage of 0.0011 mg O_3/mg SS [suspended solids]. The effects of the above conditions on COD solubilization and SS reduction were 20.8% and 13.5%, respectively. Similarly the rate of biogas production with ozone pretreatment was reported to be 0.535 L/(g VS), relatively higher than that seen with the control treatment 0.202 L/(g VS) (Sowmya et al. 2015). Based on an earlier research study, researchers chose two optimal ozone dosages of 0.1 g O_3/g TS and 0.16 g O_3/g TS. In these conditions the outcomes were estimated as COD solubilization of 18% and TS solubilization of 26%, respectively (based on a graph), with median particle size diameters of 33.2 μm and 32.6 μm, respectively; no improvement was shown when the ozone dosage was increased, and the rates of biogas production were 1.08 and 1.25 times higher, respectively, than those seen with untreated (raw) sludge (Bougrier et al. 2006). During ozone pretreatment the sludge disintegrated and harmed bacterial cells. Three ozone dosages were chosen: 25 mg O_3/g DS [dissolved solids], 50 mg O_3/g DS, and 80 mg O_3/g DS. The results of the three chosen doses were analyzed in terms of the degree of disintegration attained: merely 10.4% after 90 min, 40.3% after 105 min, and 46.7% after 105 min. The results obtained showed that 50 mg O_3/g DS produced a higher degree of disintegration than 25 mg O_3/g DS, whereas the difference was minor compared with 80 mg O_3/g DS. Based on the above outcomes the optimal ozone dose was 50 mg O_3/g DS, also considering the ozone cost. Hence the concentration of solids was decreased to 49% and VS was reduced to 45%. Similar improvements of 699% for COD solubilization, 169% for total nitrogen (TN), 2379% for total phosphorus (TP), 602% for protein, 528% for polysaccharides, and 556% for DNA were observed, but there were no significant differences in the size of particle distribution in sludge flocs (Zhang et al. 2009). An overview of ozone pretreatment is shown in Table 9.3.

9.4 Chemical Pretreatment

Chemical pretreatment is more efficient to disintegrate organic fractions accumulated in wastewater treatment plants. This pretreatment is conducted with use of strong acids or alkalis (Li et al. 2007). Hydrolysis of waste activated sludge in biodegradation processes has been recognized as effective while using strong acids or alkalis for pretreatment (Chen et al. 2007). Usually, the microbial activity and enzyme catalysis is very powerful at a neutral pH for most bacteria (Huijun et al. 2015). On the other

Table 9.3 Overview of ozone pretreatment

References	Type of sludge	Optimized conditions	Results
Sowmya et al. (2015)	Dairy waste activated sludge	Ozone dosage 0.0011 mg O_3/mg SS	COD solubilization 20.8%
		Ozone flow 0.075 L/min	SS reduction 13.5%
			Biogas production: control 0.202 L/(g VS), ozone pretreated 0.535 L/(g VS)
Bougrier et al. (2006)	Municipal waste activated sludge	Ozone dosage 0.10 g O_3/g TS	COD solubilization 18%
			TS solubilization 27%
			Median diameter of particle size 33.2 μm
			Biogas production 1.08 times that of raw sludge
		Ozone dosage 0.16 g O_3/g TS	COD solubilization 22%
			TS solubilization 25%
			Median diameter of particle size 32.6 μm
			Biogas production 1.25 times that of raw sludge
Zhang et al. (2009)	Wastewater treatment plant sludge	Ozone dosage 50 mg O_3/g DS	Degree of disintegration 40.3% after 105 min:
			Sludge solids 49.1% (reduced)
			Volatile solids 45.7% (reduced)
			SCOD 699% (increased)
			TN 169% (increased)
			TP 2379% (increased)
			Protein 602% (increased)
			Polysaccharide 528% (increased)
			DNA 556% (increased)
Salsabil et al. (2010)	Municipal waste activated sludge	0.1 g O_3/g TS_0	Specific energy 46,285 kJ/kg TS_0
			SCOD 10%
			Protein 9%
			Carbohydrate 7.1%
			TSS removal 15%
			VSS removal 19.2%

COD chemical oxygen demand, *DS* dissolved solids, *SCOD* soluble chemical oxygen demand, *SS* suspended solids, *TN* total nitrogen, *TP* total phosphorus, *TS* total solids, TS_0 initial total solids, *TSS* total suspended solids, *VS* volatile solids, *VSS* volatile suspended solids

hand, neutralization after chemical pretreatment (with strong acids or alkalis) leads to additional increases in the concentration of salts. In particular, this type of pretreatment will be more active when accomplished by expensive chemicals (Uma Rani et al. 2012a).

9.4.1 Treatment Implementation and Outcome

Chemical pretreatment involves alkaline and acidic pretreatment methods that can disrupt organic fractions through addition of alkalis or acids. Anaerobic digestion generates bioenergy from municipal sludge, which may be increased positively due to alkaline pretreatment. The effects of these pretreatments result in better biogas generation and effective solubilization of organic fractions. Citric acid has been shown to efficiently disrupt the floc structure and remove EPS at a lower concentration (0.05 g/g SS). At this concentration, the EPS removed was found to be 55 mg/L with limited cell lysis. The concentrations of the extracellular enzymes protease and amylase at this dosage were detected to be 0.052 U/mL and 0.042 U/mL, respectively. Citric acid has a probable experimental function as it is completely decomposable. The outcomes of the disintegrated process showed that 21.4% and 16.2% of SS reduction and COD solubilization, respectively, were achieved by citric acid–mediated bacterial disintegration, which were relatively greater effects than those obtained with bacterial disintegration (14.28% and 10.0%, respectively) (Kavitha et al. 2014b). The outcome of alkaline pretreatment of waste activated sludge with 40 meq/L NaOH, 34% of gas production, 41% of VS removal, and 30% of COD removal was improved in all parameters versus the control treatment (Lin et al. 2009). An overview of chemical pretreatment is shown in Table 9.4.

9.5 Thermal Pretreatment

Thermal pretreatment is a well-known physical method to decrease the excess organic matter (sludge) generated from wastewater treatment plants. In this type of pretreatment, temperature plays a major role and cell walls of pathogens are broken. The damage to the cell wall leads to release of the inner organic content into the solution (Rajesh Banu et al. 2008; Esakki Raj et al. 2013). Application of heat energy is the main part of thermal pretreatment and has been widely examined in a temperature range from 60 to 270 °C (Climent et al. 2007). A temperature range >100 °C is considered high-temperature thermal pretreatment, whereas a range <100 °C is considered low-temperature thermal pretreatment (Gavala et al. 2003). Thermal pretreatment is good for killing pathogens, which also improves the dewatering properties and boosts solubilization of sludge (Kuglarz et al. 2013).

9.5.1 Treatment Implementation and Outcome

Thermal pretreatment plays an effective role, and in one study, two different high temperatures of 170 °C and 190 °C were chosen for 30 min of pretreatment. The results were that the COD solubilization rates and TS solubilization showed positive results. At 170 °C the COD solubilization rate and TS solubilization were 48% and 41%, respectively. Similarly, at 190 °C the COD solubilization rate and TS solubilization were 49% and 38%, respectively. Hence the COD solubilization effectively

Table 9.4 Overview of thermal pretreatment

References	Type of sludge	Optimized conditions	Results
Bougrier et al. (2006)	Municipal waste activated sludge	170 °C, 30 min	CST 39 ± 1 s
			COD solubilization 48%
			TS solubilization 41%
			Median diameter of particle size 76.8 μm
			Biogas production 1.59 times that of raw sludge
		190 °C, 30 min	CST 29 ± 4 s
			COD solubilization 49%
			TS solubilization 38%
			Median diameter of particle size 77.1 μm
			Biogas production 1.59 times that of raw sludge
Salsabil et al. (2010)	Municipal waste activated sludge	90 °C, 90 min (thermic)	Specific energy 558,620 kJ/kg TS_0
			SCOD 16.8%
			Protein 45%
			Carbohydrate 37.6%
			TSS removal 15.8%
			VSS removal 21.2%
Dhar et al. (2012)	Municipal waste activated sludge	90 °C, 30 min	VSS reduction 27%
			Soluble organic nitrogen 596 mg/L
			Soluble carbohydrate 555 mg/L
			TVFA 178 mg/L
			Specific CH_4 production 368 mL CH_4/g VSS
Bougrier et al. (2008)	Municipal waste activated sludge	70 °C, 9 h	Biogas increased 30% versus raw sludge
Rashad et al. (2010)	Municipal waste activated sludge	70 °C	78% biogas and 60% methane production

COD chemical oxygen demand, *CST* capillary suction time, *SCOD* soluble chemical oxygen demand, *TS* total solids, TS_0 initial total solids, *TSS* total suspended solids, *TVFA* total volatile fatty acids, *VSS* volatile suspended solids

increased and TS solubilization slightly decreased. The rates of biogas production performance showed the same outcome at both temperatures, at 1.59 times that of raw sludge. Analysis of one more physical parameter—the median diameter of the particle size—showed that the median diameter was increased at 190 °C (77.1 μm) when compared with 170 °C (76.8 μm). Similarly the capillary suction time was decreased at 170 °C (39 ± 1) when compared with 190 °C (29 ± 4) (Bougrier et al. 2006). Thermal pretreatment at 70 °C also increased production of biogas by 78% and methane production by 60% (Rashad et al. 2010). An overview of thermal pretreatment is shown in Table 9.5.

Table 9.5 Overview of chemical pretreatment

References	Type of sludge	Optimized conditions	Results
Kavitha et al. (2014b)	Municipal waste activated sludge	Citric acid dosage 0.05 g/g SS	Raw sludge: SS reduction 6.5%, SCOD 8.5%
		Protease dosage 0.052 U/ml	Deflocculated-bacterially pretreated sludge: SS reduction 16.2%, SCOD 21.4%
		Amylase dosage 0.042 U/mL	Flocculated sludge: SS reduction 10%, SCOD 14.82%
Lin et al. (2009)	Waste activated sludge	40 meq/L NaOH	Gas production 34%
			VS removal 41%
			COD removal 30%

COD chemical oxygen demand, *SCOD* soluble chemical oxygen demand, *SS* suspended solids, *VS* volatile solids

9.6 Combination Pretreatment

9.6.1 Treatment Implementation and Outcome

9.6.1.1 Thermochemical Pretreatment

Combination thermal pretreatment, commonly with alkaline, is widely practiced and is defined as thermochemical pretreatment. Combination pretreatment disrupts the organic fraction to release the inner organic content into the solution, and the solubilization of the organic fraction increases the rate of biogas generation. A photograph of a thermochemical treatment setup is shown in Fig. 9.4. Strong alkalis such as 0.1 N NaOH, 0.1 N KOH, and 0.1 N $Ca(OH)_2$ were combined with an optimum temperature of 80 °C and a pretreatment time of 30 min. The results of this pretreatment were evaluated in terms of SCOD release (12.4%, 10.4%, and 9% for 0.1 N NaOH, 0.1 N KOH, and 0.1 N $Ca(OH)_2$, respectively) and SS reduction (9.9%, 8.1%, and 6.5%, respectively) (Uma Rani et al. 2012b). Sewage sludge was pretreated by a combination of thermal treatment at an optimized temperature of 60 °C for 90 min and chemical treatment at an optimized dosage of 4.0 N NaOH. The results shows that COD solubilization reached 68% and the VSS/TSS [volatile suspended solids/total suspended solids] ratio reached 0.69. Similarly, in optimized conditions, certain parameters warranted analysis for future study: alkalinity 5.40 ± 0.15 g $CaCO_3$/L, TCOD [total chemical oxygen demand] 22.94 ± 0.38 g/L, SCOD 15.53 ± 0.04 g/L, soluble TN 1.29 ± 0.02 g/L N, soluble TP 0.48 ± 0.01 g/L P, NH^{+}_{4}-N 0.55 ± 0.07 g/L, soluble protein 2.35 ± 0.03 g/L, and soluble carbohydrate 0.83 ± 0.03 g/L (Cho et al. 2013). An overview of thermochemical pretreatment is shown in Table 9.6.

9.6.1.2 Thermobiological Pretreatment

Heat treatment was used by Yan et al. (2008) to reduce excess sludge volume. This is a comparatively easy procedure for reduction of sludge volume. One hour of heat

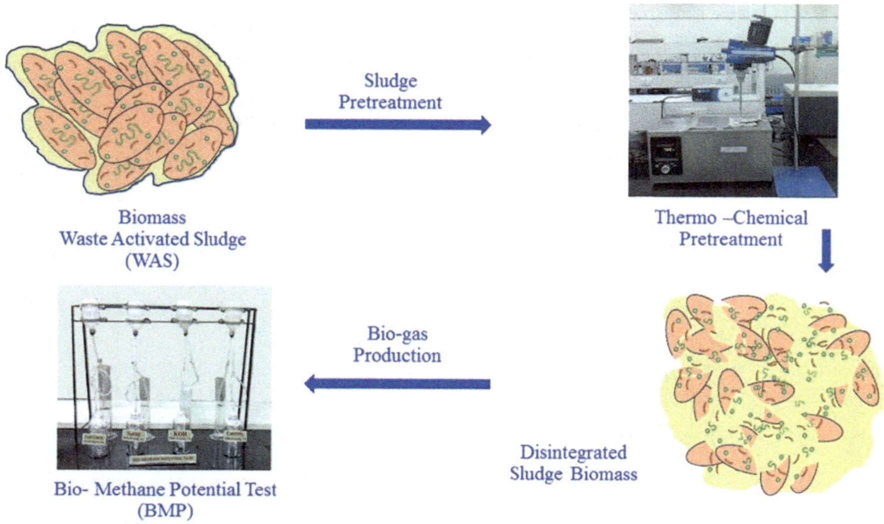

Fig. 9.4 Example of a thermochemical pretreatment setup

treatment increases protease activity in the sludge supernatant, which leads to more sludge degradation. For sludge degradation a protease-producing bacterium, *Bacillus licheniformis*, is used, which is thermophilic (55 °C). It enhances sludge degradation with SS reduction of 56% and biogas yield of 57% (Merrylin et al. 2013a).

9.6.1.3 Chemobiological Pretreatment

Many research studies have been done to explain the effects of chemicals on enzyme pretreatments. According to Kavitha et al. (2015b), NaCl induced biological disintegration (*Bacillus jerish* 03 and *Bacillus jerish* 04) with solubilization of about 23%. A citric acid dosage of 0.05 g/g SS disrupted flocs and increases sludge enzyme action with a 21.4% decrease in SS and 16.2% COD solubilization (Kavitha et al. 2014b). Use of the biosurfactant-producing bacterium *Planococcus jake* 01 is a novel method, which enhanced biogas production to 0.322 L/g VS with the help of 0.06 g/g SS of calcium chloride to disrupt sludge flocs (Kavitha et al. 2015b). An ethylenediaminetetraacetic acid (EDTA) dosage of 0.2 g/g SS removed EPS and improved bacterial pretreatment with SS reduction of 48.5% and COD solubilization of 47.3% attained in aerobic digestion (Kavitha et al. 2013). According to Ushani et al. (2016), a very low dosage (0.009 g/g SS) of surfactant (dioctyl sodium sulphosuccinate) disturbed EPS and, coupled with biological pretreatment (*Bacillus cereus*), was proven to be efficient with 20% COD solubilization. A schematic diagram of chemo-induced bacterial pretreatment is shown in Fig. 9.5.

Table 9.6 Overview of thermochemical pretreatment

References	Type of sludge	Optimized conditions	Results
Kavitha et al. (2015a)	Municipal waste activated sludge	0.1 N NaOH, 80 °C, 30 min	SCOD 12.4%, SS reduction 9.9%
		0.1 N KOH, 80 °C, 30 min	SCOD 10.4%, SS reduction 8.1%
		0.1 N $Ca(OH)_2$, 80 °C, 30 min	SCOD 9%, SS reduction 6.5%
Shehu et al. (2012)	Municipal waste activated sludge	2.29 M NaOH, 88.5 °C, 21 min	Degree of pretreatment achieved 61%
Chou et al. (2010)	Palm oil mill effluent	8.8 g/L NaOH, 32.5 °C, 41.2 h reaction time	COD solubilization 83%
Cho et al. (2013)	Sewage treatment plant sludge	4.0 N NaOH, 60 °C, 90 min	Alkalinity 5.40 ± 0.15 g $CaCO_3$/L
			TCOD 22.94 ± 0.38 g/L
			SCOD 15.53 ± 0.04 g/L
			COD solubilization 68%
			VSS/TSS 0.69
			Soluble T-N 1.29 ± 0.02 g/L N
			Soluble T-P 0.48 ± 0.01 g/L P
			NH^{+}_4-N 0.55 ± 0.07 g/L
			Soluble protein 2.35 ± 0.03 g/L
			Soluble carbohydrate 0.83 ± 0.03 g/L

COD chemical oxygen demand, *SCOD* soluble chemical oxygen demand, *SS* suspended solids, *TCOD* total chemical oxygen demand, *TN* total nitrogen, *TP* total phosphorus, *TSS* total suspended solids, *VSS* volatile suspended solids

9.6.1.4 Ultrasonic–Biological Pretreatment

Biological disintegration through sonication is an effective process. Floc destruction was efficient at specific energy of 2.45 kJ/kg TS with 23% COD solubilization attained by this ultrasonic–biological pretreatment (Kavitha et al. 2016d).

9.6.1.5 Chemoultrasonic Pretreatment

Uma Rani et al. (2013b) studied the consequence of various alkaline agents, pH, and sonic response times and obtained COD solubilization, SS reduction, and biogas generation of 59%, 46%, and 80%, respectively, at specific energy of 4172 kJ/kg TS for NaOH at pH 10. Ultrasonic pretreatment was enhanced by 0.05 g/g SS of citric acid, which removes EPS. The COD solubilization and SS reduction obtained were 22.70% and 20.28%, respectively, for specific energy of 171.9 kJ/kg TS (Gayathri et al. 2015).

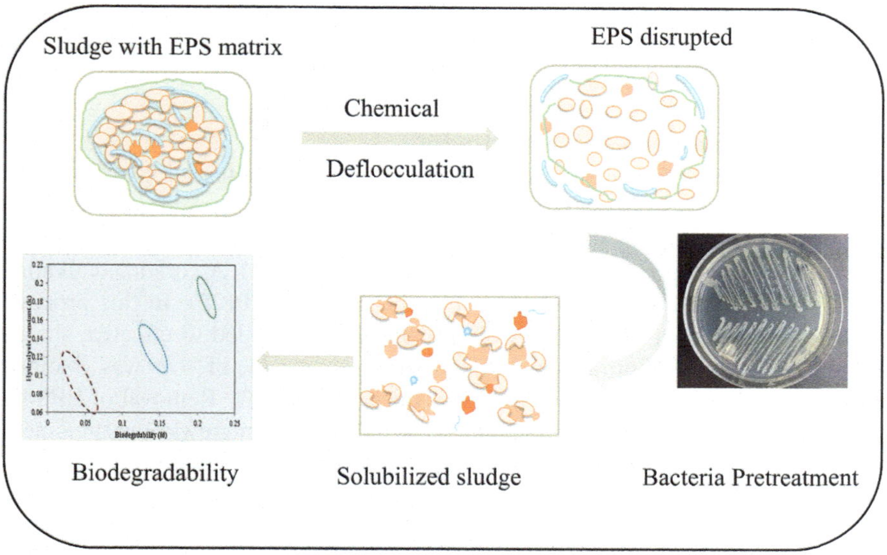

Fig. 9.5 Schematic diagram of chemo-induced bacterial pretreatment

9.6.1.6 Ultrasonic–Ozone Pretreatment

Sludge deflocculation using ultrasonication was also used followed by ozone treatment for cell disintegration. EPS removal was ultrasonically induced at an energy dosage of 76.4 kJ/kg TS. An ozone quantity of 0.0011 mg O_3/mg SS induced cell disintegration and 25.4% and 17.8% COD solubilization and SS reduction, respectively, in comparison with ozone-pretreated sludge (Sowmya et al. 2015).

9.6.1.7 Photocatalysis–Biological Pretreatment

Photocatalysis is a process of photoreactions in the presence of a catalyst to form an electron–hole pair, in turn generating hydroxyl radicals, leading to secondary reactions. The ultraviolet emission essential for the photocatalytic procedure comes from the sun or artificial sources. Dewatering of sludge is improved by using a photo-Fenton reaction. A photocatalysis reaction is used to treat pharmaceutical wastewater (Adish Kumar et al. 2012) and phenolic wastewater (Adish Kumar et al. 2014). Hospital wastewater was alleviated by using a fluidized bed solar photo-Fenton method with pH 3, 5 mM of $Fe2^+$ dosage, 50 mM of H_2O_2 dosage, and 40 g/L of silica carrier, and obtained 98% COD removal with the BOD_5/COD [5-day biochemical oxygen demand/chemical oxygen demand] ratio enhanced from 0.16 to 0.7 (Anjana Anand et al. 2015). According to Kavitha et al. (2015c), the solar photo-Fenton method obtained 4.62% COD solubilization in a 4-h time interval. A solar photocatalytic practice, using TiO_2 in an advanced oxidation process (AOP), is frequently done for management of wastewater (Rajesh Banu et al. 2008), degradation of phenolic wastewater (Adish Kumar et al. 2015), and pretreatment of sludge (Liu et al. 2014). A TiO_2 dosage of 0.03 g/g SS in a solar photocatalytic process is considered to be a superior technique for EPS destruction with negligible cell lysis, followed by bacterial treatment; in one study this approach result in SS reduction of 22.8% and COD solubilization of 22.9% (Godvin Sharmila et al. 2015).

9.6.1.8 Chemodisperser Pretreatment

Chemo-mediated disperser pretreatment is an inventive methodology. A sodium dodecyl sulfate dosage of 0.04 g/g SS disturbed EPS in sludge floc, followed by disperser pretreatment, achieving COD solubilization and SS reduction of 26% and 22.9%, respectively, at 013 kJ/kg TS (Poornima et al. 2014). A combined alkaline and disperser method was studied by Uma Rani et al. (2012b). In this process, NaOH at pH 10 with 4544 kJ kg^{-1} TS of supplied energy led to effective sludge disintegration. In this optimal condition, a high biogas value of 76% was obtained with COD solubilization of 24% and SS reduction of 23.3%. Removal of EPS by sodium citrate (0.1 g sodium citrate/g SS) and subsequent microwave pretreatment produced COD solubilization of 31% and SS reduction of 37%, with specific energy of 14,000 kJ/kg TS (Vimala et al. 2015a, b).

9.7 Conclusion

The emerging worldwide concerns about the accumulative surplus quantity of waste activated sludge, renewable energy demands, and global warming have encouraged research on the speeding up and boosting of the anaerobic digestion process. Pretreatment, coupled with anaerobic digestion, helps to reduce sludge volume, pathogens, and organic and inorganic loads. Pretreatment procedures can be categorized as biological, chemical, mechanical, thermal, or a combination of these. Enhancement of pretreatment by removal of EPS has advantages over pretreatment alone. Many methods such as deflocculation (removal of EPS) of sludge by chemicals, sonication, and dispensers have their own advantages. Among them, mechanical methods with a mild action such as low temperature, low-intensity sonication, and a low-rpm [revolutions per minute] dispenser have a great influence on EPS removal without cell lysis. Of the commonly described pretreatment methods tried at the laboratory scale, only rare mechanical, thermal, thermochemical, and ozonation methods have been effectively applied at full scale. Combinations of two or more pretreatment methods together are more effective than single pretreatment methods. These combination pretreatment methods enhance the biogas yield, improving the energy balance and economic feasibility. Therefore many research studies are ongoing to scale up the combined pretreatments for optimum usage in terms of significant energy recovery and net profit.

Opinion

Sewage sludge is an inescapable concern associated with man's activities and their growth. Auspiciously, it consists of constituents such as organic matter, nutrients, etc., which can be used beneficially. This book gives a brief overview of the ASP, pretreatment of sludge, and energy recovery from sludge in the form of biogas.

Problems occurring during pretreatment and anaerobic digestion are discussed with an aim to overcome the difficulties. Theoretical knowledge from the experimental outcomes of various research studies in the field of converting waste activated sludge (renewable biomass) to renewable energy (biogas) is elucidated in detail. This book chapter details various pretreatment methods adopted for energy generation in the form of biogas (methane) from waste biomass. Some additional points are mentioned regarding enhancement of pretreatment methods, such as removal of EPSs, leading to enhancement of pretreatment which, in turn, enhances biogas production. The book chapter not only reviews the various types of pretreatment but also includes the process mechanisms behind these pretreatment methods. Ultrasonication and thermal pretreatment methods in full-scale use are described in detail. The discussion of experimental outcomes of various pretreatment methods will be worthwhile for other researchers to recognize the advantages and limitations of individual pretreatment methods.

References

Adish Kumar S, Sivajothi S, Rajesh Banu J (2012) Coupled solar photo-Fenton process with aerobic sequential batch reactor for treatment of pharmaceutical wastewater. Desalin Water Treat 48:89–95. https://doi.org/10.1080/19443994.2012.698799

Adish Kumar S, Kanmani S, Rajesh Banu J (2014) Solar photocatalytic treatment of phenolic wastewaters: influence of chlorides, sulphates, aeration, liquid volume and solar light intensity. Desalin Water Treat 52:7957–7963. https://doi.org/10.1080/19443994.2012.69879910.1080/19443994.2013.834522

Adish Kumar S, Kanmani S, Rajesh Banu J, Yeom IT (2015) Evaluation of bench-scale solar photocatalytic reactors for degradation of phenolic wastewaters. Desalin Water Treat 57:16862–16870. https://doi.org/10.1080/19443994.2015.1083481

Anjana Anand AS, Adish Kumar S, Rajesh Banu J, Ginni G (2015) The performance of fluidized bed solar photo Fenton oxidation in the removal of COD from hospital wastewaters. Desalin Water Treat 57:9093–9100. https://doi.org/10.1080/19443994.2015.1021843

Ariunbaatar J, Panico A, Esposito G, Pirozzi F, Piet N, Lens L (2014) Pretreatment methods to enhance anaerobic digestion of organic solid waste. Appl Energy 123:143–156. https://doi.org/10.1016/j.apenergy.2014.02.035

Ayol A (2005) Enzymatic treatment effects on dewaterability of anaerobically digested biosolids—I: performance evaluations. Process Biochem 40:2427–2434. https://doi.org/10.1016/j.procbio.2004.09.023

Ayol A, Dente SK (2005) Enzymatic treatment effects on dewaterability of anaerobically digested biosolids—II: laboratory characterizations of drainability and filterability. Process Biochem 40:2435–2442. https://doi.org/10.1016/j.procbio.2004.09.024

Bougrier C, Albasi C, Delgenes JP, Carrere H (2006) Effect of ultrasonic, thermal and ozone pretreatments on waste activated sludge solubilisation and anaerobic biodegradability. Chem Eng Process 45:711–718. https://doi.org/10.1016/j.cep.2006.02.005

Bougrier C, Delgenes JP, Carrere H (2008) Effects of thermal treatments on five different waste activated sludge samples solubilization, physical properties and anaerobic digestion. J Chem Eng 139:236–244. https://doi.org/10.1016/j.cej.2007.07.099

Braguglia CM, Gianico A, Mininni G (2011) Laboratory-scale ultrasound pre-treated digestion of sludge: heat and energy balance. Bioresour Technol 102:7567–7573. https://doi.org/10.1016/j.biortech.2011.05.025

Carballa M, Manterola G, Larrea L, Ternes T, Omil F, Lema JM (2007) Influence of ozone pretreatment on sludge anaerobic digestion: removal of pharmaceutical and personal care products. Chemosphere 67:1444–1452. https://doi.org/10.1016/j.chemosphere.2006.10.004

Carrere H, Dumas C, Battimelli A, Batsone DJ, Delgenes JP, Steyer JP (2010) Pretreatment methods to improve sludge anaerobic degradability: a review. J Hazard Mater 183:1–15. https://doi.org/10.1016/j.jhazmat.2010.06.129

Chen Y, Jiang S, Yuan H, Zhou Q, Gu G (2007) Hydrolysis and acidification of waste activated sludge at different pHs. Water Res 41:683–689. https://doi.org/10.1016/j.watres.2006.07.030

Cho SK, Shin HS, Kim DH (2012) Waste activated sludge hydrolysis during ultrasonication: two step disintegration. Bioresour Technol 121:480–483. https://doi.org/10.1016/j.biortech.2012.07.024

Cho HU, Park SK, Ha JH, Park JM (2013) An innovative sewage sludge reduction by using a combined mesophilic anaerobic and thermophilic aerobic process with thermal-alkaline treatment and sludge recirculation. J Environ Manag 129:274–282. https://doi.org/10.1016/j.jenvman.2013.07.009

Chou KW, Norli I, Anees A (2010) Evaluation of the effect of temperature, NaOH concentration and pretreatment time on solubilization of palm oil mill effluent (POME) using response surface methodology (RSM). Bioresour Technol 101:8616–8622. https://doi.org/10.1016/j.biortech.2010.06.101

Climent M, Ferrer I, Baeza MM, Artola A, Vázquez F, Font X (2007) Effects of thermal and mechanical pretreatments of secondary sludge on biogas production under thermophilic conditions. J Chem Eng 133:335–342. https://doi.org/10.1016/j.cej.2007.02.020

Dhar BR, Nakhla G, Ray MB (2012) Techno-economic evaluation of ultrasound and thermal pretreatments for enhanced anaerobic digestion of municipal waste activated sludge. Waste Manag 32:542–549. https://doi.org/10.1016/j.wasman.2011.10.007

Dharmsthiti S, Kuhasuntisuk B (1998) Lipase from *Pseudomonas aeruginosa* LP602: biochemical properties and application for wastewater treatment. Ind. Microbiol Biotechnol 21:75–80. https://doi.org/10.1038/sj.jim.2900563

Erikkson T, Borrieso J, Tjerneld F (2002) Mechanism of surfactant effect in enzymatic hydrolysis of lignocelluloses. Enzym Microb Technol 31:353–364. https://doi.org/10.1016/S0141-0229(02)00134-5

Esakki Raj S, Rajesh Banu J, Kaliappan S, Yeom I-T, Adish Kumar S (2013) Effects of side-stream, low temperature phosphorus recovery on the performance of anaerobic/anoxic/oxic systems integrated with sludge, pretreatment. Bioresour Technol 140:376–384. https://doi.org/10.1016/j.biortech.2013.04.061

Fantozzi F, Buratti C (2009) Biogas production from different substrates in an experimental continuously stirred tank reactor anaerobic digester. Bioresour Technol 100:5783–5789. https://doi.org/10.1016/j.biortech.2009.06.013

Felice BD, Pontecorvo G, Carfagna M (1997) Degradation of wastewaters from olive oil mills by *Yarrowia lipolytica* ATCC 20255 and *Pseudomonas putida*. Acta Biotechnol 17:231–239. https://doi.org/10.1002/abio.370170306

Gavala HN, Yenal U, Skiadas IV, Westermann P, Ahring BK (2003) Mesophilic and thermophilic anaerobic digestion of primary and secondary sludge. Effect of pre-treatment at elevated temperature. Water Res 37:4561–4572. https://doi.org/10.1016/S0043-1354(03)00401-9

Gayathri T, Kavitha S, Adish Kumar S, Kaliappan S, Yeom IT, Rajesh Banu J (2015) Effect of citric acid induced deflocculation on the ultrasonic pretreatment efficiency of dairy waste activated sludge. Ultrason Sonochem 22:333–340. https://doi.org/10.1016/j.ultsonch.2014.07.017

Godvin Sharmila V, Kavitha S, Rajashankar K, Yeom IT, Rajesh Banu J (2015) Effects of titanium dioxide mediated dairy waste activated sludge deflocculation on the efficiency of bacterial disintegration and cost of sludge management. Bioresour Technol 197:64–71. https://doi.org/10.1016/j.biortech.2015.08.038

Gopikumar S, Arulazhagan P, Kavitha S, Adish Kumar S, Rajesh Banu J (2016) Evaluation of operational parameters for semi-continuous anaerobic digester treating pretreated waste activated sludge. Desalin Water Treat 57:9093–9100. https://doi.org/10.1080/19443994.2015.1029526

Guo JS, Xu YF (2011) Review of enzymatic sludge hydrolysis. J Bioremed Biodegr 2:1–7. https://doi.org/10.4172/2155-6199.1000130

Harrison STL (1991) Bacterial cell disruption: a key unit operation in the recovery of intracellular products. Biotechnol Adv 9:217–240. https://doi.org/10.1016/0734-9750(91)90005-G

He J-G, Xin X-D, Qiu W, Zang J, Wen Z-D, Tang J (2014) Performance of the lysozyme for promoting the waste activated sludge biodegradability. Bioresour Technol 170:108–114. https://doi.org/10.1016/j.biortech.2014.07.095

Huijun M, Chen X, Liu H, Liu H, Fu B (2015) Improved volatile fatty acids anaerobic production from waste activated sludge by pH regulation: alkaline or neutral pH? Waste Manag 48:397–403. https://doi.org/10.1016/j.wasman.2015.11.029

Jagadish HP, Raj MA, Muralidhara PL, Desai SM, Mahadeva Raju GK (2012) Kinetics of anaerobic digestion of water hyacinth using poultry litter as inoculum. Int J Environ Sci Dev 3:94–98. https://doi.org/10.7763/IJESD.2012.V3.195

Jianguo J, Shihui Y, Maozhe C, Qunfang Z (2009) Disintegration of sewage sludge with bifrequency ultrasonic treatment. Water Sci Technol 60:1445–1453. https://doi.org/10.2166/wst.2009.469

Jianguo J, Changxiu G, Jiaming W, Sicong T, Yujing Z (2014) Effects of ultrasound pre-treatment on the amount of dissolved organic matter extracted from food waste. Bioresour Technol 155:266–271. https://doi.org/10.1016/j.biortech.2013.12.064

Kavitha S, Adish Kumar S, Yogalakshmi KN, Kaliappan S, Rajesh Banu J (2013) Effect of enzyme secreting bacterial pretreatment on enhancement of aerobic digestion potential of waste activated sludge interceded through EDTA. Bioresour Technol 15:210–219. https://doi.org/10.1016/j.biortech.2013.10.021

Kavitha S, Adish Kumar S, Kaliappan S, Yeom IT, Rajesh Banu J (2014a) Improving the amenability of municipal waste activated sludge for biological pretreatment by phase-separated sludge disintegration method. Bioresour Technol 169:700–706. https://doi.org/10.1016/j.biortech.2014.07.065

Kavitha S, Jayashree C, Adish Kumar S, Kaliappan S, Rajesh Banu J (2014b) Enhancing the functional and economical efficiency of a novel combined thermo chemical disperser disintegration of waste activated sludge for biogas production. Bioresour Technol 173:32–41. https://doi.org/10.1016/j.biortech.2014.09.078

Kavitha S, Kaliappan S, Adish Kumar S, Yeom IT, Rajesh Banu J (2015a) Effect of NaCl induced floc disruption on biological disintegration of sludge for enhanced biogas production. Bioresour Technol 192:807–811. https://doi.org/10.1016/j.biortech.2015.05.071

Kavitha S, Karthika P, Rajesh Banu J, Yeom IT, Adish Kumar S (2015b) Enhancement of waste activated sludge reduction potential by amalgamated solar photo-Fenton treatment. Desalin Water Treat 57:13144–13156. https://doi.org/10.1080/19443994.2015.1055810

Kavitha S, Saranya T, Kaliappan S, Adish Kumar S, Yeom IT, Rajesh Banu J (2015c) Accelerating the sludge disintegration potential of a novel bacterial strain *Planococcus jake* 01 by CaCl induced deflocculation. Bioresour Technol 175:396–405. https://doi.org/10.1016/j.biortech.2014.10.122

Kavitha S, Yukesh Kannah R, Yeom IT, Uan DK, Rajesh Banu J (2015d) Combined thermo-chemo-sonic disintegration of waste activated sludge potential municipal waste activated sludge by ultrasonic aided bacterial disintegration. Bioresour Technol 200:161–169. https://doi.org/10.1016/j.biortech.2015.10.026

Kavitha S, Jessin Brindha GM, Sally Gloriana A, Rajashankar K, Yeom IK, Rajesh Banu J (2016a) Enhancement of aerobic biodegradability potential municipal waste activated sludge by ultrasonic aided bacterial disintegration. Bioresour Technol 200:161–169. doi:10.1016/j.biortech.2015.10.026

Kavitha S, Rajesh Banu J, Subitha G, Ushani U, Yeom IT (2016b) Impact of thermo-chemo-sonic pretreatment in solubilizing waste activated sludge for biogas production: energetic analysis and economic assessment. Bioresour Technol 219:479–486. https://doi.org/10.1016/j.biortech.2016.07.115

Kavitha S, Rajesh Banu J, Vinoth Kumar J, Rajkumar M (2016c) Improving the biogas production performances of municipal waste activated sludge via disperser induced microwave disintegration. Bioresour Technol 217:21–27. https://doi.org/10.1016/j.biortech.2016.02.034

Kavitha S, Saji Pray S, Yogalakshmi KN, Adish Kumar S, Yeom IT, Rajesh Banu J (2016d) Effect of chemo-mechanical disintegration on sludge anaerobic digestion for enhanced biogas production. Environ Sci Pollut Res 23:2402–2414. https://doi.org/10.1007/s11356-015-5461-z

Kianmehr P, Parker W, Seto P (2010) An evaluation of protocols for characterization of ozone impacts on WAS properties and digestibility. Bioresour Technol 101:8565–8572. https://doi.org/10.1016/j.biortech.2010.06.061

Kleinig AR, Middelberg APJ (1998) On the mechanism of microbial cell disruption in high-pressure homogenization. Chem Eng Sci 53:891–898. https://doi.org/10.1016/S0009-2509(97)00414-4

Kuglarz M, Karakashev D, Angelidaki I (2013) Microwave and thermal pretreatment as methods for increasing the biogas potential of secondary sludge from municipal wastewater treatment plants. Bioresour Technol 134:290–297. https://doi.org/10.1016/j.biortech.2013.02.001

Lakshmi MV, Merrylin J, Kavitha S, Adish Kumar S, Rajesh Banu J, Yeom IT (2014) Solubilization of municipal sewage waste activated sludge by novel lytic bacterial strains. Environ Sci Pollut Res 21:2733–2743. https://doi.org/10.1007/s11356-013-2228-2

Li H, Jin Y, Mahar RB, Wang Z, Nie Y (2007) Effects and model of alkaline waste activated sludge treatment. Bioresour Technol 99:5140–5144. https://doi.org/10.1016/j.biortech.2007.09.019

Lin Y, Wang D, Wu S, Wang C (2009) Alkali pretreatment enhances biogas production in the anaerobic digestion of pulp and paper sludge. J Hazard Mater 170:366–373. https://doi.org/10.1016/j.jhazmat.2009.04.086

Liu C, Shi W, Li H, Lei Z, He L, Zhang Z (2014) Improvement of methane production from waste activated sludge by on-site photocatalytic pretreatment in a photocatalytic anaerobic fermenter. Bioresour Technol 155:198–203. https://doi.org/10.1016/j.biortech.2013.12.041

Luo K, Yang Q, Li XM, Chen HB, Liu X, Yang GJ, Zeng GM (2013) Novel insights into enzymatic-enhanced anaerobic digestion of waste activated sludge by three-dimensional excitation and emission matrix fluorescence spectroscopy. Chemosphere 91:579–585. https://doi.org/10.1016/j.chemosphere.2012.12.002

Mata-Alvarez J, Mace S, Llabres P (2000) Anaerobic digestion of organic solid wastes. An overview of research achievements and perspectives. Bioresour Technol 74:3–16. https://doi.org/10.1016/S0960-8524(00)00023-7

Merrylin J, Adish Kumar S, Kaliappan S, Yeom IT, Rajesh Banu J (2013a) Biological pretreatment of non-flocculated sludge augments the biogas production in the anaerobic digestion of the pretreated waste activated sludge. Environ Technol 34:2113–2123. https://doi.org/10.1080/09593330.2013.810294

Merrylin J, Kaliappan S, Adish Kumar S, Yeom IT, Rajesh Banu J (2013b) Effect of extracellular polymeric substances on sludge reduction potential of *Bacillus licheniformis*. Int J Environ Sci Technol 10:85–92. https://doi.org/10.1007/s13762-012-0141-8

Monique R, Elisabeth GN, Etienne P, Dominique L (2008) A high yield multimethod extraction protocol for protein quantification in activated sludge. Bioresour Technol 99:7464–7471. https://doi.org/10.1016/j.biortech.2008.02.025

Ngoc TL, Carine JL, Henri D (2013) Ultrasonic sludge pretreatment under pressure. Ultrason Sonochem 20:1203–1210. https://doi.org/10.1016/j.ultsonch.2013.03.005

Nguyen MT, Yasin NHM, Miyazaki T, Maeda T (2014) Enhancement of sludge reduction and methane production by removing extracellular polymeric substances from waste activated sludge. Chemosphere 117:552–558. https://doi.org/10.1016/j.chemosphere.2014.08.055

Parawira W, Murto M, Read JS, Mattiasson B (2007) A study of two-stage anaerobic digestion of solid potato waste using reactors under mesophilic and thermophilic conditions. Environ Technol 28:1205–1216. https://doi.org/10.1080/09593332808618881

Parmar N, Singh A, Ward OP (2001) Enzyme treatment to reduce solids and improve settling of sewage sludge. J Ind Microbiol Biotechnol 26:383–386. https://doi.org/10.1038/sj.jim.7000150

Perez-Elvira SI, Nieto Diez P, Fdz-Polanco F (2006) Sludge minimization technologies. Rev Environ Sci Biotechnol 5:375–398. https://doi.org/10.1007/s11157-005-5728-9

Pilli S, Bhunia P, Yang S, Leblanc RJ, Tyagi RD, Surampalli RY (2011) Ultrasonic pretreatment of sludge. Ultrason Sonochem 18:1–18. https://doi.org/10.1016/j.ultsonch.2010.02.014

Poornima Devi T, Vimala Ebenezer A, Adish Kumar S, Kaliappan S, Rajesh Banu J (2014) Effect of deflocculation on the efficiency of disperser induced dairy waste activated sludge disintegration and treatment cost. Bioresour Technol 167:151–158. https://doi.org/10.1016/j.biortech.2014.06.004

Rajesh Banu J, Kaliappan S (2007) Treatment of tannery wastewater using hybrid upflow anaerobic sludge blanket reactor. J Environ Eng Sci 6:415–421. https://doi.org/10.1139/s06-063

Rajesh Banu J, Kaliappan S, Beck D (2006) High rate anaerobic treatment of sago wastewater using HUASB with PUF as carrier. Int J Environ Sci Technol 3:69–77. https://doi.org/10.1007/BF03325909

Rajesh Banu J, Kaliappan S, Yeom IT (2007a) Treatment of domestic wastewater using upflow anaerobic sludge blanket reactor. Int J Environ Sci Technol 4:363–370. https://doi.org/10.1007/BF03326295

Rajesh Banu J, Kaliappan S, Yeom IT (2007b) Two-stage anaerobic treatment of dairy wastewater using HUASB with PUF and PVC carrier. Biotechnol Bioprocess Eng 12:257–264. https://doi.org/10.1007/BF02931101

Rajesh Banu J, Anandan S, Kaliappan S, Yeom IT (2008) Treatment of dairy wastewater using anaerobic and solar photocatalytic methods. Sol Energy 82:812–819. https://doi.org/10.1016/j.solener.2008.02.015

Rajesh Banu J, Arulazhagan P, Adish Kumar S, Kaliappan S, Lakshmi AM (2014) Anaerobic co-digestion of chemical- and ozone-pretreated sludge in hybrid upflow anaerobic sludge blanket reactor. Desalin Water Treat 54:3269–3278. https://doi.org/10.1080/19443994.2014.912156

Rajesh Banu J, Ushani U, Merrylin J, Kaliappan S (2016) Evaluation of operational parameters for biodegration of bacterially disintegrated sludge. Desalin Water Treat 57:25018–25027. https://doi.org/10.1080/19443994.2016.1150888

Rashad R, Tjalfe GP, Nizami A-S, Zaki-ul-Zaman A, Murphy Jerry D, Kiely G (2010) Effect of thermal, chemical and thermo-chemical pretreatments to enhance methane production. Energy 35:4556–4561. https://doi.org/10.1016/j.energy.2010.07.011

Sahinkaya S, Sevimli MF (2013) Sono-thermal pretreatment of waste activated sludge before anaerobic digestion. Ultrason Sonochem 20:587–594. https://doi.org/10.1016/j.ultsonch.2012.07.006

Salsabil MR, Laurent J, Casellas M, Dagot C (2010) Techno-economic evaluation of thermal treatment, ozonation and sonication for the reduction of wastewater biomass volume before aerobic or anaerobic digestion. J Hazard Mater 174:323–333. https://doi.org/10.1016/j.jhazmat.2009.09.054

Shehu MA, Manan ZA, Alwi SR (2012) Optimization of thermo-alkaline pretreatment of sewage sludge for enhanced biogas yield. Bioresour Technol 114:69–74. https://doi.org/10.1016/j.biortech.2012.02.135

Sowmya Packyam G, Kavitha S, Adish Kumar S, Kaliappan S, Yeom IT, Rajesh Banu J (2015) Effect of sonically induced deflocculation on the efficiency of ozone mediated partial sludge disintegration for improved production of biogas. Ultrason Sonochem 26:241–248. https://doi.org/10.1016/j.ultsonch.2015.01.015

Sri Bala Kameswari K, Kalyanaraman C, Thanasekaran K (2011) Effect of ozonation and ultrasonication pretreatment processes on co-digestion of tannery solid wastes. Clean Techn Environ Policy 13:517–525. https://doi.org/10.1007/s-10098-010-0334-0

Suresh Karthik Kumar M, Krishna Kumar T, Arulazhagan P, Adish Kumar S, Yeom IT, Rajesh Banu J (2014) Effect of alkaline and ozone pretreatment on sludge reduction potential of a membrane bioreactor treating high-strength domestic wastewater. Desalin Water Treat 55:1127–1134. https://doi.org/10.1080/19443994.2014.923335

Suslick KS (1998) Sonochemistry. In: Kirk–Othmer encyclopedia of chemical technology, vol 26, 4th edn. Wiley, New York, pp 517–541. ISBN: 0-471-52696-7

Toreci I, Kennedy KJ, Droste RL (2009) Evaluation of continuous mesophilic anaerobic sludge digestion after high temperature microwave pretreatment. Water Res 43:1273–1284. https://doi.org/10.1016/j.watres.2008.12.022

Uma Rani R, Adish Kumar S, Kaliappan S, Rajesh Banu J (2012a) Combined treatment of alkaline and disperser for improving solubilization and anaerobic biodegradability of dairy waste activated sludge. Bioresour Technol 126:107–116. https://doi.org/10.1016/j.biortech.2012.09.027

Uma Rani R, Adish Kumar S, Kaliappan S, Yeom IT, Rajesh Banu J (2012b) Low temperature thermo-chemical pretreatment of dairy waste activated sludge for anaerobic digestion process. Bioresour Technol 103:415–424. https://doi.org/10.1016/j.biortech.2011.09.124
Uma Rani R, Adish Kumar S, Kaliappan S, Yeom IT, Rajesh Banu J (2013a) Impacts of microwave pretreatments on the semi-continuous anaerobic digestion of dairy waste activated sludge. Waste Manag 33:1119–1127. https://doi.org/10.1016/j.wasman.2013.01.016
Uma Rani R, Adish Kumar S, Kaliappan S, Yeom IT, Rajesh Banu J (2013b) Enhancing the anaerobic digestion potential of dairy waste activated sludge by two step sono-alkalization pretreatment. Ultrason Sonochem 21:1065–1074. https://doi.org/10.1016/j.ultsonch.2013.11.007
Ushani U, Kavtiha S, Johnson M, Yeom IT, Rajesh Banu J (2016) Upgrading the hydrolytic potential of immobilized bacterial 1 pretreatment for boosting biofuel production. Environ Sci Pollut Res 24(1):813–826. https://doi.org/10.1007/s11356-016-7819-2
Vimala Ebenezer A, Arulazhagan P, Adish Kumar S, Yeom IT, Rajesh Banu J (2015a) Effect of deflocculation on the efficiency of low-energy microwave pretreatment and anaerobic biodegradation of waste activated sludge. Appl Energy 145:104–110. https://doi.org/10.1016/j.apenergy.2015.01.133
Vimala Ebenezer A, Kaliappan S, Adish Kumar S, Yeom IT, Rajesh Banu J (2015b) Influence of deflocculation on microwave disintegration and anaerobic biodegradability of waste activated sludge. Bioresour Technol 185:194–201. https://doi.org/10.1016/j.biortech.2015.02.102
Wang BB, Peng DC, Hou YP, Li HJ, Pei LY, LF Y (2014) The important implications of particulate substrate in determining the physicochemical characteristics of extracellular polymeric substances (EPS) in activated sludge. Water Res 58:1–8. https://doi.org/10.1016/j.watres.2014.03.060
Water Environment Federation (1992) Design of municipal wastewater treatment plants, Manual of Practice No. 8. McGraw-Hill Education, New Delhi. ISBN 97800971663588
Watson SD, Akhurst T, Whiteley CG, Rose PD, Pletschke BI (2004) Primary sludge floc degradation is accelerated under biosulphidogenic conditions: enzymological aspects. Enzym Microb Technol 34:595–602. https://doi.org/10.1016/j.enzmictec.2004.01.004
Weemaes M, Grootaerd H, Simons F, Verstraete W (2000) Anaerobic digestion of ozonized biosolids. Water Sci Technol 34:2330–2336. https://doi.org/10.1016/S0043-1354(99)00373-5
Xin XD, He JG, Qiu W, Tang J, Liu TT (2015) Microbial community related to lysozyme digestion process for boosting waste activated sludge biodegradability. Bioresour Technol 175:112–119. https://doi.org/10.1016/j.biortech.2014.10.042
Yan S, Miyanaga K, Xing XH, Tanji Y (2008) Succession of bacterial community and enzymatic activities of activated sludge by heat-treatment for reduction of excess sludge. Biochem Eng J 39:598–603. https://doi.org/10.1016/j.bej.2007.12.002
Yan ST, Chu LB, Xing XH, AF Y, Sun XL, Jurcik B (2009) Analysis of the mechanism of sludge ozonation by a combination of biological and chemical approaches. Water Res 43:195–203. https://doi.org/10.1016/j.watres.2008.09.039
Zhang P, Zhang G, Wang W (2007) Ultrasonic treatment of biological sludge: floc disintegration, cell lysis and inactivation. Bioresour Technol 98:207–210. https://doi.org/10.1016/j.biortech.2005.12.002
Zhang GM, Zhang PY, Yang J, Liu HZ (2008) Energy efficient sludge sonication: power and sludge characteristics. Bioresour Technol 99:9029–9903. https://doi.org/10.1016/j.biortech.2008.04.021
Zhang G, Yang J, Liu H, Zhang J (2009) Sludge ozonation: disintegration, supernatant changes and mechanisms. Bioresour Technol 100:1505–1509. https://doi.org/10.1016/j.biortech.2008.08.041
Zhang S, Zhang P, Zhang G, Fan J, Zhang Y (2012) Enhancement of anaerobic sludge digestion by high-pressure homogenization. Bioresour Technol 118:496–501. https://doi.org/10.1016/j.biortech.2012.05.089
Zhang Y, Zhang P, Ma W, Wu H, Ma B, Zhang S, Xu X (2013) Sewage sludge solubilization by high-pressure homogenization. Water Sci Technol 67:2399–2405. https://doi.org/10.2166/wst.2013.141

Mass Production of Microalgae in Photobioreactors for Biodiesel Application: Selection, Limitations, and Optimization

10

Sanjay Pawar and Suvidha Gupta

Abstract

The algal biodiesel is produced in four steps: cultivation, harvesting and dewatering, lipid extraction, and transesterification reaction (converting lipids to fatty acid methyl esters, FAME, i.e., biodiesel). Microalgae cultivation in photobioreactor (PBR) is not only costly but also very challenging step among the steps of algal biodiesel production. In this chapter, the advantages and disadvantages of various types of PBR have been discussed in the context of mass production at outdoor conditions. Several environmental limitations which affect the performance of PBR are critically reviewed with respect to the selections of PBRs and microalgae strains. The mode of cultivation was found to affect the biomass and lipid productivity under different conditions. Thus, an overview suggesting the optimal ways to enhance the total lipid content of microalgae under different modes of cultivation is discussed in this chapter. The research work of microalgae cultivation on wastewater is tabulated and discussed in the context of feasibility of process. The quality of biodiesel strongly depends on the proportion of saturated and unsaturated fatty acids. Thus, various process parameters and strategies which affect the lipid's composition are also discussed in detail.

Keywords

Mixotrophic • Scale-up • Cetane number • Fatty acids • Biofuel

S. Pawar (✉) • S. Gupta
Environmental Biotechnology and Genomics Division (EBGD), CSIR – National Environmental Engineering Research Institute (NEERI), Nehru Marg, Nagpur 440020, Maharashtra, India
e-mail: sanjay6282@gmail.com; sb_pawar@neeri.res.in

H.J. Purohit et al. (eds.), *Optimization and Applicability of Bioprocesses*, https://doi.org/10.1007/978-981-10-6863-8_10

10.1 Introduction

Microalgae cultivation for various applications is not new to us. For several years, microalgae have been cultivated in photobioreactors (PBRs) for protein supplements, carbohydrates, and active compounds such as astaxanthin and beta-carotene (Bux and Chisti 2016). But, over the time, when the need and security for fuels were increased, it received wide attention as the possible solution for energy (fuel) scarcity because of its inherent property to accumulate a large quantity of lipids (up to 50% of its dry cell weight (DCW)) (Chisti 2007). Microalgae cultivation has many advantages such as rapid growth, higher biomass and lipid productivity, and robustness to various environmental conditions; however, the high production cost of microalgal biodiesel needs to reduce down by optimization of entire process as suggested by several previous studies (Brennan and Owende 2010; Chisti 2013; Quinn and Davis 2015).

In general, microalgae growth depends on light intensity, compositions of nutrients, carbon source, pH, temperature, and mixing, etc. (Miron et al. 2003; Mata et al. 2010). However, all these parameters are having some limitations when supplied in low or high extents than a particular value. For examples, high light intensity causes photoinhibition, poor mixing causes the photorespiration and settling of cells, nutrient (C:N:P) composition affects the biomass and lipid productivity, temperature alters the lipid composition which then directly affects the biodiesel quality, and so on (Pruvost et al. 2009, 2011; Hulatt and Thomas 2011). It is desired to get the high biomass concentrations of selected species with the main focus on getting more lipid productivity. But, after achieving certain concentration, microalgae cells start to form flocks and settle down at the bottom of the PBR, or sticking on the PBR wall becomes evident. The formation of flocks, in turn, reduces the ability of cells to capture the light, and the mutual shading effect becomes predominant. One way to overcome the above problem is mixing. Mixing can be provided by paddles such as in open raceway ponds or using air in closed vertical column PBRs (Pawar 2016b). The precautions must be taken while providing the mixing to the culture in the PBRs so that cells are not damaging because of the high shear stress (Contreras et al. 1998; Michels et al. 2010), and here hydrodynamics play a major role. The technological aspects can be manipulated with an optimal design for indoor cultivation, but outdoor cultivation becomes very challenging due to environmental constraints.

The following steps are generally considered before selecting the site for microalgae cultivation:

1. Choice of geographical location (sunlight and rainfall patterns)
2. Type of PBR and scale of production
3. Microalga strain (freshwater/marine or robustness for wastewater)
4. Mode of cultivation (autotrophic/mixotrophic)
5. Availability of wastewater treatment facility and/or thermal power station in nearby area
6. Post processing scenario (biodiesel, bioethanol, bio-oil, biogas, etc.)

The total lipid accumulation and its composition in the microalgae vary with respect to species, nutrient manipulation, and mode of cultivation (Gim et al. 2014; Gris et al. 2014; Gomma et al. 2015; Lari et al. 2016). In general, for 100 L of biodiesel (density = 840 g/L), approximately 3360 kg of microalgae with 25% (w/w) lipid content is required to be produced. With the biomass productivity of 2 g/L (15 day harvesting time), a total volume of 1680 kL culture needs to be processed in the PBR (Yang et al. 2011). These numbers show that a very massive scale is required to meet the requirement of biodiesel production per annum for a state or a country.

10.2 Choice of Geographical Location

The mapping of solar intensity and rainfall patterns at various geographical locations is prerequisite for microalgae cultivation on a large scale as pointed out by Pawar (2016a). The through estimate of the feasibility of microalgae cultivation at several geographical locations in the USA has been given by Quinn et al. (2012). The temperature-dependent growth model was used for estimation of microalgae biomass and lipid productivities. It has been reported that the geographical location plays a major role to determine the economic feasibility of the entire production system. This factor is very important due to the fact that the microalgae growth ceases at light intensity of 200 μmol/m^2/s, which is just 10–20% of the peak sunlight in a daytime. Thus, the culture may be subjected to photoinhibition with the incompatible type of PBR (Lopez et al. 2006; Pruvost et al. 2015). The high temperature also speeds up the process of water evaporation which is another main drawback. Thus, in the case of autotrophic cultivations in tropical countries, PBR having less impact on the culture instability should be selected.

Three designs of PBR are widely studied in the context of mass production of microalgae, viz., open raceway pond (ORP), tubular loop PBR, and vertical tubular PBR, as shown in Figs. 10.1, 10.2, and 10.3, respectively. ORP possesses the risk of

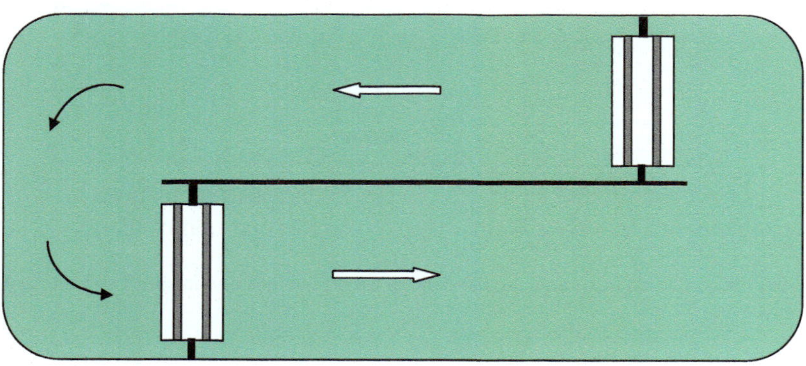

Fig. 10.1 Open raceway pond photobioreactor equipped with paddles

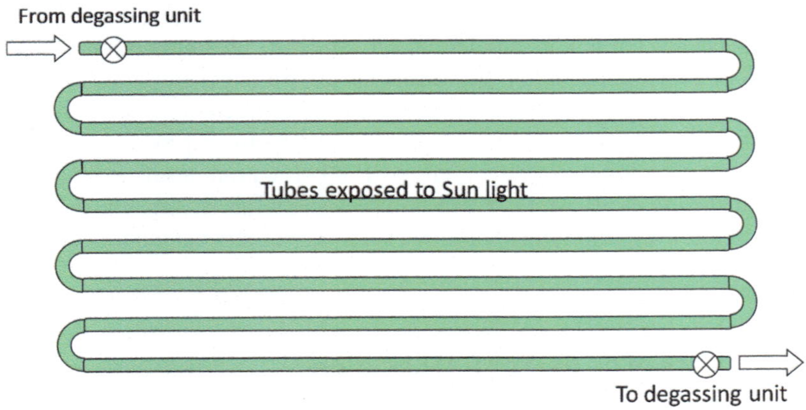

Fig. 10.2 Tubular loop photobioreactor: array of several tubes with vertical or horizontal position

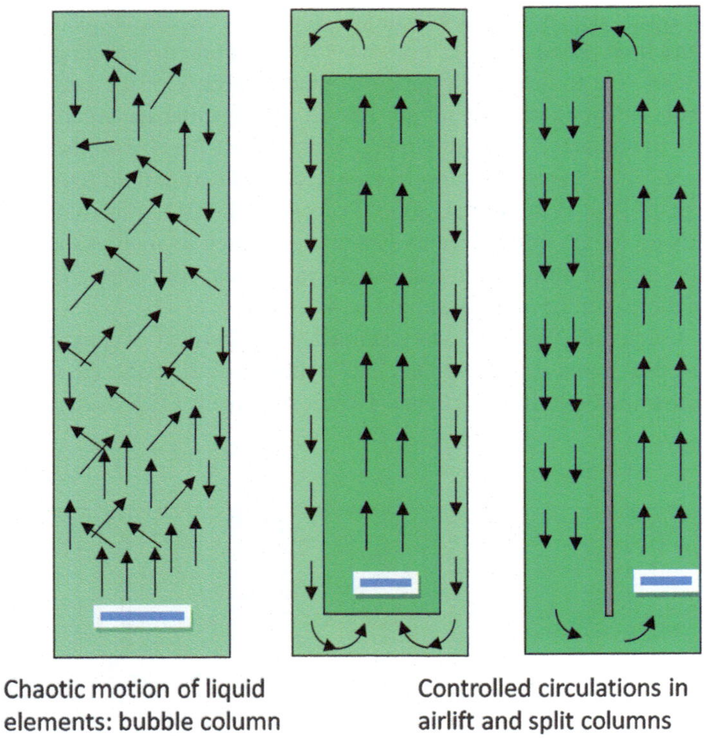

Fig. 10.3 Flow pattern in vertical tubular column PBRs: bubble column, airlift column, and split column

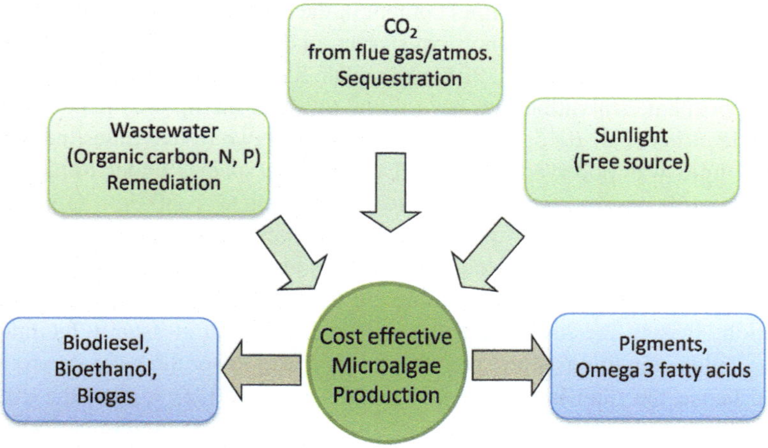

Fig. 10.4 Cost-effective strategy for mass-scale microalgae production

contamination and water loss due to evaporation at high temperatures, while tubular loop PBR is having the risk of photoinhibition as well as photorespiration due to its small diameter tubes. The outdoor cultivations were also found to change the composition of membrane lipids (Venkata Mohan et al. 2015). The lower cultivation temperatures were found to increase the unsaturated fatty acid content in the cells and decrease membrane fluidity (Venkata Mohan et al. 2015). The normal range of temperature for microalgae cultivation was found in between 25 and 30 °C. Nevertheless, at high sun intensity areas, vertical bubble column PBRs with the diameter approximately 20–30 cm can be used to avoid the photoinhibition (Miron et al. 1999; Lopez et al. 2006).

Figure 10.4 represents the cost-effective methodology for large-scale microalgae production. To make the process cost-effective, an installation near to the thermal power station can be thought a good option so that the need of CO_2 can be fulfilled by flue gas in the case of photoautotrophic cultivation or a location near central wastewater treatment facility can be opted for heterotrophic or mixotrophic cultivation (Gomez-Serrano et al. 2015; Gouveia et al. 2016). Thus, outdoor cultivation of microalgae is not only beneficial to CO_2 sequestration but also to the wastewater bioremediation besides saving in cost of biodiesel production. Moreover, the requirement of freshwater for the cultivation can be partially replaced by wastewater (Gouveia et al. 2016).

10.3 Selection of Right Microalga Strain for Biodiesel Production

There have been hundreds of microalgae strains identified so far, and still research is in pursuit to identify the most suitable strains for biodiesel application (Nascimento et al. 2013, 2014; Valdez-Ojeda et al. 2015). Different strategies have been reported in the literature to increase the lipid content in the algal cells. One common method to increase the lipid content is nutrient starvation (Vasumathi et al. 2012; Prathima Devi et al. 2013; Chandra et al. 2014; Yewalkar-Kulkarni et al. 2016). Nitrogen, as well as phosphate starvation, has been found to affect the lipid content (Benvenuti et al. 2015; Rios et al. 2015; Lari et al. 2016). About two to five times increment in the total lipid was observed for several microalgae species under nitrate-deplete condition (150 mg/L) as compared to nitrate-replete condition (1500 mg/L) (Griffiths et al. 2012). An increase in the lipid content with respect to starvation is not obvious. Several studies have observed that some microalgae strains change the lipid composition instead of an increase in the total lipid content under nutrient starvation. This is due to the fact that lipid enhancement is usually facilitated in nitrate-deplete conditions which, on the contrary, also lower the biomass productivity and hence the total lipid content. Griffiths et al. (2012) studied the effect of N-deficient and N-sufficient conditions on total lipid accumulation for several microalgae species. Among them, *Chlorella vulgaris* and *Scenedesmus* were found to accumulate total lipid over 40% of DCW in 14 days. Similarly, in another study, Benvenuti et al. (2015) examined lipid productivity of different strains under nitrogen-deprived conditions. It has been stated that the right strain for industrial application should contain the high capacity of lipid accumulation and high photosynthetic efficiency under nitrogen starvation. Among the studied strains, *Nannochloropsis* sp. was found the right strain for industrial application as it accumulated lipid up to 45% of its DCW with the productivity of 238 mg/L/d.

The total lipids in the microalgae mainly consist of phospholipids, glycolipids, and triacylglycerol (TAG). The lipid is converted to biodiesel (fatty acid methyl esters, FAME) by the transesterification reaction. The lipids contain a different proportion of saturated fatty acid (SFA), monounsaturated fatty acid (MUFA), and polyunsaturated fatty acid (PUFA). The proportion of SFA to UFA is the most important parameter to determine the biodiesel quality (Griffiths et al. 2012; Nascimento et al. 2013, 2014). Biodiesel quality is determined by certain parameters such as cetane number (CN), saponification value (SV), iodine value (IV), and cold filter plug point (CFPP) (Ramos et al. 2009). The CN value decides the ignition quality, and it should be above 51 as per the European standard of biodiesel (EN14214) and above 47 as per the standard of ASTM. The IV value suggests the oxidative stability of the FAME. IV value below 120 g I_2 /100 g is acceptable. The unsaturated component of the FAME is highly susceptible to oxidation due to the double-bond carbon chain. The CFPP value denotes the performance of the engine at lower temperatures. The higher SFA content of FAME increases the value of CFPP which in turn affects the performance of the engine in cold days.

Overall, the saturated fraction of fatty acids in FAME increases the CN number but deteriorates the cold flow property, while some unsaturated fraction of fatty acids increases the lubricity and the risk of oxidation (Nascimento et al. 2013, 2014). So, there should be a balance between the ratios of SFA and UFA contents of lipid. Thus, the strategy should be focused not only to enhance the lipid content but also getting the FAME with the desired ratio of different fatty acids. For example, microalgae species such as *Chlorella and Scenedesmus* are widely studied in the context of biodiesel production. Although the CN for these species obtained between 51 and 65, the CFPP falls in the range of 8.6–11.6 °C for *C. vulgaris* and *Sc. obliquus* (Nascimento et al. 2013, 2014). These CFPP values are not acceptable as in the winter season the engine filter blockage or cold start problems can arise. Thus, the species suitability cannot be judged on the basis of CN only, and other factors are also equally important (Ramos et al. 2009).

The lipid content and its composition for some microalgae strains are shown in Table 10.1. Various combinations of experiments were implemented to get the FAME with desired compositions of standard quality. It should be noted that many laboratory experiments are generally conducted in indoor conditions with fixed light intensity, whereas the actual light intensity falling on the surface of PBR varies a lot in the case of outdoor cultivation which may produce different results. Thus, a PBR should be taken into account in all design parameters along with biological aspects of the process which affect the final quality of biodiesel.

10.4 Mode of Cultivation

Microalgae constitute about 50% of carbon, 1–10% of nitrogen, and 1–8% of phosphorous (Lari et al. 2016). The assimilation of these elements is channeled through the metabolic pathways. In general, there are two primary mechanisms of carbon fixation in microalgae: autotrophic and heterotrophic (Venkata Mohan et al. 2015). Microalgae species can be grown in three modes of cultivation, viz., photoautotrophic, heterotrophic, and mixotrophic. The photoautotrophic cultivation of microalgae is common and widely studied. Under this mode of cultivation, microalgae use CO_2 as an inorganic carbon source and light energy for photosynthesis to generate the biomass, i.e., reduction of carbon dioxide to carbohydrate through the Calvin-Benson cycle (Lowrey et al. 2015). The factors which affect the biomass growth and lipid accumulation are as follows: CO_2 fraction in the inlet air, light intensity, nutrient sufficiency or deficiency, etc.

In the heterotrophic mode of cultivation, organic carbon is used as a substrate for microalgae growth (Ceron-Garcia et al. 2013). Organic carbon is used as a sole energy source, and light is not required for the growth of cells. Various organic carbon sources such as acetate, glucose, and glycerol have been found to support the microalgae growth in heterotrophic or mixotrophic cultivation (Bhatnagar et al. 2011; Perez-Garcia et al. 2011; Ceron-Garcia et al. 2013; Gupta et al. 2017). The factors which affect the growth of microalgae are as follows: cost for organic carbon source, substrate inhibition, risk of contamination, etc. (Chen and Johns 1995).

The mixotrophic mode of cultivation actually is a combination of photoautotrophic and heterotrophic modes of cultivation. Not all microalgae species show such kind of behavior. Some microalgae strains have the capability to assimilate simultaneously both inorganic and organic carbon in the presence and absence of light, respectively. In the mixotrophic mode of cultivation, acetyl-CoA pool is maintained by two ways, i.e., CO_2 fixation and extracellular organic carbon (Venkata Mohan et al. 2015). Mixotrophic cultivation shows reduced photoinhibition and improved

Table 10.1 Some potential microalgae strains for biodiesel production

Microalgae sp.	CN	IV	SV	DU %	CFPP (˚C)	References
Chlamydomonas sp.	64.94	27.34	220.17	28.15	−12.23	Nascimento et al. (2013)
Desmodesmus brasiliensis	53.28	87.05	205.46	86.84	−0.55	Nascimento et al. (2013)
Scenedesmus obliquus	63.63	35.38	216.04	36.63	−11.87	Nascimento et al. (2013)
Chlorella vulgaris	61.83	52.63	199.37	58.17	−10.81	Nascimento et al. (2013)
Botryococcus braunii	52.67	94.60	197.36	100.69	−7.96	Nascimento et al. (2013)
Botryococcus terribilis	59.50	66.90	193.22	69.41	−10.26	Nascimento et al. (2013)
Chlorella vulgaris	52	92	–	–	4	Griffiths et al. (2012)
Scenedesmus sp.	54	82	–	–	8	Griffiths et al. (2012)
Phaeodactylum tricornutum	53	72	–	–	9	Griffiths et al. (2012)
Nannochloropsis sp.	57	59	–	–	14	Griffiths et al. (2012)
Spirulina platensis	51	56	–	–	19	Griffiths et al. (2012)
Phaeodactylum tricornutum	48.3	–	–	116.9	−10.6	Gangadhar et al. (2016)
Nannochloropsis oculata	46.7	–	–	124.3	−10.7	Gangadhar et al. (2016)
Tetraselmis sp.	47.3	–	–	126.6	−8.5	Gangadhar et al. (2016)
Scenedesmus abundance	52.15	94	202	84.5	–	Mandotra et al. (2014)
Chlorella vulgaris	56.7	65	217.8	74.1	4.5	Francisco et al. (2010)
Scenedesmus obliquus	56.1	68.2	217.5	67.8	20.8	Francisco et al. (2010)
Scenedesmus obliquus[1]	59.3	66.36	195.4	76.53	−2.48	Wu and Miao (2014)
Scenedesmus obliquus[2]	57.51	74.51	195.14	85.74	−7.01	Wu and Miao (2014)
Chlorella pyrenoidosa[1]	43.85	133.37	198.05	132.08	−0.78	Wu and Miao (2014)
Chlorella pyrenoidosa[2]	41.38	144.29	198.17	132.08	−5.85	Wu and Miao (2014)
Scenedesmus abundans[3]	57.7	64.7	210.7	51.3	–	Gonzalez-Garcinuno et al. (2014)

(continued)

Table 10.1 (continued)

Microalgae sp.	CN	IV	SV	DU %	CFPP (°C)	References
Scenedesmus abundans[4]	54.7	82.2	202.9	89.9	–	Gonzalez-Garcinuno et al. (2014)
Chlorella ellipsoidea[3]	59.3	58.9	207.9	64.6	–	Gonzalez-Garcinuno et al. (2014)
Chlorella ellipsoidea[4]	57.5	70.2	202.2	74.2	–	Gonzalez-Garcinuno et al. (2014)

CN cetane number, *IV* iodine value, *SV* saponification value, *DU* degree of unsaturation, *CFPP* cold filter plug point
Superscripts 1 and 2 denote nitrate concentrations 0.3 and 0.6 g/L, respectively
Superscripts 3 and 4 denote nitrogen sources ammonium nitrate and sodium nitrate, respectively

growth rates and lipid accumulation over autotrophic and heterotrophic cultivations (Perez-Garcia et al. 2011; Prathima Devi and Venkata Mohan 2012).

All three modes of cultivation have been studied with the aim to achieve maximum possible biomass and lipid productivities. Kim et al. (2013) investigated the cell growth rate and organic carbon and nutrient removal efficiency of microalga *C. sorokiniana* under three modes of cultivation, viz., autotrophic, heterotrophic, and mixotrophic. A growth rate of 0.55/day was obtained in heterotrophic cultivation as compared to 0.24/day in phototrophic cultivation and 0.44/day in mixotrophic cultivation. A similar trend was also observed for removal of N and P. However, the suitability of mode totally depends on the experimental design. In other studies, mixotrophic mode of cultivation was found very suitable to deal with real wastewaters. For example, Wang et al. (2012) studied the mixotrophic cultivation of *C. pyrenoidosa* with diluted primary piggery wastewater (COD: 250–1000 mg/L). The biomass productivity was obtained in the range of 0.3–0.6/day. Similarly, Sforza et al. (2012) investigated the effect of time of sparging of excess CO_2 fraction in the inlet air (5%) and different organic substrates (ethanol, glycerol, and acetate) on biomass and lipid productivities under mixotrophic cultivation of *C. protothecoides* and *N. salina*. The maximum specific growth rate was obtained in mixotrophic cultivation for both species for 5% CO_2 in feed air supplied in the light-hours only, while glycerol concentrations were maintained at 1% w/v for respective runs. The maximum specific growth rates of 0.867 and 0.613/day were reported for *C. protothecoides* and *N. salina*, respectively. Recently, Shen et al. (2015) studied the effect of organic and inorganic carbons on microalgae growth and nutrient removal of *S. obliquus*. The initial concentrations of TOC (total organic carbon) were varied between 20 and 120 mg/L, while CO_2 concentrations were kept in the range of 0.03–15% (v/v). The maximum biomass productivity of 0.577 g/L/d was obtained for 5% (v/v) CO_2 concentration in the inlet air.

It is very interesting to note that the lipid profiling (composition of various fatty acids) is different under different modes of cultivation. Liu et al. (2011) investigated the lipid and fatty acid profiles under photoautotrophic and heterotrophic cultivation of *C. zofingiensis*. It has been found that lipids obtained under heterotrophic cultivation accumulated primarily neutral lipids up to 79.5% of total lipids (out of which about 88.7% was TAG), whereas lipids under photoautotrophic cultivation were mainly detected with membrane lipids such as glycolipids and phospholipids. In another study, similar results were obtained for the cultivation of *C. sorokiniana* (Wen et al. 2013). The strain accumulated lipids up to 56% of DCW after 7 days of cultivation under heterotrophic conditions on glucose as substrate. On the other hand, it accumulated lipid up to 19% of DCW in total 30 days under photoautotrophic growth. The relative neutral lipid content was found much higher in the heterotrophic cultivation as compared to the photoautotrophic cultivation (Wen et al. 2013). A direct comparison between the acetate and glucose as carbon source showed that acetate enhances the lipid production while maintaining the pH of the culture in the suitable range (Huang et al. 2016).

To make the process economically feasible, many authors have proposed the use of industrial wastewater as a source of the nutrients (N and P) and organic carbon for microalgae growth under heterotrophic cultivations (Perez-Garcia et al. 2011; Sankaran et al. 2014; Venkata Mohan et al. 2015), while the use of flue gas (10–18% CO_2) under the photoautotrophic cultivation has also been suggested to curb the CO_2 in the atmosphere which is a major cause of greenhouse effect (Valdes et al. 2012; Ji et al. 2015). Both said systems of microalgae cultivation find their benefits in lowering the water and air pollution, respectively. Various studies have shown that microalgae are not only reducing the COD of the wastewater significantly but also producing the valuable biomass which further can be used for biomethane production (Uggetti et al. 2016). On the other hand, previous studies have shown that microalgae can sustain at high concentration of CO_2 (about 2–10% v/v) in the inlet air/gas (Hulatt and Thomas 2011; Arudchelvam and Nirmalakhandan 2012a, b). However, the optimal concentration of CO_2 in feed air varies with respect to the species. Microalgae are successfully used for the treatment of various effluents (Table 10.2) such as food processing wastewater, municipal, piggery, dairy, and distillery/brewery wastewaters under different modes of cultivation (Hongyang et al. 2011; Abreu et al. 2012; Mata et al. 2012; Gentili 2014; Tan et al. 2014; Chu et al. 2015; Ding et al. 2015; Gupta et al. 2016). That is why the ability of microalgae to grow on wastewater and on the high fraction of CO_2 is highly applauded.

Nutrient optimization plays an important role while growing microalgae. Initial C:N:P ratio was found to affect the growth and lipid accumulation in microalgae (Silaban et al. 2014; Choi and Lee 2015; Li et al. 2016). In general, microalgae are being cultivated in the nutrient-sufficient media. The excess quantity or very low quantity of nutrients can affect the growth of the microalgae by inhibition or deprivation, respectively. Although there is no such limit defined, there are vast strains available with different composition of these elements, and the stoichiometric requirement will be different. In a previous study, for the cultivation of freshwater microalgae species *C. vulgaris*, the nitrate inhibition limit of 3.75 g/L was found, while phosphorous with a concentration above than 0.2 g/L was found to inhibit the growth (Li et al. 2016).

Table 10.2 Microalgae cultivation on wastewaters under different modes of cultivation

Mode of cultivation	Microalgae sp.	Wastewater (WW)	COD or TOC mg·L^{-1}	COD or TOC removal (%)	TN removal (%)	TP removal (%)	References
Mixotrophic	*Chlorella pyrenoidosa*	Anaerobic digested starch processing WW	702.4–1026.2	65.99	83.06	96.97	Tan et al. (2014)
Mixotrophic	*Scenedesmus* sp.	Municipal, dairy, pulp and paper WW	1905	92.7	>95 NH_4	>95 PO_4–P	Gentili (2014)
Mixotrophic	*Chlorella pyrenoidosa*	Piggery wastewater	250–1000	57.6	54.7–74.6	31–77.7	Wang et al. (2012)
Autotrophic/ Mixotrophic	*Chlorella* sp.	Anaerobic digested dairy manure WW	700–1300	27.4–38.4	75.7–82.5	62.5–74.7	Wang et al. (2010)
Heterotrophic	*Chlorella* sp.	Concentrated municipal WW	2304	90.8	89.1	80.9	Li et al. (2011)
Mixotrophic	*Chlorella pyrenoidosa*	Soybean processing WW	3000–3750	77.8	88.8	70.3	Hongyang et al. (2011)
Mixotrophic	*Chlorella pyrenoidosa*	Anaerobic digested starch processing WW	196.9–209.1 TOC	53.8–71.9 TOC	69.3–85.1	95.2–100	Chu et al. (2015)
Mixotrophic	*Scenedesmus obliquus*	Municipal waste water with food wastewater	~56–220 TOC	58–74 TOC	83–97	82–89	Ji et al. (2015)
Mixotrophic	*Euglena gracilis, C. pyrenoidosa*	Dairy effluent	2900	80–86	96–100 NH_3–N	98 PO_4–P	Zhan et al. (2015)
Mixotrophic	*Scenedesmus bijuga*	Anaerobic digested food wastewater	630.3–238.48	66.4	90.7	90.5	Su et al. (2011)
Mixotrophic	*Scenedesmus obliquus*	Synthetic brewery effluent	3635	66.8	20.8	–	Mata et al. (2010)
Mixotrophic	*Chlorella pyrenoidosa*	Swine wastewater	1064–3665	60–70	40–90 NH_3–N	–	Wang et al. (2015)
Mixotrophic	*Chlorella* sp.	Food Processing Ind. WW	1000–4000	78–91	58–90	43–62	Gupta et al. (2016)

10.5 Optimization of PBR

Several laboratory studies have shown that microalgae are one step ahead in terms of biomass productivity as compared to oilseed crops and thus proposed as the most suitable candidate for biofuel production. But, the implementation of microalgae cultivation on larger scale is not easy and straightforward. The lab scale studies cannot be extrapolated as such to a large scale as the process hydrodynamics becomes very challenging with increasing scale. When the process is to be scaled up, the first question that comes in mind is: which type of photobioreactor will serve the purpose? It is not that much difficult to identify the suitable PBR design with respect to the desired products and by-products. ORP is simplest and cheaper among the available types of PBR and widely used for microalgae cultivation (Jorquera et al. 2010; Benavides et al. 2013; Chiaramonti et al. 2013; Hadiyanto et al. 2013), while closed-type PBRs are sophisticated but expensive (Miron et al. 2000). Tubular loop PBR consists of array of several long tubes of smaller diameter in a loop arrangement connected with a common degassing unit. The culture is driven by low shear pump or by airlift column. The third category, column PBRs consist of a simple design of gas-liquid bubble column or flat plate. The airlift PBR is also a type of column PBRs but with the advantage of well directive flow which assists microalgae cells in receiving the light intermittently while passing through the light and dark zones continuously.

General desired characteristics of PBR are as follows:

1. High surface area to volume ratio
2. High biomass productivity (g/L/day)
3. Energy optimized system (sparging, pumping, paddle or circulation)
4. Cost optimization (PBR cost, nutrients, water, CO_2, light, etc.)
5. Ease of CO_2 feed, utilization, and loss minimization
6. Provision for avoidance of photoinhibition and photorespiration

No matter which design is chosen for the cultivation of microalgae, few problems associated with the PBR remain the same. The light attenuation is one of them. This single factor puts limit on desired productivity per unit land area (Wang et al. 2014). Despite of occupying larger land area, the volumetric productivity of biomass remains lower in ORP due to its shallow structure which is made purposely to receive the abundant light on the surface. The light intensity is a major requirement for the autotrophic cultivation; hence, PBRs are designed with the provision of higher surface area to volume ratio. ORP and flat-plate PBR are excellent examples of this category (Sierra et al. 2008). The other closed-type PBRs such as tubular loop and bubble column have little issue with light attenuation unless the culture is dense. But, as the main goal is to obtain the higher biomass concentration, mutual shading effect becomes dominant when the culture becomes dense within few days of operation. In such a case, the airlift PBR which is having the advantage of well directive flow/circulations can be found suitable. Photorespiration was found a major limitation in the tubular loop PBR for its implementation at larger scale

(Miron et al. 1999). Moreover, a degassing unit, i.e., gas-sparged column, is required to install for the removal of accumulated oxygen in the culture. Furthermore, the culture needs to be transferred from tubes to the degassing unit using centrifugal pump, and loss of culture can be found due to mechanical shear (Miron et al. 2003; Scarsella et al. 2012). On contrary, the limitation imposed by photorespiration can be nullified in the airlift PBR due to gentle mixing in the column using air.

The closed PBRs show certain advantages such as improved biomass and lipid yield, cell circulation along with mixing (prevention of settling of cells and sticking), high photosynthetic efficiency, less contamination, and greater areal productivity as compared to ORP (Miron et al. 1999, 2000; Hulatt and Thomas 2011; Pegallapati and Nirmalakhandan 2012; Pawar 2016b). The limitation of light penetration or its unavailability in dense culture can be overcome by mixing the culture by external means either by air circulation or paddle. The way of mixing the culture with the help of air circulation is costlier but very efficient than it imparted by a paddle. It is well known that the hydrodynamics play an important role in gas-sparged bioreactor (Luo and Al-Dahhan 2012; Pawar 2016b). The hydrodynamic parameters such as overall gas holdup, liquid circulation velocity, and shear stress play a decisive role in culture viability and growth (Contreras et al. 1998; Barbosa et al. 2004). The gas-sparged reactors are generally operated in two main regimes: homogeneous and heterogeneous. The homogeneous regime is desired in the case of shear-sensitive cells because of low turbulence in such kind of flows, whereas heterogeneous regime is characteristic of high turbulent flow which certainly damages the shear-sensitive cells, and loss of culture is inevitable.

Among the vertical column PBRs, the airlift column PBR is of great interest. It is a type of bubble column which contains two concentric tubes: the outer is known as downcomer, whereas the inner is called riser. It has a characteristic well directive liquid flow, i.e., from riser to downcomer to riser (Chisti et al. 1998). This gentle liquid circulation is very effective in the case of microalgae cultivation as it brought the cells from the dark zone (riser) to the light zone (downcomer). Thus, the light and dark cycles are maintained. Moreover, the limitation imposed by light penetration can also be nullified (Miron et al. 2000). When horizontal loop tubular PBRs are compared with vertical tubular column PBRs, the latter arrangement is more beneficial than the former due to several advantages such ease of mixing with air sparging, reduced risks of photoinhibition and photorespiration, uniform concentration gradients, suspension and local circulation of dense culture, and regulation of instant dark/light periods (Miron et al. 2003; Lopez et al. 2006).

In previous articles, the effects of various design parameters on microalgae productivity in several types of vertical closed PBRs have been critically investigated (Miron et al. 2000; Sierra et al. 2008; Luo and Al-Dahhan 2012; Fernandes et al. 2014; San Pedro et al. 2014). As the cultivation of microalgae in a gas-sparged tubular column is very sensitive to the parameters such as temperature, shear stress, irradiance intensity, concentration of O_2 and CO_2 in the culture, gas and liquid velocities, and biomass concentration, it has been suggested that comprehensive studies are required on the pilot scale before making any conclusion regarding the suitability of PBR for particular species. Thus, any modeling or simulation study should take into account the effects of light transfer, mass transfer, and mixing simultaneously (Nauha and Alopaeus 2013). Further, in the case of upscaling of airlift PBR, it should also be noted that as the volume of PBR increases, the liquid circulation velocity tends to increase which further may increase the local values of turbulence and shear stress affecting the growth of microalgae.

10.6 Cost and Feasibility of Microalgal Biodiesel Production

Biodiesel from microalgae is produced in four steps using different methods: the first step is cultivation of microalgae in PBR using inorganic or organic carbon, the second step is the harvesting of cells using centrifugation or flotation, the third step is the lipid extraction using solvents by wet (Bligh dryer) or dry (Folch) methods, and then the last step is transesterification reaction with methanol either with acid or base catalyst to obtain the biodiesel (FAME). The entire process of getting algal biodiesel is technically feasible but not economically. The first step is thoroughly discussed in the above sections. The comprehensive analysis of other steps can be found in the literature (Brennan and Owende 2010), and they are not discussed here. Lipid extraction is an important step after cultivation as the yield of biodiesel is directly proportional to the yield of lipid extraction. There are several methods used for lipid extraction which generally contain some alcohol and ethers. The efficiency of different lipid extraction methods was found to be affected by the solvent polarity. In a previous study, dichloromethane/methanol extraction and Folch extraction (methanol/chloroform) methods showed highest extraction yields, but the best FAME profile was obtained with hexane/ethanol (Soares et al. 2014). So, the choice of extraction method is also needed to manipulate according to the final desired properties of biodiesel.

Optimization of the PBR, whether open or closed system, is an inevitable part for the feasibility of the entire process. In a notable work by Ippoliti et al. (2016), the microalga strain *Tisochrysis lutea* was cultivated in an optimized pilot scale tubular PBR of 3000 L capacity with a biomass productivity of 20 g/m^2/d and lipid content of 25% of DCW. Such PBRs are susceptible to high levels of dissolved oxygen and variation in temperature gradients in long tubes and need attention to maintain the uniform culture properties. In another study (Eustance et al. 2016), the effect of depth of culture and nitrogen starvation on lipid productivity is studied for *Scenedesmus acutus* strain in a pilot scale ORP (area 30.37 m^2). The depth of culture of 9 cm was found suitable for maximum lipid productivity as compared to the depth of 20 cm.

The cost optimization of entire process is the most anxious part of study. As the PBR has major share of total cost of production (Fig. 10.5), many researchers are focusing on enhancing the biomass and lipid productivities of various microalgae strains. Various combinations of post processing operations have been also proposed (Brennan and Owende 2010). Complete life cycle analyses of microalgal biodiesel have been critically reviewed by many researchers (Davis et al. 2011; Delrue et al. 2012; Quinn and Davis 2015). The cost of biodiesel production is 2.25 $/L for ORP pathway, while with closed PBR pathway, it was reported about 4.8 $/L (Davis et al.

Fig. 10.5 The cost share of different steps for biodiesel production

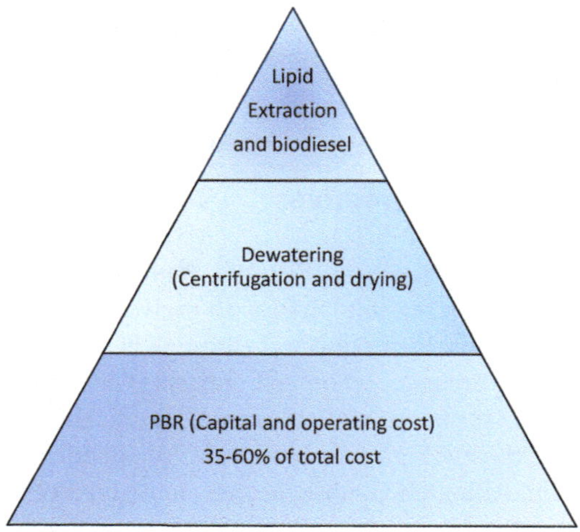

2011). On the other hand, Delrue et al. (2012) reported the cost of biodiesel production in between 1.35 and 2.36 $/L for an optimal route of studied estimates. Nevertheless, the current production cost of biodiesel with thorough estimates has been reported in the range of 2–10 $/L which is several times higher than that of petrodiesel (Brennan and Owende 2010; Mata et al. 2010; Pawar 2016a). Pawar (2016a) suggested that the costs of biomass and biodiesel production should be based on the actual operating days per annum and season-wise performance of PBR. It has been also suggested to take into account the effect of various environmental factors such as varying and high intensity sunlight (day and season wise), cloud cover (rainy season), and variable ambient temperature and associated biomass productivity before evaluating the cost of biodiesel production. However, the culture temperature could be different from the ambient temperature and may require provision for cooling or heating systems accordingly. In a previous study, for *C. pyrenoidosa*, a mean biomass productivity of 0.06, 0.15, 0.19, 0.37, and 0.09 g/L/day was obtained in winter (7–12 °C), spring (17–22 °C), autumn (23–27 °C), summer (35–39 °C), and summer (42–45 °C) seasons, respectively (Tan et al. 2014).

Different microalgae strains have different requirement of nutrients. Hence, in the literature, different media are found to be used for its cultivation. Out of that, the common media are BBM, BG11, F/2, and CHU. Nutrient management is the key to produce quality results. The tolerance of microalgae to high concentration of organic carbon saves freshwater requirement as such cultivation can be done on concentrated industrial wastewater or synthetic one without dilution or with some 10% dilution (Gupta et al. 2016, 2017). The dilution factor is also equally important for the feasibility of microalgae cultivation on wastewaters.

Beyond all above points, according to Chisti (2013), the innovative new technology is necessary in current situation for algal biofuel to be a cost-effective sustainable fuel. In pursue of that, a clear focus should be placed to genetic and metabolic engineering of microalgae production. It includes increasing the photosynthetic efficiency of microalgae, enhancement in carbon capture by the cells, and channeling of the energy absorbed by microalgae toward lipid accumulation in the cells (Chisti 2013).

10.7 Perspectives

Optimization of the PBR and manipulation of nutrients are two utmost important things of a feasible process to derive the biodiesel of standard quality. Microalgae cultivation on wastewaters (through heterotrophic or mixotrophic mode) having dissolved organic carbon and nutrients (N and P) needs to be studied thoroughly as it can lower the cost of cultivation significantly. Microalgae growth in the presence of a relatively high proportion of CO_2 is added advantage of the mixotrophic cultivation. Although glucose, acetate, and glycerol are found suitable for microalgae cultivation as organic sources of carbon under heterotrophic or mixotrophic mode, the crude glycerol can be considered as a potential cost-effective substrate as it is a major by-product of biodiesel production (transesterification reaction), and thus, its use in microalgae production can make the system more cost-effective. For autotrophic cultivation, several microalgae strains can be grown in CO_2-rich gas (up to 10% v/v), and thus suitable methodology can be determined for CO_2 sequestration. As not all standard properties of biodiesel are satisfied completely by microalgal biodiesel, mixing of lipids from various strains can be thought as a suitable strategy for future studies or estimates.

Acknowledgment The authors are thankful to Director, CSIR-National Environmental Engineering Research Institute (CSIR-NEERI) for providing necessary facilities for this work. The corresponding author is thankful to the Department of Science and Technology, New Delhi, for financial support for this research work under the scheme of DST INSPIRE Faculty Award (IFA13-ENG63).

References

Abreu AP, Fernandes B, Vicente AA, Teixeira J, Dragone G (2012) Mixotrophic cultivation of *Chlorella vulgaris* using industrial dairy waste as organic carbon source. Bioresour Technol 118:61–66. https://doi.org/10.1016/j.biortech.2012.05.055

Arudchelvam Y, Nirmalakhandan N (2012a) Energetic optimization of algal lipid production in bubble columns: part I: evaluation of gas sparging. Biomass Bioenergy 46:757–764. https://doi.org/10.1016/j.biombioe.2012.08.011

Arudchelvam Y, Nirmalakhandan N (2012b) Energetic optimization of algal lipid production in bubble columns: part II: evaluation of CO_2 enrichment. Biomass Bioenergy 46:765–772. https://doi.org/10.1016/j.biombioe.2012.08.012

Barbosa MJ, Hadiyanto WRH (2004) Overcoming shear stress of microalgae cultures in sparged photobioreactors. Biotechnol Bioeng 85:78–85. https://doi.org/10.1002/bit.10862

Benavides AMS, Torzillo G, Kopecky J, Masojidek J (2013) Productivity and biochemical composition of *Phaeodactylum tricornutum* (*Bacillariophyceae*) cultures grown outdoors in tubular photobioreactors and open ponds. Biomass Bioenergy 54:115–122. https://doi.org/10.1016/j.biombioe.2013.03.016

Benvenuti G, Bosma R, Cuaresma M, Janssen M, Barbosa MJ, Wijffels RH (2015) Selecting microalgae with high lipid productivity and photosynthetic activity under nitrogen starvation. J Appl Phycol 27:1425–1431. https://doi.org/10.1007/s10811-014-0470-8

Bhatnagar A, Chinnasamy S, Singh M, Das K (2011) Renewable biomass production by mixotrophic algae in the presence of various carbon sources and wastewaters. Appl Energy 88:3425–3431. https://doi.org/10.1016/j.apenergy.2010.12.064

Brennan L, Owende P (2010) Biofuels from microalgae – a review of technologies for production, processing, and extraction of biofuels and coproducts. Renew Sust Energ Rev 14:557–577. https://doi.org/10.1016/j.rser.2009.10.009

Bux F, Chisti Y (2016) Algae biotechnology: products and processes. Springer International Publishing, Switzerland. https://doi.org/10.1007/978-3-319-12334-9

Ceron-Garcia MC, Fernandez-Sevilla JM, Sanchez-Miron A, Garcia-Camacho F, Contreras-Gomez A, Molina-Grima E (2013) Mixotrophic growth of *Phaeodactylum tricornutum* on fructose and glycerol in fed-batch and semi-continuous modes. Bioresour Technol 147:569–576. https://doi.org/10.1016/j.biortech.2013.08.092

Chandra R, Rohit MV, Swamy YV, Venkata Mohan S (2014) Regulatory function of organic carbon supplementation on biodiesel production during growth and nutrient stress phases of mixotrophic microalgae cultivation. Bioresour Technol 165:279–287. https://doi.org/10.1016/j.biortech.2014.02.102

Chen F, Johns MR (1995) A strategy for high cell density culture of heterotrophic microalgae with inhibitory substrates. J Appl Phycol 7:43–46

Chiaramonti D, Prussi M, Casini D, Tredici MR, Rodolfi L, Bassi N, Zittelli GC, Bondioli P (2013) Review of energy balance in raceway ponds for microalgae cultivation: re-thinking a traditional system is possible. Appl Energy 102:101–111. https://doi.org/10.1016/j.apenergy.2012.07.040

Chisti Y (2007) Biodiesel from microalgae. Biotechnol Adv 25:294–306. https://doi.org/10.1016/j.biotechadv.2007.02.001

Chisti Y (2013) Constraints to commercialization of algal fuels. J Biotech 167:201–214. https://doi.org/10.1016/j.jbiotec.2013.07.020

Chisti MY, Halard B, Moo-Young M (1998) Liquid circulation in airlift reactors. Chem Eng Sci 43:451–457. https://doi.org/10.1016/0009-2509(88)87005-2

Choi HJ, Lee SM (2015) Effect of the N/P ratio on biomass productivity and nutrient removal from municipal wastewater. Bioprocess Biosyst Eng 38:761–766. https://doi.org/10.1007/s00449-014-1317-z

Chu HQ, Tan XB, Zhang YL, Yang LB, Zhao FC, Guo J (2015) Continuous cultivation of *Chlorella pyrenoidosa* using anaerobic digested starch processing wastewater in the outdoors. Bioresour Technol 185:40–48. https://doi.org/10.1016/j.biortech.2015.02.030

Contreras A, Garcia F, Molina E, Merchuk JC (1998) Interaction between CO_2 - mass transfer, light availability, and hydrodynamics stress in the growth of *Phaeodactylum tricornutum* in a concentric tube airlift photobioreactor. Biotechnol Bioeng 60:317–325

Davis R, Aden A, Pienkos PT (2011) Techno-economic analysis of autotrophic microalgae for fuel production. Appl Energy 88:3524–3531. https://doi.org/10.1016/j.apenergy.2011.04.018

Delrue F, Setier PA, Sahut C, Cournac L, Roubaud A, Peltier G, Froment AK (2012) An economic, sustainability, and energetic model of biodiesel production from microalgae. Bioresour Technol 111:191–200. https://doi.org/10.1016/j.biortech.2012.02.020

Ding J, Zhao F, Cao Y, Xing L, Liu W, Mei S, Li S (2015) Cultivation of microalgae in dairy farm wastewater without sterilization. Int J Phytoremediation 17:222–227. https://doi.org/10.1080/15226514.2013.876970

Eustance E, Wray JT, Badvipour S, Sommerfeld MR (2016) The effects of cultivation depth, areal density, and nutrient level on lipid accumulation of Scenedesmus acutus in outdoor raceway ponds. J Appl Phycol 28:1459–1469. https://doi.org/10.1007/s10811-015-0709-z

Fernandes BD, Mota A, Ferreira A, Dragone G, Teixeira JA, Vicente AA (2014) Characterization of split cylinder airlift photobioreactors for efficient microalgae cultivation. Chem Eng Sci 117:445–454. https://doi.org/10.1016/j.ces.2014.06.043

Francisco EC, Neves DB, Jacob-Lopes E, Franco TT (2010) Microalgae as feedstock for biodiesel production: carbon dioxide sequestration, lipid production and biofuel quality. J Chem Technol Biotechnol 85:395–403. https://doi.org/10.1002/jctb.2338

Gangadhar KN, Pereira H, Diogo HP, Borges dos Santos RM, Prabhavathi Devi BLA, Prasad RBN, Custodio L, Malcata FX, Varela J, Barreira L (2016) Assessment and comparison of the properties of biodiesel synthesized from three different types of wet microalgal biomass. J Appl Phycol 28:1571–1578. https://doi.org/10.1007/s10811-015-0683-5

Gentili FG (2014) Microalgal biomass and lipid production in mixed municipal, dairy, pulp and paper wastewater together with added flue gases. Bioresour Technol 169:27–32. https://doi.org/10.1016/j.biortech.2014.06.061

Gim GH, Kim JK, Kim HS, Kathiravan MN, Yang H, Jeong SH, Kim SW (2014) Comparison of biomass production and total lipid content of freshwater green microalgae cultivated under various culture conditions. Bioprocess Biosyst Eng 37:99–106. https://doi.org/10.1007/s00449-013-0920-8

Gomez-Serrano C, Morales-Amaral MM, Acien FG, Escudero R, Fernandez-Sevilla JM, Molina-Grima E (2015) Utilization of secondary-treated wastewater for the production of freshwater microalgae. Appl Microbiol Biotechnol 99:6931–6944. https://doi.org/10.1007/s00253-015-6694-y

Gomma AE, Lee SK, Sun SM, Yang SH, Chung G (2015) Improvement in oil production by increasing Malonyl-CoA and Glycerol-3-Phosphate pools in *Scenedesmus quadricauda*. Indian J Microbiol 55:447–455. https://doi.org/10.1007/s12088-015-0546-4

González-Garcinuño A, Tabernero A, Sánchez-Álvarez JM, del Valle EMM, Galán MA (2014) Effect of nitrogen source on growth and lipid accumulation in Scenedesmus abundans and Chlorella ellipsoidea. Bioresour Technol 173:334–341. https://doi.org/10.1016/j.biortech.2014.09.038

Gouveia L, Graça S, Sousa C, Ambrosano L, Ribeiro B, Botrel EP, Neto PC, Ferreira AF, Silva CM (2016) Microalgae biomass production using wastewater: treatment and costs scale-up considerations. Algal Res 16:167–176. https://doi.org/10.1016/j.algal.2016.03.010

Griffiths MJ, van Hille RP, Harrison STL (2012) Lipid productivity, settling potential and fatty acid profile of 11 microalgal species grown under nitrogen replete and limited conditions. J Appl Phycol 24:989–1001. https://doi.org/10.1007/s10811-011-9723-y

Gris B, Morosinotto T, Giacometti GM, Bertucco A, Sforza E (2014) Cultivation of Scenedesmus obliquus in photobioreactors: effects of light intensities and light–dark cycles on growth, productivity, and biochemical composition. Appl Biochem Biotechnol 172:2377–2389. https://doi.org/10.1007/s12010-013-0679-z

Gupta S, Pandey RA, Pawar SB (2016) Microalgal bioremediation of food-processing industrial wastewater under mixotrophic conditions: kinetics and scale-up approach. Front Chem Sci Eng. 10(4):499–508. https://doi.org/10.1007/s11705-016-1602-2

Gupta S, Pandey RA, Pawar SB (2017) Bioremediation of synthetic high–chemical oxygen demand wastewater using microalgal species *Chlorella pyrenoidosa*. Biorem J 21:38–51. https://doi.org/10.1080/10889868.2017.1282936

Hadiyanto H, Elmore S, Van Gerven T, Stankiewicz A (2013) Hydrodynamic evaluations in high rate algae pond (HRAP) design. Chem Eng J 217:231–239. https://doi.org/10.1016/j.cej.2012.12.015

Hongyang S, Yalei Z, Chunmin Z, Xuefei Z, Jinpeng L (2011) Cultivation of *Chlorella pyrenoidosa* in soybean processing wastewater. Bioresour Technol 102:9884–9890. https://doi.org/10.1016/j.biortech.2011.08.016

Huang A, Sun L, Wu S, Liu C, Zhao P, Xie X, Wang G (2016) Utilization of glucose and acetate by *Chlorella* and the effect of multiple factors on cell composition. J Appl Phycol 29(1):23–33. https://doi.org/10.1007/s10811-016-0920-6

Hulatt C, Thomas D (2011) Productivity, carbon dioxide uptake and net energy return of microalgal bubble column photobioreactors. Bioresour Technol 102:5775–5787. https://doi.org/10.1016/j.biortech.2011.02.025

Ippoliti D, Gonzalez DA, Martin I, Sevilla JMF, Pistocchi R, Acien FG (2016) Outdoor production of *Tisochrysis lutea* in pilot-scale tubular photobioreactors. J Appl Phycol 28:3159–3166. https://doi.org/10.1007/s10811-016-0856-x

Ji MK, Yun HS, Park YT, Kabra AN, IH O, Choi J (2015) Mixotrophic cultivation of a microalga *Scenedesmus obliquus* in municipal wastewater supplemented with food wastewater and flue gas CO_2 for biomass production. J Environ Manag 159:115–120. https://doi.org/10.1016/j.jenvman.2015.05.037

Jorquera O, Kiperstok A, Sales EA, Embirucu M, Ghirardi ML (2010) Comparative energy life-cycle analyses of microalgal biomass production in open ponds and photobioreactors. Bioresour Technol 101:1406–1413. https://doi.org/10.1016/j.biortech.2009.09.038

Kim S, Park J, Cho Y, Hwang S (2013) Growth rate, organic carbon and nutrient removal rates of *Chlorella sorokiniana* in autotrophic, heterotrophic and mixotrophic conditions. Bioresour Technol 144:8–13. https://doi.org/10.1016/j.biortech.2013.06.068

Lari Z, Moradi-kheibari N, Ahmadzadeh H, Abrishamchi P, Moheimani NR, Murry MA (2016) Bioprocess engineering of microalgae to optimize lipid production through nutrient management. J Appl Phycol 28(6):3235–3250. https://doi.org/10.1007/s10811-016-0884-6

Li C, Yu Y, Zhang D, Liu J, Ren N, Feng Y (2016) Combined effects of carbon, phosphorus and nitrogen on lipid accumulation of Chlorella vulgaris in mixotrophic culture. J Chem Technol Biotechnol 91:680–684. https://doi.org/10.1002/jctb.4623

Li Y, Chen YF, Chen P, Min M, Zhoi W, Martinez B, Zhu J, Ruan R (2011) Characterization of a microalga *Chlorella* sp. well adapted to highly concentrate municipal wastewater for nutrient removal and biodiesel production. Bioresour Technol 102:5138–5144. https://doi.org/10.1016/j.biortech.2011.01.091

Liu J, Huang J, Sun Z, Zhong Y, Jiang Y, Chen F (2011) Differential lipid and fatty acid profiles of photoautotrophic and heterotrophic *Chlorella zofingiensis*: assessment of algal oils for biodiesel production. Bioresour Technol 102:106–110. https://doi.org/10.1016/j.biortech.2010.06.017

Lopez MCG, Sanchez EDR, Lopez JLC, Fernandez FGA, Sevilla JMF, Rivas J, Guerrero MG, Grima EM (2006) Comparative analysis of the outdoor culture of *Haematococcus pluvialis* in tubular and bubble column photobioreactors. J Biotechnol 123:329–342. https://doi.org/10.1016/j.jbiotec.2005.11.010

Lowrey J, Brooks MS, McGinn PJ (2015) Heterotrophic and mixotrophic cultivation of microalgae for biodiesel production in agricultural wastewaters and associated challenges – a critical review. J Appl Phycol 27:1485–1498. https://doi.org/10.1007/s10811-014-0459-3

Luo H, Al-Dahhan MH (2012) Airlift column photobioreactors for *Porphyridium Sp* culturing: part I: effects of hydrodynamics and reactor geometry. Biotechnol Bioeng 109:932–941. https://doi.org/10.1002/bit.24361

Mandotra SK, Kumar P, Suseela MR, Ramteke PW (2014) Freshwater green microalga *Scenedesmus abundans*: a potential feedstock for high quality biodiesel production. Bioresour Technol 156:42–47. https://doi.org/10.1016/j.biortech.2013.12.127

Mata TM, Antonio AM, Nidia S (2010) Microalgae for biodiesel production and other applications: a review. Renew Sust Energ Rev 14:217–232. https://doi.org/10.1016/j.rser.2009.07.020

Mata TM, Melo AC, Simoes M, Caetano NS (2012) Parametric study of a brewery effluent treatment by microalgae *Scenedesmus obliquus*. Bioresour Technol 2012(107):151–158. https://doi.org/10.1016/j.biortech.2011.12.109

Michels MHA, van der Goot AJ, Norsker NH, Wijffels RH (2010) Effects of shear stress on the microalgae *Chaetoceros muelleri*. Bioprocess Biosyst Eng 33:921–927. https://doi.org/10.1007/s00449-010-0415-9

Miron AS, Camacho FG, Gomez AC, Grima EM, Chisti Y (1999) Comparative evaluation of compact photobioreactor for large scale monoculture of microalgae. J Biotechnol 70:249–270. https://doi.org/10.1016/S0168-1656(99)00079-6

Miron AS, Camacho FG, Gomez AC, Grima EM, Chisti Y (2000) Bubble column and airlift photobioreactor for algal culture. AICHE J 46:1872–1887. https://doi.org/10.1002/aic.690460915

Miron AS, Garcia MCC, Gomez AC, Camacho FG, Grima EM, Chisti Y (2003) Shear stress tolerance and biochemical characterization of *Phaeodactylum tricornutum* in quasi steady-state continuous culture in outdoor photobioreactors. Biochem Eng J 16:287–297. https://doi.org/10.1016/S1369-703X(03)00072-X

Nascimento IA, Marques SSI, Cabanelas ITD, Pereira SA, Druzian JI, Oliveira de Souza C, Vich DV, Carvalho GC, Nascimento MA (2013) Screening microalgae strains for biodiesel production: lipid productivity and estimation of fuel quality based on fatty acids profiles as selective criteria. Bioenergy Res 6:1–13. https://doi.org/10.1007/s12155-012-9222-2

Nascimento IA, Marques SSI, Cabanelas ITD, Carvalho GC, Nascimento MA, Souza CO, Druzian JI, Hussain J, Liao W (2014) Microalgae versus land crops as feedstock for biodiesel: productivity, quality, and standard compliance. Bioenergy Res 7:1002–1013. https://doi.org/10.1007/s12155-014-9440-x

Nauha EK, Alopaeus V (2013) Modeling method for combining fluid dynamics and algal growth in a bubble column photobioreactor. Chem Eng J 229:559–568. https://doi.org/10.1016/j.cej.2013.06.065

Pegallapati AK, Nirmalakhandan N (2012) Modeling algal growth in bubble columns under sparging with CO enriched air. Bioresour Technol 124:137–145. https://doi.org/10.1016/j.biortech.2012.08.026

Pawar S (2016a) Effectiveness mapping of open raceway pond and tubular photobioreactors for sustainable production of microalgae biofuel. Renew Sust Energ Rev 62:640–653. https://doi.org/10.1016/j.rser.2016.04.074

Pawar SB (2016b) Process engineering aspects of vertical column photobioreactors for mass production of microalgae. ChemBioEng Rev 3:101–115. https://doi.org/10.1002/cben.201600003

Perez-Garcia O, Escalante FME, de Bashan LE, Bashan Y (2011) Heterotrophic cultures of microalgae: metabolism and potential products. Water Res 45:11–36. https://doi.org/10.1016/j.watres.2010.08.037

Prathima Devi M, Venkata Mohan S (2012) CO_2 supplementation to domestic wastewater enhances microalgae lipid accumulation under mixotrophic microenvironment: effect of sparging period and interval. Bioresour Technol 112:116–123. https://doi.org/10.1016/j.biortech.2012.02.095

Prathima Devi M, Swamy YV, Venkata Mohan S (2013) Nutritional mode influences lipid accumulation in microalgae with the function of carbon sequestration and nutrient supplementation. Bioresour Technol 142:278–286. https://doi.org/10.1016/j.biortech.2013.05.001

Pruvost J, Vooren GV, Cogne G, Legrand J (2009) Investigation of biomass and lipid production with *Neochloris oleoabundans* in photobioreactor. Bioresour Technol 100:5988–5995. https://doi.org/10.1016/j.biortech.2009.06.004

Pruvost J, Vooren GV, Gouic BL, Couzinet-Mossion A, Legrand J (2011) Systematic investigation of biomass and lipid productivity by microalgae in photobioreactors for biodiesel application. Bioresour Technol 102:150–158. https://doi.org/10.1016/j.biortech.2010.06.153

Pruvost J, Cornet JF, Borgne FL, Goetz V, Legrand J (2015) Theoretical investigation of microalgae culture in the light changing conditions of solar photobioreactor production and comparison with cyanobacteria. Algal Res 10:87–99. https://doi.org/10.1016/j.algal.2015.04.005

Quinn JC, Davis R (2015) The potentials and challenges of algae based biofuels: a review of the techno-economic, life cycle, and resource assessment modeling. Bioresour Technol 184:444–452. https://doi.org/10.1016/j.biortech.2014.10.075

Quinn JC, Catton K, Wagner N, Bradley TH (2012) Current large-scale US biofuel potential from microalgae cultivated in photobioreactors. Bioenergy Res 5:49–60. https://doi.org/10.1007/s12155-011-9165-z

Ramos MJ, Fernandez CM, Casas A, Rodriguez L, Perez A (2009) Influence of fatty acid composition of raw materials on biodiesel properties. Bioresour Technol 1:261–268. https://doi.org/10.1016/j.biortech.2008.06.039

Rios LF, Klein BC, Luz LF, Filho RM, Maciel MRW (2015) Nitrogen starvation for lipid accumulation in the microalga species *Desmodesmus* sp. Appl Biochem Biotechnol 175:469–476. https://doi.org/10.1007/s12010-014-1283-6

San Pedro A, Gonzalez-Lopez CV, Acien FG, Molina-Grima E (2014) Outdoor pilot-scale production of *Nannochloropsis gaditana*: influence of culture parameters and lipid production rates in tubular photobioreactors. Bioresour Technol 169:667–676. https://doi.org/10.1016/j.biortech.2014.07.052

Sankaran K, Premalatha M, Vijayasekaran M, Somasundaram VT (2014) DEPHY project: distillery wastewater treatment through anaerobic digestion and phycoremediation – a green industrial approach. Renew Sust Energ Rev 37:634–643. https://doi.org/10.1016/j.rser.2014.05.062

Scarsella M, Torzillo G, Cicci A, Belotti G, Filippis D, Bravia M (2012) Mechanical stress tolerance of two microalgae. Process Biochem 47:1603–1611. https://doi.org/10.1016/j.procbio.2011.07.002

Sforza E, Cipriani R, Morosinotto T, BertuccoA GGM (2012) Excess CO_2 supply inhibits mixotrophic growth of *Chlorella protothecoides* and *Nannochloropsis salina*. Bioresour Technol 104:523–529. https://doi.org/10.1016/j.biortech.2011.10.025

Shen QH, Jiang JW, Chen LP, Cheng LH, XH X, Chen HL (2015) Effect of carbon source on biomass growth and nutrients removal of *Scenedesmus obliquus* for wastewater advanced treatment and lipid production. Bioresour Technol 190:257–263. https://doi.org/10.1016/j.biortech.2015.04.053

Sierra E, Acien FG, Fernandez JM, Garcia JL, Gonzalez C, Molina E (2008) Characterization of a flat plate photobioreactor for the production of microalgae. Chem Eng J 138:136–147. https://doi.org/10.1016/j.cej.2007.06.004

Silaban A, Bai R, Gutierrez-Wing MT, Negulescu II, Rusch KA (2014) Effect of organic carbon, C:N ratio and light on the growth and lipid productivity of microalgae/cyanobacteria coculture. Eng Life Sci 14:47–56. https://doi.org/10.1002/elsc.201200219

Soares AT, Costa DC, Silva BF, Lopes RG, Derner RB, Filho NRA (2014) Comparative analysis of the fatty acid composition of microalgae obtained by different oil extraction methods and direct biomass transesterification. Bioenergy Res 7:1035–1044. https://doi.org/10.1007/s12155-014-9446-4

Su Y, Mennerich A, Urban B (2011) Municipal wastewater treatment and biomass accumulation with a wastewater born and settleable algalbacterial culture. Water Res 45:3351–3358. https://doi.org/10.1016/j.watres.2011.03.046

Tan X, Chu H, Zhang Y, Yang L, Zhao F, Zhou X (2014) *Chlorella pyrenoidosa* cultivation using anaerobic digested starch processing wastewater in an airlift circulation photobioreactor. Bioresour Technol 170:538–548. https://doi.org/10.1016/j.biortech.2014.07.086

Uggetti E, Passos F, Sole M, Garfi M, Ferrer I (2016) Recent achievements in the production of biogas from microalgae. Waste Biomass Valor 8(1):129–139. https://doi.org/10.1007/s12649-016-9604-3

Valdés FJ, Hernández MR, Catalá L, Marcilla A (2012) Estimation of CO stripping/CO microalgae consumption ratios in a bubble column photobioreactor using the analysis of the pH profiles. Application to Nannochloropsis oculata microalgae culture. Bioresour Technol 119:1–6. https://doi.org/10.1016/j.biortech.2012.05.120

Valdez-Ojeda R, Gonzalez-Munoz M, Us-Vazquez R, Narvaez-Zapata J, Chavarria-Hernandez JC, Lopez-Adrian S, Barahona-Perez F, Toledano-Thompson T, Garduno-Solorzano G, RMEG M (2015) Characterization of five fresh water microalgae with potential for biodiesel production. Algal Res 7:33–44. https://doi.org/10.1016/j.algal.2014.11.009

Vasumathi KK, Premalatha M, Subramanian P (2012) Parameter influencing the design of photobioreactor for the growth of microalgae. Renew Sust Energ Rev 16:5443–5450. https://doi.org/10.1016/j.rser.2012.06.013

Venkata Mohan S, Rohit MV, Chiranjeevi P, Chandra R, Navaneeth B (2015) Heterotrophic microalgae cultivation to synergize biodiesel production with waste remediation: progress and perspectives. Bioresour Technol 184:169–178. https://doi.org/10.1016/j.biortech.2014.10.056

Wang L, Li Y, Chen P, Min M, Chen Y, Zhu J, Ruan RR (2010) Anaerobic digested dairy manure as a nutrient supplement for cultivation of oilrich green microalgae *Chlorella* sp. Bioresour Technol 101:2623–2628. https://doi.org/10.1016/j.biortech.2009.10.062

Wang H, Xiong H, Hui Z, Zeng X (2012) Mixotrophic cultivation of *Chlorella pyrenoidosa* with diluted primary piggery wastewater to produce lipids. Bioresour Technol 104:215–220. https://doi.org/10.1016/j.biortech.2011.11.020

Wang SK, Stiles AR, Guo C, Liu CZ (2014) Microalgae cultivation in photobioreactors: an overview of light characteristics. Eng Life Sci 14:550–559. https://doi.org/10.1002/elsc.201300170

Wang Y, Guo W, Yen HW, Ho SH, Lo YC, Cheng CL, Ren N, Chang JS (2015) Cultivation of *Chlorella vulgaris* JSC-6 with swine wastewater for simultaneous nutrient/COD removal and carbohydrate production. Bioresour Technol 198:619–625. https://doi.org/10.1016/j.biortech.2015.09.067

Wen Q, Chen Z, Li P, Han Y, Fend Y, Ren N (2013) Lipid production for biofuels from effluent-based culture by heterotrophic *Chlorella Protothecoides*. Bioenergy Res 6:877–882. https://doi.org/10.1007/s12155-013-9308-5

Wu H, Miao X (2014) Biodiesel quality and biochemical changes of microalgae Chlorella pyrenoidosa and Scenedesmus obliquus in response to nitrate levels. Bioresour Technol 170:421–427. https://doi.org/10.1016/j.biortech.2014.08.017

Yang J, Xu M, Zhang X, Hu Q, Sommerfeld M, Chen M (2011) Life-cycle analysis on biodiesel production from microalgae: water foot print and nutrients balance. Bioresour Technol 102:159–165. https://doi.org/10.1016/j.biortech.2010.07.017

Yewalkar-Kulkarni S, Gera G, Nene S, Pandare K, Kulkarni B, Kamble S (2016) Exploiting phosphate-starved cells of Scenedesmus sp. for the treatment of raw sewage. Indian J Microbiol 57(2):241–249. https://doi.org/10.1007/s12088-016-0626-0

Zhan J, Zhang Q, Qin M, Hong Y (2015) Selection and characterization of eight fresh water green algae strains for synchronous water purification and lipid production. Front Environ Sci Eng 10:548–558. https://doi.org/10.1007/s11783-016-0831-4

Bioflocculants and Production of Microalgal Biomass

11

Mihir C. Sarang and Anuradha S. Nerurkar

Abstract

With the increase in world population, the demand of fossil fuels has increased rapidly. With the rising concern of limited availability of the fossil fuels, the demand of alternative fuels like biodiesel has attained a considerable attention over a decade. Microalgae have shown to be one of the promising sources for the production of biodiesel, as it can be grown easily on inorganic nutrients and have a tendency to accumulate large amount of fatty acids, which can be converted to biodiesel. However, the recovery of microalgae is one of the major problems that limits its industrial production. Physical methods like centrifugation and filtration are costly. This chapter focuses on flocculation and the use of bioflocculants for the cost-effective recovery of microalgal biomass.

Keywords

Bioflocculant • Biomass • Coagulation • Microalgae • Photobioreactor

11.1 Introduction

A variety of industrial processes require separation of very fine suspended and dissolved solids where flocculation is an important process that can be used for clarification through sedimentation. The removal of such suspended particles by ordinary filtration or membrane filtration is a challenging task. On the other hand, techniques like centrifugation can be costly. Hence, flocculation is widely used for the removal of such suspended solids in waste water treatment, food and beverage, mineral,

M.C. Sarang • A.S. Nerurkar (✉)
Department of Microbiology and Biotechnology Centre, Faculty of Science,
The Maharaja Sayajirao University of Baroda, Vadodara 390002, Gujarat, India
e-mail: anuner26@yahoo.com

H.J. Purohit et al. (eds.), *Optimization and Applicability of Bioprocesses*,
https://doi.org/10.1007/978-981-10-6863-8_11

paper and pulp, and biodiesel industry. Chemical flocculants are harmful and there are strict norms about the usage of these flocculants. In the background of disadvantages of the chemical flocculants, the bioflocculants provide an economical and eco-friendly process, significant in the perspective of rising concern about environmental pollution.

Microalgae cultivation technology is attractive from the viewpoint of energy production since they provide fuel security and reduce carbon dioxide emissions. Photosynthetically derived fuels have distinct advantages from environmental viewpoint, and microalgae technology is particularly promising since saline and waste water can be used for its cultivation. Biofuel is one of the important products of microalgal technology. Larkum et al. (2012) have reviewed the strain improvement of microalgae for biofuel production and suggested that it promises to be the technology of the future. The production of biodiesel from microalgae at industrial scale involves energy and cost-intensive processes in its cultivation and particularly harvesting. Harvesting incurs 20–30% of the total cost of production. Generally centrifugation is used for harvesting at small scale; however flocculation is the method of choice at large-scale harvesting. The use of bioflocculants can reduce the cost of microalgae production (Salim et al. 2011). This chapter provides information on bioflocculants and their applications with special reference to harvesting process of microalgae production that is part of biodiesel industry.

11.2 Flocculation

The phenomenon in which particles in colloidal suspension undergo aggregation to form larger clumps is called flocculation. Flocculation is based on solid-liquid separation process, where the added flocculant interacts with the suspended colloids and removes them from the aqueous phase. The agents that cause flocculation are called flocculants. The nature of flocculants can either be synthetic chemical or natural. Flocculation can be induced by two ways, (i) flocculation by coagulation and (ii) direct flocculation (Amuda and Amoo 2007).

11.2.1 Flocculation by Coagulation

Flocculation by coagulation involves the use of inorganic metal ions that interact with the colloidal particles and neutralize their surface charge. This allows the suspended colloidal particles to come near one another and form aggregates that sediment easily. It must be noted that the microflocs generated by the phenomenon of coagulation may easily disintegrate by external forces. In order to increase the stability of small flocs generated by coagulation, polymeric flocculants are needed that increase the size as well as stability of flocs (Lee et al. 2012).

The chemical flocculants can either be organic or inorganic. The inorganic chemical flocculants take part in the phenomenon of flocculation by coagulation and include alum (aluminum sulfate, aluminum chloride, polyaluminum chloride, alum, ferrous sulfate, ferric chloride, magnesium chloride, and calcium chloride). It must be noted that the charge neutralization of suspended particles depends on the amount of salts added. Optimum flocculation is observed when the zeta potential of the particles becomes zero (isoelectric point). Excessive use of metal coagulants can cause reversal of zeta potential that finally results in re-dispersal of suspended particles (Kleimann et al. 2005).There are several disadvantages of inorganic chemical flocculant that includes their requirement in large amount for efficient flocculation, need of specific pH to induce flocculation and toxicity to humans and other animals (Polizzi et al. 2002; Ward et al. 2006; Banks et al. 2006).

11.2.2 Direct Flocculation

Direct flocculation involves the use of polymeric substances to form denser and stable flocs by bridge or patch formation that settles easily. In flocculation by coagulation, a specific pH is required for the precipitation of metal hydroxides, whereas in direct flocculation the polymer can function at a wide pH range (Chong 2012). Organic chemical flocculants are involved in direct flocculation. They are high molecular weight, linear or branched, anionic/cationic or amphoteric, water-soluble polymers synthesized by products of petroleum industry (Suopajarvi et al. 2013). For example, polyacrylamide is one of the common flocculants synthesized by polymerization of monomers of acrylamide. Other such flocculants include polyacrylic acid, polyamine, and polydiallyl dimethyl ammonium chloride (Singh et al. 2000). Majority of chemically synthesized polymers are not safe for the environment and are resistant to degradation (Brostow et al. 2009). The monomeric forms of chemical flocculants such as acrylamide, ethyleneimine, epichlorohydrin, and dimethylamine are extremely toxic and can act as a potent carcinogen. The release of such toxic monomers into the environment can have adverse effect on living organisms by entering the food chain (Sharma et al. 2006).

11.2.2.1 Mechanism of Direct Flocculation

The interaction of the flocculant with the suspended particles in direct flocculation can either include electrostatic patch or bridging mechanism. The electrostatic patch mechanism involves the binding of charged low molecular weight polymer with the suspended particles having opposite charge. As the polymer binds the suspended particles, the surface charge of the particle gets partially neutralized resulting in formation of small cationic patches on the particle (Blanco et al. 2002).The cationic polymer on the particle not only decreases electrostatic repulsion between particles but also attracts other suspended particles with anionic charge to form large aggregates. This aggregated material becomes dense and settles as a floc. The flocs formed by this mechanism are stable than the flocs formed by coagulation, but less stable than the flocs formed by bridge mechanism (Fig. 11.1a).

In flocculation by bridge mechanism, the polymer with high molecular weight gets adsorbed on suspended particles and facilitates its charge neutralization. Interaction of particles associated with flocculant with other free particles results in

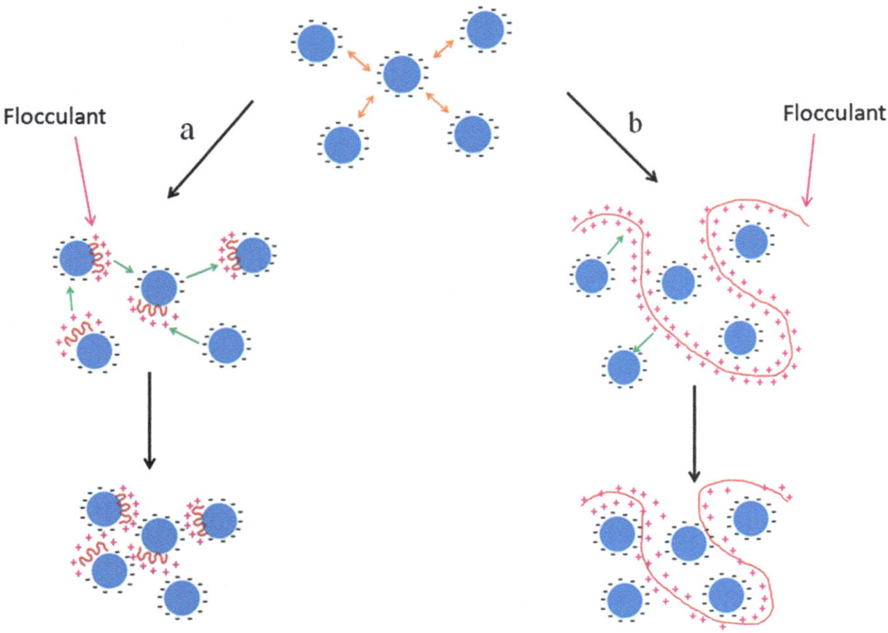

Fig. 11.1 Mechanism of flocculation by bioflocculants: (**a**) flocculation by patching; (**b**) by bridge formation

formation of large networks and bridges resulting in better flocculation efficiency as compared to low molecular weight polymer. The ability to form bridge between the particles depends on particle density of the colloids, collision frequency, and binding properties of the polymer. The bridging mechanism finally results in the formation of large aggregated flocs that are stable and settle easily there by facilitating separation (Biggs et al. 2000; Blanco et al. 2002) (Fig. 11.1b).

11.3 Bioflocculants

The rising problem of environmental pollution and health hazards of chemical flocculants has shifted the interest to bioflocculants. Bioflocculants are the products of living organisms that can interact with the colloidal particles in suspension resulting in their aggregation. The bioflocculants can either be obtained from microorganisms or plants.

11.3.1 Microbial Bioflocculants

Microbial flocculants are produced by various bacteria and fungi. Most of these are abundantly found in the flocs of the activated sludge. Studies have shown that there is a strong correlation between the production of extracellular polymeric substances by the bacteria, cell aggregation, and bioflocculation. The extracellular polymeric substances (EPS) are produced by a variety of microorganisms including bacteria, yeast, fungi, and algae. The microbial EPS can either be made up of polysaccharides, proteins, lipids, glycolipids, or glycoproteins. The bioflocculant EPS synthesized by the bacteria can either be capsular (cell bound), be slime type (loosely bound to the cells), or secreted in the extracellular environment as dissolved organic matter. These polymers are sticky form matrix around the suspended colloids resulting in large and stable flocs. Besides these, transport exopolymeric particles (TEP) have been reported in bacteria and diatoms. TEP are the products of microbial EPS, produced under abiotic conditions like shearing forces. In the natural conditions like marine environment, TEP have a major role in the aggregation of diatoms and other marine algae. The flocculation activity of TEP can either be due to electrostatic patch mechanism or by bridge formation (Passow 2002).

Microbial bioflocculants are high molecular weight heteropolysaccharides with branched or unbranched chains. The heteropolysaccharide is mainly composed of sugars like glucose, mannose, galactose, fucose, ribose, and rhamnose (Suh et al. 1997). The presence of arabinose in bioflocculant imparts good cell aggregation and biofilm-forming activity. Besides sugars uronic acids, sulfate esters, pyruvates, and proteins may also be present. The presence of sulfate in polysaccharide bioflocculant makes the polymer hydrophilic, imparting gel-like nature. Sulfate is also known to increase the flocculation activity of the bioflocculant (Zhou et al. 2012). Table 11.1 summarizes the chemical nature and properties of bioflocculants produced by different microorganisms.

11.3.2 Plant-Based Bioflocculants

Besides microbial based bioflocculants, several naturally existing polymers such as starch, chitosan, cellulose, natural gums, and mucilages of plant origin also have the ability to flocculate the suspended colloids in suspension (Renault et al. 2009). The efficiency of starch as a flocculant is however low, and it can be easily hydrolyzed by the amylases secreted by bacteria. Cellulose chemically modified to sodium carboxymethyl cellulose has been used for the treatment of drinking water (Khiari et al. 2010). Natural gums and mucilages have also been employed for the treatment of industrial effluents (Gupta and Ako 2005). Most of these polymers act by the mechanism of matrix and bridge formation around the suspended particles, resulting in large aggregates. The plant-based bioflocculants have gained considerable interest along with microbial flocculants as they are environmentally friendly, biodegradable, nontoxic to humans and other animals, and easily available.

Most of the plant-based bioflocculants are heteropolysaccharides containing different monosaccharides, like arabinose, galactose, glucose, mannose, xylose, rhamnose, and uronic acid (Jani et al. 2009; Mirhosseini and Amid 2012). The extraction of bioflocculants from the plants is mainly done by solvent extraction method. Most of the bioflocculants from okra, psyllium, tamarind, and fenugreek have been isolated by this method (Agrawal et al. 2001; Mishra and Bajpai 2005).

Table 11.1 Bioflocculants produced by different bacteria, their chemical nature, and properties

Organism	Growth media	Coagulant aid	Chemical composition	Functional groups in flocculation	Ability to withstand stress	References
Bacillus sp. A-5A	Glucose and ammonium nitrate, pH 6	Ca^{+2}	Uronic acid, neutral sugar, and protein	FTIR analysis – methoxy, hydroxyl, and amino groups	Thermal stable bioflocculant that retains 65% activity after heating at 100 °C for 25 min	Ugbenyenn et al. (2014)
Halomonas sp. and *Micrococcus* sp. consortia	Glucose, ammonium sulfate, and urea	Ca^{+2} Mn^{+2}, Al^{+3}	Uronic acid protein, and sugar	FTIR analysis – methoxy, hydroxyl, and amino groups	Thermal stable bioflocculant that retains 70% activity at 80 °C for 30 min	Okaiyeto et al. (2013)
Bacillus licheniformis	Minimal media	Ca^{+2}	Proteoglycan, neutral sugar, amino-sugar carbohydrate, and protein	Carboxy, hydroxy, and amino groups	Stable below 80 °C and pH range 3–8	Xiong et al. (2010)
Pseudoalteromonas sp.	Minimal media	Ca^{+2} and Fe^{+2}		Acetyl groups	Work in saline conditions with 35% NaCl, work at low temperature below 15 °C	Li et al. (2008)
Methylobacterium sp.	Minimal media	Ca^{+2}		Carboxyl and hydroxyl groups	Thermostable, retains 70% activity at 100 °C for 30 min	Luvuyo et al. (2013)
Pseudomonas sp.	Minimal media		Carbohydrates, proteins, and glycol-proteins	Carboxyl, hydroxyl, and methoxyl groups	Thermally stable between 5 and 100 °C	Zaki et al. (2011)

Besides using these natural polymers directly for flocculation, they can be chemically modified by the addition of more branches of another polymer to give a grafted polymer (Kumar et al. 2012; Kalia et al. 2013). As the grafted polymers have a high molecular weight, high hydrodynamic volume, and large surface area for the adsorption of colloidal particles, they are considered as very efficient flocculants (Ghosh et al. 2010; Pal et al. 2012).

11.4 Factors Affecting Bioflocculant Production

A variety of factors like carbon source, pH, temperature, metal ions, aeration, etc. have shown to affect bioflocculant production. Studies with *Sorangium cellulosum* have shown that the bioflocculant production in the presence of soluble starch accounted about 14.8 g/l compared to 1 g/l in glucose. On the other hand, combination of starch and glucose showed maximum 17.5 g/l yield of bioflocculant (Zhang et al. 2002). *Bacillus licheniformis* isolated from contaminated Luria broth medium have shown maximum bioflocculant production in medium consisting sucrose, glucose, and soluble starch as carbon sources and urea as nitrogen source (Xiong et al. 2010). *Bacillus* spp. isolated from the marine environment have shown maximum bioflocculant production utilizing glucose and ammonium nitrate as carbon and nitrogen source, respectively (Cosa et al. 2013).

The role of initial pH of the media on bioflocculant production varies with different isolates. In case of *Bacillus licheniformis*, maximum flocculation activity was observed at initial pH 7.5 (Xiong et al. 2010). In case of mixed cultures of *Methylobacterium* sp. and *Actinobacterium* sp., maximum bioflocculant production is observed at pH 9, while production is minimum at neutral and acidic pH (Luvuyo et al. 2013). *Bacillus* spp. isolated from the marine environment show maximum bioflocculant production at a wide pH range for 3–9 with optimum production at pH 6. Metal ions may stimulate or inhibit the bioflocculant production in the culture media. In case of *Bacillus* spp., Na^{+}, K^{+}, Li^{+}, Mn^{+2}, and Al^{+3} favor bioflocculant production with maximum production observed with Ca^{+2} ions. On the other hand, Fe^{+3} had inhibitory effect (Cosa et al. 2013).

11.5 Factors Affecting Flocculation Activity

Different factors such as molecular weight of the bioflocculant, pH, temperature, and concentration of metal ions affect flocculation activity. High molecular weight bioflocculant provides large binding sites and forms a netlike matrix around the suspended particles that increases the flocculation activity. On the other hand, low molecular weight polymer shows flocculation by electrostatic patch mechanism. The metal ions aid flocculation, by neutralizing the surface charge of the suspended particles allowing them to come closer and interact with the bioflocculant added. This phenomenon is known as flocculation by coagulation as described earlier. Increase in the flocculation activity has been observed by the addition of Al^{+3}, Fe^{+3},

Ca^{+2}, and Mg^{+2} ions. These metal ions induce coagulation at specific concentration and pH. It must be noted that excessive addition of metal coagulants can cause reversal of zeta potential that can decrease the stability of flocs and overall flocculation efficiency.

The pH should play a decisive role in flocculation by bioflocculants. Carboxyl, hydroxyl and amino are common functional groups present on bioflocculants involved in flocculation. A decrease in pH can cause protonation of the functional groups on bioflocculant imparting net positive charge. On the other hand, increase in pH imparts net negative charge on the polymer.

11.6 Microalgae Cultivation and Harvesting

Microalgae are the unicellular microorganisms that live in water bodies like lakes, rivers, and oceans. They utilize nitrogen and phosphates present in water to generate biomass by the process of photosynthesis. With the increase in world population and demand of fossil fuels, the demand of biofuels like biodiesel has increased over the last decade (Foley et al. 2011). Studies have shown that microalgae can be a promising source of generating biofuel. Most of the microalgae act as a good source of triglycerides that can be extracted to form biodiesel by transesterification reaction with methanol. Acids, alkali, and enzymes like lipase aid the transesterification reaction. Most of the transesterification reaction is catalyzed by alkali, as the use of enzymes like lipase can be costly (Chisti 2008; Posten and Schaub 2009; Wijffels and Barbosa 2010). Besides these the algal biomass can be used to produce bioplastic, methane in anaerobic digester, biofertilizer, and as a feed for fish or livestock. The production of microalgae at large scale is a challenging task. Cost-effective production and harvesting are the two major factors that must be taken into consideration while cultivating microalgae at a large scale.

11.6.1 Cost-Effective Production of Microalgae

Cultivation of microalgae requires nutrients like nitrates, phosphates, and inorganic nutrients along with CO_2 and sunlight. Different methods such as growth in open ponds, bioreactor, suspended cultures, immobilized cultures, and biofilm-based methods have been employed for the large-scale production.

11.6.2 Open Ponds

They are commonly known as high rate algal ponds (HRAP) that are commonly used for microalgal growth. These are artificially created shallow ponds with an impeller at its bottom to provide mixing of nutrients and algal biomass. In this method the availability of atmospheric carbon (CO_2) via diffusion can be a growth-limiting factor. However providing CO_2 externally has shown an increase in the biomass as well as lipid production in the cells (Chiu et al. 2009; Vandamme et al. 2013).Continuous CO_2 sparging may not be economically feasible as the retention time of CO_2 bubbles can be short and they finally end up in the atmosphere (Mata et al. 2010). Besides these, control of other parameters like pH and temperature is the major drawback of open pond system.

11.6.3 Photobioreactors

Photobioreactors are the most efficient for the growth of microalgae. The design of photobioreactors may be horizontal, vertical, and helical (Chisti 2007). The photobioreactors have several advantages over open ponds, which include better bulk mixing, pH and temperature control, minimal evaporative loss, and better growth rate (Mata et al. 2010). One of the most commonly used reactors is modified air-lift reactors. These bioreactors are transparent and are equipped with a riser column, sparger to provide aeration, and pH probes. As the air/CO_2 bubbles are introduced in the riser column, the media and the algal biomass begin to move upward in the riser column and downward in the downcomer column. The continuous sparging not only provides carbon source but also causes mixing of nutrients as well as biomass in the reactor (Merchuk et al. 2007). The distribution of light in the reactor might be a growth-limiting factor. To solve this problem, specially designed transparent impellers are attached with a perforated rod in the middle of reactor. Nutrients as well as CO_2 are allowed to diffuse through the perforated rod, whereas light passing through this rod illuminates the transparent impeller, giving enough photons for the growth of algae. One of the limitations of photobioreactor is the accumulation of oxygen generated by photosynthesis that can cause photooxidative damage to the cells, which may decrease the growth rate (Carvalho et al. 2006). Also the maintenance of tubular bioreactors can be costly (Grima et al. 1999).

11.6.4 Flat-Plate Bioreactors

These are rectangular reactors constructed by joining two transparent sheets together. A typical flat-plate reactor has dimensions: height 1.5 m, length 2.5 m, and with 7 cm. As the width of the reactor is less, there are no dark zones in the reactor. This provides enough illumination for the algal biomass growing in the reactor. Aeration is provided by the sparger that provides aeration, mass transfer, and bulk mixing. Recently, Offshore Membrane Enclosure for Growing Algae (OMEGA) system was developed by NASA for waste water treatment as well as cultivation of microalgae. This technique uses specially developed plastic bags that are partially permeable to oxygen and carbon dioxide. The microalgae in the bags utilize the nutrients present in the waste water to generate biomass, and the treated water leaves the bag via osmosis (Trent et al. 2013).

11.6.5 Immobilized Algae in Matrix

The use of immobilized microalgae in matrix-like alginate has been successful in the waste water treatment. As algae are immobilized, they utilize the nutrients in waste water for their metabolism (Chevalier et al. 2000). The immobilized microalgae in alginate beads have shown better accumulation of lipids and pigments as compared to free algae when coimmobilized with *Azospirillum brasilense* (De Bashan et al. 2002). However, it must be noted that the cost of immobilized matrix makes this method economically futile. Most of the works on immobilized algae have been done at bench scale only.

11.6.6 Algal Biofilms

Besides suspended cultures microalgae also have a tendency to form biofilms. It has been found that if the surface area is large enough, then the amount of algal biomass produced can be greater than the suspended cultures. The algal biofilm cannot only remove the nutrients of waste water more efficiently but also provide advantage on recovery of algal cells. With respect to waste water treatment, trickling filters and rotating biological contactors are bacterial biofilm-based reactors that are widely used for large-scale waste water treatment. The formation of biofilms depends on several factors such as the presence of cations and proteins, ability of bacteria to produce EPS, as well as availability of organic nutrients in the waste water. The development of algal biofilms can be aided by the EPS-producing bacteria present in the waste water (Hodoki 2005). Integrated use of open ponds and algal trickling filters has shown better removal of waste water nutrients as well as removal of algal cells by biofilm formation (Hoffmann 1998).

11.7 Harvesting of Microalgae

Besides cultivation, the recovery of algal biomass following cultivation is a major challenge (Uduman et al. 2010). The harvesting method depends on microalgal size, density, and value of the product. Generally two main stages, viz., bulk harvesting and thickening, are involved. In the former separation of biomass from the bulk, suspension is carried out, and in the latter concentration of the slurry is done (Brennan and Owende 2010). The common technique used for microalgae harvesting is centrifugation which is costly at an industrial scale. One of the alternative and cost-effective methods for harvesting is by flocculation, which is part of the bulk harvesting stage. The media used for microalgal cultivation is rich in phosphates and nitrates. Addition of calcium ions to the broth causes precipitation of phosphates as calcium phosphates. As the microalgal cells have net negative charge, the precipitates with net positive charge can neutralize them resulting in flocculation. However this phenomenon requires high concentration of phosphate ions (Christenson and Sims 2011). Besides calcium, magnesium ions that precipitate at alkaline pH also have a tendency to flocculate microalgae (Vandamme et al. 2012). Other metal salts like alum and ferric chloride have been used for the harvesting of microalgae *Dunaliella*.

Recent studies have shown that cationic, magnetic nanoparticles made up of Fe_2O_3 can be used for the microalgae harvesting with application of external magnetic field. These nanoparticles can also be attached to a cationic polymer making the harvesting process more efficient (Lim et al. 2012). It should be noted that biomass recovered by this method contains high concentration of metal ions. The removal of metal ions from the biomass becomes difficult. The residual metal ions can hinder the extraction of lipids and carotenoids from the biomass (Rwehumbiza et al. 2012). Besides metal salts, organic chemical flocculants like polyacrylamide have shown effective recovery of microalgae; however the use of microalgal biomass recovered in this way, for animal feed, can be a major concern due to the presence of chemicals (Brennan and Owende 2010).

The algal blooms in lakes show spontaneous flocculation caused by the extracellular polymers made by the algae (Vandamme et al. 2013). Bioflocculation generally involves electrostatic attraction between the positively charged bioflocculant and negatively charged microalgal cells, resulting in large stable aggregates (Zhou et al. 2012; Zhang and Hu 2012). Biopolymer such as chitosan, a derivative of chitin, is a cationic, high molecular weight polymer capable of flocculating microalgae via bridge mechanism. However, flocculation by chitosan requires a relatively low pH. Besides chitin cationic starch with quaternary ammonium groups has shown to flocculate microalgae, at a broader pH range.

Bacteria as well as fungi have a tendency to cause flocculation of microalgae (Zhang and Hu 2012; Zhou et al. 2012). Extracellular poly-γ-glutamic acid produced by *Bacillus subtilis* has shown effective flocculation of microalgae. Cocultivation of algal and bacterial cells is one of the methods that allows growth of both algal and bacterial biomass. As the bacterial biomass increases, it interacts with the algal cells resulting in bacterial algal flocs. It must be noted that additional carbon source must be added to the media to support bacterial growth. Besides media components, the salts present in the media and pH influence the flocculation process. Studies with *Chlorococcum nivale*, *Chlorococcum ellipsoideum*, and *Scenedesmus* sp. have shown that flocculation activity of algae increases with decrease in the pH, with optimum pH of 4. Charge neutralization of organic acid functional groups on the algal surface, by H^+ ions, is mainly responsible for the aggregation of algal cells. Similar results were observed when poly-γ-glutamic acid purified from *Bacillus licheniformis* was used for the flocculation of *Desmodesmus* sp. The flocculation activity was increased from 44% to 98% when the pH was dropped from 7.2 to 3 (Liu et al. 2013). Most of the microalgae are known to produce significant amounts of extracellular polysaccharides that can interfere with the flocculant added.

The exopolymeric substance produced by *Scenedesmus quadricauda* has shown to recover the algal biomass up to 86% in presence of Zn^{+2} ions. This EPS is chemically composed of carbohydrate and protein and has hydroxyl, carboxyl, and amino groups that help in zeta potential reduction and bridge formation with the algal cells causing flocculation (Bibi et al. 2016). Cyanobacteria like *Synechocystis salina* also have the ability to flocculate *Chlorella vulgaris* and *Pseudokirchneriella subcapitata* by decreasing the zeta potential on cell surface. Studies on cell surface physiochemical properties of *S. salina* revealed that it possesses a hydrophobic surface that allows it to form aggregates. Cocultivation of *S. salina* with other microalgae resulted in more hydrophobic algal suspension. This increase in hydrophobicity causes the algal cells to repel water molecules and form aggregates, which settle as flocs. Besides hydrophobicity the ionic strength of the media and pH also plays an important role in governing the flocculation of microalgae and cyanobacteria (Gonçalves et al. 2015). Studies with marine isolate *Cobetia marina* show recovery of *Chlorella vulgaris* by flocculation – flotation phenomenon. In this process, unlike sedimentation the flocs float on the surface of liquid phase, due to its low density. This bioflocculant was composed of carbohydrate and protein and showed maximum recovery of *Chlorella* up to 92% in the presence of $CaCl_2$ as a coagulating aid (Lei et al. 2015). Bioflocculants from halotolerant, alkaliphilic *Bacillus agaradhaerens* have shown 80% flocculating activity with *Chlorella minutissima*. Chemical analysis showed the presence of polysaccharides, proteins, and nucleic acids in the bioflocculant (Liu et al. 2015).

Fungi such as *Aspergillus niger* have a tendency to bind and aggregate microalgal cells facilitating efficient removal of algal biomass (Zhang and Hu 2012). The formation of algal-fungal pellets is mainly regulated by the pH which in turn can be regulated by the addition of glucose in the growth media. Besides pH the number of fungal spores used as an initial inoculum also plays an important role. The algal-fungal pellets can be further reused for the removal of phosphates and nitrates from the waste water (Zhou et al. 2012). Besides *Aspergillus niger*, *Aspergillus lentulus* isolated from effluents of textile industry have shown nearly 100% recovery of *Chroococcus* spp. within 6 h, with fungal-to-algal ratio 1:3. The flocculation process involved electrostatic interaction between negatively charged algal cells and positively charged fungal mycelia. The cellulase produced by the fungi also helps in the breakdown of algal cell wall into soluble sugars that increased the methane production up to 50%, during anaerobic digestion. The recovery of algal biomass with fungi can have several applications like production of biomethane, biohydrogen, and biodiesel (Prajapati et al. 2016).

Recently plant-based flocculant isolated from *Strychnos potatorum* commonly known as clearing nut, has shown 99% recovery of *Chlorella vulgaris*. The isolation and purification of this bioflocculant are easy and cost-effective as compared to microbial bioflocculant (Razack et al. 2015). Derivatives of chitin like chitosan have shown promising results for the replacement of chemical flocculant like polyacrylamide, for harvesting of *Chlamydomonas reinhardtii*. The efficiency of harvesting *Chlamydomonas* was 35 kg per kg of chitosan used. However cultivation of *Chlamydomonas* in the presence of chitosan showed slight decrease in the growth rate, but was not lethal. Attempts have also been made to recycle the flocculant present in the supernatant, after the harvesting is done (Delrue et al. 2015).

In this section flocculation as a technique for separation of algal biomass has been focused; however, a variety of techniques are used and there is no standard technique as such. Microalgae are an attractive feedstock for production of biofuel since its productivity is high and it does not require arable landlike agricultural feedstock. As a source of biodiesel, it can compete with the fossil fuel provided it is grown in a cost-effective way. This is feasible since it requires simple nutrients for its cultivation. However, harvesting step is cost intensive in addition to bulk cultivation process. Application of advanced methods of improvement of microalgal stocks, cultivation process, and harvesting process should make it economically viable. The use of bioflocculants in harvesting is one of the ways in which it can be made cost-effective. Environmental concerns are going to make bioflocculants important in this application as it is projected that microalgal biodiesel will be sustainable process in near future.

References

Agarwal M, Srinivasan R, Mishra A (2001) Study on flocculation efficiency of okra gum in sewage waste water. Macromol Mater Eng 286:560–563. https://doi.org/10.1002/1439-2054(20010901)286

Amuda OS, Amoo IA (2007) Coagulation/flocculation process and sludge conditioning in beverage industrial wastewater treatment. J Hazard Mater 141:778–783. https://doi.org/10.1016/j.jhazmat.2006.07.044

Banks WA, Niehoff ML, Drago D, Zatta P (2006) Aluminum complexing enhances amyloid β protein penetration of blood–brain barrier. Brain Res 1116:215–221. https://doi.org/10.1016/j.brainres.2006.07.112

Bibi R, Ahmad Z, Imran M, Hussain S, Ditta A, Mahmood S, Khalid A (2016) Algal bioethanol production technology: a trend towards sustainable development. Renew Sustain Energy Rev 71:976–985. https://doi.org/10.1016/j.rser.2016.12.126

Biggs S, Habgood M, Jameson GJ (2000) Aggregate structures formed via a bridging flocculation mechanism. Chem Eng J 80:13–22. https://doi.org/10.1016/S1383-5866(00)00072-1

Blanco A, Negro C, Tijero J (2002) Flocculation monitoring: focused beam reflectance measurement as a measurement tool. Can J Chem Eng 80:1–7. https://doi.org/10.1002/cjce.5450800403

Brennan L, Owende P (2010) Biofuels from microalgae—a review of technologies for production, processing, and extractions of biofuels and coproducts. Renew Sust Energ Rev 14:557–577. https://doi.org/10.1016/j.rser.2009.10.009

Brostow W, Lobland HH, Pal S, Singh RP (2009) Polymeric flocculants for wastewater and industrial effluent treatment. J Mater Educ 31:157–166. https://doi.org/10.1016/j.watres.2007.03.012

Carvalho AP, Meireles LA, Malcata FX (2006) Microalgal reactors: a review of enclosed system designs and performances. Biotechol Progr 22:1490–1506. https://doi.org/10.1021/bp060065r

Chevalier P, Proulx D, Lessard P, Vincent WF, De la Noüe J (2000) Nitrogen and phosphorus removal by high latitude mat-forming cyanobacteria for potential use in tertiary wastewater treatment. J Appl Phycol 12:105–112. https://doi.org/10.1023/A:1008168128654

Chisti Y (2007) Biodiesel from microalgae. Biotechnol Adv 25:294–306. https://doi.org/10.1016/j.biotechadv.2007.02.001

Chisti Y (2008) Biodiesel from microalgae beats bioethanol. Trends Biotechnol 26:126–131. https://doi.org/10.1016/j.tibtech.2007.12.002

Chiu SY, Kao CY, Tsai MT, Ong SC, Chen CH, Lin CS (2009) Lipid accumulation and CO_2 utilization of Nannochloropsis oculata in response to CO_2 aeration. Bioresour Technol 100:833–838. https://doi.org/10.1016/j.biortech.2008.06.061

Chong MF (2012) Direct flocculation process for wastewater treatment. In: Sharma (ed) Advances in water treatment and pollution prevention, pp 201–230. ISBN: 978-94-007-4203-1. https://doi.org/10.1007/978-94-007-4204-8_8

Christenson L, Sims R (2011) Production and harvesting of microalgae for wastewater treatment, biofuels, and bioproducts. Biotechnol Adv 29:686–702. https://doi.org/10.1016/j.biotechadv.2011.05.015

Cosa S, Ugbenyen AM, Mabinya LV, Rumbold K, Okoh AI (2013) Characterization and flocculation efficiency of a bioflocculant produced by a marine Halobacillus. Environ Technol 34(2671):2679. https://doi.org/10.1080/09593330.2013.786104

de Bashan LE, Bashan Y, Moreno M, Lebsky VK, Bustillos JJ (2002) Increased pigment and lipid content, lipid variety, and cell and population size of the microalgae *Chlorella spp.* when co-immobilized in alginate beads with the microalgae-growth-promoting bacterium *Azospirillum brasilense*. Can J Microbiol 48:514–521. https://doi.org/10.1139/W02-051

Delrue F, Imbert Y, Fleury G, Peltier G, Sassi JF (2015) Using coagulation–flocculation to harvest Chlamydomonas reinhardtii: coagulant and flocculant efficiencies, and reuse of the liquid phase as growth medium. Algal Res 9:283–290. https://doi.org/10.1016/j.algal.2015.04.004

Foley JA, Ramankutty N, Brauman KA, Cassidy ES, Gerber JS, Johnston M, Mueller ND, O'Connell C, Ray DK, West PC, Balzer C (2011) Solutions for a cultivated planet. Nature 478:337–342. https://doi.org/10.1038/nature10452

Ghosh S, Sen G, Jha U, Pal S (2010) Novel biodegradable polymeric flocculant based on polyacrylamide-grafted tamarind kernel polysaccharide. Bioresour Technol 101:9638–9644. https://doi.org/10.1016/j.biortech.2010.07.058

Gonçalves AL, Ferreira C, Loureiro JA, Pires JC, Simões M (2015) Surface physicochemical properties of selected single and mixed cultures of microalgae and cyanobacteria and their relationship with sedimentation kinetics. Bioresour Bioprocess 2:21. https://doi.org/10.1186/s40643-015-0051-y

Grima EM, Fernández FA, Camacho FG, Chisti Y (1999) Photobioreactors: light regime, mass transfer, and scaleup. J Biotechnol 70:231–247. https://doi.org/10.1016/j.biortech.2011.11.105

Gupta BS, Ako JE (2005) Application of guar gum as a flocculant aid in food processing and potable water treatment. Eur Food Res Technol 221:746–751. https://doi.org/10.1007/s00217-005-0056-4

Hodoki Y (2005) Bacteria biofilm encourages algal immigration onto substrata in lotic systems. Hydrobiologia 539:27–34. https://doi.org/10.1007/s10750-004-3082-5

Hoffmann JP (1998) Wastewater treatment with suspended and nonsuspended algae. J Phycol 34:757–763. https://doi.org/10.1046/j.1529-8817.1998.340757.x

Jani GK, Shah DP, Prajapati VD, Jain VC (2009) Gums and mucilages: versatile excipients for pharmaceutical formulations. Asian J Pharm Sci 4:309–323

Kalia S, Sabaa MW, Kango S (2013) Polymer grafting: a versatile means to modify the polysaccharides. In: Polysaccharide based graft copolymers. Springer, Berlin. https://doi.org/10.1007/978-3-642-36566-91

Khiari R, Dridi-Dhaouadi S, Aguir C, Mhenni MF (2010) Experimental evaluation of eco-friendly flocculants prepared from date palm rachis. J Environ Sci 22:1539–1543. https://doi.org/10.1016/S1001-0742(09)60286-2

Kleimann J, Gehin-Delval C, Auweter H, Borkovec M (2005) Super-stoichiometric charge neutralization in particle-polyelectrolyte systems. Langmuir 21:3688–3698. https://doi.org/10.1021/la046911u

Kumar R, Setia A, Mahadevan N (2012) Grafting modification of the polysaccharide by the use of microwave irradiation—a review. Int J Recent Adv Pharm Res 2:45–53. ISSN: 2230-9306

Larkum AW, Ross IL, Kruse O, Hankamer B (2012) Selection, breeding and engineering of microalgae for bioenergy and biofuel production. Trends Biotechnol 30:198–205. https://doi.org/10.1016/j.tibtech.2011.11.003

Lee KE, Morad N, Teng TT, Poh BT (2012) Development, characterization and the application of hybrid materials in coagulation/flocculation of wastewater: a review. Chem Eng J 203:370–386. https://doi.org/10.1016/j.cej.2012.06.109

Lei X, Chen Y, Shao Z, Chen Z, Li Y, Zhu H, Zhang J, Zheng W, Zheng T (2015) Effective harvesting of the microalgae Chlorella vulgaris via flocculation–flotation with bioflocculant. Bioresour Technol 198:922–925. https://doi.org/10.1016/j.biortech.2015.08.095

Li WW, Zhou WZ, Zhang YZ, Wang J, Zhu XB (2008) Flocculation behavior and mechanism of an exopolysaccharide from the deep-sea psychrophilic bacterium *Pseudoalteromonas sp.* SM9913. Bioresour Technol 99:6893–6899. https://doi.org/10.1016/j.biortech.2008.01.050

Lim JK, Chieh DCJ, Jalak SA, Toh PY, Yasin NM, Ng BW, Ahmad AL (2012) Rapid magnetophoretic separation of microalgae. Small 8:1683–1692. https://doi.org/10.1002/smll.201102400

Liu C, Wang K, Jiang JH, Liu WJ, Wang JY (2015) A novel bioflocculant produced by a salt-tolerant, alkaliphilic and biofilm-forming strain *Bacillus agaradhaerens* C9 and its application in harvesting *Chlorella minutissima* UTEX2341. Biochem Eng J 93:166–172. https://doi.org/10.1016/j.bej.2014.10.006

Liu J, Zhu Y, Tao Y, Zhang Y, Li A, Li T, Sang M, Zhang C (2013a) Freshwater microalgae harvested via flocculation induced by pH decrease. Biotechnol Biofuels 6:1. https://doi.org/10.1186/1754-6834-6-98

Luvuyo N, Nwodo UU, Mabinya LV, Okoh AI (2013) Studies on bioflocculant production by a mixed culture of *Methylobacterium sp. Obi* and *Actinobacterium sp.* Mayor. BMC Biotechnol 13:62. https://doi.org/10.1186/1472-6750-13-62

Mata TM, Martins AA, Caetano NS (2010) Microalgae for biodiesel production and other applications: a review. Renew Sust Energ Rev 14:217–232. https://doi.org/10.1016/j.rser.2009.07.020

Merchuk JC, Rosenblat Y, Berzin I (2007) Fluid flow and mass transfer in a counter-current gas–liquid inclined tubes photo-bioreactor. Chem Eng Sci 62:7414–7425. https://doi.org/10.1016/j.ces.2007.08.058

Mirhosseini H, Amid BT (2012) A review study on chemical composition and molecular structure of newly plant gum exudates and seed gums. Food Res Int 46:387–398. https://doi.org/10.1016/j.foodres.2011.11.017

Mishra A, Bajpai M (2005) Flocculation behaviour of model textile wastewater treated with a food grade polysaccharide. J Hazard Mater 118:213–217. https://doi.org/10.1016/j.jhazmat.2004.11.003

Okaiyeto K, Nwodo UU, Mabinya LV, Okoh AI (2013) Characterization of a Bioflocculant produced by a consortium of *Halomonas sp. Okoh* and *Micrococcus sp.* Leo. Int J Environ Res Public Health 10:5097–5110. https://doi.org/10.3390/ijerph10105097

Pal S, Sen G, Ghosh S, Singh RP (2012) High performance polymeric flocculants based on modified polysaccharides—microwave assisted synthesis. Carbohydr Polym 87:336–342. https://doi.org/10.1016/j.carbpol.2011.07.052

Passow U (2002) Transparent exopolymer particles (TEP) in aquatic environments. Prog Oceanogr 55:287–333. https://doi.org/10.1016/S0079-6611(02)00138-6

Polizzi S, Pira E, Ferrara M, Bugiani M, Papaleo A, Albera R, Palmi S (2002) Neurotoxic effects of aluminium among foundry workers and Alzheimer's disease. Neuro Toxicol 23:761–774. https://doi.org/10.1016/S0161-813X(02)00097-9

Posten C, Schaub G (2009) Microalgae and terrestrial biomass as source for fuels—a process view. J Biotechnol 142:64–69. https://doi.org/10.1016/j.jbiotec.2009.03.015

Prajapati SK, Bhattacharya A, Kumar P, Malik A, Vijay VK (2016) A method for simultaneous bioflocculation and pretreatment of algal biomass targeting improved methane production. Green Chem 18:5230–5238. https://doi.org/10.1039/C6GC01483F

Razack SA, Duraiarasan S, Shellomith AS, Muralikrishnan K (2015) Statistical optimization of harvesting *Chlorella vulgaris* using a novel biosource, *Strychnos potatorum*. Biotechnol Rep 7:150–156. https://doi.org/10.1016/j.btre.2015.06.006

Renault F, Sancey B, Badot PM, Crini G (2009) Chitosan for coagulation/flocculation processes–an eco-friendly approach. Eur Polym J 45:1337–1348. https://doi.org/10.1016/j.eurpolymj.2008.12.027

Rwehumbiza VM, Harrison R, Thomsen L (2012) Alum-induced flocculation of preconcentrated *Nannochloropsis salina*: residual aluminium in the biomass, FAMEs and its effects on microalgae growth upon media recycling. Chem Eng J 200:168–175. https://doi.org/10.1016/j.cej.2012.06.008

Salim S, Bosma R, Vermuë MH, Wijffels RH (2011) Harvesting of microalgae by bio-flocculation. J Appl Phycol 23:849–855. https://doi.org/10.1007/s10811-010-9591-x

Sharma BR, Dhuldhoya NC, Merchant UC (2006) Flocculants—an ecofriendly approach. J Polym Environ 14:195–202. https://doi.org/10.1007/s10924-006-0011-x

Singh RP, Karmakar GP, Rath SK, Karmakar NC, Pandey SR, Tripathy T, Panda J, Kanan K, Jain SK, Lan NT (2000) Biodegradable drag reducing agents and flocculants based on polysaccharides: materials and applications. Polym Eng Sci 40:46–60. https://doi.org/10.1002/pen.11138

Suh HH, Kwon GS, Lee CH, Kim HS, HM O, Yoon BD (1997) Characterization of bioflocculant produced by *Bacillus sp.* DP-152. J Ferment Bioeng 84:108–112. https://doi.org/10.1016/S0922-338X (97)82537-8

Suopajärvi T, Liimatainen H, Hormi O, Niinimäki J (2013) Coagulation–flocculation treatment of municipal wastewater based on anionized nanocelluloses. Chem Eng J 231:59–67. https://doi.org/10.1016/j.cej.2013.07.010

Trent JD, Gormly SJ, Embaye TN, Delzeit LD, Flynn MT, Liggett TA, Buckwalter PW, Baertsch R (2013) US Patent No. 8,409,845. Washington, DC: U.S. Patent and Trademark Office

Uduman N, Qi Y, Danquah MK, Forde GM, Hoadley A (2010) Dewatering of microalgal cultures: a major bottleneck to algae-based fuels. J Renew Sust Energ 2:012701. https://doi.org/10.1063/1.3294480

Ugbenyen AM, Cosa S, Mabinya LV, Okoh AI (2014) Bioflocculant production by *Bacillus sp.* Gilbert isolated from a marine environment in South Africa. Appl Biochem Microbiol 50:49–54. https://doi.org/10.1134/S0003683814010104

Vandamme D, Foubert I, Muylaert K (2013) Flocculation as a low-cost method for harvesting microalgae for bulk biomass production. Trends Biotechnol 31:233–239. https://doi.org/10.1016/j.tibtech.2012.12.005

Vandamme D, Foubert I, Fraeye I, Meesschaert B, Muylaert K (2012) Flocculation of *Chlorella vulgaris* induced by high pH: role of magnesium and calcium and practical implications. Bioresour Technol 105:114–119. https://doi.org/10.1016/j.biortech.2011.11.105

Ward RJ, McCrohan CR, White KN (2006) Influence of aqueous aluminium on the immune system of the freshwater crayfish *Pacifastacus leniusculus*. Aquat Toxicol 77:222–228. https://doi.org/10.1016/j.aquatox.2005.12.006

Wijffels RH, Barbosa MJ (2010) An outlook on microalgal biofuels. Sci 329:796–799. https://doi.org/10.1126/science.1189003

Xiong Y, Wang Y, Yu Y, Li Q, Wang H, Chen R, He N (2010) Production and characterization of a novel bioflocculant from *Bacillus licheniformis*. Appl Environ Microbiol 76:2778–2782. https://doi.org/10.1128/AEM.02558-09

Zaki S, Farag S, Elreesh GA, Elkady M, Nosier M, El Abd D (2011) Characterization of bioflocculants produced by bacteria isolated from crude petroleum oil. Int J Environ Sci Technol 8:831–840. https://doi.org/10.1007/BF03326266

Zhang J, Wang R, Jiang P, Liu Z (2002) Production of an exopolysaccharide bioflocculant by *Sorangium cellulosum*. Lett Appl Microbiol 34:178–181. https://doi.org/10.1046/j.1472-765x.2002.01068.x

Zhang J, Hu B (2012) A novel method to harvest microalgae via co-culture of filamentous fungi to form cell pellets. Bioresour Technol 114:529–535. https://doi.org/10.1016/j.biortech.2012.03.054

Zhou W, Cheng Y, Li Y, Wan Y, Liu Y, Lin X, Ruan R (2012) Novel fungal pelletization-assisted technology for algae harvesting and wastewater treatment. Appl Biochem Biotechnol 167:214–228. https://doi.org/10.1007/s12010-012-9667-y

Biohydrogen Production: An Outlook of Fermentative Processes and Integration Strategies

12

G.N. Nikhil, Omprakash Sarkar, and S. Venkata Mohan

Abstract

The emanations upon combustion of petroleum fuels cause grave pessimistic impact on surrounding ambiance and global climate change. Therefore, the contemporary stance of scientific fraternity is to generate energy and mercantile products through biological methods with waste as a resource. Biohydrogen production from wastewater seems to be a promising green option for sustainable renewable energy. The process is feasible from practical point of view and can be operated under ambient conditions. It has been attracting attention due to its applicability to different types of wastewaters, and the production costs of biohydrogen can compete economically with other traditional methods. However, the crucial challenges like enhancing rate and yield for sustainable biohydrogen production still persist. During fermentation process, the undissociated volatile fatty acids (VFAs) and alcohols accumulate in the system leading to inhibition and redundancy in substrate degradation. Employing integration strategies with other bioprocesses like photo-fermentation or bio-electrochemical systems is the sanguine option to make the process frugally possible.

Keywords

Biohydrogen • Fermentation • Renewable • Energy • Photosynthetic

G.N. Nikhil • O. Sarkar • S. Venkata Mohan (✉)
Bioengineering and Environmental Sciences Lab, EEFF Department, CSIR-Indian Institute of Chemical Technology (CSIR-IICT), Hyderabad 500 007, Telangana, India
e-mail: vmohan_s@yahoo.com; svmohan@iict.res.in

H.J. Purohit et al. (eds.), *Optimization and Applicability of Bioprocesses*,
https://doi.org/10.1007/978-981-10-6863-8_12

12.1 Present Scenario

Paucity of worldwide petroleum reserves and apprehensions concerning environmental pollution prompt for the advancement of green energy substitutes (Arimi et al. 2015). Bioenergy has the sustainable potential that is appealing great curiosity amongst researchers across the world (Rama Mohan 2015). Recently, biohydrogen is cited to be one of the prospective areas of interest (Venkata Mohan 2009). In 2010, the market demand of international hydrogen production was 53 million metric tons and is anticipated to rise with a growth rate of 5.6% per annum. Worldwide commercial H_2 production presently is produced majorly from natural gas, oil and coal and to certain extent by water electrolysis (Markets and Markets 2011; Bhaskar et al. 2013; Rama Mohan 2015). Alternatively, H_2 from waste biomass/wastewater using biological methods as renewable resource is gaining much interest. By this way negative-valued organic waste is transformed to green energy abating the pollution (Ghimire et al. 2015). The governing prerequisite for wastewater treatment is an ideal prospect to produce biohydrogen. By this way, it diminishes the overall expenditure towards effluent treatment, and additional revenue could be incurred from biohydrogen as a fuel (Nissilä et al. 2014; Thi et al. 2016).

12.2 Biohydrogen Production

There are several routes for H_2 production, viz. physical, chemical, biological and thermochemical (Fig. 12.1). BioH_2 production is majorly contributed by anaerobic fermentation which is categorised as light-independent and light-dependent (Venkata Mohan et al. 2007a; Hallenbeck 2013). Light-independent fermentation process is usually referred to as dark fermentation that hires both strict anaerobes and facultative bacteria. This process results a higher rate of H_2 and formation of

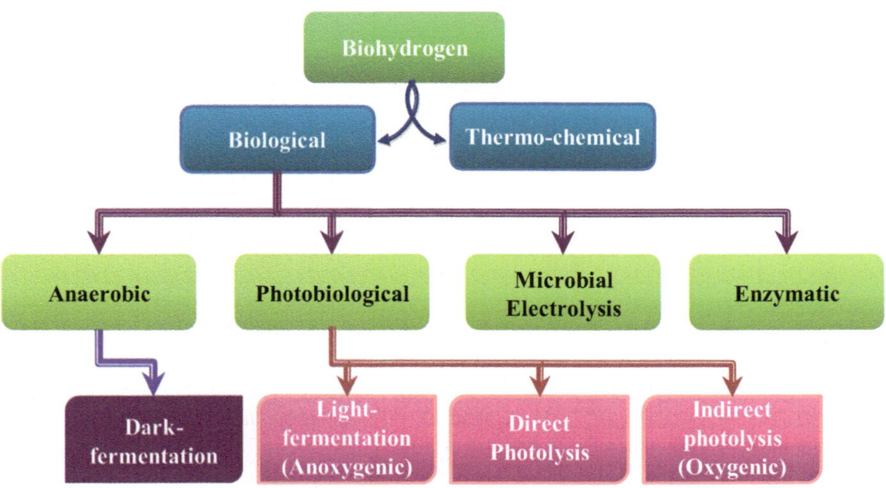

Fig. 12.1 Flowchart representing various routes for biohydrogen production

soluble metabolic products (SMPs), viz. volatile fatty acids, solvents, etc. Acidogenic anaerobes are incompetent of consuming these undissociated organic acids resulting in the accumulation of these acids leading to a drop in pH value. Subsequently, H_2 production is inhibited and, finally, results in low H_2 yields (Guo et al. 2010; Venkata Mohan et al. 2017). Light-dependent processes used by cyanobacteria and green algae are direct and indirect biophotolysis or via photo-fermentation mediated by photosynthetic bacteria. H_2 production by photosynthetic bacteria occurs by photo-fermentation of inorganic/organic acids (Lam and Lee 2013). The biochemistry varies considerably based on the biocatalyst, operational conditions, microenvironment and substrate (Lin et al. 2012).

12.2.1 Photo-Fermentation

Photosynthetic microorganisms, viz. cyanobacteria, photosynthetic bacteria and algae, have varied photosynthetic machinery (anoxygenic or oxygenic) that functions for H_2 production (Allakhverdiev et al. 2010). The process involves generation of a proton gradient by energy from sunlight and electron generation either by direct photolysis of water (Fig. 12.2a) or by indirect photolysis involving a parallel photosystem II (PSII)-independent process (Fig. 12.2b). In direct photolysis, the sunlight functions as a powerhouse for PSII ensuing production of oxidising equivalents that are used for the water oxidation into protons, electrons and O_2 (Krassen et al. 2009). Then, ferredoxin in reduced state is the electron donor for [FeFe]-hydrogenases in both types of photolysis. Thereafter, these reducing equivalents are transported to the chlorophyll α-dimer (P700) in photosystem I (PSI). The P700 is energised upon absorption of sunlight and then discharges electrons to the acceptor site of PSI containing iron–sulphur clusters via the electron transport chain (Hallenbeck 2013). Later, PSII is driven by water splitting where in the electrons are transported to reducing equivalents which then finally reduced by the [FeFe]-hydrogenase (*Hyd*A) to H_2. Hydrogenase enzyme performs as a discharge valve in presence of protons and electrons from the reduced ferredoxin to produce H_2. The activity of hydrogenase enzyme relies on the number of reducing equivalents from either of two photosynthetic processes (Constant and Hallenbeck 2013).

Microalgae are outstanding for their ability of high translation of solar energy to molecular H_2 (12–14%) by the oxidation of water molecules (Sambusiti et al. 2015). In direct biophotolysis, O_2 is generated as by-product of PSII that is a major suppressor of the hydrogenase enzyme; therefore, it can be operated for short periods of time (Blankenship et al. 1995). In indirect photolysis, the generated electrons and protons are generated and stored as starch during photosynthesis. Under certain stress conditions, the electrons and protons are fed into the plastoquinone pool and then onto *Hyd*A via PSI resulting in H_2 production. In case of microalgae and cyanobacteria, compounds for reserve accumulate during the Calvin cycle. At night, the compounds in reserve act as source energy for the cell activities. The change from aerobic to anaerobic condition is complemented by termination of a photosynthetic light reaction and generation of a surplus reductant, which is finally converted to H_2 by hydrogenase (Allakhverdiev et al. 2010). Under anaerobic conditions,

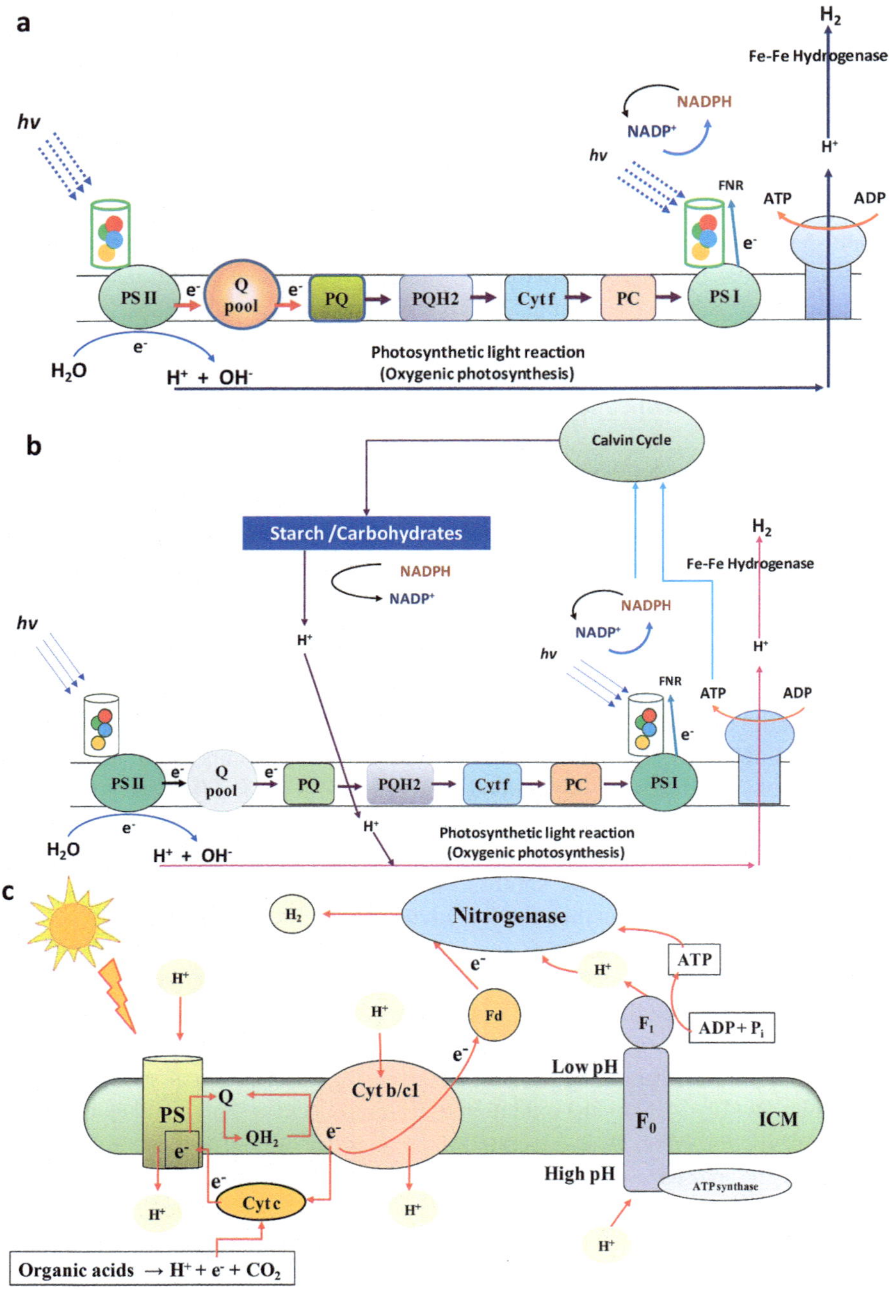

Fig. 12.2 Schematic illustration of the photosynthetic mechanisms for biohydrogen production – (**a**) direct photolysis involved in oxygenic photosynthesis; (**b**) indirect photolysis involved in oxygenic photosynthesis; (**c**) anoxygenic mechanism involved with photosynthetic bacteria

[FeFe]-hydrogenases (*Hyd*A1 and *Hyd*A2) catalyse the molecular H_2 by proton reduction. In cyanobacteria, H_2 is produced during nitrogen fixation by action of nitrogenase enzyme (Greenbaum 1988a, b; Melis et al. 2000).

Photosynthetic bacteria utilise the visible and/or near-infrared spectrum to produce H_2 by catabolising organic molecules by anoxygenic photosynthesis (Fig. 12.2c). In these bacteria, the [FeFe]-hydrogenase can easily evade the distress of oxygen, since they do not use water as source for electron generation (Melis and Happe 2001; Krassen et al. 2009). In this process, the bacteriochlorophyll (*BChl*) elicits for the formation of bacteriopheophytin (*BPh*) upon light absorption. The electrons from *BPh* are transferred to quinine pool (QA) and then to the cytochrome subunit of the reaction centre that generates a proton gradient for ATP synthesis and finally to H_2 (Hallenbeck and Benemann 2002). The efficacy of light energy translation to H_2 is significantly higher in photosynthetic bacteria than cyanobacteria since the quantum of light energy requirement is less than the water photolysis. Thus, the photosynthetic bacteria make photo-fermentation process more feasible, viable and its adaptability in consuming a variety of substrates (Chandra and Venkata Mohan 2011, 2014).

12.2.2 Dark Fermentation

Dark fermentation involves a multitude of biochemical metabolic reactions, viz. hydrolysis, acidogenesis, acetogenesis and methanogenesis (Lin et al. 2012). The composite organic matter is catabolised to simple molecules during hydrolysis and consequently fermented to organic acids during acidogenesis (Dahiya et al. 2015). As a special case, H_2 production is feasible from acetic acid by microbial consortia of acetogens and homoacetogens. Further, the acetoclastic methanogens translate the volatile fatty acids to methane during methanogenesis (Thi et al. 2016). Most often, acetogens nurture in syntrophic alliance with the hydrogenotrophic methanogens, thereby retaining low partial pressure to allow acidogenesis to become thermodynamically favourable. Therefore, the methanogenic activity is suppressed to increase the yields of bioH_2 as a sole metabolic by-product (Venkata Mohan et al. 2008a, b; Goud and Venkata Mohan 2012a; Sarkar et al. 2016).

The organic sugars are first metabolised during the glycolysis to pyruvate, a key precursor for subsequent microbial fermentation (Venkata Mohan et al. 2007b). Consequently, pyruvate can result in a variety of short-chain organic fatty acids, and H_2 is also produced during the metabolism (Fig. 12.3a). Anaerobes that are facultative metabolise pyruvate to acetyl-CoA and then to formate catalysed by pyruvate formate lyase and subsequently to hydrogen by formate hydrogen lyase (Nikhil et al. 2014b). Anaerobes that are obligate oxidise pyruvate to acetyl-CoA through pyruvate ferredoxin oxidoreductase. Metabolites formed during anaerobic substrate metabolism raise the accessibility of reducing equivalents within the bacteria (Wong et al. 2014). Protons from the redox mediators separate in presence NADH dehydrogenase enzyme and subsequently reduced to H_2 with electrons offered by the oxidised ferredoxin upon action of hydrogenase enzyme. The membrane-bound NADH dehydrogenase, cytochrome complex and other carrier proteins channel the

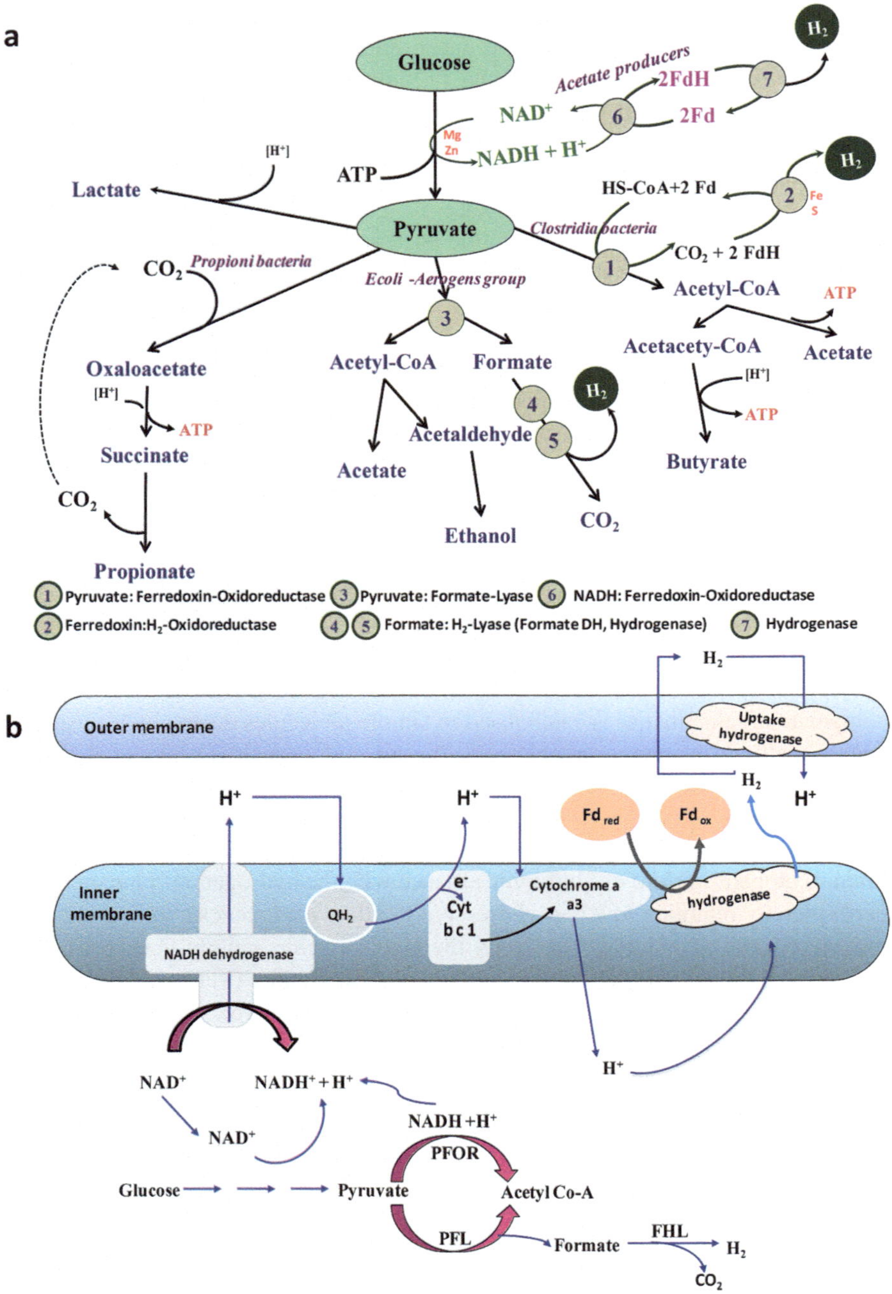

Fig. 12.3 (**a**) Different biochemical steps involved in the dark fermentation of glucose (organic matter); (**b**) schematic illustration of substrate conversion and H_2 production mechanism during dark fermentation

electrons via the quinone (Q) pool (Fig. 12.3b). Uninterrupted interconversions of quinone and protons assist in the transport of electrons to cytochrome b–c_1 complex and then to cytochrome aa_3. Ultimately, the electrons are transferred from cytochrome aa_3 to iron-containing protein ferredoxin. The ferredoxin in reduced state transfers electrons to the catalytic site of hydrogenase where protons are combined to form H_2 (Saratale et al. 2013).

Dark fermentation has gained significant credit as a practically feasible scheme amongst the other biological ways for H_2 production, especially when wastewater is considered as a substrate and assorted bacterial consortium as a biocatalyst (Agler et al. 2011; Nikhil et al. 2014a). The prominent features of this process are: process simplicity, less energy intensive, less carbon footprints, use of broad spectrum of organics and operation at ambient temperatures. Besides, the process is economically steady and robust for large scale of H_2 production (Venkata Mohan 2009). If the fermentation favours for the acetic acid synthesis, then the stoichiometric yield of H_2 is 4 mol per mol of glucose, whereas if the fermentation pathway is butyric acid, the yield of H_2 is 2 mol per mol of glucose. Yet, the experimental H_2 yield is lower than the stoichiometric yield, since a fraction of the substrate is consumed for bacterial cell mass and in few cases by-products might reduce the H_2 yield. Strategies are developed to attain superior H_2 yields at better rates, viz. novel reactor designs, variable mode of operation, inoculum type and pretreatment, nature and pretreatment of substrate and many more (Venkata Mohan et al. 2008a; Monlau et al. 2015; Sarkar et al. 2016).

The potential sources for anaerobic cultures are found almost in all types of natural environments. Anaerobic bacteria under the class of *Firmicutes, which* are apt biocatalysts for H_2 production (Venkata Mohan et al. 2011). A mixture of bacterial flora is largely preferred for continuous H_2 production; however, the microbial culture can show a divergence in H_2 production competence because of the co-occurrence of H_2-consuming bacteria in overall microbial diversity. In such conditions, conditioning of inoculum is performed to augment H_2 producers due to their compliance in hostile environments (Sarkar et al. 2013; Goud et al. 2017). Moreover, proton fluxes in/out bacteria affects the enzyme activity, biochemical pathways and substrate decomposition. pH is also acute to sustain ample cellular ATP levels, since surplus H^+ ions are pumped out using ATP to ensure cell neutrality (Srikanth and Venkata Mohan 2012). The initial pH can impact the duration of lag phase involved in spore germination and enzyme synthesis. The operational organic load of bioreactor can affect several functional issues which include accumulation of undissociated VFA and pH variations. This subsequently changes the diversity of microbial flora with subsequent amendments of the allied metabolic pathways (Van Ginkel and Logan 2005a; Goud and Venkata Mohan 2012b). Commercial biohydrogen production still poses certain process impede like the inhibitory effect caused by the undissociated VFA (Van Ginkel and Logan 2005b; Goud et al. 2014; Srikanth and Venkata Mohan 2014; Sarkar et al. 2017). Improving H_2 production rate and yield are the grave challenges to sustain economic production. In this regard, various strategies were reported in the literature. A few of them are selection and pretreatment of microbial consortia, immobilisation of consortia, statistical techniques for process optimisation, sequencing of bioreactors, bio-electrochemical treatment, multiple process integration and bio-augmentation (Pasupuleti and Venkata Mohan 2015a, b). Commercial operation still poses certain process impedes like inhibitory effect of undissociated VFA (Van Ginkel and Logan 2005b; Goud et al. 2014; Srikanth and Venkata Mohan 2014; Sarkar et al. 2017).

12.2.3 Integration Strategies

12.2.3.1 Biohythane

A proportionate mixture of hydrogen and methane is commercially available as Hythane™, HCNG or methagen (Eden 2010). The recommended composition of H_2 with CH_4 ranges between 10% (v/v) and 25% (v/v) (Liu et al. 2013). Conventional methods could be unsustainable due to their reliance on exhausting petroleum reserves. Yet, the contemporary trends in bioenergy exploit the different biological routes for the production of both methane and hydrogen from waste organic matter. Literature reports are available on two-step fermentation process to produce a mixture of H_2 and CH_4 from wastewater (Venkata Mohan et al. 2008b; Mohanakrishna and Venkata Mohan 2013). This biological design assures the composition of biogas mixture (CH_4 + H_2) by regulating the environments of microbial fermentation. The usual operation is a two-step process that has a few advantages, but operation of two bioreactors is frugally not that sustainable (Cavinato et al. 2012; Willquist et al. 2012). A recent investigation reported biohythane from single-stage bioreactor using spent-wash wastewater (Pasupuleti and Venkata Mohan 2015b). In another study, a unique strategy was reported that augments acidogenesis for the production of biohythane (Sarkar and Venkata Mohan 2016, 2017).

12.2.3.2 Hybrid Dark–Photo Fermentation

A blend of both dark and photo-fermentation is a credible option to accomplish a yield of 12 mol H_2/mol glucose. Integrative fermentation can be considered as an operational and competent option to harness maximum H_2 yield with simultaneous wastewater treatment (Srikanth et al. 2009; Laurinavichene et al. 2012). A variety of photosynthetic bacteria are capable of H_2 production which can uptake the volatile fatty acids as carbon source and light as energy sources. Studies have been carried out with two-stage integration of heterotrophic dark with photo-fermentation for bioH_2 production (Chandra et al. 2015). It was noticed that photo-fermentation of acid-rich effluents is considered to be complicated due to light infiltration complications, multifaceted nutritional requirements, obligatory operational conditions, substrate (VFA and ammonia) inhibitions and vulnerability for impurity (Wang et al. 2009; Chandra and Venkata Mohan 2014).

12.2.3.3 Microbial Electrolysis

Microbial electrolysis is an electrically driven H_2 production process wherein the exchange of electron equivalents in carbon-based composites to H_2 gas involves bio-electrochemical reactions (Call and Logan 2008). A microbial electrochemical cell (MEC) has a prerequisite of additional potential to assist the metabolism of undissociated organic acids into H_2 which is obligatory to cross the endothermic barrier (Cheng and Logan 2011). Therefore, the protons and electrons are driven with the supplementation of an external voltage from anode to cathode which is reduced to H_2. The typical redox potential for H_2 reduction is −0.414 V, and a voltage >0.11 V facilitates H_2 formation at the cathode. These applied potentials are moderately low compared to the applied external potential of 1.23 V for electrolysis of water. For example, H_2 (−0.414 V) from acetate (−0.279 V) at the cathode occurs upon appliance of voltage (−0.135 V). In reality, a relatively higher voltage is required than the prerequisite; owing to system overpotentials aroused by physicochemical and microbial influences.

It is reported that the applied potentials selectively enrich the growth of electroactive bacteria that efficiently reduce electrons (Arunasri et al. 2016). Operating an MEC can recover more than 90% of H_2 as against 33% by the dark fermentation. Thus, the application of MEC can be seen with the usage of acidogenic effluents rich in short-chain fatty acids for recovery of additional H_2/CH_4 production (Wagner et al. 2009; Modestra et al. 2015). The prospects of MEC over conventional water electrolysis are low-energy consumption, appreciable H_2 yields and wastewater treatment. Nevertheless, biocatalyst diversity, electrode materials, membrane, applied potential, substrate loading rate and reactor configuration critically affect the functioning of MEC (Nam et al. 2011). To begin with, studies were carried out in double chamber that allows separate capture of H_2 at cathode and prevents fouling by anodic bacteria (Pisciotta et al. 2012). But, separation leads to acidification at the anode chamber; so, removing the partition creates a single-chamber MEC that reduces the applied potential attenuating the pH and energy losses. Electrodes coated with platinum are commonly used as chemical catalysts for H_2 evolution as it significantly diminishes the cathode overpotential. Conventional methods use platinum which is costly, consumable and vulnerable to contamination by constituents in the effluents. Recently, research on biocathodes is of prime attention as they are eco-friendly and renewable (Jeremiasse et al. 2010; Nikhil et al. 2015b).

During the operation of MEC, the hydrogen evolution reaction (HER) is linked to the pH. The buffer capacity of wastewater is typically low resulting in accumulation of protons resulting in acidification at anode (Hamelers et al. 2010). The use of chemical buffers could be a possible option that alleviates the pH gradient. The efficacy of a buffer relies on its dissociation constant (pKa) and diffusivity. Buffers with pKa slightly above the operating pH (near neutral) aid to sustain the internal pH close to the external pH (Zhu et al. 2009; Liang et al. 2014). Inappropriate use of buffer (above 300 mM) affects the system and unsuitable for the effluent discharge; besides addition of salts would escalate the operating economics (Ambler and Logan 2011). It was reported that the external voltage exhibited a dual consequence over process performance by regulating the in situ buffering capacity using a biocathode (Lenin Babu et al. 2013b; Nikhil et al. 2015a). Integrating MEC with other processes to harvest biohydrogen generation is the state-of-the-art research amongst the scientific fraternity (Escapa et al. 2014). Small additional voltage with a high current density will be a crucial challenge for scaling-up of MEC to mercantile applications (Lenin Babu et al. 2013a; Venkata Mohan et al. 2014).

12.2.3.4 Bio-based Products

The dark-fermentative effluents are rich in VFA, and these can be used as possible substrate for production of polyhydroxyalkanoates (PHAs) commonly referred to as 'bioplastics' (Venkata Mohan et al. 2010; Sarkar et al. 2016). PHAs are biopolyesters of hydroxyalkanoates that amass as cellular storage materials produced under surplus carbon and nutrient-deprived environments (Patel et al. 2012; Fradinho et al. 2013). Commercial scale of PHA is produced using plant-derived resources and pure cultures that rise the production costs. Compared to other substrates (carbohydrates or proteins), the use of VFA facilitates PHA synthesis depriving the participation of glycolysis and β-oxidation pathways. Reports are available on PHA production from synthetic VFA and biohydrogen effluents (VFA-rich) under an anoxic microenvironment using a mixed bacterial culture (Venkateswar Reddy and Venkata Mohan 2012). If acetate and butyrate compositions are relatively higher amounts, then polyhydroxybutyrate (PHB) is the product type, and if propionate and valerate are relatively higher amounts, then polyhydroxyvalerate (PHV) is the product type. The combined production of H_2 and PHA followed by methanogenesis is possible option to make the whole process economical and sustainable (Patel et al. 2012; Venkateswar Reddy et al. 2014).

12.3 Waste Treatment vs Biohydrogen Production

Ideally, a unit kilogram of chemical oxygen demand (COD) equals to 5.2 mol of glucose, and upon dark fermentation each mole of glucose results in 4 mol and 2 mol of H_2 depending upon acetic acid and butyric acid pathways, respectively (Venkata Mohan et al. 2007c). On the other hand, photo-fermentation yields 12 mol of H_2 per mol of glucose. H_2 produced by acetate pathway of dark fermentation yields 89.6 l of H_2 per mol of glucose which is equivalent to 466.6 l of H_2 per kg COD. Assuming with only 40% of COD removal, then H_2 yield in dark and photo-fermentation is 125 g and 16.6 g, respectively. For example, if an industry discharging ~3×10^{11} l of effluent per annum containing a typical COD of 20 g/l, then it amounts to 6000×10^6 g of COD for that year. The yields of H_2 with dark and photo-fermentation are about 5×10^6 kg and 300×10^6 kg per year that accounts for many million dollars/year (at a rate of $ 4/kg H_2) (Markets 2011).

Rapid industrialisation is resulting in immense amounts of waste. Conventional treatment of wastewater is an energy-exhaustive process. Probing means to produce or harness valuable products from wastewater remediation are significantly pursued in recent times (Angenent et al. 2004; Venkata Mohan et al. 2010). The prerequisite requirement for their management markedly makes wastewater a choice for biohydrogen production, besides abating the overall treatment cost (Guo et al. 2010; Elsharnouby et al. 2013). Till date, considerable efforts were put on the application of various wastewaters for the production of biohydrogen through fermentation processes (both light-dependent and light-independent) (Hallenbeck 2013). Both wastewater treatment and H_2 production are equally important, and a balance should exist between technical expertise and substrate uptake with operating circumstances (Show and Lee 2013). Assessing these conditions is particularly important in

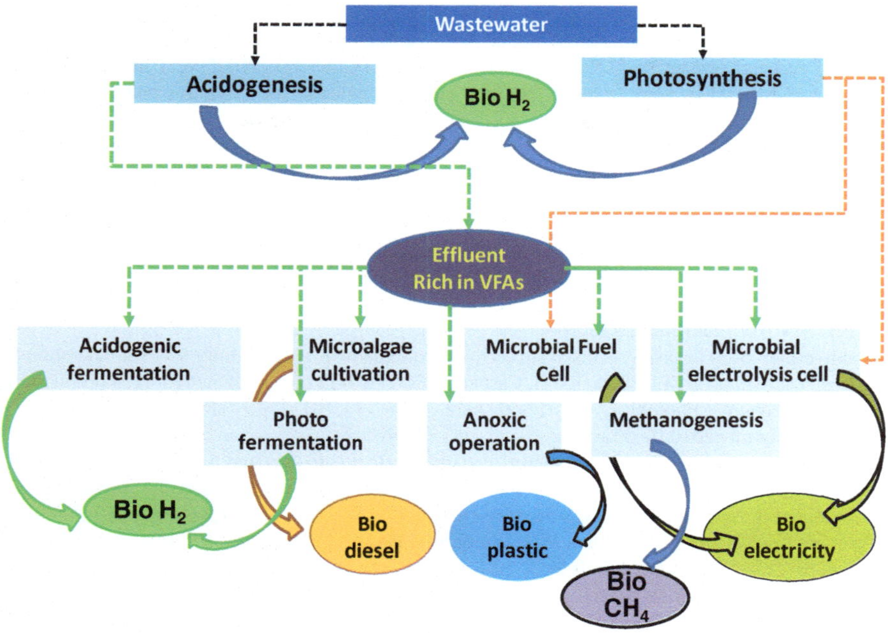

Fig. 12.4 Schematic details of various strategic routes possible for effective H_2 production and other value-added products

sustaining the financial possibility and environmental competence of the process (Arimi et al. 2015). An illustrative plan of strategies to recover energy and products for value additions along with wastewater treatment is represented in Fig. 12.4.

12.4 Future Perspectives

Hydrogen is an imperative substitute for energy domain and can be produced from a variety of production technologies (Dunn 2002; Nouni 2012). Presently, investigations are being dedicated on viable and eco-friendly biohydrogen from biomass to substitute fossil fuels (Hallenbeck and Ghosh 2009; Rama Mohan 2015). Waste biomass can be deliberated as the paramount choice and has the foremost prospective that encounters energy supplies and could assure fuel stock in the upcoming future. Utilisation of remaining organic portion after acidogenesis is also the most significant aspect to be paid momentous consideration (Mohanakrishna et al. 2010; Ghimire et al. 2015). Photobiological processes with acidogenic wastewater as substrate are comparatively less studied and need further exploitation. Multi-process combination is effectively evaluated for economic viability and commercialisation (Mohanakrishna and Venkata Mohan 2013; Wang and Ren 2013). Overall, a biorefinery concept with closed-loop approach visualises negative-valued waste as a potential renewable feedstock that will pave new opportunities for the growth of bio-based society with circular economy (Venkata Mohan et al. 2016).

Acknowledgements Funding received from the Ministry of New and Renewable Energy (MNRE), Government of India and Council for Scientific and Industrial Research (CSIR) in the form of research grants as MNRE Project No. 103/131/2008-NT, XII 5-year network project (SETCA (CSC-0113), respectively. GNN and OS acknowledge the CSIR for providing Senior Research Fellowship.

References

Agler MT, Wrenn BA, Zinder SH, Angenent LT (2011) Waste to bioproduct conversion with undefined mixed cultures: the carboxylate platform. Trends Biotechnol 29:70–78. https://doi.org/10.1016/j.tibtech.2010.11.006

Allakhverdiev SI, Thavasi V, Kreslavski VD, Zharmukhamedov SK, Klimo VV, Ramakrishna S, Los DA, Mimurod M, Nishiharae H, Carpentier R (2010) Photosynthetic hydrogen production. J Photochem Photobiol C: Photochem Rev 11:101–113. https://doi.org/10.1016/j.jphotochemrev.2010.07.002

Ambler JR, Logan BE (2011) Evaluation of stainless steel cathodes and a bicarbonate buffer for hydrogen production in microbial electrolysis cells using a new method for measuring gas production. Int J Hydrog Energy 36:160–166. https://doi.org/10.1016/j.ijhydene.2010.09.044

Angenent LT, Karim K, Al-Dahhan MH, Wrenn BA, Domíguez-Espinosa R (2004) Production of bioenergy and biochemicals from industrial and agricultural wastewater. Trends Biotechnol 22:477–485. https://doi.org/10.1016/j.tibtech.2004.07.001

Arimi MM, Knodel J, Kiprop A, Namango SS, Zhang Y, Geißen S-U (2015) Strategies for improvement of biohydrogen production from organicrich wastewater: a review. Biomass Bioenergy 75:101–118. https://doi.org/10.1016/j.biombioe.2015.02.011

Arunasri K, Modestra JA, Yeruva DK, Krishna KV, Venkata Mohan S (2016) Polarized potential and electrode materials implication on electrofermentative di-hydrogen production: microbial assemblages and hydrogenase gene copy variation. Bioresour Technol 200:691–698. https://doi.org/10.1016/j.biortech.2015.10.091

Bhaskar T, Balagurumurthy B, Singh R, Poddar MK (2013) Thermochemical route for biohydrogen production. In: Pandey A, Chang JS, Hallenbeck P, Larroche C (eds) Biohydrogen. Elsevier, Amsterdam, pp 285–316. ISBN: 978-0-444-59555-3

Blankenship RE, Olson JM, Miller M (1995) Antenna complexes from green photosynthetic bacteria. In: Blankenship RE, Madigan Bauer MT (eds) Anoxygenic photosynthetic bacteria. Springer, Dordrecht. ISBN: 978-0-7923-3681-5, pp 399–435. https://doi.org/10.1007/0-306-47954-0_20

Call D, Logan BE (2008) Hydrogen production in a single chamber microbial electrolysis cell lacking a membrane. Environ Sci Technol 42:3401–3406. https://doi.org/10.1021/es8001822

Cavinato C, Giuliano A, Bolzonella D, Pavan P, Cecchi F (2012) Bio-hythane production from food waste by dark fermentation coupled with anaerobic digestion process: a long-term pilot scale experience. Int J Hydrog Energy 37:11549–11555. https://doi.org/10.1016/j.ijhydene.2012.03.065

Chandra R, Venkata Mohan S (2011) Microalgal community and their growth conditions influence biohydrogen production during integration of dark-fermentation and photo-fermentation processes. Int J Hydrog Energy 36:12211–12219. https://doi.org/10.1016/j.ijhydene.2011.07.007

Chandra R, Venkata Mohan S (2014) Enhanced bio-hydrogenesis by co-culturing photosynthetic bacteria with acidogenic process: augmented dark-photo fermentative hybrid system to regulate volatile fatty acid inhibition. Int J Hydrog Energy 39:7604–7615. https://doi.org/10.1016/j.ijhydene.2014.01.196

Chandra R, Nikhil GN, Venkata Mohan S (2015) Single-stage operation of hybrid dark-photo fermentation to enhance biohydrogen production through regulation of system redox condition: evaluation with real-field wastewater. Int J Mol Sci 16:9540–9556. https://doi.org/10.3390/ijms16059540

Cheng S, Logan BE (2011) High hydrogen production rate of microbial electrolysis cell (MEC) with reduced electrode spacing. Bioresour Technol 102:3571–3574. https://doi.org/10.1016/j.biortech.2010.10.025

Constant P, Hallenbeck PC (2013) Hydrogenase. In: Pandey A, Chang JS, Hallenbeck P, Larroche C (eds) Biohydrogen. Elsevier, Amsterdam, pp 75–102. ISBN: 978-0-444-59555-3

Dahiya S, Sarkar O, Swamy Y, Venkata Mohan S (2015) Acidogenic fermentation of food waste for volatile fatty acid production with cogeneration of biohydrogen. Bioresour Technol 182:103–113. https://doi.org/10.1016/j.biortech.2015.01.007

Dunn S (2002) Hydrogen futures: toward a sustainable energy system. Int J Hydrog Energy 27:235–264. https://doi.org/10.1016/S0360-3199(01)00131-8

Eden (2010) Annual Report. Available from: www.edenenergy.com.au

Elsharnouby O, Hafez H, Nakhla G, El Naggar MH (2013) A critical literature review on biohydrogen production by pure cultures. Int J Hydrog Energy 38:4945–4966. https://doi.org/10.1016/j.ijhydene.2013.02.032

Escapa A, San Martin MI, Moran A (2014) Potential use of microbial electrolysis cells in domestic wastewater treatment plants for energy recovery. Front Energy Res 2:1–10. https://doi.org/10.3389/fenrg.2014.00019

Fradinho J, Domingos J, Carvalho G, Oehmen A, Reis M (2013) Polyhydroxyalkanoates production by a mixed photosynthetic consortium of bacteria and algae. Bioresour Technol 132:146–153. https://doi.org/10.1016/j.biortech.2013.01.050

Ghimire A, Frunzo L, Pirozzi F, Trably E, Escudie R, Lens PNL, Esposito G (2015) A review on dark fermentative biohydrogen production from organic biomass: process parameters and use of by-products. Appl Energy 144:73–95. https://doi.org/10.1016/j.apenergy.2015.01.045

Goud RK, Venkata Mohan S (2012a) Acidic and alkaline shock pretreatment to enrich acidogenic biohydrogen producing mixed culture: long term synergetic evaluation of microbial inventory, dehydrogenase activity and bio-electro kinetics. RSC Adv 2:6336–6353. https://doi.org/10.1039/C2RA20526B

Goud RK, Venkata Mohan S (2012b) Regulating biohydrogen production from wastewater by applying organic load-shock: change in the microbial community structure and bio-electrochemical behavior over long-term operation. Int J Hydrog Energy 37:17763–17777. https://doi.org/10.1016/j.ijhydene.2012.08.124

Goud RK, Sarkar O, Chiranjeevi P, Venkata Mohan S (2014) Bioaugmentation of potent acidogenic isolates: a strategy for enhancing biohydrogen production at elevated organic load. Bioresour Technol 165:223–232. https://doi.org/10.1016/j.biortech.2014.03.049

Goud RK, Arunasri K, Yeruva DK, Krishna KV, Dahiya S, Venkata Mohan S (2017) Impact of selectively enriched microbial communities on longterm fermentative biohydrogen production. Bioresour Technol. https://doi.org/10.1016/j.biortech.2017.03.147

Greenbaum E (1988a) Energetic efficiency of hydrogen photoevolution by algal water splitting. Biophys J 54:365–368. https://doi.org/10.1016/S0006-3495(88)82968-0

Greenbaum E (1988b) Interfacial photoreactions at the photosynthetic membrane interface: an upper limit for the number of platinum atoms required to form a hydrogen-evolving platinum metal catalyst. J Phys Chem 92:4571–4574. https://doi.org/10.1021/j100327a001

Guo XM, Trably E, Latrille E, Carrère H, Steyer J-P (2010) Hydrogen production from agricultural waste by dark fermentation: a review. Int J Hydrog Energy 35:10660–10673. https://doi.org/10.1016/j.ijhydene.2010.03.008

Hallenbeck PC (2013) Photofermentative biohydrogen production. In: Pandey A, Chang JS, Hallenbeck P, Larroche C (eds) Biohydrogen. Elsevier, Amsterdam, pp 145–159. ISBN: 978-0-444-59555-3

Hallenbeck PC, Benemann JR (2002) Biological hydrogen production; fundamentals and limiting processes. Int J Hydrog Energy 27:1185–1193. https://doi.org/10.1016/S0360-3199(02)00131-3

Hallenbeck PC, Ghosh D (2009) Advances in fermentative biohydrogen production: the way forward? Trends Biotechnol 27:287–297. https://doi.org/10.1016/j.tibtech.2009.02.004

Hamelers HV, Ter Heijne A, Sleutels TH, Jeremiasse AW, Strik DP, Buisman CJ (2010) New applications and performance of bioelectrochemical systems. Appl Microbiol Biotechnol 85:1673–1685. https://doi.org/10.1007/s00253-009-2357-1

Jeremiasse AW, Hamelers HVM, Buisman CJN (2010) Microbial electrolysis cell with a microbial biocathode. Bioelectrochemistry 78:39–43. https://doi.org/10.1016/j.bioelechem.2009.05.005

Krassen H, Schwarze A, Friedrich B, Ataka K, Lenz O, Heberle J (2009) Photosynthetic hydrogen production by a hybrid complex of photosystem I and [NiFe]-hydrogenase. ACS Nano 3:4055–4061. https://doi.org/10.1021/nn900748j

Lam MK, Lee KT (2013) Biohydrogen production from algae. In: Pandey A, Chang JS, Hallenbeck P, Larroche C (eds) Biohydrogen. Elsevier, Amsterdam, pp 161–184. ISBN: 978-0-444-59555-3

Laurinavichene TV, Belokopytov BF, Laurinavichius KS, Khusnutdinova AN, Seibert M, Tsygankov AA (2012) Towards the integration of darkand photo-fermentative waste treatment. 4. Repeated batch sequential dark- and photofermentation using starch as substrate. Int J Hydrog Energy 37:8800–8810. https://doi.org/10.1016/j.ijhydene.2012.01.132

Lenin Babu M, Sarma P, Venkata Mohan S (2013a) Microbial electrolysis of synthetic acids for biohydrogen production: influence of biocatalyst pretreatment and pH with the function of applied potential. J Microb Biochem Technol S 6:2. https://doi.org/10.4172/1948-5948.S6-003

Lenin Babu M, Subhash GV, Sarma PN, Venkata Mohan S (2013b) Bio-electrolytic conversion of acidogenic effluents to biohydrogen: an integration strategy for higher substrate conversion and product recovery. Bioresour Technol 133:322–331. https://doi.org/10.1016/j.biortech.2013.01.029

Liang D, Liu Y, Peng S, Lan F, Lu S, Xiang Y (2014) Effects of bicarbonate and cathode potential on hydrogen production in a biocathode electrolysis cell. Front Environ Sci Eng 8:624–630. https://doi.org/10.1007/s11783-013-0584-2

Lin C-Y, Lay C-H, Sen B, Chu C-Y, Kumar G, Chen C-C, Chang J-S (2012) Fermentative hydrogen production from wastewaters: a review and prognosis. Int J Hydrog Energy 37:15632–15642. https://doi.org/10.1016/j.ijhydene.2012.02.072

Liu Z, Zhang C, Lu Y, Wu X, Wanf L, Wang L, Han B, Xing QH (2013b) States and challenges for high-value biohythane production from waste biomass by dark fermentation technology. Bioresour Technol 135:292–303. https://doi.org/10.1016/j.biortech.2012.10.027

Markets and Markets (2011) Hydrogen generation market-by merchant and captive type, distributed and centralized generation, application and technology-trends and global forecasts (2011–2016). Report code: EP1708. http://www.marketsandmarkets.com/Market-Reports/hydrogengeneration-market-494.html

Melis A, Happe T (2001) Hydrogen production. Green algae as a source of energy. Plant Physiol 127:740–748. https://doi.org/10.1104/pp.010498

Melis A, Zhang L, Forestier M, Ghirardi ML, Seibert M (2000) Sustained photobiological hydrogen gas production upon reversible inactivation of oxygen evolution in the green alga *Chlamydomonas reinhardtii*. Plant Physiol 122:127–136. https://doi.org/10.1104/pp.122.1.127

Modestra JA, Babu ML, Venkata Mohan S (2015) Electro-fermentation of real-field acidogenic spent wash effluents for additional biohydrogen production with simultaneous treatment in a microbial electrolysis cell. Sep Purif Technol 150:308–315. https://doi.org/10.1016/j.seppur.2015.05.043

Mohanakrishna G, Venkata Mohan S (2013) Multiple process integrations for broad perspective analysis of fermentative H production from wastewater treatment: technical and environmental considerations. Appl Energy 107:244–254. https://doi.org/10.1016/j.apenergy.2013.01.085

Mohanakrishna G, Venkata Mohan S, Sarma PN (2010) Utilizing acid-rich effluents of fermentative hydrogen production process as substrate for harnessing bioelectricity: an integrative approach. Int J Hydrog Energy 35:3440–3449. https://doi.org/10.1016/j.ijhydene.2010.01.084

Monlau F, Kaparaju P, Trably E, Steyer J-P, Carrere H (2015) Alkaline pretreatment to enhance one-stage CH and two-stage H /CH production from sunflower stalks: mass, energy and economical balances. Chem Eng J 260:377–385. https://doi.org/10.1016/j.cej.2014.08.108

Nam J-Y, Tokash JC, Logan BE (2011) Comparison of microbial electrolysis cells operated with added voltage or by setting the anode potential. Int J Hydrog Energy 36:10550–10556. https://doi.org/10.1016/j.ijhydene.2011.05.148

Nikhil GN, Venkata Mohan S, Swamy YV (2014a) Behavior of acidogenesis during biohydrogen production with formate and glucose as carbon source: substrate associated dehydrogenase expression. Int J Hydrog Energy 39:7486–7495. https://doi.org/10.1016/j.ijhydene.2013.12.169

Nikhil GN, Venkata Venkata Mohan S, Swamy YV (2014b) Systematic approach to assess biohydrogen potential of anaerobic sludge and soil rhizobia as biocatalysts: influence of crucial factors affecting acidogenic fermentation. Bioresour Technol 165:323–331. https://doi.org/10.1016/j.biortech.2014.02.097

Nikhil GN, Subhash GV, Yeruva DK, Venkata Mohan S (2015a) Synergistic yield of dual energy forms through biocatalyzed electrofermentation of waste: stoichiometric analysis of electron and carbon distribution. Energy 88:281–291. https://doi.org/10.1016/j.energy.2015.05.043

Nikhil GN, Venkata Mohan S, Swamy YV (2015b) Applied potentials regulate recovery of residual hydrogen from acid-rich effluents: influence of biocathodic buffer capacity over process performance. Bioresour Technol 188:65–72. https://doi.org/10.1016/j.biortech.2015.01.084

Nissilä ME, Lay C-H, Puhakka JA (2014) Dark fermentative hydrogen production from lignocellulosic hydrolyzates – a review. Biomass Bioenergy 67:145–159. https://doi.org/10.1016/j.biombioe.2014.04.035

Nouni M (2012) Hydrogen energy and fuel cell technology: recent developments and future prospects in India. Renew Energy. Akshay Urja 5:10–14

Pasupuleti SB, Venkata Mohan S (2015a) Acidogenic hydrogen production from wastewater: process analysis with the function of influencing parameters. Int J Energy Res 39:1131–1141. https://doi.org/10.1002/er.3322

Pasupuleti SB, Venkata Mohan S (2015b) Single-stage fermentation process for high-value biohythane production with the treatment of distillery spent-wash. Bioresour Technol 189:177–185. /10.1016/j.biortech.2015.03.128

Patel SK, Singh M, Kumar P, Purohit HJ, Kalia VC (2012) Exploitation of defined bacterial cultures for production of hydrogen and polyhydroxybutyrate from pea-shells. Biomass Bioenergy 36:218–225. https://doi.org/10.1016/j.biombioe.2011.10.027

Pisciotta JM, Zaybak Z, Call DF, Nam JY, Logan BE (2012) Enrichment of microbial electrolysis cell biocathodes from sediment microbial fuel cell bioanodes. Appl Environ Microbiol 78:5212–5219. https://doi.org/10.1128/AEM.00480-12

Rama Mohan S (2015) Structure and growth of research on biohydrogen generation using wastewater. Int J Hydrog Energy 40:16056–16069. https://doi.org/10.1016/j.ijhydene.2015.08.072

Sambusiti C, Bellucci M, Zabaniotou A, Beneduce L, Monlau F (2015) Algae as promising feedstocks for fermentative biohydrogen production according to a biorefinery approach: a comprehensive review. Renew Sustain Energy Rev 44:20–36. https://doi.org/10.1016/j.rser.2014.12.013

Saratale GD, Saratale RG, Chang J-S (2013) Biohydrogen from renewable resources. In: Pandey A, Chang JS, Hallenbeck P, Larroche C (eds) Biohydrogen. Elsevier, Amsterdam, pp 185–221. ISBN: 978-0-444-59555-3

Sarkar O, Venkata Mohan S (2016) Deciphering acidogenic process towards biohydrogen, biohythane, and short chain fatty acids production: multi-output optimization strategy. Biofuel Res J 3:458–469. https://doi.org/10.18331/BRJ2016.3.3.5

Sarkar O, Venkata Mohan S (2017) Pre-aeration of food waste to augment acidogenic process at higher organic load: valorizing biohydrogen, volatile fatty acids and biohythane. Bioresour Technol. https://doi.org/10.1016/j.biortech.2017.05.053

Sarkar O, Goud RK, Subhash GV, Venkata Mohan S (2013) Relative effect of different inorganic acids on selective enrichment of acidogenic biocatalyst for fermentative biohydrogen production from wastewater. Bioresour Technol 147:321–331. https://doi.org/10.1016/j.biortech.2013.08.021

Sarkar O, Kumar AN, Dahiya S, Krishna KV, Yeruva DK, Venkata Mohan S (2016) Regulation of acidogenic metabolism towards enhanced short chain fatty acid biosynthesis from waste: metagenomic profiling. RSC Adv 6:18641–18653. https://doi.org/10.1039/C5RA24254A

Sarkar O, Butti SK, Venkata Mohan S (2017) Acidogenesis driven by hydrogen partial pressure towards bioethanol production through fatty acids reduction. Energy 118:425–434. https://doi.org/10.1016/j.energy.2016.12.017

Show K-Y, Lee D-J (2013) Bioreactor and bioprocess design for biohydrogen production. In: Larroche AP-SCCH (ed) Biohydrogen. Elsevier, Amsterdam. In: Pandey A, Chang JS, Hallenbeck P, Larroche C (eds) Biohydrogen, pp 317–337. Elsevier, Amsterdam. ISBN: 978-0-444-59555-3

Srikanth S, Venkata Mohan S (2012) Regulatory function of divalent cations in controlling the acidogenic biohydrogen production process. RSC Adv 2:6576–6589. https://doi.org/10.1039/C2RA20383A

Srikanth S, Venkata Mohan S (2014) Regulating feedback inhibition caused by the accumulated acid intermediates during acidogenic hydrogen production through feed replacement. Int J Hydrog Energy 39:10028–10040. https://doi.org/10.1016/j.ijhydene.2014.04.152

Srikanth S, Venkata Mohan S, Prathima Devi M, Peri D, Sarma PN (2009) Acetate and butyrate as substrates for hydrogen production through photo-fermentation: process optimization and combined performance evaluation. Int J Hydrog Energy 34:7513–7522. https://doi.org/10.1016/j.ijhydene.2009.05.095

Thi NBD, Lin C-Y, Kumar G (2016) Waste-to-wealth for valorization of food waste to hydrogen and methane towards creating a sustainable ideal source of bioenergy. J Clean Prod 122:29–41. https://doi.org/10.1016/j.jclepro.2016.02.034

Van Ginkel S, Logan BE (2005a) Inhibition of biohydrogen production by undissociated acetic and butyric acids. Environ Sci Technol 39:9351–9356. https://doi.org/10.1021/es0510515

Van Ginkel SW, Logan B (2005b) Increased biological hydrogen production with reduced organic loading. Water Res 39:3819–3826. https://doi.org/10.1016/j.watres.2005.07.021

Venkata Mohan S (2009) Harnessing of biohydrogen from wastewater treatment using mixed fermentative consortia: process evaluation towards optimization. Int J Hydrog Energy 34:7460–7474. https://doi.org/10.1016/j.ijhydene.2009.05.062

Venkata Mohan S, Lalit Babu V, Sarma PN (2007a) Anaerobic biohydrogen production from dairy wastewater treatment in sequencing batch reactor (AnSBR): effect of organic loading rate. Enzym Microb Technol 41:506–515. https://doi.org/10.1016/j.enzmictec.2007.04.007

Venkata Mohan S, Vijaya Bhaskar Y, Murali Krishna P, Chandrasekhara Rao N, Lalit Babu V, Sarma PN (2007b) Biohydrogen production from chemical wastewater as substrate by selectively enriched anaerobic mixed consortia: influence of fermentation pH and substrate composition. Int J Hydrog Energy 32:2286–2295. https://doi.org/10.1016/j.ijhydene.2007.03.015

Venkata Mohan S, Vijaya Bhaskar Y, Sarma PN (2007c) Biohydrogen production from chemical wastewater treatment in biofilm configured reactor operated in periodic discontinuous batch mode by selectively enriched anaerobic mixed consortia. Water Res 41:2652–2664. https://doi.org/10.1016/j.watres.2007.02.015

Venkata Mohan S, Babu VL, Sarma P (2008a) Effect of various pretreatment methods on anaerobic mixed microflora to enhance biohydrogen production utilizing dairy wastewater as substrate. Bioresour Technol 99:59–67. https://doi.org/10.1016/j.biortech.2006.12.004

Venkata Mohan S, Mohanakrishna G, Sarma P (2008b) Integration of acidogenic and methanogenic processes for simultaneous production of biohydrogen and methane from wastewater treatment. Int J Hydrog Energy 33:2156–2166. https://doi.org/10.1016/j.ijhydene.2008.01.055

Venkata Mohan S, Venkateswar Reddy M, Venkata Subhash G, Sarma PN (2010) Fermentative effluents from hydrogen producing bioreactor as substrate for poly (β -OH) butyrate production with simultaneous treatment: an integrated approach. Bioresour Technol 101:9382–9386. https://doi.org/10.1016/j.biortech.2010.06.109

Venkata Mohan S, Agarwal L, Mohanakrishna G, Srikanth S, Kapley A, Purohit HJ, Sarma PN (2011) Firmicutes with iron dependent hydrogenase drive hydrogen production in anaerobic bioreactor using distillery wastewater. Int J Hydrog Energy 36:8234–8242. https://doi.org/10.1016/j.ijhydene.2011.04.021

Venkata Mohan S, Velvizhi G, Vamshi Krishna K, Lenin Babu M (2014) Microbial catalyzed electrochemical systems: a bio-factory with multi-facet applications. Bioresour Technol 165:355–364. https://doi.org/10.1016/j.biortech.2014.03.048

Venkata Mohan S, Nikhil GN, Chiranjeevi P, Reddy CN, Rohit MV, Kumar AN, Sarkar O (2016) Waste biorefinery models towards sustainable circular bioeconomy: crit rev future perspect. Bioresour Technol 215:2–12. https://doi.org/10.1016/j.biortech.2016.03.130

Venkata Mohan S, Srikanth S, Nikhil GN (2017) Augmentation of bacterial homeostasis by regulating in situ buffer capacity: significance of total dissolved salts over acidogenic metabolism. Bioresour Technol 225:34–39. https://doi.org/10.1016/j.biortech.2016.11.027

Venkateswar Reddy M, Venkata Mohan S (2012) Influence of aerobic and anoxic microenvironments on polyhydroxyalkanoates (PHA) production from food waste and acidogenic effluents using aerobic consortia. Bioresour Technol 103:313–321. https://doi.org/10.1016/j.biortech.2011.09.040

Venkateswar Reddy M, Chitanya DNSK, Nikhil GN, Venkata Mohan S, Sarma PN (2014) Influence of co-factor on enhancement of bioplastic production through wastewater treatment. Clean Soil Air Water 42:809–814. https://doi.org/10.1002/clen.201300105

Wagner RC, Regan JM, S-E O, Zuo Y, Logan BE (2009) Hydrogen and methane production from swine wastewater using microbial electrolysis cells. Water Res 43:1480–1488. https://doi.org/10.1016/j.watres.2008.12.037

Wang H, Ren ZJ (2013) A comprehensive review of microbial electrochemical systems as a platform technology. Biotechnol Adv 31:1796–1807. https://doi.org/10.1016/j.biotechadv.2013.10.001

Wang B, Wan W, Wang J (2009) Effect of ammonia concentration on fermentative hydrogen production by mixed cultures. Bioresour Technol 100:1211–1213. https://doi.org/10.1016/j.biortech.2008.08.018

Willquist K, Nkemka VN, Svensson H, Pawar S, Ljunggren M, Karlsson H, Murto M, Hulteberg C, van Niel EWJ, Liden G (2012) Design of a novel biohythane process with high H and CH production rates. Int J Hydrog Energy 37:17749–17762. https://doi.org/10.1016/j.ijhydene.2012.08.092

Wong YM, TY W, Juan JC (2014) A review of sustainable hydrogen production using seed sludge via dark fermentation. Renew Sustain Energy Rev 34:471–482. https://doi.org/10.1016/j.rser.2014.03.008

Zhu H, Parker W, Basnar R, Proracki A, Falletta P, Béland M, Seto P (2009) Buffer requirements for enhanced hydrogen production in acidogenic digestion of food wastes. Bioresour Technol 100:5097–5102. https://doi.org/10.1016/j.biortech.2009.02.066

Development of Dry Anaerobic Technologies of Bio-waste and Unlock the Barriers for Valorization

13

Ahmed Tawfik and Mohamed Elsamadony

Abstract

Solid waste (SW) is highly produced, wherever there is human habitation. Uncollected and improper management of SW in developing countries creates sanitary hazardous and sources of public nuisance. Fortunately, biodegradable organic fractions represent approximately 60–70% of SW, which could be easily converted into energy in the form of H_2 via dry and wet anaerobic digestion process. Moreover, the digestate is rich with plant nutrients (N&P), which could be used for crop production. Factors affecting H_2 production such as pH levels, moisture content, temperature, food to microorganism (F/M) ratio, N/P ratio, mixing mode, sludge residence time (SRT), hydraulic retention time (HRT), organic loading rate (OLR), particle size (PS), and initial ammonia concentration are addressed in this chapter. Moreover, enhancement of H_2 production using mixtures of different substrates via co-digestion bioprocess is extensively discussed to unlock the barriers for valorization of organic fraction of municipal solid waste (OFMSW). This will certainly provide the stakeholders and community the beneficial use of OFMSW to contribute closing the cycles of rural–urban–industrial nutrient and renewable energy in a sustainable way.

Keywords

Bio-waste • Anaerobic digestion • Biodegradation • Hydrogen

A. Tawfik (✉)
Environmental Engineering Department, Egypt – Japan University of Science and Technology (E-JUST), P.O. Box 179, New Borg El Arab City, 21934 Alexandria, Egypt

Water Pollution Research Department, National Research Centre, P.O. Box 12622, Giza, Egypt
e-mail: ahmed.tawfik@ejust.edu.eg; Tawfik8@hotmail.com

M. Elsamadony
Public Works Engineering Department, Faculty of Engineering, Tanta University, 31521 Tanta City, Egypt

H.J. Purohit et al. (eds.), *Optimization and Applicability of Bioprocesses*,
https://doi.org/10.1007/978-981-10-6863-8_13

13.1 Introduction

Solid waste (SW) production is a serious issue in low-income countries. Financial support, contribution of private and public authorities, is very limited for solid waste management. Total solid waste of about 88% is mainly dumped in open areas, where natural decomposition takes place creating severe environmental problems and pollution of groundwater. Burning up of solid waste is the worst option, where smog a formation of black cloud in the sky results in air pollution. SW is composed of carbohydrate, proteins, fats, and other components and represents approximately 60–70% of the total municipal solid waste (MSW) (Tawfik and El-Qelish 2012). Anaerobic digestion (AD) process represents one of the most promising techniques for conversion of organic fraction of municipal solid waste (OFMSW) into renewable energy. OFMSW is free of charge and abundant and considered as a superior substrate for H_2 production using anaerobes (Dias et al. 2014). The importance of AD process for energy production is a promising approach as a substitute for conventional fossil fuels (Elsamadony et al. 2015a). Proliferation of biological H_2 from OFMSW via AD bioprocess would result in a clean environment, and it also represents a sustainable biofuel source for countries lacking sufficient reservoirs of fossil fuels. Moreover, H_2 produces only water as a combustion by-product and generates energy yields of 122 kJ/g (2.75 times higher than hydrocarbon fuels) (Soltan et al. 2016). Several technologies have been used for H_2 production from SW such as thermo-catalytic reformation process and water electrolysis (Irmak and Öztürk 2010). However, these methods consume electricity derived from combustion of the fossil fuels. Therefore, AD of OFMSW is preferable for H_2 production due to its simplicity, low-cost technology, and degradation of a wide range of bio-wastes (Table 13.1). The main obstacles and advantages of AD bioprocess are listed in Table 13.1, which should be simultaneously addressed and considered.

Wet, semidry, and dry AD process is classified based on solid content, i.e., ≤10%, 10–20%, and >20%, respectively (Elsamadony et al. 2015c). Wet AD is not an appropriate technology for conversion of OFMSW, where dilution by water is

Table 13.1 Advantages and disadvantages of anaerobic digestion (AD) process

Advantages
1. Reduction of the volume of bio-waste
2. The digestate is rich with N and P and could be used for crop production
3. Sustain high organic loading rate (OLR)
4. Low operating and maintenance costs
5. Long starvation period and thus it is suitable for the treatment of seasonal wastes
Disadvantages
1. Slow degradation process where long sludge residence time and large reactor volume are required
2. Long start-up period due to slow growth rate of anaerobes
3. Buffer reagent addition may be required to maintain an acceptable pH level
4. Produce odors and corrosive volatile sulfur compounds

Table 13.2 H_2 yield from different substrates via dry anaerobic digestion (DAD) process

Substrate	Initial pH value	Operation mode	Temperature (°C)	H_2 yield	References
OFMSW + slaughterhouse waste	5–6	Semicontinuous	34	52.5–71.3 mL/g $VS_{removed}$	Gómez et al. (2006)
OFMSW	5.7	Continuous	38	127 mL/g$VS_{removed}$	Alzate-Gaviria et al. (2007)
				99 mL/g $VS_{removed}$	
Food waste + olive mill waste	6.5	Batch	39	44 mL/g VS_{added}	De Gioannis et al. (2013)
Food waste				81 mL/g VS_{added}	
				114 mL/g VS_{added}	
				58 mL/g VS_{added}	
			36	101 mL/g $COD_{removed}$	Chen et al. (2006)
	5.5	Semicontinuous	55	205 mL/g VS_{added}	Chu et al. (2008)
	5.5		37	332 mL/g VS_{added}	Elbeshbishy et al. (2011b)
	5–6		34	27-28 mL/gVS_{added}	Gómez et al. (2009)
Vegetable waste	6.0–7.0		55	0.48 mmol/g COD_{added}	Lee et al. (2008)

required resulting in a large volume of reactors, which should be avoided. Dry AD (DAD) of such wastes is a promising option, which provides a high volumetric H_2 production and positive energy balance (Elsamadony and Tawfik 2015a; Romero Aguilar et al. 2013). Li et al. (2013a, b) found that DAD achieved a high volumetric energy production of 14.2 $L_{CH4}/L_{reactor}$ from co-fermentation of corn stover and chicken manure at a ratio of 1:1, which was 3.2- and 1.4-fold greater than those obtained from wet and semidry AD processes, respectively. Table 13.2 exhibits high H_2 generation from OFMSW and food waste (FW) via AD process as compared to other substrates (Elbeshbishy et al. 2011a; Hwang et al. 2011).

13.1.1 Pathway of H_2 Production

Anaerobes can easily utilize the OFMSW (carbohydrates, protein, and fats) through successive bio-processes. Hydrolysis process is carried out by hydrolytic bacteria to degrade proteins, carbohydrate, and fats into simple monomers (Fig. 13.1). The products are utilized further by acidogenesis bacteria to produce organic acids, H_2, and CO_2. Acetate is generated from a combination of H_2 and CO_2 by homoacetogenic bacteria through the Wood–Ljungdahl pathway. Acetoclastic/hydrogenotrophic methanogens generate CH_4 and CO_2 from conversion of organic acids and H_2

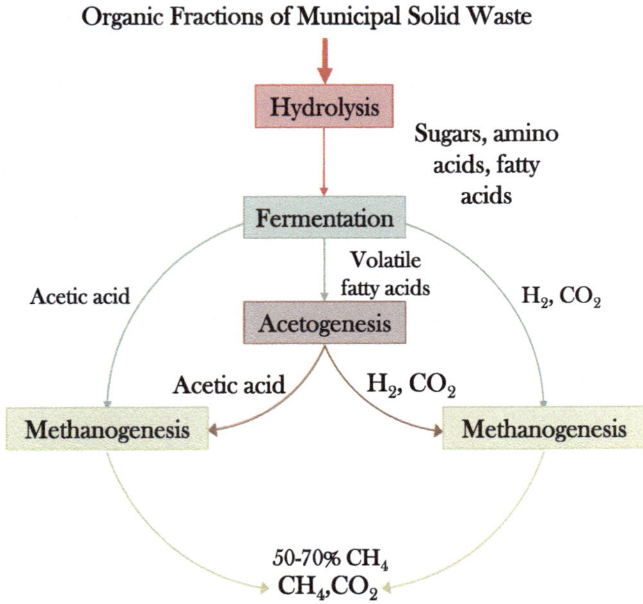

Fig. 13.1 Anaerobic pathway degradation of OFMSW containing carbohydrate, lipids, and protein

(Hallenbeck and Centre-ville 2009). Moreover, the presence of sulfate and/or nitrate in the substrate facilitated the utilization of H_2 (Saady 2013).

13.1.2 Anaerobes Producing Hydrogen

Strict and facultative bacteria are the main anaerobes that able to produce H_2 via DAD process. Strict anaerobes such as *Clostridium* sp. need the complete absence of oxygen to grow and could efficiently generate H_2 under mesophilic and thermophilic conditions at a wide range of pH levels. Facultative microbes (*Enterobacter* sp.) have the ability to generate hydrogenase enzyme, which is responsible for H_2 production (Li and Fang 2007) and has the merit of generation of adenosine triphosphate (ATP) under aerobic and anaerobic respiration that outperform strict microbes. However, *Clostridium* sp. produces 4 mole $H_2/mol_{glucose}$ (Eq. 13.1) as compared to *Enterobacter* sp., which generates only 1 mole $H_2/mol_{glucose}$ (Chin et al. 2003). Butyrate production is accompanied by 2 moles of H_2 per mole of glucose conversion (Eq. 13.2)

$$\text{Glucose} + 2H_2O \rightarrow 2\text{Acetate} + 2CO_2 + 4H_2 \quad (13.1)$$

$$\text{Glucose} \rightarrow \text{Butyrate} + 2CO_2 + 2H_2 \quad (13.2)$$

Co-cultures of *Clostridium sp.* with *Enterobacter* sp. provided a superior H_2 yield as compared to single culture (Chong et al. 2009). Seppälä et al. (2011) found that H_2 yield of 1.45 and 2.09 mole $H_2/mol_{glucose}$ was achieved using sole culture of *E. coli* and *C. butyricum*, respectively. However, co-cultures of *E. coli* and *C. butyricum* provided H_2 yield of 1.65 mole $H_2/mol_{glucose}$ and achieved maximum cumulative volumetric H_2 production. Yokoi et al. (2002) observed that co-culture of *C. butyricum* (strict anaerobe) with *E. aerogenes* (facultative bacteria) achieved an H_2 yield of 2.7 mol $H_2/mol_{glucose}$, which exceeded 2 mol $H_2/mol_{glucose}$ provided by pure culture (*Clostridium* sp.).

13.2 Key Factors Affecting H_2 Production

Several factors such as pH levels, moisture content, temperature, food to microorganism (F/M) ratio, N/P ratio, mixing operation mode, sludge residence time (SRT) and hydraulic retention time (HRT), organic loading rate (OLR), particle size, and initial ammonia concentration affect biological H_2 production and anaerobic metabolic activities.

13.2.1 pH Value

The pH value significantly affects H_2 evaluation via DAD, where it influences the activity of hydrogenase enzyme as well as the metabolic pathway. H^+ ion concentration is a critical factor for maintaining sufficient ATP levels in the AD process. However, high levels of H^+ ion consume ATP to sustain cell neutrality rather than to produce H_2 (Mohd Yasin et al. 2011). Continuous DAD process produces considerable amounts of volatile fatty acids (VFAs) resulting in a drop in pH value. Thus, it requires continuous supplementation of the digester with sodium carbonate to increase the buffering capacity (Ghimire et al. 2015). The optimal pH value for H_2 fermentative largely varies from 4.5 to 9 (Khanal 2003; Luo et al. 2010; Wu et al. 2010). The optimal pH value is close to 5.5, which enables maximum H_2 production due to augmentation of hydrogenase activity (Andersch et al. 1983).

13.2.2 Total Solid Content

Moisture and total solids (TS) content are the major parameters, which affect the DAD process (Abbassi-Guendouz et al. 2012; Karthikeyan and Visvanathan 2012). TS content exceeding 30% (w/w) negatively affect the H_2 production due to accumulation of VFAs (Li et al. 2013a, b). Abbassi-Guendouz et al. (2013) found that increasing the TS content inhibited the activity and growth of *Clostridium* sp., which is responsible for H_2 production. The H_2 production was substantially decreased at increasing TS content from 21% to 35% and from 10% to 35% (w/w) (Valdez-Vazquez and Poggi-Varaldo 2009; Robledo-Narváez et al. 2013). On the other hand, insufficient TS content could limit the energy production resulting in low-energy/mass transfer bioprocess (Banerjee et al. 2005; Valdez-Vazquez and Poggi-Varaldo 2009).

13.2.3 Temperature

Temperature is a critical factor affecting the behavior of H_2-producing bacteria as well as the activity of enzymes and coenzymes. Fortunately, H_2-producing bacteria could flourish under psychrophilic, mesophilic, thermophilic, and hyperthermophilic conditions where temperature varied from 10 to 110°C (Fig. 13.2). However, temperature exceeding 55 C stimulates metabolic, specific growth, and destruction rates of substrate and enhances H_2 production (Wang and Wan 2008). Moreover, thermophilic conditions are presumed to enhance the activity of hydrogenase enzyme (Valdez-Vazquez et al. 2005). Solubilization of solids under thermophilic conditions is higher as compared to mesophilic one (Lee et al. 2006; Appels et al. 2008). The optimum temperature for biological H_2 production ranges from 37 (mesophilic) to 55 °C (thermophilic) (Li and Fang 2007; Appels et al. 2008).

13.2.4 Food to Microorganism (F/M)Ratio

The initial substrate concentration (F) provides a carbon source for biosynthesis, and the microorganisms (M) are responsible for degradation of feedstock. The metabolic pathway of H_2 production is mainly affected by changing F/M ratio (Soest

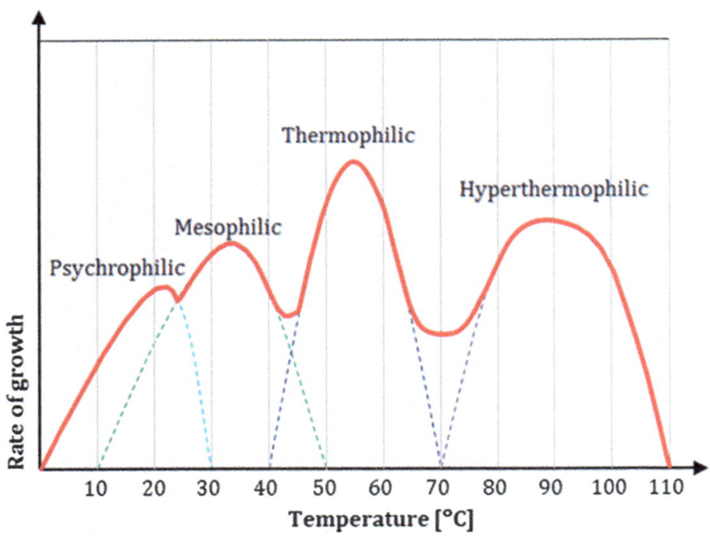

Fig. 13.2 Bacterial growth rate under psychrophilic, mesophilic, thermophilic, and hyperthermophilic conditions

and Wine 1967). A proper range of F/M ratio would enhance the capability of H_2 producers (Sreela-or et al. 2011a). H_2 production increased at higher F/M ratio of 4–6.6 g_{COD}/g_{VSS}.d (Hafez et al. 2010). Pan et al. (2008) found that biological H_2 production from food waste (FW) was the optimum at F/M ratio of 6 g VS_{feed}/g VS_{seed}. The highest H_2 yield from FW occurred at F/M ratio of 7.8 g_{COD}/g_{VSS} and a temperature of 36 °C (Chen et al. 2006).

13.2.5 Carbon to Nitrogen (C/N) Ratio

Nitrogen is an essential element for growth of H_2 producers. However, high and low C/N ratio would negatively affect the H_2 production. Anaerobes have the ability to utilize carbon source 25–30 times faster than nitrogen content, and the optimum C/N ratio varied from 20 to 30 (Elsamadony et al. 2015a; Kim 2004). Co-digestion processes supply sufficient C/N ratio of 30 for anaerobic consortium bacterium (Yadvika et al. 2004). The optimal C/N ratio for H_2 production from co-digestion of rice straw and sewage sludge was 25:1 (Kim et al. 2012). Sreela-or et al. (2011b) found that H_2 fermentation of the combination of food waste and sewage sludge was maximum at C/N ratio of 33.

13.2.6 Mixing Operation Modes

Mixing is an essential process to sustain adequate contact between the microbial consortia and the substrate (Elsamadony and Tawfik 2015a). Moreover, flocculation of suspended solids, distribution of nutrients and trace elements for anaerobes, and guarantee of a uniform temperature along the digester occurred due to mixing process (Weiland 2010). The latter plays an important role for transfer the substrate from bulk liquid to biomass, which facilitates the degradation process. Ong et al. (2002) found that mixing promoted energy production and overcomes accumulation of gases inside the sludge blanket. Furthermore, sufficient agitation is introduced to the AD process to avoid sludge washout. Nevertheless, 29–54% of the total power generated from anaerobic digester could be consumed for operation process. The efficiency of mixing process is dependent on the shape and type of paddles, as well as the operating period per day. Improper mechanical mixing devices provide dead zones and reduce the contact time between substrate and bacterium consortium, negatively influencing the performance of the AD process (Ghanimeh et al. 2012). On the other hand, proper design of mixing device and its intensity is necessary in order to maximize the net gain energy and improve the efficiency of DAD process. However, at high mixing intensity, shear force destroys the flocs and breaks down the granules, which deteriorate the H_2 production (Hoffmann et al. 2008; Lindmark et al. 2014). In addition, microbial shift could take place in the digester resulting in slow degradation process. Mixing is important for AD process particularly at high loading rate and solid contents (Karim et al. 2005). Furthermore, intermittent mixing mode is preferred to enhance the H_2 generation, as well as to minimize the energy consumption (Kaparaju et al. 2008).

13.2.7 Sludge Residence Time (SRT) and Hydraulic Retention Time (HRT)

Sludge residence time (SRT) is the average of the retained biomass in the anaerobic digester. The SRT is an important design parameter for AD process and expressed in days. Moreover, short SRT (3–5 days) and HRT (6–12 h) are preferred for H_2 producers. On the other hand, long SRT and HRT encourage the growth of H_2 consumers and promote the methanogens, which should be avoided. HRT is the time (h or days) for feedstock to pass through the digester. Fortunately, SRT is an equal HRT in DAD process (Appels et al. 2008; Fernández-Rodríguez et al. 2014).

13.2.8 Organic Loading Rate (OLR)

Organic loading rate (OLR) is the quantity of daily organic content of the feedstock per reactor volume and expressed as g COD/L/day or g volatile solids/L/day. OLR exceeding 20 g COD/L/day is preferred for H_2 producers (Hafez et al. 2010). However, overloading the anaerobic digester would increase the accumulation of VFAs and deteriorate the H_2 production process and yield (Fernández-Rodríguez et al. 2014).

13.2.9 Particle Size (PS)

Particle size (PS) exerts a significant effect on the efficiency of DAD process. Anaerobic consortium bacteria utilize particulate organic matter for H_2 production via three main steps, namely, hydrolysis, acidogenesis, and acetogenesis processes. However, hydrolysis of organic particulate matter is the rate-limiting step in DAD, which affects both the mass transfer and the substrate accessibility (Izumi et al. 2010). Hydrolytic enzymes are mainly responsible for the hydrolysis of organics into the particulate form (Angelidaki and Sanders 2004). Reduced particle size exhibits a perfect pretreatment technique for efficient conversion of wastes-rich particulate organic matter (Barakat et al. 2013). The reduction of particle size would increase the surface area which enhances the degradation process using hydrolytic enzymes resulting an efficient substrate uptake rate and quite high H_2 yield (Monlau et al. 2013; Agyeman and Tao 2014). Wen (2004) found that reducing the particle size of animal manure from 840–590 to 590–350 μm improved glucose yield by 29%. However, further reduction of particle size exhibited no effect. The highest substrate uptake rate occurred on reducing the particle size of food waste from 2.14 to 1.02 mm (Kim et al. 2000). This undoubtedly emphasized the importance of the reduction of particle size prior DAD bioprocess.

13.3 Upgrading of DAD Process

13.3.1 Substrate Pretreatment

Mechanical, thermal, microwave, and alkali pretreatment methods are mainly used to increase the solubility of organics and facilitate the anaerobic conversion of solid wastes (Elbeshbishy et al. 2011a, b; Ma et al. 2011). Surfactants, particularly non-ionic type, would enhance H_2 yield from OFMSW and improve the enzymatic hydrolysis of particulate organic substances (Qing et al. 2010; Elsamadony et al. 2015c). Surfactant molecule could also change the structure of organics and allow the enzymes to penetrate into its surface (Leaño and Babel 2012). Moreover, surfactant has a positive effect on the interaction between enzyme and substrate (Sipos et al. 2011). Recently, fungal species were tested for enhancement of the H_2 generation from agricultural waste such as cornstalks (Cheng and Liu 2012).

13.3.2 Co-fermentation Process

Mono-fermentation of OFMSW showed some difficulties, i.e., insufficient trace elements content, high C/N ratio, and low buffering capacity and suffered from accumulation of VFAs in the digester (Lin et al. 2013; Zhou et al. 2013). Co-fermentation of various substrates can overcome such drawbacks and provide nutrient balance and buffering capacity to improve H_2 production (Razaviarani and Buchanan 2014). Table 13.3 shows the enhancement ratio of H_2 production from co-fermentation of OFMSW and FW with other substrates. The highest H_2 yield was achieved from co-fermentation of OFMSW with gelatinous solid waste (GSW) and paper mill sludge (PMS) (Elsamadony and Tawfik 2015b). PMS and GSW are rich in calcium ions, which play an important role for enhancing the growth of anaerobes and

Table 13.3 Co-digestion of various substrates

Substrates	Ratio	Operational conditions	Enhancement ratio[a]	References
OFMSW + sewage sludge	1: 1	Batch, 55 °C, pH 5.5	1.20	Tyagi et al. (2014)
FW + lime mud	1: 0.03	Batch, 55 °C, pH 6.5–6.9	1.51	Zhang et al. (2013)
FW + white mud from ammonia soda process	1: 0.015	Batch, 55 °C, pH 6.6–7.4	1.54	Zhang and Wang (2013)
OFMSW + cellulose	1: 0.2	Batch, 37 °C	1.35	Ponsá et al. (2011)
OFMSW + protein	1: 0.2		1.54	
OFMSW + gelatin solid waste + Paper mill sludge	0.7:0.2:0.1	Batch, 55 °C, pH 6.0	1.62	Elsamadony and Tawfik (2015b)

[a]Enhancement ratio is calculated based on the difference between mono-fermentation OFMSW and co-fermentation with other substrates

subsequently H_2 production and yield (Chen et al. 2008; Zhang and Wang 2013). Moreover, Ca^{++} acts as a coagulant aids to improve the biomass accumulation inside the reactor (Yuan et al. 2010). Furthermore, GSW is protein-rich feedstock, which provides a proper C/N ratio during co-fermentation of OFMSW (Kim 2004). C/N ratio not only augments the nutrients content but also increases the bacterial activity and H_2 productivity (Zhou et al. 2013). Moreover, co-fermentation of OFMSW with GSW and PMS provides a sufficient buffering capacity and improves the stability of the AD bioprocess (Zhang et al. 2014). Zhou et al. (2013) and Kim (2004) found that C/N ratio increased from 26 to 31 for co-fermentation of food waste (FW) with sewage sludge. Several researchers (Zhang et al. 2011, 2013, 2014; Wang et al. 2013; Zhang and Wang 2013) showed that co-digestion process increases the buffering capacity, trace element content, and balanced C/N ratio resulting in an improved H_2 productivity.

13.3.3 Trace Elements Addition

Bio-hydrogen production requires essential micronutrients for bacterial metabolism during fermentation process. However, high concentration of metals would inhibit the activity of H_2-producing anaerobes (Srikanth and Mohan 2012; Wu et al. 2015). Sodium ions enable better anaerobic metabolism (Lin et al. 2013). Iron and nickel ions are the backbone of [Fe], [Ni–Fe], and [Ni–Fe–Se] hydrogenase enzymes, which are responsible for H_2 production (Srikanth and Mohan 2012). Potassium ions facilitate nutrients transportation by increasing cell wall permeability, while coenzyme utilizes cobalt to stimulate acetogenesis (acetate-forming) activity (Wu et al. 2015). Magnesium and manganese facilitate substrate utilization rate, while little amount of zinc is required for nutrition of anaerobes (Srikanth and Mohan 2012). The highest cumulative H_2 production was achieved with iron, nickel, magnesium, manganese, and zinc at concentrations of 100, 16, 600, 15, and 12 mg/L, respectively (Srikanth and Mohan 2012). Iron, nickel, and cobalt concentrations of 1233, 6.1, and 2.5 mg/kg, respectively, were essential for H_2-producing bacteria (Schmidt et al. 2014). Moreover, Lin et al. (2013) found that sodium concentration of 100–200 mg/L was acceptable for H_2 producers. In addition, Wu et al. (2015) investigated the effect of combination of different metals on AD process and found that the optimal values of iron, cobalt, nickel, molybdenum, potassium, magnesium, zinc, and manganese were 349.0, 37.2, 165.6, 7.3, 824.5, 62.4, 0.8, and 22.7 mg/L, respectively. Optimal calcium concentration of 3 g/L was recorded to provide the best fermentation performance (Ahn et al. 2006).

13.3.4 Enzymes Enrichment

Addition of enzymes plays a key role for hydrolysis of organic matter during AD process (Mongkolthanaruk and Dharmsthiti 2002). The COD removal increased from 13% to 87% with pretreatment of residues containing yeast cells with lytic enzymes (Mallick et al. 2010). In addition, Masse et al. used pancreatic lipase for pretreatment of pork fat to reduce the particle size from 359 to 68 μm (Masse et al. 2001). These findings indicate the ability of external enzymes for enhancement of hydrolysis process and increase the solubilization of solid waste. However, the enzymes supplementation is not economically feasible, due to high costs. Therefore, trials were made to find out a cheap natural source of enzymes such as *Carica papaya* (Elsamadony et al. 2015b). *C. papaya* could enhance the H_2 productivity from degradation of protein and lipids; however, it had a limited effect on the conversion of carbohydrate fraction. Carica papaya's latex is well known as a rich source of the four cysteine endopeptidases: papain, chymopapain, glycyl endopeptidase, caricain, and lipases (Azarkan et al. 2003; Abdelkafi et al. 2011). Chaiwut et al. (2010) extracted proteases from papaya peels which could be used for hydrolysis of protein.

13.4 Conclusions

DAD process is a promising approach for developing countries suffering from lack of conventional fuels. Huge amounts of municipal solid waste are produced in tropical and subtropical countries without a proper management, due to lack of financial support. Fortunately, the major portions of these wastes are biodegradable organics (60–70%), which can be easily converted into H_2-using anaerobes. However, the H_2 production and yield from OFMSW are mainly affected by several parameters such as pH value, solid content, cultivation temperature, F/M ratio, C/N ratio, mixing operation mode, and particle size, which have been well addressed in this chapter. Moreover, enhancement of H_2 production from bio-waste has been extensively discussed.

References

Abbassi-Guendouz A, Brockmann D, Trably E, Dumas C, Delgenès J-P, Steyer J-P, Escudié R (2012) Total solids content drives high solid anaerobic digestion via mass transfer limitation. Bioresour Technol 111:55–61. https://doi.org/10.1016/j.biortech.2012.01.174

Abbassi-Guendouz A, Trably E, Hamelin J, Dumas C, Steyer JP, Delgenè J-P, Escudié R (2013) Microbial community signature of high-solid content methanogenic ecosystems. Bioresour Technol 133:256–262. https://doi.org/10.1016/j.biortech.2013.01.121

Abdelkafi S, Barouh N, Fouquet B, Fendri I, Pina M, Scheirlinckx F, Villeneuve P, Carrière F (2011) Carica papaya lipase: a naturally immobilized enzyme with interesting biochemical properties. Plant Foods Hum Nutr 66:34–40. https://doi.org/10.1007/s11130-010-0206-0

Agyeman FO, Tao W (2014) Anaerobic co-digestion of food waste and dairy manure: effects of food waste particle size and organic loading rate. J Environ Manag 133:268–274. https://doi.org/10.1016/j.jenvman.2013.12.016

Ahn J-H, Do TH, Kim SD, Hwang S (2006) The effect of calcium on the anaerobic digestion treating swine wastewater. Biochem Eng J 30:33–38. https://doi.org/10.1016/j.bej.2006.01.014

Alzate-Gaviria LM, Sebastian PJ, Pérez-Hernández A, Eapen D (2007) Comparison of two anaerobic systems for hydrogen production from the organic fraction of municipal solid waste and synthetic wastewater. Int J Hydrog Energy 32:3141–3146. https://doi.org/10.1016/j.ijhydene.2006.02.034

Andersch W, Bahl H, Gottschalk G (1983) Microbiology and biotechnology level of enzymes involved in acetate, butyrate, acetone and butanol formation by clostridium acetobutylicum. Eur J Appl Microbiol Biotechnol 3:327–332
Angelidaki I, Sanders W (2004) Assessment of the anaerobic biodegradability of macropollutants. Rev Environ Sci Biotechnol 3:117–129. https://doi.org/10.1007/s11157-004-2502-3
Appels L, Baeyens J, Degrève J, Dewil R (2008) Principles and potential of the anaerobic digestion of waste-activated sludge. Prog Energy Combust Sci 34:755–781. https://doi.org/10.1016/j.pecs.2008.06.002
Azarkan M, El Moussaoui A, van Wuytswinkel D, Dehon G, Looze Y (2003) Fractionation and purification of the enzymes stored in the latex of Carica papaya. J Chromatogr B 790:229–238. https://doi.org/10.1016/S1570-0232(03)00084-9
Banerjee R, Mukherjee G, Patra KC (2005) Microbial transformation of tannin-rich substrate to gallic acid through co-culture method. Bioresour Technol 96:949–953. https://doi.org/10.1016/j.biortech.2004.08.004
Barakat A, de Vries H, Rouau X (2013) Dry fractionation process as an important step in current and future lignocellulose biorefineries: a review. Bioresour Technol 134:362–373. https://doi.org/10.1016/j.biortech.2013.01.169
Chaiwut P, Pintathong P, Rawdkuen S (2010) Extraction and three-phase partitioning behavior of proteases from papaya peels. Process Biochem 45:1172–1175. https://doi.org/10.1016/j.procbio.2010.03.019
Chen W, Chen S, Kumarkhanal S, Sung S (2006) Kinetic study of biological hydrogen production by anaerobic fermentation. Int J Hydrog Energy 31:2170–2178. https://doi.org/10.1016/j.ijhydene.2006.02.020
Chen Y, Cheng JJ, Creamer KS (2008) Inhibition of anaerobic digestion process: a review. Bioresour Technol 99:4044–4064. https://doi.org/10.1016/j.biortech.2007.01.057
Cheng X-Y, Liu C-Z (2012) Fungal pretreatment enhances hydrogen production via thermophilic fermentation of cornstalk. Appl Energy 91:1–6. https://doi.org/10.1016/j.apenergy.2011.09.014
Chin H-L, Chen Z-S, Chou CP (2003) Fedbatch operation using *Clostridium acetobutylicum* suspension culture as biocatalyst for enhancing hydrogen production. Biotechnol Prog 19:383–388. https://doi.org/10.1021/bp0200604
Chong M-L, Sabaratnam V, Shirai Y, Hassan MA (2009) Biohydrogen production from biomass and industrial wastes by dark fermentation. Int J Hydrog Energy 34:3277–3287. https://doi.org/10.1016/j.ijhydene.2009.02.010
Chu C-F, Li Y-Y, K-Q X, Ebie Y, Inamori Y, Kong H-N (2008) A pH- and temperature-phased two-stage process for hydrogen and methane production from food waste. Int J Hydrog Energy 33:4739–4746. https://doi.org/10.1016/j.ijhydene.2008.06.060
De Gioannis G, Muntoni A, Polettini A, Pomi R (2013) A review of dark fermentative hydrogen production from biodegradable municipal waste fractions. Waste Manag 33:1345–1361. https://doi.org/10.1016/j.wasman.2013.02.019
Dias T, Fragoso R, Duarte E (2014) Anaerobic co-digestion of dairy cattle manure and pear waste. Bioresour Technol 164:420–423. https://doi.org/10.1016/j.biortech.2014.04.110
Elbeshbishy E, Hafez H, Dhar BR, Nakhla G (2011a) Single and combined effect of various pretreatment methods for biohydrogen production from food waste. Int J Hydrog Energy 36:11379–11387. https://doi.org/10.1016/j.ijhydene.2011.02.067
Elbeshbishy E, Hafez H, Nakhla G (2011b) Ultrasonication for biohydrogen production from food waste. Int J Hydrog Energy 36:2896–2903. https://doi.org/10.1016/j.ijhydene.2010.12.009
Elsamadony M, Tawfik A (2015a) Potential of biohydrogen production from organic fraction of municipal solid waste (OFMSW) using pilot-scale dry anaerobic reactor. Bioresour Technol 196:9–16. https://doi.org/10.1016/j.biortech.2015.07.048
Elsamadony M, Tawfik A (2015b) Dry anaerobic co-digestion of organic fraction of municipal waste with paperboard mill sludge and gelatin solid waste for enhancement of hydrogen production. Bioresour Technol 191:157–165. https://doi.org/10.1016/j.biortech.2015.05.017

Elsamadony M, Tawfik A, Suzuki M (2015a) Surfactant-enhanced biohydrogen production from organic fraction of municipal solid waste (OFMSW) via dry anaerobic digestion. Appl Energy 149:272–282. https://doi.org/10.1016/j.apenergy.2015.03.127

Elsamadony M, Tawfik A, Danial A, Suzuki M (2015b) Optimization of hydrogen production from organic fraction of municipal solid waste (OFMSW) dry anaerobic digestion with analysis of microbial community. Int J Energy Res 39:929–940. https://doi.org/10.1002/er.3297

Elsamadony M, Tawfik A, Danial A, Suzuki M (2015c) Use of Carica Papaya enzymes for enhancement of H production and degradation of glucose, protein, and lipids. Energy Procedia 75:975–980. https://doi.org/10.1016/j.egypro.2015.07.308

Fernández-Rodríguez J, Pérez M, Romero LI (2014) Dry thermophilic anaerobic digestion of the organic fraction of municipal solid wastes: solid retention time optimization. Chem Eng J 251:435–440. https://doi.org/10.1016/j.cej.2014.04.067

Ghanimeh S, El M, Saikaly P (2012) Bioresource technology mixing effect on thermophilic anaerobic digestion of source-sorted organic fraction of municipal solid waste. Bioresour Technol 117:63–71. https://doi.org/10.1016/j.biortech.2012.02.125

Ghimire A, Valentino S, Frunzo L, Trably E, Escudié R, Pirozzi F, Lens PNL, Esposito G (2015) Biohydrogen production from food waste by coupling semi-continuous dark-photofermentation and residue post-treatment to anaerobic digestion: a synergy for energy recovery. Int J Hydrog Energy 40:16045–16055. https://doi.org/10.1016/j.ijhydene.2015.09.117

Gómez X, Morán A, Cuetos MJ, Sánchez ME (2006) The production of hydrogen by dark fermentation of municipal solid wastes and slaughterhouse waste: a two-phase process. J Power Sources 157:727–732. https://doi.org/10.1016/j.jpowsour.2006.01.006

Gómez X, Cuetos MJ, Prieto JI, Morán A (2009) Bio-hydrogen production from waste fermentation: mixing and static conditions. Renew Energy 34:970–975. https://doi.org/10.1016/j.renene.2008.08.011

Hafez H, Nakhla G, El Naggar MH, Elbeshbishy E, Baghchehsaraee B (2010) Effect of organic loading on a novel hydrogen bioreactor. Int J Hydrog Energy 35:81–92. https://doi.org/10.1016/j.ijhydene.2009.10.051

Hallenbeck PC, Centre-ville S (2009) Fermentative hydrogen production: principles, progress, and prognosis. Int J Hydrog Energy 34:7379–7389. https://doi.org/10.1016/j.ijhydene.2008.12.080

Hoffmann RA, Garcia ML, Veskivar M, Karim K, Al-Dahhan MH, Angenent LT (2008) Effect of shear on performance and microbial ecology of continuously stirred anaerobic digesters treating animal manure. Biotechnol Bioeng 100:38–48. https://doi.org/10.1002/bit.21730

Hwang J-H, Choi J-A, Abou-Shanab RAI, Min B, Song H, Kim Y, Lee ES, Jeon B-H (2011) Feasibility of hydrogen production from ripened fruits by a combined two-stage (dark/dark) fermentation system. Bioresour Technol 102:1051–1058. https://doi.org/10.1016/j.biortech.2010.08.047

Irmak S, Öztürk İ (2010) Hydrogen rich gas production by thermocatalytic decomposition of kenaf biomass. Int J Hydrog Energy 35:5312–5317. https://doi.org/10.1016/j.ijhydene.2010.03.081

Izumi K, Okishio Y, Nagao N, Niwa C, Yamamoto S, Toda T (2010) Effects of particle size on anaerobic digestion of food waste. Int Biodeterior Biodegrad 64:601–608. https://doi.org/10.1016/j.ibiod.2010.06.013

Kaparaju P, Buendia I, Ellegaard L, Angelidakia I (2008) Effects of mixing on methane production during thermophilic anaerobic digestion of manure: lab-scale and pilot-scale studies. Bioresour Technol 99:4919–4928. https://doi.org/10.1016/j.biortech.2007.09.015

Karim K, Hoffmann R, Thomas Klasson K, Al-Dahhan MH (2005) Anaerobic digestion of animal waste: effect of mode of mixing. Water Res 39:3597–3606. https://doi.org/10.1016/j.watres.2005.06.019

Karthikeyan OP, Visvanathan C (2012) Bio-energy recovery from high-solid organic substrates by dry anaerobic bio-conversion processes: a review. Rev Environ Sci Biotechnol 12:257–284. https://doi.org/10.1007/s11157-012-9304-9

Khanal S (2003) Biological hydrogen production: effects of pH and intermediate products. Int J Hydrog Energy 29:1123–1131. https://doi.org/10.1016/j.ijhydene.2003.11.002

Kim S (2004) Feasibility of biohydrogen production by anaerobic co-digestion of food waste and sewage sludge. Int J Hydrog Energy 29:1607–1616. https://doi.org/10.1016/j.ijhydene.2004.02.018

Kim IS, Kim DH, Hyun S (2000) Effect of particle size and sodium ion concentration on anaerobic thermophilic food waste digestion. Water Sci Technol 41:67–73

Kim M, Yang Y, Morikawa-Sakura MS, Wang Q, Lee MV, Lee D-Y, Feng C, Zhou Y, Zhang Z (2012) Hydrogen production by anaerobic codigestion of rice straw and sewage sludge. Int J Hydrog Energy 37:3142–3149. https://doi.org/10.1016/j.ijhydene.2011.10.116

Leaño EP, Babel S (2012) The influence of enzyme and surfactant on biohydrogen production and electricity generation using Palm Oil Mill Effluent. J Clean Prod 31:91–99. https://doi.org/10.1016/j.jclepro.2012.02.026

Lee K, Lin P, Chang J (2006) Temperature effects on biohydrogen production in a granular sludge bed induced by activated carbon carriers. Int J Hydrog Energy 31:465–472. https://doi.org/10.1016/j.ijhydene.2005.04.024

Lee Z-K, Li S-L, Lin J-S, Wang Y-H, Kuo P-C, Cheng S-S (2008) Effect of pH in fermentation of vegetable kitchen wastes on hydrogen production under a thermophilic condition. Int J Hydrog Energy 33:5234–5241. https://doi.org/10.1016/j.ijhydene.2008.05.006

Li C, Fang HHP (2007) Fermentative hydrogen production from wastewater and solid wastes by mixed cultures. Crit Rev Environ Sci Technol 37:1–39. https://doi.org/10.1080/10643380600729071

Li A, Chu Y, Wang X, Ren L, Yu J, Liu X, Yan J, Zhang L, Wu S, Li S (2013a) A pyrosequencing-based metagenomic study of methaneproducing microbial community in solid-state biogas reactor. Biotechnol Biofuels 6:3. https://doi.org/10.1186/1754-6834-6-3

Li Y, Zhang R, Chen C, Liu G, He Y, Liu X (2013b) Biogas production from co-digestion of corn stover and chicken manure under anaerobic wet, hemi-solid, and solid state conditions. Bioresour Technol 149:406–412. https://doi.org/10.1016/j.biortech.2013.09.091

Lin Y, Wu S, Wang D (2013) Hydrogen-methane production from pulp & paper sludge and food waste by mesophilic–thermophilic anaerobic codigestion. Int J Hydrog Energy 38:15055–15062. https://doi.org/10.1016/j.ijhydene.2012.01.051

Lindmark J, Eriksson P, Thorin E (2014) The effects of different mixing intensities during anaerobic digestion of the organic fraction of municipal solid waste. Waste Manag 34:1391–1397. https://doi.org/10.1016/j.wasman.2014.04.006

Luo G, Xie L, Zou Z, Zhou Q, Wang J-Y (2010) Fermentative hydrogen production from cassava stillage by mixed anaerobic microflora: effects of temperature and pH. Appl Energy 87:3710–3717. https://doi.org/10.1016/j.apenergy.2010.07.004

Ma J, Duong TH, Smits M, Verstraete W, Carballa M (2011) Enhanced biomethanation of kitchen waste by different pre-treatments. Bioresour Technol 102:592–599. https://doi.org/10.1016/j.biortech.2010.07.122

Mallick P, Akunna JC, Walker GM (2010) Anaerobic digestion of distillery spent wash: influence of enzymatic pre-treatment of intact yeast cells. Bioresour Technol 101:1681–1685. https://doi.org/10.1016/j.biortech.2009.09.089

Masse L, Kennedy K, Chou S (2001) Testing of alkaline and enzymatic hydrolysis pretreatments for fat particles in slaughterhouse wastewater. Bioresour Technol 77:145–155. https://doi.org/10.1016/S0960-8524(00)00146-2

Mohd Yasin NH, Rahman NA, Man HC, Mohd Yusoff MZ, Hassan MA (2011) Microbial characterization of hydrogen-producing bacteria in fermented food waste at different pH values. Int J Hydrog Energy 36:9571–9580. https://doi.org/10.1016/j.ijhydene.2011.05.048

Mongkolthanaruk W, Dharmsthiti S (2002) Biodegradation of lipid-rich wastewater by a mixed bacterial consortium. Int Biodeterior Biodegradation 50:101–105. https://doi.org/10.1016/S0964-8305(02)00057-4

Monlau F, Barakat A, Trably E, Dumas C, Steyer J-P, Carrère H (2013) Lignocellulosic materials into biohydrogen and biomethane: impact of structural features and pretreatment. Crit Rev Environ Sci Technol 43:260–322. https://doi.org/10.1080/10643389.2011.604258

Ong HK, Greenfield PF, Pullammanappallil PC (2002) Effect of mixing on biomethanation of cattle-manure slurry. Environ Technol 23:1081–1090. https://doi.org/10.1080/09593332308618330

Pan J, Zhang R, Elmashad H, Sun H, Ying Y (2008) Effect of food to microorganism ratio on biohydrogen production from food waste via anaerobic fermentation. Int J Hydrog Energy 33:6968–6975. https://doi.org/10.1016/j.ijhydene.2008.07.130

Ponsá S, Gea T, Sánchez A (2011) Anaerobic co-digestion of the organic fraction of municipal solid waste with several pure organic cosubstrates. Biosyst Eng 108:352–360. https://doi.org/10.1016/j.biosystemseng.2011.01.007

Qing Q, Yang B, Wyman CE (2010) Impact of surfactants on pretreatment of corn stover. Bioresour Technol 101:5941–5951. https://doi.org/10.1016/j.biortech.2010.03.003

Razaviarani V, Buchanan ID (2014) Reactor performance and microbial community dynamics during anaerobic co-digestion of municipal wastewater sludge with restaurant grease waste at steady state and overloading stages. Bioresour Technol 172:232–240. https://doi.org/10.1016/j.biortech.2014.09.046

Robledo-Narváez PN, Muñoz-Páez KM, Poggi-Varaldo HM, Ríos-Leal E, Calva-Calva G, Ortega-Clemente LA, Rinderknecht-Seijas N, Estrada- Vázquez C, Ponce-Noyola MT, Salazar-Montoya JA (2013) The influence of total solids content and initial pH on batch biohydrogen production by solid substrate fermentation of agroindustrial wastes. J Environ Manag 128:126–137. https://doi.org/10.1016/j.jenvman.2013.04.042

Romero Aguilar MA, Fdez-Güelfo LA, Álvarez-Gallego CJ, Romero García LI (2013) Effect of HRT on hydrogen production and organic matter solubilization in acidogenic anaerobic digestion of OFMSW. Chem Eng J 219:443–449. https://doi.org/10.1016/j.cej.2012.12.090

Saady NMC (2013) Homoacetogenesis during hydrogen production by mixed cultures dark fermentation: unresolved challenge. Int J Hydrog Energy 38:13172–13191. https://doi.org/10.1016/j.ijhydene.2013.07.122

Schmidt T, Nelles M, Scholwin F, Pröter J (2014) Trace element supplementation in the biogas production from wheat stillage – optimization of metal dosing. Bioresour Technol 168:80–85. https://doi.org/10.1016/j.biortech.2014.02.124

Seppälä JJ, Puhakka JA, Yli-Harja O, Karp MT, Santala V (2011) Fermentative hydrogen production by Clostridium butyricum and *Escherichia coli* in pure and cocultures. Int J Hydrog Energy 36:10701–10708. https://doi.org/10.1016/j.ijhydene.2011.05.189

Sipos B, Szilágyi M, Sebestyén Z, Perazzini R, Dienes D, Jakab E, Crestini C, Réczey K (2011) Mechanism of the positive effect of poly(ethylene glycol) addition in enzymatic hydrolysis of steam pretreated lignocelluloses. C R Biol 334:812–823. https://doi.org/10.1016/j.crvi.2011.06.005

Soest PJV, Wine RH (1967) Use of detergents in the analysis of fibrous feeds. IV. Determination of plant cell-wall constituents. J AOAC 50:50–55. https://doi.org/10.1016/j.ijhydene.2012.08.110

Soltan M, Elsamadony M, Tawfik A (2016) Biological hydrogen promotion via integrated fermentation of complex agro-industrial wastes. Appl Energy. https://doi.org/10.1016/j.apenergy.2016.10.002

Sreela-or C, Imai T, Plangklang P, Reungsang A (2011a) Optimization of key factors affecting hydrogen production from food waste by anaerobic mixed cultures. Int J Hydrog Energy 36:14120–14133. https://doi.org/10.1016/j.ijhydene.2011.04.136

Sreela-or C, Plangklang P, Imai T, Reungsang A (2011b) Co-digestion of food waste and sludge for hydrogen production by anaerobic mixed cultures: statistical key factors optimization. Int J Hydrog Energy 36:14227–14237. https://doi.org/10.1016/j.ijhydene.2011.05.145

Srikanth S, Mohan SV (2012) Regulatory function of divalent cations in controlling the acidogenic biohydrogen production process. RSC Adv 2:6576–6589. https://doi.org/10.1039/c2ra20383a

Tawfik A, El-Qelish M (2012) Continuous hydrogen production from co-digestion of municipal food waste and kitchen wastewater in mesophilic anaerobic baffled reactor. Bioresour Technol 11:270–274. https://doi.org/10.1016/j.biortech.2012.02.016

Tyagi VK, Angériz Campoy R, Álvarez-Gallego CJ, Romero García LI (2014) Enhancement in hydrogen production by thermophilic anaerobic codigestion of organic fraction of municipal solid waste and sewage sludge – optimization of treatment conditions. Bioresour Technol 164:408–415. https://doi.org/10.1016/j.biortech.2014.05.013

Valdez-Vazquez I, Poggi-Varaldo HM (2009) Alkalinity and high total solids affecting H production from organic solid waste by anaerobic consortia. Int J Hydrog Energy 34:3639–3646. https://doi.org/10.1016/j.ijhydene.2009.02.039

Valdez-Vazquez I, Riosleal E, Esparzagarcia F, Cecchi F, Poggivaraldo H (2005) Semi-continuous solid substrate anaerobic reactors for H production from organic waste: mesophilic versus thermophilic regime. Int J Hydrogen Energy 30:1383–1391. https://doi.org/10.1016/j.ijhydene.2004.09.016

Wang J, Wan W (2008) Effect of temperature on fermentative hydrogen production by mixed cultures. Int J Hydrog Energy 33:5392–5397. https://doi.org/10.1016/j.ijhydene.2008.07.010

Wang W, Xie L, Luo G, Zhou Q (2013) Enhanced fermentative hydrogen production from cassava stillage by co-digestion: the effects of different co-substrates. Int J Hydrog Energy 38:6980–6988. https://doi.org/10.1016/j.ijhydene.2013.04.004

Weiland P (2010) Biogas production: current state and perspectives. Appl Microbiol Biotechnol 85:849–860. https://doi.org/10.1007/s00253-009-2246-7

Wen Z (2004) Hydrolysis of animal manure lignocellulosics for reducing sugar production. Bioresour Technol 91:31–39. https://doi.org/10.1016/S0960-8524(03)00166-4

Wu X, Yao W, Zhu J (2010) Effect of pH on continuous biohydrogen production from liquid swine manure with glucose supplement using an anaerobic sequencing batch reactor. Int J Hydrog Energy 35:6592–6599. https://doi.org/10.1016/j.ijhydene.2010.03.097

L-J W, Kobayashi T, Kuramochi H, Li Y-Y, K-Q X (2015) Effects of potassium, magnesium, zinc, and manganese addition on the anaerobic digestion of de-oiled grease trap waste. Arab J Sci Eng 41:2417–2427. https://doi.org/10.1007/s13369-015-1879-3

Yadvika Santosh Sreekrishnan TR, Kohli S, Rana V (2004) Enhancement of biogas production from solid substrates using different techniques – a review. Bioresour Technol 95:1–10. https://doi.org/10.1016/j.biortech.2004.02.010

Yokoi H, Maki R, Hirose J, Hayashi S (2002) Microbial production of hydrogen from starch-manufacturing wastes. Biomass Bioenergy 22:389–395. https://doi.org/10.1016/S0961-9534(02)00014-4

Yuan Z, Yang H, Zhi X, Shen J (2010) Increased performance of continuous stirred tank reactor with calcium supplementation. Int J Hydrog Energy 35:2622–2626. https://doi.org/10.1016/j.ijhydene.2009.04.018

Zhang J, Wang Q (2013) Buffering and nutrient effects of white mud from ammonia–soda process on thermophilic hydrogen fermentation from food waste. Int J Hydrog Energy 38:13564–13571. https://doi.org/10.1016/j.ijhydene.2013.08.047

Zhang L, Lee Y-W, Jahng D (2011) Anaerobic co-digestion of food waste and piggery wastewater: focusing on the role of trace elements. Bioresour Technol 102:5048–5059. https://doi.org/10.1016/j.biortech.2011.01.082

Zhang J, Wang Q, Jiang J (2013) Lime mud from paper-making process addition to food waste synergistically enhances hydrogen fermentation performance. Int J Hydrogen Energy 38:2738–2745. https://doi.org/10.1016/j.ijhydene.2012.12.048

Zhang J, Wang Q, Zheng P, Wang Y (2014) Anaerobic digestion of food waste stabilized by lime mud from papermaking process. Bioresour Technol 170:270–277. https://doi.org/10.1016/j.biortech.2014.08.003

Zhou P, Elbeshbishy E, Nakhla G (2013) Optimization of biological hydrogen production for anaerobic co-digestion of food waste and wastewater biosolids. Bioresour Technol 130:710–718. https://doi.org/10.1016/j.biortech.2012.12.069

Modelling for Anaerobic Process

14

J. Rajesh Banu, S. Kavitha, and K. Tamilarasan

Abstract

Anaerobic biodegradation is an exceptional performance for generation of energy-rich biogas from different types of waste biomass. Experimental techniques were usually employed for upgrading AD processes; however, it requires building of costly trial methods and protracted experimentation. Preferably, modelling is believed to explain the various phases of biological mechanisms that include dynamic studies and rheological mechanics in various digesters that explain the environmental and functional circumstances and state. As a result of its complicated pathway and reaction mechanism, the optimising parameters of the overall process and further system improvement of this technique has to be predicted through development of numerical models for designing and prediction. In the recent times, various numerical models have developed and improved to evaluate system complication. This chapter reviews and covers the benefits and needs for anaerobic process modelling and highlights the recent developments in designing criteria. The main important predicted and organized forms are elaborated, and its efficiency was significantly discussed. The significance of the accessibility of complex models and simulation as well as prediction studies explaining its suitability in running the anaerobic digesters and consideration about instability of AD processes are explained in detail. In addition, prediction studies of digester performance and functioning state are elaborated.

Keywords

Anaerobic • Biodegradation • Designing • Digesters • Biofuels • Modelling

J. Rajesh Banu (✉) • S. Kavitha • K. Tamilarasan
Department of Civil Engineering, Regional Centre of Anna University, Tirunelveli 627007, Tamil Nadu, India
e-mail: rajeshces@gmail.com

H.J. Purohit et al. (eds.), *Optimization and Applicability of Bioprocesses*,
https://doi.org/10.1007/978-981-10-6863-8_14

14.1 Introduction

The growing attention of global warming and expanding energy outlay has led to an increasing awareness for renewable energy in modern years. The process of anaerobic degradation (AD) is employed as general method in stabilization of different types of waste resulting in energy recovery upturn in the form of energy-rich biogas (Kythreotou et al. 2014; Kavitha et al. 2014a, 2015a; Sun et al. 2015). Biogas is a fuel energy reserve that presents the outlook of substituting remnants for transport division as well as restraining greenhouse gas release concerned in weather alteration. AD is an exceptionally multifaceted process that relies on the mutual connections among the different assemblage of microbes involved (Banu et al. 2006, 2007a, b, 2015, 2016; Banu and Kaliappan 2007). A well-stable equilibrium among these microbial populations is essential for efficient degradation providing the prevalent transformation of organics to methane. During AD process, the substrate degradation and biogas generation are distressed by numerous factors such as pH, alkalinity, volatile fatty acids (VFA), substrate characterization, reactor configuration, nutritional requirements, carbon-to-nitrogen ratio, biogas production, gas liquid transfer, organic input, mixing and toxic substances.

In the current years, the relevance of AD process for the digestion of waste biomass has developed enormously, with a yearly development rate of 25%. AD process has conventionally been designated as a black box system owing to complication of the method. To comprehend and to analyse the mechanism of processes taking place inside the digestion system, modelling analysis is essential for fine digester design. Through kinetic analysis, operational conditions, stable conditions of system and characteristics of influent can be predicted (Gavala et al. 2003; Batstone 2006; Bravo et al. 2011; Kacprzak et al. 2012; Yu et al. 2013). For designing, a model must spotlight on accurateness as well as prediction (Batstone 2006). Different models have been built to give improved knowledge about the reaction affecting biodegradation system. Mathematical models can be used as useful implements for proper understanding of multifaceted process and to assist functioning and designing of the system. By predicting the performance of the system, the process can be optimised, and failure of the system can be barred. Modelling of AD process has been extensively expanded 1970s onwards (Mairet et al. 2011; Palanichamy and Palani 2014; Antonopoulou et al. 2015; Arumugam et al. 2015; Mejdoub and Ksibi 2015), from uncomplicated models (e.g. designating single restrictive reaction) (Donoso-Bravo et al. 2010) or dual reactions Bernard et al. 2001) to additional rational demonstration (e.g. the ADM1 model – (Batstone et al. 2002a) with 19 biochemical results).

Preferably, modelling is hypothetical for explaining the characteristic features of biochemical processes that include hydrodynamic reaction and growth and development of microbial community that occurs in various ecological and functional circumstances. Conversely, the chore of acquiring suitable analytical parameters is convoluted through reality that AD is a complicated energetic method that demands the intensive participation of several bacterial communities. The quantity of such population differs in an indefinite way with alteration in retaining time, substrate, temperature, digester design and other functioning parameters. The lack of information pertaining to metabolic reactions of specific bacteria involved in AD process limits the predicting influence of AD. As a result, present efforts to predict a viable AD model usually depend on hypothesis that initiates the nonidealities. The main objective of this section is to reassess the existing literature of AD system modelling as it paves the way for designing models that describe kinetic analysis of digester optimisation, operating parameters, upgrade of process and functioning. In this chapter, the main structure of the model, rate equations of kinetics, review of parameter validation and estimation of derived parameters are completely discussed.

14.2 Overview of Various AD Models

There are various existing models of AD process describing about optimisation of functioning balance and proficiency of the process. The various existing models describing about modelling of AD process are described in the following section. Table 14.1 summarizes the various models, processes and its various applications.

14.2.1 Pathways and Reaction Mechanism of Anaerobic Degradability

The biochemical process taking place during AD process is a very multifaceted reaction mediated through a group of microbes that comprise hydrolytic, hydrogen-utilizing, hydrogen-generating, acidogenic and methanogenic bacterial groups (Appels et al. 2008). These complicated sequential reactions lead to the generation and discharge of intermediary compounds, their solubilization into liquid phase and their biochemical transformation into simpler molecules. The cleavage of cell walls and lysis of biomass (waste-activated sludge (WAS) or algal cells) are required before the anaerobic degradation of waste biomass or biosolids (Merrylin et al. 2013a; Esakki Raj et al. 2013; Suresh Karthik Kumar et al. 2015; Kavitha et al. 2016a; Eswari et al. 2016). Soluble organics in the WAS (proteins, carbohydrates, lipids, cellulose) (Merrylin et al. 2013b, 2014a, b; Veera Lakshmi et al. 2014; Uan et al. 2009, 2012, 2013a, b) are biochemically hydrolysed through hydrolytic enzymes (protease, amylase, cellulase and lipase) (Kavitha et al. 2016b) to simpler molecules that can be easily moved through bacterial cell membranes and are subjected to different metabolic reactions inside the cell. They are then biochemically transformed into intermediary products such as fatty acids, amino acids and acetic acids and finally transformed into methane and carbon dioxide.

14.2.1.1 A Comprehensive Model of Anaerobic Biotransformation

The production and decomposition pathways of intermediary metabolites and its connected reactions detailed by Angelidaki et al. (1993a, b) are grouped as comprehensive model of AD process. It is an energetic outline explaining about the anaerobic biodegradation of complicated wastes and co-digestion of such wastes. This

Table 14.1 Summarization of various models, process involved and applications

S.No	Model references	Process involved	Kinetics analysis	Application
1	Hill (1982)	Acidogenesis	Monod kinetics	Accounted for inhibition of VFA
		Acetogenesis	Monod kinetics	
		Methanogenesis	Monod kinetics	
2	Bryers (1985)	Hydrolysis	First order	For insoluble organic matter
3	Costello et al. (1991)	Acidogenesis	Monod kinetics	pH products
		Acetogenesis	Monod kinetics	pH products
		Methanogenesis	Monod kinetics	pH parameters
4	Siegrist et al. (2002)	Hydrolysis	First order	Insoluble substrates
		Acetate conversion to methane	Monod kinetics	pH, free NH3
5	Angelidaki et al. (1999)	Hydrolysis	First order	For insoluble organic matter
		Acidogenesis	Monod kinetics	
		Acetogenesis	Monod kinetics	Accounted for inhibition of VFA
		Methanogenesis	Monod kinetics	
6	ADM1 model Batstone et al. (2002a)	Hydrolysis	First order	For insoluble organic matter and protein-rich substrates
				Parameter estimation and uncertainty identification of biodegradability and hydrolysis with 95% confidence interval
				Accurate prediction
7	Biochemical model Yu et al. (2013)	Hydrolysis	First-order rate-limiting approaches	Suitable for particulate organic material
		Acetogenesis		
		Methanogenesis		
		Hydrolysis	Monod kinetics	Complex substrates at high organic loading
		Methanogenesis	Monod kinetics	
				Soluble substrates
				Inhibition of VFA
8	Black-box models Holubar et al. (2000)	Degradation by microbial population	Artificial neural networks	Effluent quality studies
				Measurement of multitude of parameter to study the operating conditions and performance of anaerobic degradability treatment process

(continued)

Table 14.1 (continued)

S.No	Model references	Process involved	Kinetics analysis	Application
9	Ideal models (phenomenological models). Vavillin et al. (2007)	Anaerobic digester configurations	Fundamental kinetics and biomass balance relationships	Mixing and recirculation ratio studies in reactors
10	Nonideal models (phenomenological models). Vlyssides et al. (2007)	Reactor configurations for nonidealities	Modelling of sludge bed and blanket	Design, scale up and optimisation studies
				Effective mixing strategies studies

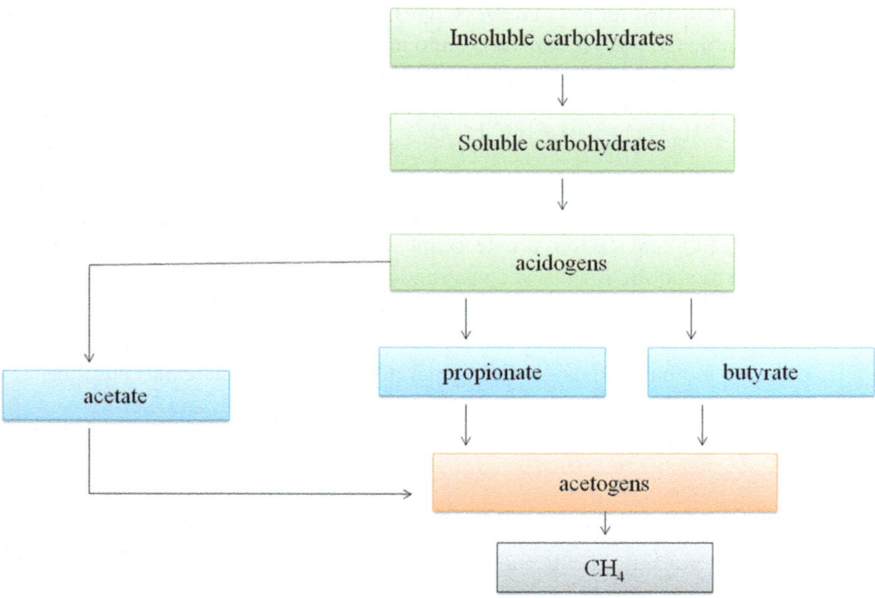

Fig. 14.1 Flowchart of Angelidaki model

comprehensive model of Angelidaki describes different phases of AD process (Fig. 14.1). The design elucidates the lytic conversion of insoluble organics. The hydrolysis process was mediated by eight groups of bacteria. In this model, proteins, CHO, NH_3, phosphates and other intermediaries such as volatile fatty acids, etc. are considered to be key model parameters. The hydrolytic action permits an energetic alteration in the reaction by changing the composition of the substrate. This has been motivated by altering the input data of substrate.

The model also elucidates the comprehensive details of pH and temperature. In this model, the following assumptions were made:

- NH_3 is believed to hamper methanogenesis.
- CH_3COOH is believed to hamper acetogenesis.

- Total volatile fatty acids are believed to hamper acidogenesis.
- The μmax of the bacterial population and ionising property of NH_3 rely on the temperature and pH of the medium.

Based on the above-mentioned assumptions, to attain realistic precision with no definite substantial experiments, the assumed substrate constituents are suitably predicted and described by the subsequent components that are normally estimated.

Carbohydrates

The model categorizes the carbohydrates ($C_6H_{10}O_5$) as dissolvable, indissolvable and inert $C_6H_{10}O_5$. Insoluble $C_6H_{10}O_5$ $(C_6H_{10}O_5)_{insol}$ is degraded through hydrolytic enzyme to soluble $C_6H_{10}O_5$ $(C_6H_{10}O_5)_{sol}$, and inert $C_6H_{10}O_5$ $(C_6H_{10}O_5)_{inert}$ is explained in the following equation (Angelidaki et al. 1993a):

$$\left(C_6H_{10}O_5\right)_{insol} \rightarrow F_c\left(C_6H_{10}O_5\right)_{sol} + \left(1-F_c\right)\left(C_6H_{10}O_5\right)_{inert} \quad (14.1)$$

where F_c is the portion of $C_6H_{10}O_5$ that can be hydrolyzable.

$(C_6H_{10}O_5)_{sol}$ is then decomposed to $C_2H_4O_2$, $C_3H_6O_2$ and $C_4H_8O_2$ in the acidogenic pace. $C_3H_6O_2$ and $C_4H_8O_2$ are decomposed to $C_2H_4O_2$ and H_2 in the acetogenic pace. The two metabolic reactions concerned in the formation of CH_4 in the methanogenesis pace are as follows:

- H_2-utilizing methanogenesis – CH_4 generated from propionate and butyrate
- Acetoclastic methanogenesis – $C_2H_4O_2$ decomposed to CH_4 and CO_2

Proteins

The gelatin ($CH_{2.03}O_{0.6}N_{0.3}\ S_{0.001}$) is assumed to be a proteinaceous molecule in this model. Proteins are initially decomposed to amino acids and are explained in the following equation:

$$\left(\text{Proteins}\right)_{insol} \rightarrow F_p\left(\text{Amino acids}\right) + \left(1-F_p\right)\left(\text{Protein}\right)_{inert} \quad (14.2)$$

where F_p is a portion of protein that can be hydrolyzable.

Amino acids are further decomposed to VFA through acidogenic bacteria. Acids formed during this phase include $C_2H_4O_2$, $C_3H_6O_2$, $C_4H_8O_2$ and $C_5H_{10}O_2$.

- Acetoclastic methanogenesis – $C_2H_4O_2$ decomposed to CH_4 and CO_2
- Acetogenesis and H_2 utilizing – CH_4 generated from propionate and methanogenesis H_2

This model explains the performance of reactors fed with manure. VFA build-up diminishes the pH and leads to a decrement in free NH_3 and hinders the methanogenesis. The system is thus auto controlled; if not, the extent of the interruption is bigger than the process can endure. When this happens, the pH falls considerably, and the drop leads to digester collapse.

14.2.2 Anaerobic Digestion Model No. 1 (ADM1)

The most extensively employed model for AD process is the ADM1, designed through a chore group of professionals in International Water Association (IWA) (Batstone et al. 2002a; Fedorovich et al. 2003). The greatly controlled model contains several phases explaining chemical and physical reactions; it includes the four phases: hydrolysis, acidogenesis, acetogenesis and methanogenesis, and the variation of above-mentioned phases for input substrates. It is extremely precise; however, it needs input of numerous factors and an elevated extent of complication. This is applied as an outline model allowing researchers to contemplate on change for definite reasons. This comprises the kinetic analysis describing the breakdown of organic materials to CHO, other organics and subsequent degradation of the organics to simpler ones. The breakdown of particulate organic matter in a process is often rate restricting and assumed to follow first-order kinetic analysis. In acidogenic and acetogenic phases, the dynamics of $C_2H_4O_2$, $C_3H_6O_2$ and $C_4H_8O_2$ are involved. In methanogenic phase, conversion of $C_2H_4O_2$ and H_2/CO_2 to biogas is involved. The inhibitory behaviour of the metabolic reaction of NH_3, pH, $C_2H_4O_2$, H_2 and N constraints is involved. The modelling moreover explains gaseous-aqueous transformation and ion organization and separation. This model has been broadly employed as well as authenticated to simulate AD process of many solubles such as industrialized waste waters (Batstone et al. 2002b), waste-activated sludge (Blumensaat and Keller 2005) and municipal sludge (Batstone et al. 2002a), and olive mill dissipates (Feng et al. 2006).

14.2.2.1 ADM1 Implementation and Simulating Pathway

ADM1 implementation includes a scheme of discrepancy equations which comprises 32 quantitative state parameters and a group of discrepancy and numerical equations, and 26 state parameters and 8 inherent numerical equations are employed (Batstone et al. 2002a, b). Examples for such implementation are formation of lactic acid, reduction of sulphate and nitrate and competitive utilization of hydrogen and carbon dioxide (Blumensaat and Keller 2005). ADM1 is not broadly applied owing to the devoid of execution in an industrial programme and because of the limited investigation (Batstone et al. 2002a). Batstone et al. (2002a) have described two instances of ADM1 demonstration that elucidates the benefits of modelling. Alteration of the system in the pilot scale was assessed, and calculations are observed to have high precision. ADM1 model has implementation:

- Implementation of paradigm model in a mixed-tank reactor which was authorized by the ADM1 systematic and technological information (Batstone et al. 2002b) regularly to observe precise methods
- Implementation in hypothetical investigation for innovative dispersed factors

The outlined ADM1 model suggests chief progression linked with transforming multifaceted soluble organics into methane, carbon dioxide and static derivatives (Fig. 14.2). The ADM1 model predicts the decomposition of multifaceted

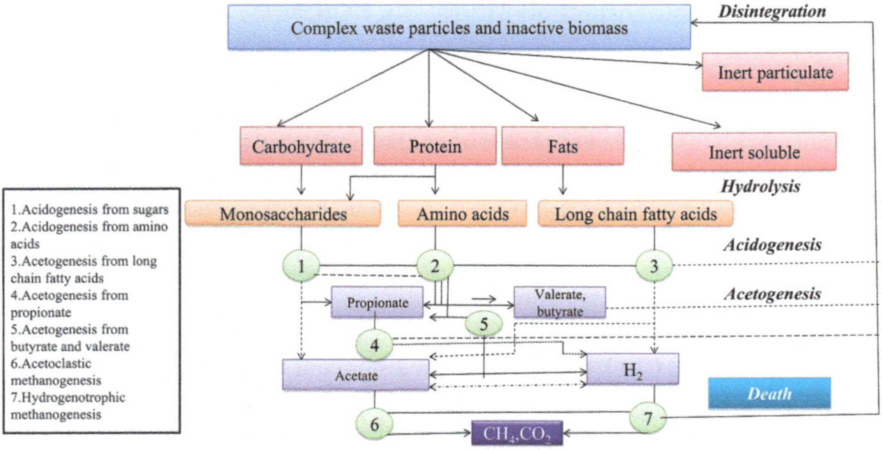

Fig. 14.2 Main pathways of ADM1 modelling

components into soluble organic matters. The decomposed components are further degraded into very simpler molecules. VFA and hydrogen gas are produced through anaerobic fermentation of proteins and CHO. $C_2H_4O_2$ and hydrogen gas are in addition produced by means of oxygenation of LCFA or transformation of $C_3H_6O_2$, $C_4H_8O_2$ and $C_5H_{10}O_2$ through acetogenesis. Methane is generated by breakage of acetate by acetoclastic methanogenesis as well as by conversion of carbon dioxide via hydrogen through hydrogenotrophic methanogenesis. The outlined ADM1 model suggests chief progression linked with transforming multifaceted soluble organics into methane, carbon dioxide and static derivatives (Fig. 14.2) (Batstone et al. 2002a).

Complicated models such as ADM1 are compatible for system designing. On the other hand, it is significantly inadequate while it is employed for system management and optimisation. The ADM1 model presumes to suggest a stable capacity, entirely blended process. As a result, the ADM1 model was designated as an extensively employed, fundamental line or consistent model that in addition meets other obligatory alterations of definite processes.

14.2.3 Biochemical Kinetics

The hypothesis of uninterrupted culturing of microbes has been formerly employed to precisely symbolize kinetic analysis of biological methods (Metcalf and Eddy 2003; Lubken et al. 2010). Biokinetics is performed mainly on the basis of fundamental growth kinetic analysis of microbes and substrate utilization kinetic analysis. Organic nutrients are presumed to be substances which are provided at surplus quantity. Subsequent expressions are various general modelling terms explaining biodegradation and successive transformation of biodegradable substrates to simpler molecules and generation of methane.

First-order kinetic model (Yu et al. 2013):

$$\mu = \frac{K_X, \max X}{X_0 - X} - b\frac{-\mathrm{d}X}{\mathrm{d}t} = K_X, \max XS = \frac{S_0}{1 + K_X, \max Xt_{\mathrm{XR}}} \tag{14.3}$$

Monod-type kinetics (Vavilin et al. 2008):

$$\mu = \frac{\mu_{\max}}{[X] + [\mathrm{Ks}]} \tag{14.4}$$

where μ = definite growth rate, $\mu_{\max}$ = maximal definite growth rate, $[X]$ = substrate quantity and Ks is the substrate saturation constant Haldane kinetics (Yu et al. 2013).

$$\mu = \frac{\mu \max}{1 + \frac{\mathrm{Ks}}{[S]} + \frac{[S]}{K_1}} \tag{14.5}$$

These expressions systematically signify the basic structure for AD process modelling.

14.2.3.1 Rate-Limiting Approaches

AD process includes four crucial biochemical pathways and involves a series of reactions. Sequential AD process relies on the assumption that one phase was much sluggish than the others. The slower phase was referred to as rate-determining or rate-restricting phase. The various rate-restricting approaches of AD process are investigated and are explained elaborately in the following section.

14.2.3.1.1 Hydrolysis

Many AD rate-restricting models directed towards the rate-determining degradation of complicated soluble organics such as waste-activated sludge (Vavilin et al. 2008; Gopi et al. 2012; Ma et al. 2013a, b; Gopi et al. 2015). Particulate organics have to be degraded to soluble ones proficient to be energetically or inactively transformed through the cell walls prior to microbial metabolic reaction. Hydrolytic process can be explained in two modes of actions (Batstone et al. 2002a; Chen et al. 2009):

- The biomass cells ooze out extracellular enzymes into the liquid phase, and these enzymes in turn are adsorbed on the soluble component or act with a soluble organics.
- The microbial biomass sticks to the component and utilizes liquefied substances produced by processes mediated by exoenzymes that are generally secreted.

The following are the expressions that describe the kinetic modelling of hydrolysis reactions that are already discussed in the literature:

- The first-order kinetics of various organics breakdown: The highest predictable representation to suggest lysis process entails first-order model on biodegradable organic components basis (Kavitha et al. 2015a):

$$\frac{dX}{dt} = -KS\frac{dC}{dt} = \alpha KS \tag{14.6}$$

where C is the amount of produced product; S is the reactant quantity; K is the reaction constant; and α is the substrate-to-product transformation coefficient. First-order kinetics ignores biodegradability-allied analysis and employs simply when the rate restriction is owing to the particulate substrate surface.

- First-order kinetic model (non-linear regression): This model can be applied to asses biodegradability and hydrolysis rate constant (Kavitha et al. 2015a, c):

$$\text{Methane}_{(t)} = \text{Methane}_{(\text{fd})} \cdot \left(1 - e\left(^{-k}{}_{\text{hyd}}{}^{t}\right)\right) \tag{14.7}$$

where $\text{Methane}_{(t)}$ is the methane yield at degradation period, t is the days, $\text{Methane}_{(\text{fd})}$ is CH_4 generation through organic degradation (portion of the organic substances which can be transformed to CH_4), k_{hyd} is degradation constant (day^{-1}) and t is the degradation period (days). The model implemented in Matlab 2012a version and the respective coding used for prediction and analysis are illustrated as Figs. 14.3 and 14.4.

The parameter assessment and improbability evaluation used were calculated with a 95% confidence region for precision investigation and factor improbability analysis.

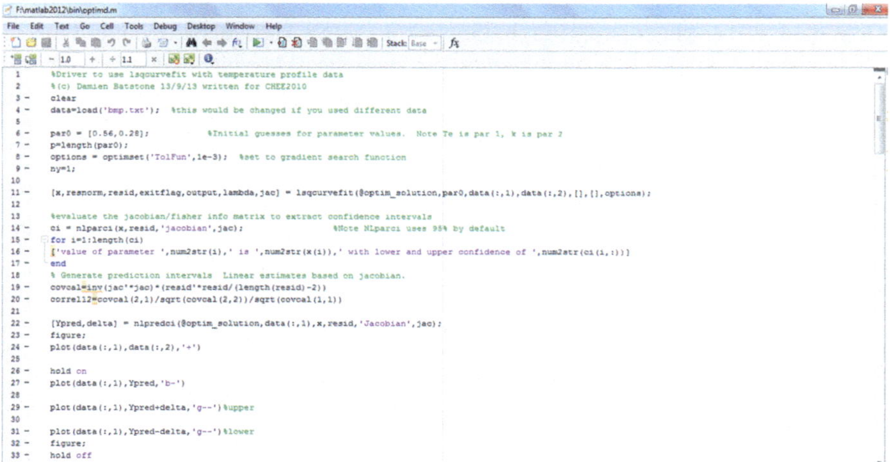

Fig. 14.3 Matlab code to obtain biodegradability parameters (Optim)

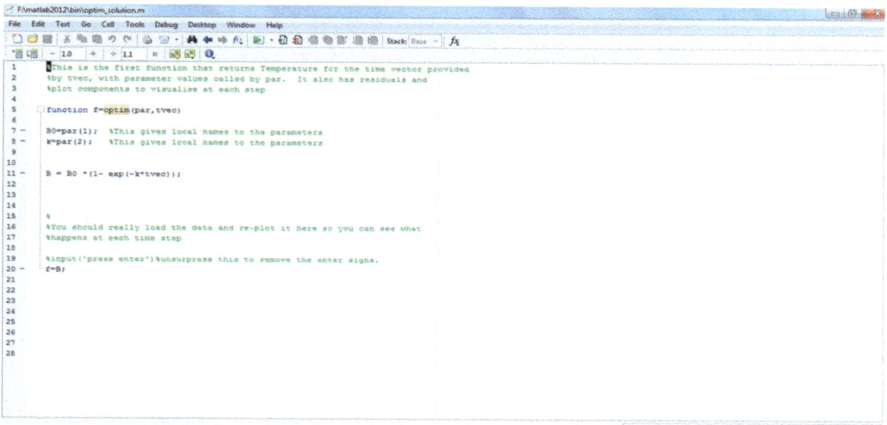

Fig. 14.4 Matlab code that gives functional input parameters (Optim solution)

14.2.4 Modified Gompertz Model

The biodegradability process assessment and evaluation and the biogas production potential can be predicted through modified Gompertz model (Budiyono et al. 2010; Kavitha et al. 2014a, b). Assuming the biogas production to the growth of methanogenic microbes, many of the researchers have used this model to calculate the biogas production. Kafle et al. (2012) have proposed that the estimation of biogas yield through this modelling resulted in appropriate error of about 0.7–13.7% and was found to be more accurate. The modified Gompertz expression employed to assess the biogas generation (Patil et al. 2012) is as follows:

$$B_t = B * \exp\left[-\exp\left[R_b / B * \exp(\lambda - t) + 1\right]\right] \tag{14.8}$$

where B is the biogas generation potential, R_b is the ultimate biogas generation extent, and λ is the lag period (minimal period spent to generate gas). The constants B, R_b, and λ can be estimated through the non-linear regression modelling in Polymath software. Figure 14.5 depicts the biogas production of various samples modelled through modified Gompertz expression and fit modelling. This equation was considered to be the finest to illustrate numerical accurateness and exactitude as compared to other functional parameters (Adiga et al. 2012). This expression is used to circumvent correlation difficulties at primary time.

14.2.5 Siegrist Model

Siegrist et al. (2002) have invented a shortened model named Siegrist model. The major discrepancies are segregation of $C_5H_{10}O_2$ and $C_4H_8O_2$ as inputs. Siegrist model parameters were executed on the basis of performance of experimentation. This modelling is developed with laboratory experiments and corroborated by

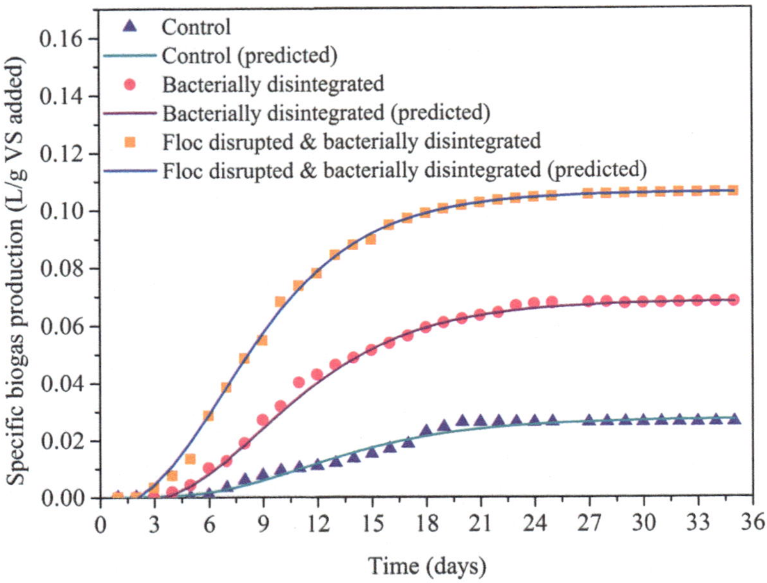

Fig. 14.5 Plot showing Gompertz model fitting of various samples for biogas production

pilot-level experimentation. Siegrist et al. (2002) performed the modelling with municipal sludge that is a mixture of primary, secondary and tertiary sludge taken from wastewater treatment plant. Biodegradable organics that are simulated in this design are fractionated into simpler molecules. The amount of feed may still be described in stoichiometry of disintegration.

14.2.5.1 Model Assumptions

- Rise in temperature can raise the upholding and therefore modify the stoichiometry.
- The K_La value in the mass transfer theory relies only upon the temperature, however, in certainty influenced by several factors.
- The digester is believed to be entirely suspended, and the sludge retention time is equivalent to the hydraulic retention time.

14.2.5.2 Limitations

- Due to alteration in biomass communities, the hasty alteration in temperature can't be calculated.
- Inclusion of size distribution study of feed in the model is denial.

Framework of Anaerobic Process Modelling

The outline for AD process modelling is a six-step procedure described as follows:

1. **Model selection step** – an exchange must be processed among precision and model complication.
2. **Selection of parameter for calibration** – this is done based on the assessment of identifiability of the precise factors. Lack of examining the parameter leads to parameter sensitivity analysis.
3. **Data collection** – this specifies experimental measurements.
4. **Accuracy measurements** – for this uncertainty is determined.
5. **Accurate plan of research** – this is to assess the factors.
6. **Confirmation procedure** – this is the calibration of achieved data described on the basis of correlation coefficient (R^2).

14.2.6 Anaerobic Digestion Reactors

Various reactors are employed in the degradation process. The perpendicular mixed-tank reactor is the most predictable digester unit employed in damp fermentation studies. The major drawback of this reactor is washing of untreated materials, the live microbes and in addition high-energy consumption (Nielsen and Villadsen 2008). Recently, various kinds of digesters have been configured to overcome the above-mentioned limitations. There are various models simulating the anaerobic digesters, and they have been described in the subsequent section:

14.2.6.1 Black-Box Models

Various types of reactors are used to enhance the performance of AD system. Wet fermentation techniques are assessed congenitally by incessant stirred-tank digester; also it produces enormous quantity of biogas. As a result, it can be relevant in biogas plants in current years. A disadvantage of the system is the elimination of some amount of active microbes at elevated organic load. This limitation occurs majorly due to insufficient mixing and more energy utilization. Batch and plug-flow style digesters have similar schematic features to activate AD process. In the present years, existing reactor configurations, suitable functioning and energy-efficient production are modified through a different sophisticated research (Perendeci et al. 2008; Yu et al. 2013). In AD system, the decomposition process carried out through microbial action is modelled through totally controlled network investigation of non-linear mathematical models called black-box models (Holubar et al. 2000). Without existing information of the model, prediction of correlation among key variables in computer structural design system is not easy. This network system needs all available input variables. This neural structural design scheme comprises of five essential divisions (Qdais et al. 2010). These divisions are essential for fuzzy logic system (Cakmakci 2007; Behera et al. 2015). Fine-tuned input and output variables are considered essential to run perfect AD process. The process performance of output variables received from regularly noticed data is also integrated in outline database arrangement.

Another imperative factor to assess the anaerobic biodegradability is the biochemical methane potential (BMP) determination (Uma et al. 2012a, b, 2013a, b;

Merrylin et al. 2013a; Poornima et al. 2014; Gayathri et al. 2015; Godvin Sharmila et al. 2015; Sowmya et al. 2015) and can be simulated using the black-box model. BMP is a fascinating method to obtain biofuel like methane as a sustainable energy source and also enhance the competence of the biodegradation (Angelidaki et al. 2009; Kavitha et al. 2013, 2014a, b, c, 2015a, b, c, d, 2016a, c, d, e, f; Ebenezer et al. 2015a, b). The major defect that diminishes the proficiency of AD process is the foaming. This is considered to be the major limitation in AD system (Wang et al. 2013). This foaming occurs when the digester is fed with activated sludge that comes from secondary treatment process. Foaming may lead to drop in methane production efficiency and liquefaction of suspended solid particles and troubling the operational function of equipment like recent sensors and propellers. The several factors that were the root cause of foaming are listed below: (1) Floc-forming microorganisms reside in activated sludge, (2) VFA, (3) proteins, (4) inorganic matter and (5) temperature (Ganidi et al. 2009). Ultimately, the black-box model approach is employed to assess the extent of foaming. The above said problems are considered to be the main root cause for generation of foaming in a digestion unit of AD process. At the present time, newest scanning image processing set-up was developed to visualize the precise properties of microbial population that are responsible for the formation of foaming in the digestion unit. Various partial expressions are employed to predict and design the settleability nature of floc-forming microbial biomass in the digester through architectural computer network system (Jenne et al. 2006; Shen et al. 2007; Bornhoft et al. 2013).

14.2.6.2 Phenomenological Models

14.2.6.2.1 Ultimate Models

Many numerical modelling expressions are designed for different ultimate digesters, for instance, continuous stirred-tank reactor (CSTR) as well as plug-flow reactor (PFR) based on mechanistic modelling and microbial community accumulation equilibrium. CSTR reactor modelling was employed to simulate numerous mathematical computation mechanisms in complicated AD process (Batstone et al. 2002b, 2015; Lauwers et al. 2013). Through this modelling, the enhancement of mixing in CSTR system process and increment of the extent of hydrolysis and methanogenesis phases in the process of anaerobic digestion with reactor encompassing either continuous or fed batch (Taricska et al. 2007; Vavilin et al. 2008; Terashima et al. 2009; Yu et al. 2011; Wu 2012a, b; Grant and Lawrence 2014) can be studied. Anaerobic fluidized-bed bioreactors (AFBR) are developed on the basis of plug flow and CSTR digester configurations. The working trend of the CSTR digester was assumed that the remixing proportion of the solids in the digester was much greater. Because of the above said reasons, trouble may occur, by reducing the recycling proportion thereby minimizing the efficiency of bed in AFBR digesters. In current years, several abundant estimation models are existing to improve the bed feature and remixing proportion in AFBR reactors. With these ideal conditions, several high-rate anaerobic digesters are developed. Mostly anaerobic degradation of $C_6H_{12}O_6$ will be functioned through periodic anaerobic baffled reactor (PAFB). This

PAFB model reactor comprises of sequential connections of complete stirred tank reactors. Another model simulated in moving bed reactors combined with complete stirred tank reactors containing well-densified municipal solid waste (MSW) through Monod kinetics and by employing Fick's law.

14.2.6.2.2 Non-ultimate Models

Nonideal elevated rate digesters are developed by taking into account the characteristic features and digester configurations of complete stirred tank and plug-flow reactors. There are various zones that are built in upflow anaerobic sludge blanket (UASB) digester. Dead zone in UASB can be employed for modelling purposes, whereas the flow in reactor should surpass the CSTR and PFR dead zone capacity. The UASB reactor is configured with three different parts that include sludge bed, layer and sludge suspension unit. The inflow liquid passes through the zone present in anaerobic sludge blanket. If this inflow liquid contains enormous quantity of biomass, then this will affect the working proficiency of digester process. These assumptions are taken into account while modelling. The CSTR have the compartment of sludge bed; blanket and PFR have the compartment of sludge settler. In such case, sludge bed and blanket in UASB reactor were modelled by nonideal condition, but the reactor configurations are fully considered by serially connected CSTR model with departed sector and bypass flow. On whole, the UASB reactor is separated by two important regions such as ideal mixers and recirculation tank which were modelled through nonideal modelling. Most of the modelling studies are contributed in the field of sludge bed and blanket (Vlyssides et al. 2007). To upgrade and optimise the working conditions of industrial digesters in field-oriented application are mainly done in the design of CSTR and PFR only. These nonideal models were simulated by various alterations and sequence-linked axial-dispersion models. In such conditions, the working function of digester is not much greater. This is entirely focussed on the retention time of liquid and transportation of liquid from inlet to outlet. The above said data analysis is sufficient to shun short circuits, dead zones and remixing proportion.

14.3 Conclusion

This chapter compiles the description of different design and prediction studies that were developed to evaluate the performance of different digester facilities during AD process. Relatively ease and practicable pace-restricting designs were emphasized in detail. For instance, the description of rate-restricting designs and its wide application concerning about the failure of system performance through the accumulation of particular nutrients were elucidated in detail. In addition, these designs were employed for recognizing pace-restricting stages which were complicated for degradation of complex substances. The recognition of those hydrolytic compounds was imperative in evaluating reactor potential. ADM1 embodies the widespread design configuration dealing with anabolic and catabolic reactions that happen within the cell and relations among anaerobes. Besides, this design could be modified for a unique AD technique concerning various anabolic and catabolic sequential processes. In order to study the complicated processes in AD, black-box models were employed for designing of treatment facilities in addition to ADM1. Since these designs ignored configuration methods, it could be more suitable for control of AD processes. Apart from these, other models afford the basis for designing various mode reactors. Non-ultimate phenomenological design was altered to explain the behaviour of fluids inside the reactors.

Opinion

This paper reviews various anaerobic models and describes that the models developed hitherto deal with numerous features that are chiefly significant for explaining the performance of anaerobic digesters. These models are essential in simulating the digester function, malfunction and probable solutions. These models can be used to design and effectively run the digesters. It was understand that recognition of rate-limiting phases at various digesting processes was hard for digestion of complicated substrates. So design and prediction studies are imperative for better understanding of the digestion process, scale up studies and run the digesters in a profitable manner. It was also observed that recognition of intermediary substances was very imperative to measure the capacity of the reactors. In addition, ADM1 corresponds to the presently most complete design of biodegradability studies that function to be a foundation to upcoming progress of modelling. In addition, more innovative aspects on the microbial community analysis will lead to the incorporation of these technologies in numerical modelling. Fluid behaviour and transportation of influent and effluent are the essential factors that possibly will entail more design analysis. However, as per our view, some essential parameters still needed to be explored such as hydrolysis, pH, temperature, inhibitors and heterogenous processes.

Acknowledgement The authors are appreciative to the DST, India, for affording economic support for this project (SR/FTP/ETA-0021/2010) in aid of Young Scientist Scheme.

References

Adiga S, Ramya R, Shankar BB, Patil JH, Geetha CR (2012) Kinetics of anaerobic digestion of water hyacinth, poultry litter, cow manure and primary sludge: a comparative study. In: Proceeding of the 2nd international conference on biotechnology and environment management. https://doi.org/https://doi.org/10.7763/IPCBEE.2012

Angelidaki I, Ellegaard L, Ahring BK (1993a) A mathematical model for dynamic simulation of anaerobic digestion of complex substrates: focusing on ammonia inhibition. Biotechnol Bioeng 42:159–166. https://doi.org/10.1002/bit.260420203

Angelidaki I, Ellegaard L, Ahring BK (1993b) A comprehensive model for anaerobic bioconversion of complex substrates to biogas. Biotechnol Bioeng 63:363–372. https://doi.org/10.1002/(SICI)1097-0290(19990505)63:33.0.CO;2-Z

Angelidaki I, Ellegaard L, Ahring BK (1999) A comprehensive model of anaerobic bioconversion of complex substrates to biogas. Biotechnol Bioeng 63:363–372

Angelidaki I, Alves M, Bolzonella D, Borzacconi L, Campos J, Guwy A (2009) Defining the biomethane potential (BMP) of solid organic wastes and energy crops: a proposed protocol for batch assays. Water Sci Technol 259:927–934. https://doi.org/10.2166/wst.2009.040

Antonopoulou G, Alexandropoulou M, Lytras C, Lyberatos G (2015) Modeling of anaerobic digestion of food industry wastes in different bioreactor types. Waste Biomass Valor 6:335–341. https://doi.org/10.1007/s12649-015-9362-7

Appels L, Baeyens J, Degreve J, Dewil R (2008) Principles and potential of the anaerobic digestion of waste activated sludge. Prog Energy Combust Sci 34:755–781. https://doi.org/10.1016/j.pecs.2008.06.002

Arumugam T, Parthiban L, Rangasamy P (2015) Two-phase anaerobic digestion model of tannery solid waste: experimental investigation and modeling with ANFIS. Arab J Sci Eng 40:279–288. https://doi.org/10.1007/s13369-014-1408-9

Banu RJ, Kaliappan S (2007) Treatment of tannery wastewater using hybrid upflow anaerobic sludge blanket reactor. J Environ Eng Sci 6:415–421. doi:10.1139/s06-063

Banu RJ, Kaliappan S, Beck D (2006) High rate anaerobic treatment of sago wastewater using HUASB with PUF as carrier. Int J Environ Sci Technol 3:69–77. https://doi.org/10.1007/BF03325909

Banu RJ, Kaliappan S, Yeom IT (2007a) Treatment of domestic wastewater using upflow anaerobic sludge blanket reactor. Int J Environ Sci Technol 4:363–370. https://doi.org/10.1007/BF03326295

Banu RJ, Kaliappan S, Yeom IT (2007b) Two-stage anaerobic treatment of dairy wastewater using HUASB with PUF and PVC carrier. Biotechnol Bioprocess Eng 12:257–264. https://doi.org/10.1007/BF02931101

Banu RJ, Arulazhagan P, Adish Kumar S, Kaliappan S, Lakshmi AM (2015) Anaerobic co-digestion of chemical- and ozone-pretreated sludge in hybrid upflow anaerobic sludge blanket reactor. Desalin Water Treat 54:3269–3278. https://doi.org/10.1080/1944399.2014.912156

Banu RJ, Ushani U, Merrylin J, Kaliappan S (2016) Evaluation of operational parameters for biodegradation of bacterially disintegrated sludge. Desalin Water Treat 57:25018–25027. https://doi.org/10.1080/19443994.2016.1150888

Batstone DJ (2006) Mathematical modelling of anaerobic reactors treating domestic waste water: rational criteria for model use. Rev Environ Sci Biotechnol 5:57–71. https://doi.org/10.1007/s11157-005-7191-z

Batstone DJ, Keller J, Angelidaki I, Kalyuzhnyi SV, Pavlostathis SG, Rozzi A, Sanders WTM, Siegrist H, Vavilin VA (2002a) The IWA anaerobic digestion model no 1 (ADM1). Water Sci Technol 45:65–73. https://doi.org/10.1.1.466.9027

Batstone DJ, Keller J, Angelidaki I, Kalyuzhnyi SV, Pavlostathis SG, Rozzi A, Sanders WTM, Siegrist H, Vavilin VA (2002b) Anaerobic digestion model no. 1 (ADM1). In: Scientific and technical report No 13 IWA task group for mathematical modelling of anaerobic digestion processes (ed). IWA Publishing, London, p 80. ISBN:1900222787, 9781900222785

Batstone DJ, Puyol D, Alsina XF, Rodriguez J (2015) Mathematical modelling of anaerobic digestion processes: applications and future needs. Rev Environ Sci Biotechnol 14:595–613. https://doi.org/10.1007/s11157-015-9376-4

Behera SK, Meher SK, Park HS (2015) Artificial neural network model for predicting methane percentage in biogas recovered from a landfill upon injection of liquid organic waste. Clean Techn Environ Policy 17:443–453. https://doi.org/10.1007/s10098-014-0798-4

Bernard O, HadjSadok Z, Dochain D, Genovesi A, Steyer J (2001) Dynamical model development and parameter identification for an anaerobic wastewater treatment process. Biotechnol Bioeng 75:424–438. https://doi.org/10.1002/bit.10036

Blumensaat F, Keller J (2005) Modelling of two-stage anaerobic digestion using the IWA anaerobic digestion model no 1 (ADM1). Water Res 39:171–183. https://doi.org/10.1016/j.watres.2004.07.024

Bornhoft A, Rauschenbach RH, Sundmacher K (2013) Steady-state analysis of the anaerobic digestion model no 1 (ADM1). Nonlinear Dynam 73:535–549. https://doi.org/10.1007/s1107-0127-0807-x

Bravo AD, Mailier J, Martin C, Rodriguez J, Lara CAA, Wouwer AV (2011) Model selection, identification and validation in anaerobic digestion: a review. Water Res 45:5347–5364. https://doi.org/10.1016/j.watres.2011.08.059

Bryers JD (1985) Structured modelling of the anaerobic digestion of biomass particulates. Biotechnol Bioeng 27:638–649. https://doi.org/10.1002/bit.260270514

Budiyono IN, Widiasa J, Sunarso S (2010) The kinetic of biogas production rate from cattle manure in batch mode. Int J Chem Biol Eng 3:39–44. doi:10.1.1.309.2870

Cakmakci M (2007) Adaptive neuro-fuzzy modelling of anaerobic digestion of primary sedimentation sludge. Bioprocess Biosyst Eng 30:349–357. https://doi.org/10.1007/s00449-0032-0131-2

Chen Z, HU D, Zhang Z, Ren N, Zhu H (2009) Modelling of two- phase anaerobic process treating traditional Chinese medicine wastewater with the IWA anaerobic digestion mod model no 1. Bioresour Technol 100:4623–4631. https://doi.org/10.1016/j.biotech.2009.04.066

Costello DJ, Greenfield PF, Lee PL (1991) Dynamic modelling of a single-stage high-rate anaerobic reactor – I. Model derivation. Water Res 25:847–858. https://doi.org/10.1016/004313543690166-N

Donoso-Bravo A, Perez-Elvira S, Fdz-Polanco F (2010) Application of simplified models for anaerobic biodegradability tests. Evaluation of pretreatment processes. Chem Eng J 160:607–614. https://doi.org/10.1016/j.cej.2010.03.082

Ebenezer VA, Arulazhagan P, Adish Kumar S, Yeom IT, Rajesh Banu J (2015a) Effect of deflocculation on the efficiency of low-energy microwave pretreatment and anaerobic biodegradation of waste activated sludge. Appl Energy 145:104–110. https://doi.org/10.1016/j.apenergy.2015.01.133

Ebenezer VA, Kaliappan S, Adish Kumar S, Yeom IT, Rajesh Banu J (2015b) Influence of deflocculation on microwave disintegration and anaerobic biodegradability of waste activated sludge. Bioresour Technol 185:194–201. https://doi.org/10.1016/j.biortech.2015.02.102

Esakki Raj S, Rajesh Banu J, Kaliappan S, Yeom IT, Adish Kumar S (2013) Effects of side-stream, low temperature phosphorus recovery on the performance of anaerobic/anoxic/oxic systems integrated with sludge pretreatment. Bioresour Technol 140:376–384. https://doi.org/10.1016/j.biortech.2013.04.061

Eswari P, Kavitha S, Kaliappan S, Yeom IT, Rajesh Banu J (2016) Enhancement of sludge anaerobic biodegradability by combined microwave-H_2O_2 pretreatment in acidic conditions. Environ Sci Pollut Res 23:13467–13479. https://doi.org/10.1007/s11356-016-6543-2

Fedorovich V, Lens P, Kalyuzhnyi S (2003) Extension of anaerobic digestion model no. 1 with processes of sulfate reduction. Appl Biochem Biotechnol 109:33–45. https://doi.org/10.1385/ABAB:109:1-3:33

Feng Y, Behrendt J, Wendland C, Otterpohl R (2006) Parameter analysis of the IWA anaerobic digestion model no. 1 for the anaerobic digestion of black water with kitchen refuse. Water Sci Technol 54:139–147. https://doi.org/10.2166/wst.2006.535

Ganidi N, Tyrrel S, Cartmell E (2009) Anaerobic digestion foaming causes – a review. Bioresour Technol 100:5546–5554. https://doi.org/10.1016/j.biortech.2009.06.024

Gavala HN, Angelidaki I, Ahring BK (2003) Kinetics and modeling of anaerobic digestion process. In: Biomethanation I. Springer, Berlin, pp 57–93. ISBN: 978-3-540-45839-5. https://doi.org/10.1007/3-540-45839-5_3

Gayathri T, Kavitha S, Adish Kumar S, Kaliappan S, Yeom IT, Rajesh Banu J (2015) Effect of citric acid induced deflocculation on the ultrasonic pretreatment efficiency of dairy waste activated sludge. Ultrason Sonochem 22:333–340. https://doi.org/10.1016/j.ultsonch.2014.048017

Godvin Sharmila V, Kavitha S, Rajashankar K, Yeom IT, Rajesh Banu J (2015) Effects of titanium dioxide mediated dairy waste activated sludge deflocculation on the efficiency of bacterial disintegration and cost of sludge management. Bioresour Technol 197:64–71. https://doi.org/10.1016/j.biortech.2015.08.038

Gopi KS, Merrylin J, Kaliappan S, Adish Kumar S, Yeom IT, Rajesh Banu J (2012) Effect of cation binding agents on sludge solubilisation potential of bacteria. Biotechnol Bioprocess Eng 17:346–352. https://doi.org/10.1007/s12257-011-0465-0

Gopi KS, Arulazhagan P, Kavitha S, Adish Kumar S, Rajesh Banu J (2015) Evaluation of operational parameters for semi-continuous anaerobic digester treating pretreated waste activated sludge. Desalin Water Treat 4:1–8. https://doi.org/10.1080/19443996.2015.1029553

Grant WD, Lawrence TM (2014) A simplified method for the design and sizing of anaerobic digestion systems for smaller farms. Environ Dev Sustain 16:345–360. https://doi.org/10.1007/s10668-013-9480-y

Hill DT (1982) A comprehensive dynamic model for animal waste methanogenesis. Trans ASAE 25:1374–1380. https://doi.org/10.13031/2013.33730

Holubar P, Zani L, Hagar M, Froschl W, Radak Z (2000) Modeling of anaerobic digestion using self-organizing maps and artificial neural networks. Water Sci Technol 41:149–156. doi:41/12/149

Jenne R, Banadda E, Gins G, Deurinck J, Smets I, Geeraerd A (2006) Use of image analysis for sludge characterization studying the relation between floc shape and sludge settleability. Water Sci Technol 54:167–174. https://doi.org/10.2166/wst.2006.384

Kacprzak A, Krzystek L, Pazdzior K, Ledakowicz S (2012) Investigation of kinetics of anaerobic digestion of canary grass. Chem Pap 66:550–555. https://doi.org/10.2478/s11696-012-0136-4

Kafle GK, Kim SH, Sung KI (2012) Ensiling of fish industry waste for biogas production: a lab scale evaluation of Biochemical Methane Potential (BMP) and kinetics. Bioresour Technol 127:326–336. https://doi.org/10.1016/j.biortech.2012.09.032

Kavitha S, Adish Kumar S, Yogalakshmi KN, Kaliappan S, Rajesh Banu J (2013) Effect of enzyme secreting bacterial pretreatment on enhancement of aerobic digestion potential of waste activated sludge interceded through EDTA. Bioresour Technol 150:210–219. https://doi.org/10.1016/j.biortech.2013.10.021

Kavitha S, Adish Kumar S, Kaliappan S, Yeom IT, Rajesh Banu J (2014a) Improving the amenability of municipal waste activated sludge for biological pretreatment by phase-separated sludge disintegration. Bioresour Technol 169:700–706. https://doi.org/10.1016/j.biortech.2014.07.065

Kavitha S, Jayashree C, Adish Kumar S, Kaliappan S, Rajesh Banu J (2014b) Enhancing the functional and economical efficiency of a novel combined thermo chemical disperser disintegration of waste activated sludge for biogas production. Bioresour Technol 173:32–41. https://doi.org/10.1016/j.biortech.2014.09.078

Kavitha S, Jayashree C, Adish Kumar S, Yeom IT, Rajesh Banu J (2014c) The enhancement of anaerobic biodegradability of waste activated sludge by surfactant mediated biological pretreatment. Bioresour Technol 168:159–166. https://doi.org/10.1016/j.biortech.201464118

Kavitha S, Adish Kumar S, Kaliappan S, Yeom IT, Rajesh Banu J (2015a) Achieving profitable biological sludge disintegration through phase separation and predicting its anaerobic biodegradability by non linear regression model. Chem Eng J 279:478–487. https://doi.org/10.1016/j.cej.2015.05.051

Kavitha S, Kaliappan S, Adish Kumar S, Yeom IT, Rajesh Banu J (2015b) Effect of NaCl induced floc disruption on biological disintegration of sludge for enhanced biogas production. Bioresour Technol 192:807–811. https://doi.org/10.1016/j.biortech.2015.05.071

Kavitha S, Saranya T, Kaliappan S, Adish Kumar S, Yeom IT, Rajesh Banu J (2015c) Accelerating the sludge disintegration potential of a novel bacterial strain Planococcus jake 01 by CaCl induced deflocculation. Bioresour Technol 175:396–405. https://doi.org/10.1016/j.biortech.2014.10.122

Kavitha S, Yukesh Kannah R, Yeom IT, Uan DK, Rajesh Banu J (2015d) Combined thermo-chemo-sonic disintegration of waste activated sludge for biogas production. Bioresour Technol 197:383–392. https://doi.org/10.1016/j.biortech.2015.08.131

Kavitha S, Catherine Stella PB, Kaliappan S, Yeom IT, Rajesh Banu J (2016a) Enhancement of anaerobic degradation of sludge biomass through surfactant-assisted bacterial hydrolysis. Process Saf Environ Prot 99:207–215. https://doi.org/10.1016/j.psep.2015.11.009

Kavitha S, Jessin Brindha GM, Sally Gloriana A, Rajashankar K, Yeom IT, Rajesh Banu J (2016b) Enhancement of aerobic biodegradability potential of municipal waste activated sludge by ultrasonic aided bacterial disintegration. Bioresour Technol 200:161–169. https://doi.org/10.1016/j.biortech.2015.10.026

Kavitha S, Rajesh Banu J, Ivin Shaju CD, Kaliappan S, Yeom IT (2016c) Fenton mediated ultrasonic disintegration of sludge biomass: biodegradability studies, energetic assessment, and its economic viability. Bioresour Technol 221:1–8. https://doi.org/10.1016/j.biortech.2016.09.012

Kavitha S, Rajesh Banu J, Subitha G, Ushani U, Yeom IT (2016d) Impact of thermo-chemo-sonic pretreatment in solubilizing waste activated sludge for biogas production: energetic analysis and economic assessment. Bioresour Technol 219:479–486. https://doi.org/10.1016/j.biortech.2016.07.115

Kavitha S, Rajesh Banu J, Vinoth Kumar J, Rajkumar M (2016e) Improving the biogas production performance of municipal waste activated sludge via disperser induced microwave disintegration. Bioresour Technol 217:21–27. https://doi.org/10.1016/j.biortech.2016.02.034

Kavitha S, Saji Pray S, Yogalakshmi KN, Adish Kumar S, Yeom IT, Banu R (2016f) Effect of chemo-mechanical disintegration on sludge anaerobic digestion for enhanced biogas production. Environ Sci Pollut Res 23:2402–2414. https://doi.org/10.1007/s11356-015-5461-z

Kythreotou N, Florides G, Tassou SA (2014) A review of simple to scientific models for anaerobic digestion. Renew Energy 71:701–714. https://doi.org/10.1016/j.renene.2014.05.055

Lauwers J, Appels L, Thompson IP, Degreve J, Van Impe JF, Dewil R (2013) Mathematical modelling of anaerobic digestion of biomass and waste: power and limitations. Prog Energy Combust Sci 39:383–402. https://doi.org/10.1016/j.pecs.2013.03.0

Lubken M, Gehring T, Wichern M (2010) Microbiological fermentation of lignocellulosic biomass: current state and prospects of mathematical modeling. Appl Microbiol Biotechnol 85:1643–1652. https://doi.org/10.1007/s00253-009-2365-1

Ma J, Frear C, Wang Z, Yu L, Zhao Q, Li X, Chen S (2013a) A simple methodology for rate limiting step determination for anaerobic digestion of complex substrates and effect of microbial community ratio. Bioresour Technol 134:391–395. https://doi.org/10.1016/j.biortech.2013.02.014

Ma J, Yu L, Frear C, Zhao Q, Li X, Chen S (2013b) Kinetics of psychrophilic anaerobic sequencing batch reactor treating flushed dairy manure. Bioresour Technol 131:6–12. https://doi.org/10.1016/j.biortech.2012.11.147

Mairet F, Bernard O, Masci P, Lacour T, Sciandra A (2011) Modelling neutral lipid production by the microalga Isochrysis affinis galbana under nitrogen limitation. Bioresour Technol 102:142–149. https://doi.org/10.1016/j.biortech.2010.06.138

Mejdoub H, Ksibi H (2015) Regulation of biogas production through waste water anaerobic digestion process: modeling and parameters optimization. Waste Biomass Valor 6:29–35. https://doi.org/10.1007/s12649-014-9324-5

Merrylin J, Adish Kumar S, Kaliappan S, Yeom IT, Rajesh Banu J (2013a) Biological pretreatment of non-flocculated sludge augments the biogas production in the anaerobic digestion of the pretreated waste activated sludge. Environ Technol 34:13–14. https://doi.org/10.1080/09593330.2013.810294

Merrylin J, Adish Kumar S, Kaliappan S, Yeom IT, Rajesh Banu J (2013b) Effect of extracellular polymeric substances (EPS) on sludge reduction potential of *Bacillus licheniformis*. Int J Environ Sci Technol 8:85–92. https://doi.org/10.1007/s13762-012-0141-8

Merrylin J, Adish Kumar S, Kaliappan S, Yeom IT, Rajesh Banu J (2014a) Effect of EPS removal on the sludge reduction potential of B. licheniformis on its optimized pH conditions. Water Environ J 28:95–103. https://doi.org/10.1111/wej.12014

Merrylin J, Adish Kumar S, Kaliappan S, Yeom IT, Rajesh Banu J (2014b) Enhancing aerobic digestion potential of municipal waste activated sludge through removal of EPS. Environ Sci Pollut Res 21:1112–1123. https://doi.org/10.1007/s11356-013-1976-3

Metcalf EI (2003) Wastewater engineering: treatment disposal and reuse. McGraw-Hill, New York. 1848p. https://doi.org/10.1036/0070418780

Nielsen J, Villadsen J (2008) Bioreactors: description and modelling. In: Rehm HJ, Reed G (eds) Biotechnology: bioprocessing, Wiley-VCH Verlag Gmb H, Weinheim. ISBN: 9783527283132. https://doi.org/https://doi.org/10.1002/9783527620845.ch5

Palanichamy J, Palani S (2014) Simulation of anaerobic digestion processes using stochastic algorithm. J Environ Health Sci Eng 12:121. https://doi.org/10.1186/s40201-014-0121-7

Patil JH, Raj MA, Muralidhara PL, Desai SM, Mahadeva Raju GK (2012) Kinetics of anaerobic digestion of water hyacinth using poultry litter as inoculum. Int J Environ Sci Dev 3:94–98. https://doi.org/10.7763/ijesd.2012.V3.195

Perendeci A, Arslan S, Çelebic SS, Tanyolaç A (2008) Prediction of effluent quality of an anaerobic treatment plant under unsteady state through ANFIS modeling with on-line input variables. Chem Eng J 145:78–85. https://doi.org/10.1016/j.cej.2008.03.008

Poornima DT, Ebenezer VA, Adish Kumar S, Kaliappan S, Rajesh Banu J (2014) Effect of deflocculation on the efficiency of disperser induced dairy waste activated sludge disintegration and treatment cost. Bioresour Technol 167:151–158. https://doi.org/10.1016/j.biort ech.2014.06.004

Qdais HA, Hani KB, Shatnawi N (2010) Modeling and optimization of biogas production from a waste digester using artificial neural network and genetic algorithm. Resour Conserv Recycl 54:359–363. https://doi.org/10.1016/j.resconrec.2009.08.012

Shen S, Premie GC, Guwy A, Dinsdale R (2007) Bifurcation and stability analysis of an anaerobic digestion model. Nonlinear Dynam 48:391–408. https://doi.org/10.1007/s11071-006-9093-1

Siegrist H, Vogt D, Garcia-Heras JL, Gujer W (2002) Mathematical model for meso- and thermophilic anaerobic sewage sludge digestion. Environ Sci Technol 36:1113–1123. https://doi.org/10.1021/es010139p

Sowmya GP, Kavitha S, Adish Kumar S, Kaliappan S, Yeom IT, Rajesh Banu J (2015) Effect of sonically induced deflocculation on the efficiency of ozone mediated partial sludge disintegration for improved production of biogas. Ultrason Sonochem 26:241–248. https://doi.org/10.1016/j.ultsonch.2015.01.015

Sun C, Cao W, Liu R (2015) Kinetics of methane production from swine manure and buffalo manure. Appl Biochem Biotechnol 177:985–995. https://doi.org/10.1007/s12010-015-1792-y

Suresh Karthik Kumar M, Krishna Kumar T, Arulazhagan P, Adish Kumar S, Yeom IT, Rajesh Banu J (2015) Effect of alkaline and ozone pretreatment on sludge reduction potential of membrane bioreactor treating high-strength domestic wastewater. Desalin Water Treat 55:1127–1185. https://doi.org/10.1080/19443994.2014.923335

Taricska JR, Long DA, Paul Chen J, Hung YT, Zou SW (2007) Anaerobic digestion. In: Biosolids treatment processes. Humana Press, Totowa, pp 135–176. ISBN- 978-1-59259-996-7. https://doi.org/https://doi.org/10.1007/978-1-59259-996-7_5

Terashima M, Goel R, Komatsu K, Yasui H, Takahashi H, Li YY, Noike T (2009) CFD simulation of mixing in anaerobic digesters. Bioresour Technol 100:2228–2233. https://doi.org/10.1016/j.biortech.2008.07.069

Uan DK, Rajesh Banu J, Chung IJ, Yeom IT (2009) Effect of thermochemical sludge pretreatment on sludge reduction and on performances of anoxic-aerobic membrane bioreactor treating low strength domestic wastewater. J Chem Technol Biotechnol 84:1350–1355. https://doi.org/10.1002/jctb.2189

Uan DK, Banu RJ, Son DH, Yeom IT (2012) Influence of ferrous sulfate on thermochemical sludge disintegration and on performances of wastewater treatment in a new process: anoxic–oxic membrane bioreactor coupled with sludge disintegration step. Biochem Eng J 66:20–26. https://doi.org/10.1016/j.bej.2012.04.013

Uan DK, Banu RJ, Kaliappan S, Yeom IT (2013a) Influence of the thermochemical sludge pretreatment on the nitrification of A/O reactor with the removal of phosphorus by simultaneous precipitation. Biotechnol Bioprocess Eng 18:313–320. https://doi.org/10.1007/s12257-012-0492-5

Uan DK, Yeom IT, Arulazhagan P, Rajesh Banu J (2013b) Effects of sludge pretreatment on sludge reduction in a lab-scale anaerobic/anoxic/oxic system treating domestic wastewater. Int J Environ Sci Technol 10:495–502. https://doi.org/10.1007/s13762-012-0120-0

Uma RR, Adish Kumar S, Kaliappan S, Rajesh Banu J (2012a) Combined treatment of alkaline and disperser for improving solubilization and anaerobic biodegradability of dairy waste activated sludge. Bioresour Technol 126:107–116. https://doi.org/10.1016/j.biortech.2012.09

Uma RR, Adish Kumar S, Kaliappan S, Yeom IT, Rajesh Banu J (2012b) Low temperature thermo-chemical pretreatment of dairy waste activated sludge for anaerobic digestion process. Bioresour Technol 103:415–424. https://doi.org/10.1016/j.biortech.2011.09.124

Uma RR, Adish Kumar S, Kaliappan S, Yeom IT, Rajesh Banu J (2013a) Impacts of microwave pretreatments on the semi-continuous anaerobic digestion of dairy waste activated sludge. Waste Manag 33:1119–1127. https://doi.org/10.1016/j.wasman.2013.01.016

Uma RR, Adish Kumar S, Kaliappan S, Yeom IT, Rajesh Banu J (2013b) Enhancing the anaerobic digestion potential of dairy waste activated sludge by two step sono-alkalization pretreatment. Ultrason Sonochem 21:1065–1074. https://doi.org/10.1016/j.ultsonch.2013.11.007

Vavilin VA, Lokshina LY, Flotats X, Angelidaki I (2007) Anaerobic digestion of solid material: multidimensional modeling of continuous flow reactor with non uniform influent concentration distributions. Biotechnol Bioeng 97:354–366. https://doi.org/10.1002/bit.21239

Vavilin VA, Fernandez B, Palatsi J, Flotats X (2008) Hydrolysis kinetics in anaerobic degradation of particulate organic material: an overview. Waste Manag 28:939–951. https://doi.org/10.1016/j.wasman.2007.03.028

Veera Lakshmi M, Merrylin J, Kavitha S, Adish Kumar S, Yeom IT, Rajesh Banu J (2014) Solubilisation of municipal sewage waste activated sludge by novel lytic bacterial strains. Environ Sci Pollut Res 21:2733–2743. https://doi.org/10.1007/s11356-013-2228-2

Vlyssides A, Barampouti EM, Mai S (2007) An alternative approach of UASB dynamic modeling. AICHE J 53:3269–3276. https://doi.org/10.1002/aic.11342

Wang H, Kalchev B, Tian Y, Simeonov I, Christov N (2013) Modelling and composed recursive model free control for the anaerobic digestion process. In: Advances in intelligent control systems and computer science. Springer, Berlin, pp 265–278. ISBN- 978-3-642-32548-9, https://doi.org/10.1007/978-3-642-32548-9_19

Wu B (2012a) CFD simulation of mixing for high-solids anaerobic digestion. Biotechnol Bioeng 109:2116–2126. https://doi.org/10.1002/bit.24482

Wu B (2012b) Integration of mixing, heat transfer, and biochemical reaction kinetics in anaerobic methane fermentation. Biotechnol Bioeng 109:2864–2874. https://doi.org/10.1002/bit.24551

Yu L, Ma J, Chen S (2011) Numerical simulation of mechanical mixing in high solid anaerobic digester. Bioresour Technol 102:1012–1018. https://doi.org/10.1016/j.biortech.2010.09.079

Yu L, Christian PW, Jingwei MA, Shulin C (2013) Mathematical modeling in anaerobic digestion (AD). Bioremed Biodeg 4:1–12. https://doi.org/10.4172/2155-6199.S4-003

Biofilm Microenvironments: Modeling Approach

15

Saheli Ghosh, Asifa Qureshi, and Hemant J. Purohit

Abstract

Biofilms are resilient and complex microbial communities or populations encased in a self-producing extracellular matrix existing in nature, which account for their robustness against a wide variety of stresses. In nature, multispecies biofilms have been found to occur frequently, whereas monospecies biofilms have been rare and found only when developed. By understanding the sociobiology of biofilm bacteria and harnessing the knowledge based on known biochemical and physical parameters, the underlying complexity of biofilm microenvironment could be better predicted by state-of-the-art advanced modeling approaches. In the present review, in-depth advanced modeling approaches based on the action of substrates on developmental biofilm physiology have been discussed. Equations for forces and adherence, quorum sensing, fluid flow dynamics, the social arrangement of community members in their favorable places, stratification, and pattern formation inside the biofilm have been elaborated. Equations to predict biofilm dispersal stages have been also discussed. Different models which uncover microbial biofilm life cycle stages to predict their complexity and recommended future variables like "omics" for magnification have been reported. This chapter will help to provide additional platform related to the genetic complex networking of biofilm bacteria for conceptual underpinnings of biofilm microenvironment.

Keywords

Biofilms • Models • Simulation • Omics • Quorum sensing • Fluid flow

S. Ghosh • A. Qureshi (✉) • H.J. Purohit
Environmental Biotechnology and Genomics Division, CSIR – National Environmental Engineering and Research Institute (NEERI), Nehru Marg, Nagpur 440020, Maharashtra, India
e-mail: a_qureshi@neeri.res.in

H.J. Purohit et al. (eds.), *Optimization and Applicability of Bioprocesses*,
https://doi.org/10.1007/978-981-10-6863-8_15

Abbreviations

CE-TOF	Capillary electrophoresis-time of flight
CLSM	Confocal laser scanning microscopy
CP-AFM	Conductive-probe atomic force microscopy
ESI-MS	Electrospray ionization mass spectrometry
FBA	Flux balance analysis
FISH	Fluorescent in situ hybridization
HPAEC	High-pH anion-exchange chromatography
IMS	Imaging mass spectrometry
MALDI-TOF	Matrix-associated laser desorption ionization-time of flight
MS-DESI	Mass spectrometry-desorption electrospray ionization
PAD	Pulsed amperometric detection
PCR	Polymerase chain reaction
STXM	Scanning transmission X-ray microscopy
TERS	Tipped enhanced Raman spectroscopy
XM	X-ray microscopy

15.1 Introduction

The developmental model of microbial biofilms has played dominant role in biofilm study, but always challenged by research community for its paradigmatic value. Based on the recent knowledge generated on biofilms, the criteria required for predicting the complex process have been presented. Biofilms represent microbial population cemented in a self-synthesizing matrix made up of nucleic acids, polysaccharides and proteins having a sequential planktonic and surface associated stages of life cycle in the environment (Halan et al. 2012). Criteria for biofilms involved adherence to a surface, microcolony formation, the growth of that microcolony and dispersal. In natural settings, bacteria have been able to form biofilms possessing different physiological stages (Molloy 2013; Qureshi et al. 2015). In fact, it is the default mode of bacterial life (Lear and Lewis 2012). The complexity of the microbial colonies in biofilms made its exhaustive analysis difficult (Koul and Kalia 2017). Biofilm concept has been envisaged by scientists by using conventional tools like biochemical assays and physical methods to understand it. But these conventional methods could provide only the qualitative information regarding the presence or absence of a particular molecule in the biofilm but cannot predict dynamic behavior. Therefore, more advanced tools to analyze biofilms had come into existence, but each method has its own limitations and benefits. The combination of different tools has to be used to gain a better understanding of biofilm complexity (Ehret and Bol 2013). Nowadays, researchers have been using different omics approaches (Zarraonaindia et al. 2013). Some mathematical modeling approaches have been partially serving the purpose. Nearly three decades were employed to understand biofilm behavior which has been still in the paradigmatic state (Moustaid et al. 2013). So it has been envisaged through the present review that combination of omics data and modeling may suffice in the future for complex biofilm behavior. Biofilms result from synchrony between competition and cooperation, and it has been compiled in the present review. Such balance has been central to building complete and predictive models of biofilms. The review focus on the magnification, i.e., in-depth understanding of biofilms through a combination of multilevel tools and simulation models based on mathematical equations or physical parameters. Various stages in biofilm formation like adhesion, fluid flow in their EPS, metabolism in biofilms, a network of microbial community pathways, cross-talks leading to co-survival, and autoaggregation and coaggregation to predict complex behavior of biofilms have been discussed understate-of-the-art perspective in the present review.

15.2 Different Biochemical and Physical Methods for Biofilm Development Process

Biofilm process could be evaluated qualitatively and quantitatively using conventional methods (Table 15.1). A number of assays for the qualitative and quantitative evaluation of bacterial biofilm colonizing on abiotic substrates together with advantages and limitations of each method have been described and compared, which revealed that some methods were suitable to quantify biofilm matrix while others were capable of evaluating both living and dead cells or quantifying exclusively viable cells in the biofilm. In particular, colorimetric methods to evaluate biofilm matrix (crystal violet,1,9-dimethyl methylene blue and fluorescein-di-acetate methods) or viable cells (LIVE/DEAD BacLight, BioTimer Assay, resazurin, tetrazolium hydroxide salt methods) were reported, and advanced microscopic techniques (mass spectrometry; confocal laser scanning microscopy; Raman spectroscopy and Electron microscopy) for detailed understanding were reported (Cerca et al. 2012; Pantanella et al. 2013; Shang et al. 2014; Kalia et al. 2017). Molecular biology techniques combined with phenotype studies also gave a broad insight into biofilm formation mechanisms (Gokcen et al. 2013; Sharma and Lal 2017). But all the reports have drawn the same conclusion that one technique alone cannot reveal the whole picture of biofilm. Thus some more advanced and improved techniques are needed to draw the bigger picture, which has been comprehensively reviewed.

15.3 Simulation Models for Predicting the Development of Bacteria in Biofilms

Biofilms have been ubiquitous in nature and industries (Kaur et al. 2014), although they have been involved in nearly 80% of infectious diseases in our body, also they have been potential candidates for bioremediation (applying microbes in the degradation of persistent compounds (Qureshi et al. 2009, 2012; Das et al. 2017) and hazardous toxic compounds (Ghosh et al. 2017). Reports were available to understand the behavior of bacteria in biofilms at each stage in their life cycle through

Table 15.1 Comparative methods for biofilm analysis

Sl no	Type of method	Name of the method	Principle	Advantages	Disadvantages	References
1	Biochemical	Crystal violet binding assay	Uses crystal violet to stain adhered cells	Semiquantitative method measures amount of adhered cells	Stains both eps and adherent cells	O'Toole (2011)
		Congo red agar method	Uses Congo red to stain adhered cells	Semiquantitative method measures amount of adhered cells	Stains both eps and adherent cells	Aparna and Yadav (2008)
		XTT assay	Uses redox indicator (XTT) to enumerate viable cells in a biofilm spectrophotometrically		Heterogeneity in biofilm hinders release of XTT	Pantanella et al. (2013)
		DMMB assay	Quantification for *S. aureus* biofilms by using a dye DMMB complexes specifically with PIA of *S. aureus* biofilms and the amount of DMMB dye released by a decomplexation solution spectrophotometrically	No. of viable cells in the biofilm can be enumerated	Not suitable for multispecies biofilm detection or any other species biofilm not made up of PIA	Pantanella et al. (2013)
		Resazurin assay	Colorimetric blue dye (resazurin) reduced by viable cells to pink fluorescent compound measured by a spectrophotometer	Specific detection for *S. aureus* biofilms	Species and strain related not suitable for multispecies biofilms	Welch et al. (2012)
		BioTimer assay	Uses a reagent containing phenol red and time elapsed in color change from red to yellow is elucidated	Viability and growth of cells in a biofilm Viability of cells inside biofilm can be determined	Not suitable for multispecies biofilm	Pantanella et al. (2013)
		LIVE/DEAD BacLight assay	Uses two nucleic acid binding dyes such as green fluorescent Syto9 crosses membranes and binds DNA and 2nd is red propidium iodide which passes onto dead cells	Viability of cells in a biofilm can be determined	Provides semiquantitative results	Cesar et al. (2013) and Pantanella et al. (2013)

2	Physical	Air-liquid interface assay	Studies attachment of bacteria to the surface	Specific for monospecies biofilms	Not suitable for multispecies biofilm	Judith et al. (2011)
		CP-AFM (conductive-probe atomic force microscopy)	Current/voltage measurements by applying voltage to putative electrical structures in a biological system	To detect electrically conductive nanowires projecting from a biofilm	Not suitable for multispecies biofilm	Golus et al. (2013)
		XM/CTXM	Microscopy based on X-ray focused radiation either mono- or polychromatic	Viewing composition of bacterial cells and biofilm	Resolution up to 25 nm	Pantanella et al. (2013)
		TERS	Combines SEM with Raman spectroscopy	High resolution	Allows only sample surface and an inner portion close to it	Pantanella et al. (2013)
		HPAEC-PAD-ESI	Detection of biofilm matrix degraded products	Specific determination of matrix degraded products of a single species biofilm	The only matrix can be analyzed	Gokcen et al. (2013)
		CLSM (confocal laser scanning microscopy)				
		Particle image velocimetry and particle tracking	Flow visualization techniques	Live cells under flowing conditions can be analyzed	Interactions between the cells cannot be analyzed	Yoon et al. (2013)
		Attenuated total reflection microscopy (ATR)	To visualize conditioning films that harbor early colonizers	Early colonizers of biofilm can be studied	Interactions between the cells cannot be analyzed	Aparna and Yadav (2008)

Table 15.2 List of mathematical modeling approaches used to study biofilm life cycle

Mathematical modeling approaches	Function at biofilm stages	References
Low-dimensional continuum model	Based on kinematics and translational diffusion of substrates in biofilm maturation stages	Wang and Zhang (2010)
Solid surface diffusion model	Solute transport in biofilms; maturation stage	Shen et al. (2012)
Empirical models	Solute transport: maturation stage	Ni and Yu (2010)
Half-order kinetics	Diffusion and reaction of solutes in the flocculated state: fully matured biofilms	Perez et al. (2005)
Two-dimensional model based on diffusion reaction equations	Coupled mass transport, biochemical reaction, and biomass growth in photosynthetic biofilm photoreactor: during maturation stage	Liao et al. (2012)
Particle-based multidimensional multispecies biofilm model	Diffusion-reaction mass balances coupled with microbial growth and spreading of biomass by hard spherical particles: maturation and dispersal stage	Picioreanu et al. (2004)
Diffusion and reaction models	Penetration of gas in a toluene-degrading biofilm	Roberts and Stewart (2004)
Mathematical models on protection of cells from antimicrobials	Protection from antimicrobial agents through the formation of persister cells: matured stage	Wang et al. (2012)
Hydrolysate diffusion and utilization model	The growth of cells in a biofilm and secretion of hydrolysate in a bulk fluid	Chakraborty et al. (2011)
Classical substrate uninhibited Monod model for differential system equations	Prediction of biofilm reactor (anaerobic) behavior treating mustard oil-rich wastewater	Frederick et al. (2011)
Prototype density-dependent diffusion-reaction biofilm model	Measures quorum sensing mediated EPS production in biofilms: microcolony formation. Based on extended Newtonian, rubber elastic gel and viscoelastic models. Measures the amount of EPS network produced in a biofilm: during maturation of biofilm	Zhang et al. (2008)
Phase field theory of modeling biofilms: Cahn-Hilliard equations	Measures changes in pH through metabolism of cells in a biofilm	Llie et al. (2012)
One-dimensional time-dependent model based on Dibdin's model and Nerst-Planck equations	Combines mass transport of solutes and hydrodynamics with biofilm development and detachment in spiral wound RO membranes leading to loss of performance	Radu et al. (2010)
2-D mathematical microscale model	Describes cell death in a biofilm: dispersal stage	Fagerlind et al. (2012)

(continued)

Table 15.2 (continued)

Mathematical modeling approaches	Function at biofilm stages	References
Dynamic model of cell death	Describes concentration profile of substrates in a fixed bed reactor degrading BTEX compounds	DeMello et al. (2010)
Phenomenological model using single differential equation	Measures biofilm growth with respect to cells attached in EPS: EPS development and growth	Friedman et al. (2014)
Model based on stokes and reaction-diffusion equation	Describes kinetics of adsorption and biodegradation of reactive black 5 in *F. trogii* biofilms also measures amount of suspended and attached cells	Lin et al. (2013)
Adomian decomposition method	Measures steady-state kinetics of substrate consumption and microbial death rate in a biofilm: later maturation stage	Muthukaruppan et al. (2013)
Mathematical modeling of dynamic competition in biofilms	Describes competition during bacterial adhesion to surfaces and reveals rate of attachment affects biofilm thickness: initial stage	D'Acunto et al. (2013)

various computational and mathematical models (Table 15.2). Simulation models applied recently to uncover different stages of the biofilm life cycle are listed (Fig. 15.1) as:

1. Models for adhesion of biofilms
2. Models for quorum sensing and fluid flow in biofilms
3. Models for bacterial growth, interactions, and heterogeneity in biofilm microenvironment
4. Models for dispersal of cells from biofilm into planktonic mode

15.3.1 Models for Adhesion of Biofilms

Simulation models have been developed in various stages of biofilm formation (Purkait et al. 2016).The first stage of biofilm involved adherence of planktonic bacteria to a surface which remains crucial step. To understand adherence phenomenon, a **K-S type model** was developed by Moustaid et al. (2013), which described bacterial adherence to a surface based on its biased random movement in response to chemicals, i.e., chemotaxis. It is described by two differential equations.

First equation describes bacterial density diffusion, chemotaxis, growth, and death. Bacterial chemotaxis is represented by the term

$$\frac{-\partial J_c}{\partial x},$$

where J_c is chemotactic flux given by

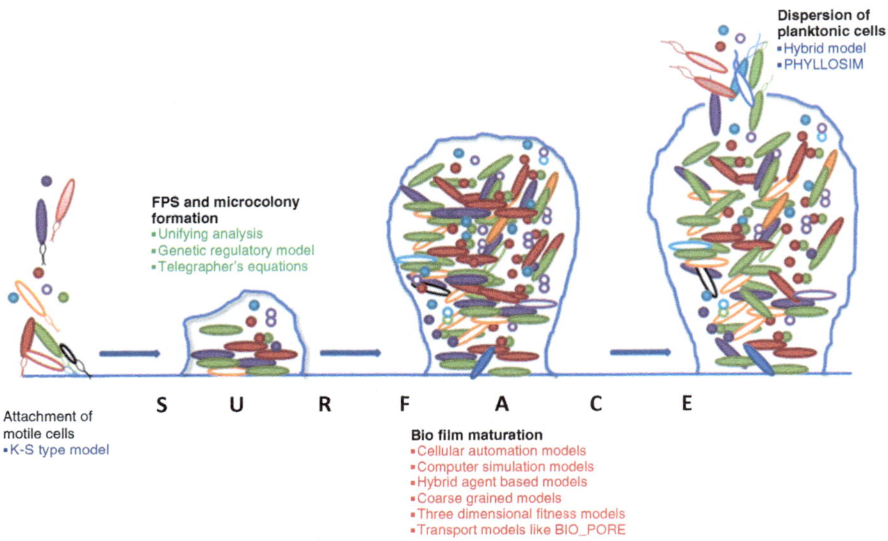

Fig. 15.1 Modeling approaches applied to study different stages of biofilm life cycle

χ= chemotactic coefficient and S=sensing molecules concentration.

$$J_c = \chi^b \frac{\partial s}{\partial x}$$

Second equation specifies about chemoattractant and chemorepellent production, diffusion, and degradation

$$\frac{\partial s}{\partial t} = \frac{\mathrm{Ds}\partial}{r^2}\left(\frac{r^2}{\partial r}\frac{\partial r}{\partial r}\right) - \lambda S + \alpha B$$

where Ds= sensing molecules diffusivity, s= sensing molecules concentration, and B= bacterial density. Testing different conditions through such model revealed that significant concentration of chemotactic signals brings microbial adherence to the surface (Moustaid et al. 2013). Swimming motility affected the transient cell-to-cell interactions "stickiness." So decreasing swimming speed corresponds to increase in the cell-to-cell interactions, i.e., clumping which is revealed by a **physical model** based on Brownian dynamics (Qi et al. 2013).

15.3.2 Modeling Approaches Based on Quorum Sensing and Fluid Flow Inside Biofilms

Quorum sensing (QS) has been an intercellular signaling system by which bacteria secrete some signaling molecules to communicate with each other to sense density of cells or "quorum" in order to initiate a particular function (Kalia and Purohit

2011; Kalia 2014). Changes in microenvironment evoked intercellular signaling which allowed bacteria to modify their environment for other members in a population to alter their behavior in response to changes in number and type of species present in a community (Ghosh et al. 2017). Bacteria coordinate their behavior like a "society" (Gui et al. 2014; Arya and Princy 2016). Bacteria call members of their population in order to deal with a difficult task which cannot be done in the unicellular state (Koul et al. 2016). Recently, research has shown that QS signals produced by bacteria enhanced biodegradation of oil, and heavy metals contaminated soil and biofouling prevention (Liao and Li 2013). Three types of quorum sensing systems have been studied in *P. aeruginosa* such as *rhl* (regulates expression of rhamnolipids, pyocyanin, etc.) (Shang et al. 2014), *las* (Pal et al. 2016), and *pqs* gene system (Lee et al. 2013). It has been reported that deleting the *rhl* QS reduced the efficiency of biodegradation of aromatic compounds such as phenol, benzoate, and phenanthrene, and the addition of butyryl homoserine lactone complemented biodegradation, which revealed that *rhl* QS systems positively regulated biodegradation process (Yong and Zhong 2013).

In context to quorum, bacteria senses signal concentration, which has been described by **unifying analysis** in terms of threshold volume occupied by cells. Unifying analysis proposed that quorum sensing allowed cells to assess themselves when critical enclosure volume (volume occupied by accumulation of numerous cells in a given space or fewer cells in a small space) has reached and introduced a dimensionless term "V" which has been threshold enclosure place expressed in number of cell volumes occupied, where $V = V_{e,c}/V_c$, $V_{e,c}$= Critical enclosure volume, and V_c= volume of the cell. Below the threshold, cells transit from freely floating to aggregated state (Nadell et al. 2013).

Bacteria exhibit behavioral variation in unicellular and aggregated state (Nadell et al. 2013). In aggregated state bacteria oscillate and to understand their behavior, a **GRN (genetic regulatory network) model** based on numerical simulation and bifurcation analysis was reported, where the population of cells remained immobilized on microfluidic chamber has been developed by researchers. The cells before transformation were analyzed, where three genes *luxI, aiiA, and yem GFP* were introduced into two different plasmids under the regulation of same single promoter *li-P. LuxI* encoded AHL which interacted with LuxR which was considered constant in GRN model. It activated the promoter *li-P*, which in turn activated the expression of selected three genes. AHL could diffuse freely across the cell membranes and coupling occurred. The situation has been modeled using GRN which revealed that the oscillatory behavior of cells was stable in the population which aroused due to high concentration of quorum sensing signals AHL which in turn arose due to Brownian diffusion. The oscillations affected the population size in the biofilm.

High cell densities mediate cellular coupling. Highly coupled cells oscillate with the same phase in synchrony as compared to weakly coupled cells (Mina et al. 2013).The competition and synergy in biofilms varied with flow conditions. When microbes form biofilms in its maturity, it formed numerous structures which lead to a network of fluid channels inside the biofilm. As stated by Davit et al. (2013), transport phenomena including flow and advection regulate the cellular behavior of biofilm members. Transport phenomena could be achieved by regulating nutrient supply, their release of waste materials, and penetration of antimicrobial agents.

Mathematical modeling approaches were used where analysis of fluid flow in biofilms was started with channel-scale description of mass transport and volume. Averaging method was used to derive a set of equations specifically for biofilm level whose thickness was greater than the channels width and has shown that solute transport could be described by two partial differential equations called **telegrapher's equations** for the average concentration and has used these equations to understand fluid transport in porous biofilms, which showed that these equations could explain the nondiffusivity of some solutes in cell clusters (Davit et al. 2013).

Kim et al. (2013) proposed another model of biofilm development of *P. aeruginosa* in microfluidic channels. The model was based on Reynolds number (Re) and dimension of the channel(r) and consequent variables (A_b/h^2 & A_b/W_{cs}^2), where A_b= 2D area of biofilm, h=channel height, and W_{cs}=consequent flow width. Analysis of this model showed that flow velocity inside the biofilms has been affected by flow width. At low Re and high flow rate inside the channels, biofilm growth was encouraged. While at high Re and high flow rate, biofilm growth was suppressed. Re is an important component of this model which regulated biofilm development. Flow rates affected biofilm growth in the study of fluid dynamics of biofilms. It has been studied in a spiral wound membrane biofilm reactor to simulate autotrophic denitrification, where the data revealed nonuniform and dense growth at the bottom of the reactor which was characterized as a minimum shear zone with low biofilm formation and denitrification (Martin et al. 2013). The results suggested that shearing force affected biofilm formation.

Zhang et al. (2013) reported that external flow conditions affected interactions between members of a biofilm community in *Pseudomonas* and *Flavobacterium* biofilms. When the supply of nutrients was low under gradual inflow conditions, two species competed with each other for survival. Under such conditions, *Flavobacteria* attained faster specific growth rate than *Pseudomonas* thereby could access favorable sites by outcompeting *Pseudomonas* in the dual species biofilm. Hence, the fate of *Pseudomonas* allowed it to confine itself to the nutritionally depleted sites of the biofilm. On the contrary, higher inflow conditions allowed a greater influx of nutrients of the growth medium where both the organisms could attain significant growth rate in the dual species biofilm. In this way, two organisms get arranged in the mixed species biofilm (Zhang et al. 2013) and structured in a single species biofilm. (Moreira et al. 2013).

15.3.3 Models for Bacterial Growth, Interactions, and Heterogeneity in Biofilm Microenvironment

In microbial biofilms when aggregates were formed, bacteria grew inside the extracellular polysaccharide (EPS); hence, the biofilms grew and heterogeneity of chemicals arouses in it. To capture such scenario, some modeling approaches such as

cellular automata have been proposed by Deygout et al. (2013), where space was divided into microscopic size boxes and evolution and development of each box were described by involving neighboring cell contents and then individual bacteria was studied. The model followed its own sets of deterministic and stochastic rules. Interactions between the cells could also be explored. In structured biofilms, channels were formed, nutrient exchange and metabolic flow occurs through channels to sustain the cells present in it (Davit et al. 2013). Proteomics combined with **computer modeling** techniques were used to elucidate the mechanism of biofilm development of *Pseudomonas aeruginosa*. Molecular modeling studies revealed that it produced siderophore called pyoverdine. During biofilm development, bacteria suffered from a variety of stresses like osmotic stress (Bazire et al. 2007), antimicrobial compounds stresses (Stewart and Costerton 2001), etc. Reports revealed various modeling studies elucidating the mechanism of stress resistance in bacterial biofilms.

Modeling techniques were nowadays being used to understand the mechanism of drug complexation in biofilms. Reports suggested that several aminoglycosides (Stewart and Costerton 2001), as well as beta-lactam antibiotics (Chang et al. 2009), enhanced biofilm formation, but the mechanism behind the process was not well understood. Reports revealed that TcaR was a repressor of ica gene locus which has been responsible for the synthesis of N-acetyl glucosamine, one of the components of EPS matrix of *S. epidermidis*. Here, the molecular modeling technique like **SOLVE** was used to understand the mechanism of TcaR operation. Data suggested that TcaR dimers bind with a 33 bp segment of the *ica* promoter region and inhibited the binding of other transcriptional factors. TcaR contained TTNNAA consensus sequence embedded in 16 bp sequence responsible for binding with the salicylates, beta-lactam, and aminoglycoside antibiotics. Binding of antibiotics induced a huge conformational change in the DNA-binding region as compared to the free protein which could not allow the proteins to bind to the DNA (Chang et al. 2009).

Various mathematical models were applied to understand toxic compounds and drug resistance in biofilms. Some mathematical models suggested that antimicrobial tolerance in biofilms was due to insufficient penetration of the drug likely in *P. aeruginosa* (Biggs and Papin 2013). Here, three mathematical models described different mechanisms of antimicrobial resistance. These mathematical models were derived by combining basic planktonic disinfection model with transport limitation and physiological heterogeneity, which depend on biofilm thickness and age. **Physiological limitation model** described the effect of antibiotic tobramycin on a 2-day-old biofilm. It predicted biphasic killing of biofilms where cell death was directly proportional to the product of the reaction of antimicrobials with the biofilm cells in a dose-dependent manner. The biofilm was segregated at the bulk liquid interface and persister cells at the substratum. The position of the antibiotic inside the biofilm described differential killing. Here, the antibiotic penetrated fully in the biofilm (D'Acunto et al. 2013). **Stoichiometric transport model** described the killing of biofilm by hypochlorites. It predicted little killing rate than physiological limitation model. The penetration of the antibiotic depends on the stoichiometric reaction of the neutralizing molecules and antibiotic which was independent of the cell viability. The reaction proceeds from a narrow zone in the biofilm. It produced a concave down killing curve. In both models, biofilm was destroyed with time (Shen et al. 2012). **Catalytic transport model** described the killing of *P. aeruginosa* biofilm by hydrogen peroxide. Catalytic reaction occurred with the antimicrobial agent and the biofilm. Biofilm dispersal was not proportional to the dose and product of the reaction between the cells and the antimicrobial agent. The reaction followed rhythmic balance between diffusion and reaction of the antibiotic instead. The balance was characterized by Thiele modulus; if the Thiele modulus >1, the antimicrobial agent never penetrated the biofilm. Hence, the biofilm was not destroyed with time in such mechanism (Dodds et al. 2000).

The main difference between the killing mechanisms of planktonic and biofilm models was that biofilm models generated nonlinear killing curve, while planktonic disinfection model generated a linear killing curve (Dodds et al. 2000; Roberts and Stewart 2004). Some reports on mathematical models predicted antimicrobial resistance in biofilms, which was also due to nutrient limitation, slow growth, and formation of hibernating or dormant cells. The mathematical model developed by Roberts and Stewart (2004) explained the statement. The concept behind the model was the dependency of antimicrobial killing with the growth rate of cells inside the biofilm, where the cellular growth inside the biofilm was measured by calculating the confined concentration of the growth-limiting substrate coupled with Monod kinetics. The model predicted that when the biofilm was challenged by an antimicrobial agent, the cellular growth inside the biofilms was stratified with no growth zones in the deep interiors. The cellular killing rate of the compound was slow as it could not penetrate much due to increased biofilm thickness initially upon the antimicrobial compound exposure. Even if it penetrated, the killing rate was decreased nonlinearly when the concentration of the growth-limiting substrate fueling the biofilm was decreased; at that time, tolerance was maximum due to mass transfer resistance. This tolerance also developed due to the formation of “dormant” or “damaged” cell, which continued to consume substrate without growth. These dormant cells further retarded antimicrobial killing (Roberts and Stewart 2004). Various modeling techniques could explain the mechanism behind persister or dormant cell formation in biofilms. Formation of these dormant or persister cells required reduced metabolism. This scenario in biofilms was captured by various models. **Hybrid agent-based metabolic model**, a multiscale model was built using a novel tool called **Mat Net**, which captured metabolic states of *P. aeruginosa* cells in terms of cellular growth and revealed aerobic metabolic reactions. It was confined to the topmost layers of cells at biofilm surface and reduced at lower layers, while addition of nitrate in the in silico media restored and enhanced cellular growth anaerobically via; denitrification where nitrate was the terminal electron acceptor (Biggs and Papin 2013).

Coarse-grained model described bacterial growth and substrate uptake in both biofilm and planktonic state in the box. In the diffusion of the substrate in each box, it diffused horizontally. Bacteria in each box use substrate one after another which increased their mass. It has a mass higher than the divisional threshold (m_{dfv}) it divided into two. The utilization of substrate by bacteria which has its own mass “m” followed Monod’s kinetics, where bacterial growth was directly proportional to this amount of substrate consumed, multiplied by conversion yield factor y, at time *t* box *i* the increase in mass dm followed this equation (Deygout et al. 2013).

$$\mathrm{dm} = \frac{\gamma \mu_j s_t, i}{k_j + S_t, i}$$

where

μ_j = maximum growth rate for either attached or planktonic bacteria
K_j= half saturation constant for either attached or planktonic bacteria

Houry et al. (2012) proposed that some planktonic bacteria moved deep inside irrigating the biofilm, resulting in pore formation, hence increased the transfer of nutrients, but it also allowed toxic entrance thereby killing biofilm cells. Cellular activity in mono- and multispecies community altered surrounding microenvironment which in turn affected cellular behavior. To explain cellular activity scenario, a **3D model** was developed by Watrous et al. (2013). These models were reported to be produced by MALDI-TOF followed by IMS analysis of cross sections of bacterial colonies growing in deep agar and entering the data sets into MATLAB for a model generation and then matched with the photographs of colonies itself. Results pointed out that this model could capture metabolic states deep beneath the agar surface. Metabolic exchange factors regulated interspecies interactions between *C. albicans* and *P. aeruginosa.* **Three-dimensional fitness model** built by Momeni et al. (2013) envisaged ecological rules, viz., competition and cooperation in a microbial community. The model revealed the ecological interactions between two distinct obligatory cooperative systems such as yeast and methane-producing community which affected patterning and strong cooperation via redox coupling. It rendered fitness benefits to both of the communities which lead to partner intermixing, and the two appeared stacked over each other. In multispecies communities, strong cooperation was the major patterning force which led to intermixing (Momeni et al. 2013).

Transport and **BIO_PORE models** were developed by using COMSOL Multiphysics™ platform to describe the functioning of constructed wetlands (CW), which were used as an alternative wastewater treatment, where a pilot horizontal CW was planted with *Phragmites australis* and fed with granular media consisting of fine gravel fed with urban wastewater. It is comprised of four sub-models such as **hydraulic** (which described abundant flow through an anisotropic porous environment), **reactive transport** (described the fates and transport of components, like dissolved oxygen), **biofilm** (described the growth and development of cells), and **plant** (release and uptake of nutrients by plant roots).This model could predict the functioning of constructed wetlands for longer time scale by limiting space and cellular growth in a biofilm. The granular media contained pores which were regarded as micro-reactor, where space and substrate limited bacterial growth.

Bacteria formed biofilms in pores which have its own space and growth limitations when microbial biomass concentration attained certain value. Substrate availability was limited due to external boundary diffusivity of biofilm. So bacteria harboring at film-water interface get more substrate and grow faster than bacteria at the interior of the granular media particles (Samso and Garcia 2013).

15.3.4 Models for Dispersal of Cells from Biofilm

After fully matured biofilms were formed and when the concentrations of cells inside was increased, they need to disperse to colonize new surfaces. To understand this process, a **hybrid model** was proposed, which was based on individual and microscopic approach as it combined three scales approach such as **microscopic approach** which described individual bacterial growth in the biofilm which followed Monod's law, a **macroscopic approach** which represented advection and diffusion of a substrate inside the exopolysaccharide as a continuous concentration in the whole biofilm, and a **mesoscopic approach** which represented discrete boxes which described coincident environment of individual bacteria in each box. It related attachment and detachment of the planktonic bacteria with the confined concentration of attached biomass. It takes concentrations of substrate (t, z), planktonic bacteria C_p (t, z), and attached bacteria C_a(t, z) into account. The attached bacteria were measured in terms of weight per surface unit. To convert the concentration of attached bacteria C_a(t, z) to planktonic ones C_p(t, z), a coefficient ∂ is used,

$$\partial = \frac{\text{Ratio of tube surface}}{\text{Ratio of tube volume}}$$

$$\partial = \frac{\pi D^4}{\frac{\pi D^2}{4,}}$$

$$\partial = \frac{4}{D}$$

The partial differential equation used as follows:

$$\frac{\partial C_a}{\partial t} = C_a\left[(F_a)G(C_a) - K_a - \beta\right] + \alpha C_p(1 - C_a)\partial^{-1}$$

where $G(C_a)$ is the fraction of daughter cells of the attached bacteria finding spaces to attach on the wall, which further depends on the free surface sites available for attachment c_A,∞ with $C_A = c_A/c_A$, ∞. So it should decrease in order to get attached on the surface until $G(C_a)$ becomes O as soon as it finds its appropriate surface C_A becomes infinite (Deygout et al. 2013). Detachment of bacteria from clusters leads to cluster size variation in the leaf surface revealed by an individual-based model **PHYLLOSIM**, which consisted of a 2-D grid having 1 mm^2 area into which virtual cells were inoculated. A range of detachment probabilities 2.5%, 5%, and 10% were simulated, which revealed an increase in cluster size ranges. Calculations with F, a measure of fit of simulated versus observed cluster sizes of different modeled conditions, were done, which resulted in low F values for those conditions having no detachment (VanderWal et al. 2013).

15.4 Future Variables for Complex Biofilm Network Models: "OMICS"

"Omics" is a modern approach to understanding cellular processes in biofilms and modify the metabolism of the individual living cell/microcolonies to exploit in industrial applications via metabolic engineering (Zarraonaindia et al. 2013). "Omics" tools range from the sequencing of genomes (genomics) (Kapley and Purohit 2009; Piubeli et al. 2015), measurement of gene transcription (transcriptomics), protein abundance (proteomics) to the product of substrate concentration (metabolomics), and metabolic flux analysis (fluxomics) (Winter and Kromer 2013). All such multilevel omics tools could be exploited in future for harnessing complex and functional genetic network of biofilm microbial communities as input variables for upcoming and future developmental biofilm models.

15.5 Perspectives

Biofilm microenvironment complexity was elaborated using modeling approaches and derived simulation models predicted the behavior of biofilm bacteria in terms of their cooperation and competition, stress tolerance, etc. Related mathematical and computational models derived from biological and physical parameters such as motions for aggregation, fluid flow dynamics, regulation in cellular behavior, growth, the interaction between the community members, pattern formation, alteration of surrounding and finally equations for dispersal could enlighten the understanding of biofilm microenvironment. It could be recommended in the future to add variables for modelling using OMICS. Omics could uncover new genes, functionally and industrially important enzymes, as well as microbial community diversified in different environments.

Acknowledgment The authors are grateful to CSIR-Senior Research Fellowship for Ms. Saheli Ghosh (19-06/2011(i) EU-IV). Funds from DST, New Delhi (DST/TM/WTI/2K15/225(G)-A), and CSIR-NEERI for 12th plan network project (ESC0306 Activity No 3.4.2) on "Clean Water: sustainable options" are acknowledged. All the authors are thankful to Director of CSIR-National Environmental Engineering Research Institute (CSIR-NEERI) for constant support and inspiration and providing infrastructural facilities [CSIR-NEERI/KRC/2017/July/EBGD/14].

References

Aparna MS, Yadav S (2008) Biofilms: microbes and disease. Braz J Infect Dis 12:526–530. https://doi.org/10.1590/S1413-86702008000600016

Arya R, Princy SA (2016) Exploration of modulated genetic circuits governing virulence determinants in *Staphylococcus aureus*. Indian J Microbiol 56:19–27. https://doi.org/10.1007/s12088-015-0555-3

Bazire A, Diab F, Jebbar M, Haras D (2007) Influence of high salinity on biofilm formation and benzoate assimilation by *Pseudomonas aeruginosa*. J Ind Microbiol Biotechnol 34:5–8. https://doi.org/10.1007/s10295-006-0087-2

Biggs MB, Papin JA (2013) Novel multiscale modeling tool applied to *Pseudomonas aeruginosa* biofilm formation. PLoS One 8:e78011. https://doi.org/10.1371/journal.pone.0078011

Cerca N, Gomes F, Pereira S, Teixeira P, Oliveira R (2012) Confocal laser scanning microscopy analysis of *S. epidermidis* biofilms exposed to farnesol, vancomycin and rifampicin. BMX Res Notes 5:244. https://doi.org/10.1186/1756-0500-5-244

Cesar N, Reffuveille F, Kathryn EF, Hancock REW (2013) The effect of nitroxides on swarming motility and biofilms, multicellular behaviors in Pseudomonas aeruginosa. Antimicrob Agents Chemother 57:4877–4881. https://doi.org/10.1128/AAC.01381-13

Chakraborty C, Chowdhury R, Bhattacharya P (2011) Experimental studies and mathematical modelling of an up flow biofilm reactor treating mustard oil-rich wastewater. Bioresour Technol 102:5596–5601. https://doi.org/10.1016/j.biortech.2011.02.010

Chang YM, Jeng WY, Ko TP, Yeh YJ, Chen CKM, Wang AHJ (2009) Structural study of TcaR and its complexes with multiple antibiotics from *Staphylococcus epidermidis*. PNAS 0913302107:1–6. https://doi.org/10.1073/pnas.0913302107

D'Acunto B, Esposito G, Frunzo L, Mattei MR, Pirozzi F (2013) Analysis and simulations of the initial phase in multispecies biofilm formation. Comm Appl Ind Math 4:e-448. https://doi.org/10.1685/journal.caim.448

Das K, Rajawat MVS, Saxena AK, Prasanna R (2017) Development of *Mesorhizobium ciceri*-based biofilms and analyses of their antifungal and plant growth promoting activity in chickpea challenged by *Fusarium* wilt. Indian J Microbiol 57:48. https://doi.org/10.1007/s12088-016-0610-8

Davit Y, Byrne H, Osborne J, Pitt-Francis J, Gavaghan D, Quintard M (2013) Hydrodynamic dispersion within porous biofilms. Phys Rev E Stat Nonlinear Soft Math Phys 87:012718. https://doi.org/10.1103/PhysRevE.87.012718

DeMello JMM, Brandao HL, De Souza AAU, Da Silva A, DeSouza SMAGU (2010) Biodegradation of BTEX compounds in a biofilm reactormodeling and simulation. J Pet Sci Eng 70:131–139. https://doi.org/10.1016/j.petrol.2009.11.004

Deygout C, Lesnec A, Campilloa F, Rapaport A (2013) Homogenized model linking microscopic and macroscopic dynamics of a biofilm: application to growth in a plug flow reactor. Ecol Model 250:15–24. https://doi.org/10.1016/j.ecolmodel.2012.10.020

Dodds MG, Grobe KJ, Stewart PS (2000) Modeling biofilm antimicrobial resistance. Biotechnol Bioeng 68:456–465. https://doi.org/10.1002/(SICI)1097-0290(20000520)68

Ehret AE, Böl M (2013) Modeling mechanical characteristics of microbial biofilms by network theory. J R Soc Interface 10:20120676. https://doi.org/10.1098/rsif.2012.0676

Fagerlind MG, Webb JS, Barraud N, McDougald D, Jansson A, Nilsson P, Harlen M, Kjelleberg S, Rice SA (2012) Dynamic modeling of cell death during biofilm development. J Theor Biol 295:23–36. https://doi.org/10.1016/j.jtbi.2011.10.007

Frederick MR, Kuttler C, Hense BA, Eberl HJ (2011) A mathematical model of quorum sensing regulated EPS production in biofilm communities. Theory Biol Med Model 8:8. https://doi.org/10.1186/1742-4682-8-8

Friedman A, Hu B, Xue C (2014) On a multiphase multicomponent model of biofilm growth. Arc Rat Mec Anal 211:257–300. https://doi.org/10.1007/s00205-013-0665-1

Ghosh S, Qureshi A, Purohit HJ (2017) Enhanced expression of catechol 1,2 dioxygenase gene in biofilm forming *Pseudomonas mendocina* EGD-AQ5 under increasing benzoate stress. Int Biodeterior Biodegrad 118:57–65. https://doi.org/10.1016/j.ibiod.2017.01.019

Gokcen A, Vilcinskas A, Wiesner J (2013) Methods to identify enzymes that degrade the main extracellular polysaccharide component of *Staphylococcus epidermidis* biofilms. Virulence 4:260–270. https://doi.org/10.4161/viru.23560

Golus J, Stankevic M, Sawicki R, Los R, Malm A, Ginalska G (2013) Quantitative analysis of biofilm formed on vascular prostheses by *Staphylococcus epidermidis* with different *ica* and *aap* genetic status. Int J Artif Organs 36:105–112. https://doi.org/10.5301/ijao.5000157

Gui Z, Wang H, Ding T, Zhu W, Zhuang X, Chu W (2014) Azithromycin reduces the production of α-hemolysin and biofilm formation in Staphylococcus aureus. Indian J Microbiol 54:114–117. https://doi.org/10.1007/s12088-013-0438-4

Halan B, Buehler K, Schmid A (2012) Biofilms as living catalysts in continuous chemical synthesis. Trends Biotechnol 30:453–465. https://doi.org/10.1016/j.tibtech.2012.05.003

Houry A, Gohar M, Deschamps J, Tischenko E, Aymerich S, Gruss A, Briandet R (2012) Bacterial swimmers that infiltrate and take over the biofilm matrix. PNAS 109:13088–13093. https://doi.org/10.1073/pnas.1200791109

Judith HM, Daniel EK, O'Toole GA (2011) Growing and analyzing static biofilms. Curr Protoc Microbiol. Unit.1B.1. https://doi.org/10.1002/9780471729259.mc01b01s22

Kalia VC (2014) Microbes, antimicrobials and resistence: the battle goes on. Indian J Microbiol 54:1–2. https://doi.org/10.1007/s12088-013-0443-7

Kalia VC, Purohit HJ (2011) Quenching the quorum sensing system: potential antibacterial drug targets. Crit Rev Microbiol 37:121–140. https://doi.org/10.3109/1040841X.2010.532479

Kalia VC, Prakash J, Koul S, Ray S (2017) Simple and rapid method for detecting biofilm forming bacteria. Indian J Microbiol 57:109–111. https://doi.org/10.1007/s12088-016-0616-2

Kapley A, Purohit HJ (2009) Genomic tools in bioremediation. Indian J Microbiol 49:108–113. https://doi.org/10.1007/s12088-009-0012-2

Kaur J, Niharika N, Lata P, Rup L (2014) Biofilms: united we stand, divided we fall. Indian J Microbiol 54:246–247. https://doi.org/10.1007/s12088-014-0460-1

Kim J, Kim HS, Han S, Lee JY, Oh JE, Chung S, Park HD (2013) Hydrodynamic effects on bacterial biofilm development in a microfluidicMicrobial biofilms: current research and applications environment. Lab Chip 13:1846–1849. https://doi.org/10.1039/c3lc40802g

Koul S, Kalia VC (2017) Multiplicity of quorum quenching enzymes: a potential mechanism to limit quorum sensing bacterial population. Indian J Microbiol 57:100–108. https://doi.org/10.1007/s12088-016-0633-1

Koul S, Prakash J, Mishra A, Kalia VC (2016) Potential emergence of multi-quorum sensing inhibitor resistant (MQSIR) bacteria. Indian J Microbiol 56:1–18. https://doi.org/10.1007/s12088-015-0558-0

Lear G, Lewis GD (2012) Microbial biofilms: current research and applications. Caister Academic press, Wymondham. ISBN: 978-1-904455-96-7

Lee J, Wu J, Deng Y, Wang J, Wang C, Wang J, Chang C, Dong Y, Paul W, Zhang LH (2013) A cell-cell communication signal integrates quorum sensing and stress response. Nat Chem Biol 9:339–343. https://doi.org/10.1038/nchembio.1225

Liao N, Li H (2013) Conceivable bioremediation techniques based on quorum sensing. Appl Mech Mater 295:39–44. https://doi.org/10.4028/www.scientific.net/AMM.295-298.39

Liao Q, Wang YJ, Wang YZ, Chen R, Zhu X, Pu Y, Lee D (2012) Two-dimensional mathematical modeling of photosynthetic bacterial biofilm growth and formation. Int J Hydrog Energy 37:15607–15615. https://doi.org/10.1016/j.ijhydene.2012.03.056

Lin YH, Lin WF, Zhang KN, Lin PY, Lee MC (2013) Adsorption with biodegradation for decolorization of reactive black 5 by *Funalia trogii* 200800 on a fly ash chitosan medium in a fluidized bed bioreactor kinetic model and reactor performance. Biodegradation 24:137–152. https://doi.org/10.1007/s10532-012-9565-6

Llie O, van Loosedrechet MCM, Picioreanu C (2012) Mathematical modeling of tooth demineralization and pH profiles in dental plaques. J Theor Biol 309:159–175. https://doi.org/10.1016/j.jtbi.2012.05.024

Martin KJ, Piciorean C, Nerenberg R (2013) Multidimensional modeling of biofilm development and fluid dynamics in a hydrogen-based, membrane biofilm reactor (MBfR). Water Res 47:4739–4751. https://doi.org/10.1016/j.watres.2013.04.031

Mina P, di Bernardo M, Savery NJ, Atanasova KT (2013) Modelling emergence of oscillations in communicating bacteria: a structured approach from one to many cells. J R Soc Interface 10:20120612. https://doi.org/10.1098/rsif.2012.0612
Molloy S (2013) Biofilm microanatomy. Nat Rev Microbiol 11:300–301. https://doi.org/10.1038/nrmicro3024
Momeni B, Brileya KA, Fields MW, Shou W (2013) Strong inter-population cooperation leads to partner intermixing in microbial communities. Elife 2:e00230. https://doi.org/10.7554/eLife.00230
Moreira JMR, Teodósio JS, Silva FC, Simões M, Melo LF, Mergulhão FJ (2013) Influence of flow rate variation on the development of *Escherichia coli* biofilms. Bioprobiosys Eng 36:1787–1796. https://doi.org/10.1007/s00449-013-0954
Moustaid F, Eladdadi A, Uys L (2013) Modeling bacterial attachment to surfaces as an early stage of biofilm development. Math Biosci Eng 10:821–842. https://doi.org/10.3934/mbe.2013.10.821
Muthukaruppan S, Eswari A, Rajendran L (2013) Mathematical modeling of a biofilm: the Adomian decomposition method. Nat Sci 5:456–462. https://doi.org/10.4236/ns.2013.54059
Nadell CD, Bucci V, Drescher K, Levin SA, Bassler BL, Xavier JB (2013) Cutting through the complexity of cell collectives. Proc R Soc B 280:2012–2770. https://doi.org/10.1098/rspb.2012.2770
Ni BJ, Yu HQ (2010) Mathematical modeling of aerobic granular sludge: a review. Biotechnol Adv 28:895–909. https://doi.org/10.1016/j.biotechadv.2010.08.004
O'Toole GA (2011) Microtiter dish biofilm formation assay. J Vis Exp 30:2437. https://doi.org/10.3791/2437
Pal S, Qureshi A, Purohit HJ (2016) Antibiofilm activity of biomolecules: gene expression study of bacterial isolates from brackish and fresh water biofouled membranes. Biologia 71:239–246. https://doi.org/10.1515/biolog-2016-0045
Pantanella F, Valenti P, Natalizi T, Passeri D, Berlutti F (2013) Analytical techniques to study microbial biofilm on abiotic surfaces: pros and cons of the main techniques currently in use. Ann Ig 25:31–42. https://doi.org/10.7416/ai.2013.1904
Perez J, Picioreanu C, van Loosdrecht M (2005) Modeling biofilm and floc diffusion process based on analytical solution of reaction-diffusion equations. Water Res 39:1311–1323. https://doi.org/10.1016/j.watres.2004.12.020
Picioreanu C, Kreft JU, Vanloosedrecht MCM (2004) Particle based multidimensional multispecies biofilm model. Appl Environ Microbiol 70:3024–3040. https://doi.org/10.1128/AEM.70.5.3024-3040.2004
Piubeli F, de Lourdes Moreno M, Kishi LT, Silva FH, Garcia MT, Mellado M (2015) Phylogenetic profiling and diversity of bacterial communities in the death valley, an extreme habitat in the Atacama Desert. Indian J Microbiol 55:392–399. https://doi.org/10.1007/s12088-015-0539-3
Purkait D, Ahuja A, Bhattacharjee U, Singha A, Rhetso K, Dey TK, Das S, Sanjukta RK, Puro K, Shakuntala I, Banerjee A, Sharma I, Bhatta RS, Mawlong M, Guha C, Pradhan NR, Ghatak S (2016) Molecular characterization and computational modelling of New Delhi metallo-β -lactamase-5 from an *Escherichia coli* isolate (KOEC3) of bovine origin. Indian J Microbiol 56:182–189. https://doi.org/10.1007/s12088-016-0569-5
Qi X, Nellas RB, Byrn MW, Russell MH, Bible AN, Alexandre G, Shen T (2013) Swimming motility plays a key role in the stochastic dynamics of cell clumping. Phys Biol 10:026005. https://doi.org/10.1088/1478-3975/10/2/026005
Qureshi A, Mohan M, Kanade GS, Kapley A, Purohit HJ (2009) In-situ bioremediation of organochlorine pesticide contaminated microcosm soil and evaluation by gene probe. Pest Manag Sci 65:798–804. https://doi.org/10.1002/ps.1757
Qureshi A, Kapley A, Purohit HJ (2012) Degradation of 2,4,6-trinitrophenol (TNP) by *Arthrobacter* sp. HPC1223 isolated from effluent treatment plant. Indian J Microbiol 52:642–647. https://doi.org/10.1007/s12088-012-0288-5
Qureshi A, Pal S, Ghosh S, Kapley A, Purohit HJ (2015) Antibiofouling biomaterials. Int J Recent Adv Multidiscip Res (IJRAMR) 2:677–684

Radu AI, Vrouwenvelder JS, van Loosedrecht MCM, Picioreanu C (2010) Modeling the effect of biofilm formation on reverse osmosis performance: flux, feed channel pressure drop and solute passage. J Membr Sci 365:1–15. https://doi.org/10.1016/j.memsci.2010.07.036

Roberts ME, Stewart PS (2004) Modeling antibiotic tolerance in biofilms by accounting for nutrient limitation. Antimicrob Agents Chemother 48:48–52. https://doi.org/10.1128/AAC.48.1.48-52.2004

Samso R, Garcia J (2013) BIO_PORE, a mathematical model to simulate biofilm growth and water quality improvement in porous media: application and calibration for constructed wetlands. Ecol Eng 54:116–127. https://doi.org/10.1016/j.ecoleng.2013.01.021

Shang Z, Wang H, Zhou S, Chu W (2014) Characterization of N-acyl-homoserine lactones (AHLs)-deficient clinical isolates of *Pseudomonas aeruginosa*. Indian J Microbiol 54:158–162. https://doi.org/10.1007/s12088-014-0449-9

Sharma A, Lal R (2017) Survey of (meta)genomic approaches for understanding microbial community dynamics. Indian J Microbiol 57:23–38. https://doi.org/10.1007/s12088-016-0629-x

Shen L, Lu Y, Liu Y (2012) Mathematical modeling of biofilm-covered granular activated carbon: a review. J Chem Technol Biotechnol 87:1513–1520. https://doi.org/10.1002/jctb.3884

Stewart PS, Costerton JW (2001) Antibiotic resistance of bacteria in biofilms. Lancet 358:135–138. https://doi.org/10.1016/S0140-6736(01)05321-1

VanderWal A, Tecon R, Kreft JU, Mooij WM, Leveau JHJ (2013) Explaining bacterial dispersion on leaf surfaces with an individual-based model (PHYLLOSIM). PLoS One 8:e75633. https://doi.org/10.1371/journal.pone.0075633

Wang Q, Zhang T (2010) Review of mathematical models for biofilms. Sol St Comm 150:1009–1022. https://doi.org/10.1016/j.ssc.2010.01.021

Wang ZW, Hamilton-Brehm SD, Lochner A, Elkins JG, Morrell-Falvey JL (2012) Mathematical modeling of hydrolysate diffusion and utilization in cellulolytic biofilms of extreme thermophile *Caldicellulosiruptor obsidiansis*. Bioresour Technol 102:3155–3162. https://doi.org/10.1016/j.biortech.2010.10.104

Watrous JD, Phelan VV, Hsu CC, Moree WJ, Duggan BM, Alexandrov T, Dorrestein PC (2013) Microbial metabolic exchange in 3D. ISME J 7:770–780. https://doi.org/10.1038/ismej.2012.155

Welch K, Cai Y, Stromme M (2012) A method for quantitative determination of biofilm viability. J Funct Biomater 3:418–431. https://doi.org/10.3390/jfb3020418

Winter G, Krömer JO (2013) Fluxomics – connecting 'omics analysis and phenotypes. Environ Microbiol 15:1901–1916. https://doi.org/10.1111/1462-2920.12064

Yong YC, Zhong JJ (2013) Regulation of aromatics biodegradation by rhl quorum sensing system through induction of catechol meta-cleavage pathway. Bioresour Technol 136:761–765. https://doi.org/10.1016/j.biortech.2013.03.134

Yoon S, Kim K, Kim JK (2013) Live-cell imaging of swarming bacteria in a fluidic biofilm formed on a soft agar gel substrate. J Vis 16:123–131. https://doi.org/10.1007/s12650-013-0159-0

Zarraonaindia I, Smith DP, Gilbert JA (2013) Beyond the genome: community-level analysis of the microbial world. Biol Philos 28:261–282. https://doi.org/10.1007/s10539-012-9357-8

Zhang T, Cogan N, Wang Q (2008) Phase field models for biofilms. I. Theory and one-dimensional simulations. SIAM J Appl Math 69:641–669. https://doi.org/10.1137/070691966

Zhang W, Sileika T, Packman AI (2013) Effects of fluid flow conditions on interactionsbetween species in biofilms. FEMS Microbiol Ecol 84:344–354. https://doi.org/10.1111/1574-6941.12066

Microbial CO_2 Fixation Bioprocesses and Desert as Future Carbon Sink

16

Leena Agarwal, Nishant A. Dafale, and Hemant J. Purohit

Abstract

Increasing levels of carbon dioxide (CO_2) in the atmosphere are causing serious effect on climatic changes. Thereby the most concerned environmental issue is global warming these days. The same culprit carbon dioxide is also involved in the most important process called autotrophy, which supports life on earth. Autotrophy could be one of the probable answers for the management of atmospheric CO_2 in an eco-friendly manner. Plants contribute a major part in CO_2 fixation but often ignored microbes are also involved in CO_2 fixation by different CO_2 fixation pathways. Since desert represents sparse green cover, we cannot ignore the microbes present in such soil, which would be contributing for carbon fixation. Oligotrophic soil like desert harbours microbes that are capable of surviving in low-carbon conditions, thus opening an area of research for the study of CO_2 fixation by oligotrophs. Since microbes require carbon for their growth, and hence under carbon starvation, they might have the mechanism for autotrophy. By exploring CO_2 fixation through oligotrophs of desert, we could probably enrich the desert soil by carbon in near future. Such microbe-based CO_2 fixation studies could help us to channelize CO_2 for the synthesis of some important bioproducts.

Keywords

Desert • CO_2 fixation • Autotroph • Oligotrophs • Carbon sink • Carbon cycles

L. Agarwal • N.A. Dafale • H.J. Purohit (✉)
Environmental Biotechnology and Genomics Division, CSIR – National Environmental Engineering and Research Institute (CSIR-NEERI), Nehru Marg, Nagpur 440020, Maharashtra, India
e-mail: hj_purohit@neeri.res.in; hemantdrd@hotmail.com

H.J. Purohit et al. (eds.), *Optimization and Applicability of Bioprocesses*,
https://doi.org/10.1007/978-981-10-6863-8_16

16.1 Introduction

Climate change is a universal question associated with an increase in CO_2 levels. This question is now of scientific interest since increased CO_2 poses serious climate change effects. Climate change is a global environmental challenge that needs targeted research and development. Gaining a greater knowledge of how carbon cycles operate in ecosystems and potential change that might occur due to increase in anthropogenic greenhouse gases will help us to develop solutions to reduce the future increase in CO_2 and other greenhouse gases. Increased atmospheric CO_2 has crossed its safety limit of 350 ppm since early 1988 (Fig. 16.1). The world's atmospheric CO_2 measurements are reported by the National Oceanic and Atmospheric Administration (NOAA) and Scripps Institution of Oceanography situated at the Mauna Loa Observatory (MLO) in Hawaii. Being CO_2 a gas, it gets uniformly mixed in the atmosphere, so observations of its concentrations from a single site like the MLO represent world atmospheric CO_2 trends. The average annual atmospheric CO_2 concentration (reported by Mauna Loa Observatory) was 407.57 ppm in 2016 and 403.45 ppm in 2015 (https://www.co2.earth/ data as on 22.05.16) (Table 16.1).

16.2 Biological CO_2 Fixation

CO_2 fixation is one of the most important bioprocesses since all ecosystems depend directly or indirectly on the organic carbon made available by autotrophic CO_2 fixation either by plants or by microbes. In addition to photosynthetic CO_2 fixation by green plants, algae and cyanobacteria, chemolithoautotrophs and Archaea also contribute to CO_2 fixation. Chemolithoautotrophy is distributed amongst both archaea and bacteria with highly diverse phylogeny, metabolic activities and ecological

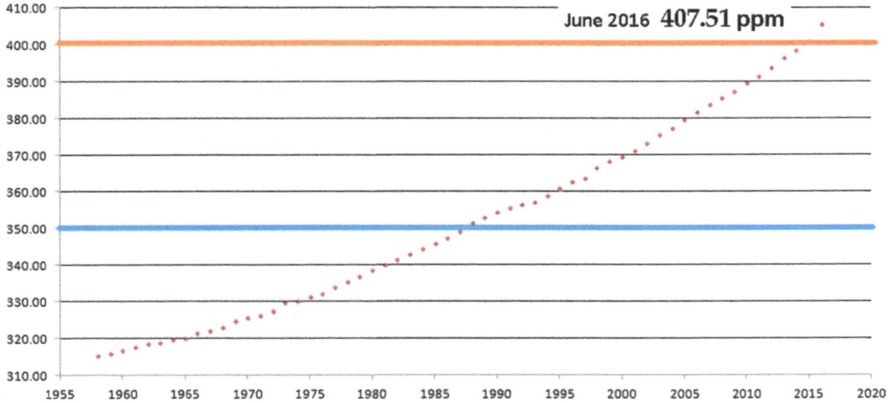

Fig. 16.1 Global status of atmospheric CO_2 (in ppm) since 1958–2016 (https://www.co2.earth/)

Table 16.1 CO_2 fixation pathway, key enzymes and their ecological distribution

SN.	Pathways	Enzymes[a]	Microbes	Carbon species	Identified organisms
1	Calvin-Bensoncycle	A – Rubisco, phosphoribulokinase	Plants, algae, Cyanobacteria and most of the proteobacteria	CO_2	Cyanobacteria – *Synechocystis* sp. (Yang et al. 2002), *Anabaena variabilis* (Berberoglu et al. 2008), *Anacystis nidulans*, *Synechococcus* sp., *Prochlorococcus* sp. (Badger and Bek 2008), *Oscillochloris* trichoides (Kuznetsov et al. 2011);
		B –rubisco			*α-proteobacteria* – *Oligotropha carboxidovorans* (Fuhrmanna et al. 2003); *Rhodobacter capsulatus*, *Rhodobacter sphaeroides*, *Xanthobacter autotrophicus Py2, Xanthobacter flavus*, *Alcaligenes eutrophus*, *Paracoccus denitrificans 1222*, *Sinorhizobium meliloti 1021*, *Bradyrhizobium japonicum USDA 110*, *Nitrobacter winogradskyi*, *Nitrobacter hamburgensis*, *Rhodospirillum rubrum ATCC 11170*, *Acidophilum cryptum JF-5*, *Rhodopseudomonas sphaeroides*, *R. palustris*, *Rhodopseudomonas acidophila* (Shively et al. 1998);
					β-proteobacteria – *Aquaspirillum autotrophicum* (*Herbaspirillum autotrophicum*) (Ding and Yokota 2004), *Rhodospirillum rubrum* and *Ralstonia eutropha*, *Cupriavidus metallidurans CH34*, *Thiomonas intermedia K12*, *Thiobacillus denitrificans* sp., *Nitrosospira multiformis*, *Nitrosomonas europaea*, *Nitrosomonas eutropha C91*, *Hydrogenophaga pseudoflava*, *Burkholderia xenovorans LB400*, *Ralstonia eutropha H16* (Badger and Bek 2008);
					γ-proteobacteria – *Allochromatium vinosum*,*Nitrosococcus oceani*, *Halorhodospira halophila*, *Alkalilimnicola ehrlichii*, *Nitrococcus mobilis*, *Halothiobacillus neapolitanus*, *Methylococcus capsulatus*, *Thiomicrospira crunogena*, *Hydrogenovibrio marinus*, *Acidithiobacillus ferrooxidans* (Badger and Bek 2008)

(continued)

Table 16.1 (continued)

SN.	Pathways	Enzymes[a]	Microbes	Carbon species	Identified organisms
2	Reductive TCA cycle	A – ATP citrate lyase, 2-oxoglutarate synthase	Green sulphur bacteria (*Chlorobiales*), sulphur-reducing bacteria, hydrogen-oxidizing bacteria and archaea	CO_2	*Chlorobiales* – *Chlorobium limicola* and *Chlorobium tepidum* (*Chlorobaculum tepidum*) (Campbell and Cary 2004)
		B – 2-oxoglutarate synthase, isocitrate dehydrogenase, pyruvate synthase, PEP carboxylase		HCO_3^-	Crenarchaeota – *Thermoproteus neutrophilus* (Schafer et al. 1989) and *Pyrobaculum islandicum* (Hu and Holden 2006)
					α-proteobacteria: *Magnetotactic coccus strain MC-1* (Williams et al. 2006)
					Aquificales – *Aquifex aeolicus* (Williams et al. 2006), *Aquifex pyrophilus* (Beh et al. 1993), *Hydrogenobacter thermophilus* (Campbell and Cary 2004), *Hydrogenobacter hydrogenophilus* (*Thermocrinis ruber*), *Sulfurihydrogenibium subterraneum*, *Thermovibrio ammonificans*, *Thermovibrio ruber*, *Desulfurobacterium crinifex*, *Desulfurobacterium thermolithotrophum* (Hügler et al. 2007)
					δ-proteobacteria (*Desulfobacterales*) –*Desulfobacter hydrogenophilus* (Campbell and Cary 2004);
					Campylobacterales (*ε-proteobacteria*) – *Thiomicrospira denitrificans* and *Candidatus Arcobacter sulfidicus* (Hügler et al. 2005)
3	Reductive acetyl CoA pathway	A – acetyl CoA synthase-CO dehydrogenase	*Firmicutes*, *Planctomycetes*, *Deltaproteobacteria*, *Spirochaeta*, Euryarchaeota	CO_2	Autotrophic acetogens and methanogens
		B- acetyl CoA synthase-CO dehydrogenase, formylmethanofuran dehydrogenase, pyruvate synthase		HCO_3^-	*Desulfobacterium autotrophicum*, *Desulfovibrio baarsii*, *Methanobacterium thermoautotrophicum*
					Methanosarcina barkeri, *Methanopyrus*
					Methanococcus, and *Methanothermus* (Berg 2011; Berg et al. 2010), *Clostridium thermoaceticum*, *Clostridium formicoaceticum*, *Acetobacterium woodii* (Saini et al. 2011)

4	3-hydroxypropionate bicycle	A – malonyl-CoA reductase, propionyl-CoA synthase and malyl-CoA lyase	Chloroflexaceae	HCO_3^-	Phototrophic green non-sulphur bacterium *Chloroflexus aurantiacus* OK-70, (Berg et al. 2010; Hügler and Sievert 2011; Berg 2011); some autotrophic archaea such as *Sulfolobus metallicus*, *Metallosphaera sedula*, *Acidianus brierleyi* and *Acidianus infernus* (Saini et al. 2011)
		B – acetyl CoA carboxylases, propionyl-CoA carboxylases			
5	3-hydroxypropionate-4-hydroxybutyrate cycle	A – acetyl CoA carboxylase, propionyl-CoA carboxylase	Crenarchaeota	HCO_3^-	Aerobic autotrophic members of Sulfolobales; *Metallosphaera sedula*, *Acidianus brierleyi* (Hügler and Sievert 2011)
		B – acetyl CoA carboxylase, propionyl- CoA carboxylase			
6	Dicarboxylate-4-hydroxybutyrate cycle	A–4-hydroxybutyryl-CoA dehydratase	Crenarchaeota	CO_2	*Ignicoccus hospitalis* (Desulfurococcales), *Thermoproteus neutrophilus* (Thermoproteales) (Hügler and Sievert 2011; Berg et al. 2010)
		B – pyruvate synthase, PEP carboxylase		HCO_3^-	

[a]A – key enzyme/gene; B – CO_2-fixing enzyme/gene

variants (Hügler and Sievert 2011). Bacterial chemolithotrophs have been investigated in diverse environments (Campbell et al. 2003; Ferrera et al. 2007; Hügler and Sievert 2011). In plants, CO_2 fixation occurs by Calvin-Benson-Bassham (CBB) cycle, while in prokaryotes other than CBB cycle, there are five alternative pathways for CO_2 fixation, viz. (a) reductive acetyl CoA (rACA) pathway, (b) reductive tricarboxylic acid (rTCA) cycle, (c) 3-hydroxypropionate (3-HP) bicycle, (d) 3-hydroxypropionate/4-hydroxybutyrate (3-HP/4-HB) cycle and (e) dicarboxylate/4-hydroxybutyrate cycle (Berg et al. 2010; Hügler and Sievert 2011).

16.2.1 Microbial Autotrophy

Microbial autotrophy annually accounts for roughly 4% of the total CO_2 fixed by terrestrial ecosystem and desert are the substantial compartment of the terrestrial ecosystem, which is often ignored in microbial ecology. The skill of using CO_2 as a carbon source for survival is called as *autotrophy*, which exists in many groups of prokaryotes. Based on energy sources used for CO_2 assimilation, growth is of two types – photosynthetic and chemosynthetic. Chemosynthetic growth refers to the oxidation of simple inorganic substances, whilephotosynthetic growth refers to light. For chemoautotrophic growth, the electron donor could be H_2, reduced nitrogen as NH^{4+} and NO_2^-, sulphur as $S_2O_3^{2-}$ and H_2S, metals as Fe^{2+} and Mn^{2+} and carbon as CO, CH_4 and CH_3OH compounds. Microorganisms fix CO_2 through the six known CO_2 fixation pathways. The ecological distribution of CO_2 fixation pathways within different groups of organisms (Fig. 16.2) depends on phylogeny, energy source, available reducing compounds, type of metals, coenzymes and the enzyme sensitivity to oxygen (Berg 2011).

The six known microbial CO_2 fixation pathways with their marker genes are described below.

16.2.1.1 Calvin-Benson-Bassham cycle

This is the most widespread and well-studied pathway for CO_2 fixation (Fig. 16.2), found in aerobic or aerotolerant bacteria (Shively et al. 1998; Berg 2011). Globally significant CBB cycle is limited to organisms with high-energy yield requiring nine ATP molecules and six NADH to synthesize one molecule of triose phosphate by using three molecules of CO_2. Besides many photosynthetic eukaryotes (plants and algae), numerous prokaryotes have also been found to rely on the CBB cycle for CO_2 fixation. The CBB cycle is restricted to few thermophiles since its upper temperature limit for operation is ~70 to 75 °C. It is also reported in purple sulphur and non-sulphur bacteria, cyanobacteria, hydrogen-oxidizing and other chemoautotrophic bacteria. The two unique enzymes of the cycle are ribulose 1,5-bisphosphate carboxylase (rubisco) and phosphoribulokinase, while other enzymes of the cycle are shared with gluconeogenesis and pentose phosphate pathway (Shively et al. 1998). In the cycle, carboxylation reaction is catalysed only by enzyme rubisco. Microbial autotrophs are difficult to culture, and very few studies reported for rubisco activity in soil (Yuan et al. 2012).

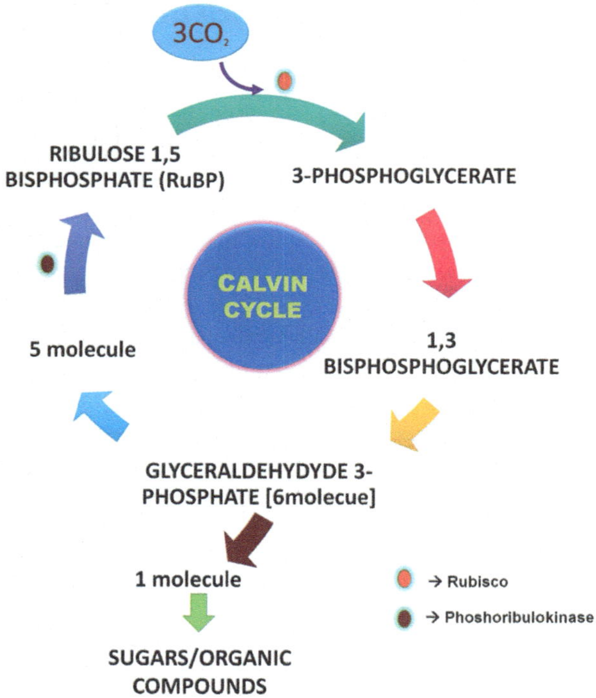

Fig. 16.2 Calvin-Benson-Bassham cycle represents fixation of three molecules of CO_2 and the key enzymes of the cycle

(a) *Ribulose 1.5-bisphosphate carboxylase (Rubisco)*

Rubisco is a copious protein (enzyme) on the planet and is the major entryway for inorganic carbon (CO_2) to enter biosphere. Two molecules of 3-phosphoglycerate (3-PGA) are produced from a single molecule of ribulose 1,5-bisphosphate (RuBP), CO_2 and H_2O, while the rest 12 reactions of the cycle regenerate RuBP. The cycle fixes three molecules of CO_2 to form three molecules of RuBP and six molecules of 3-PGA. Out of six PGA molecules, only one is used for the biosynthesis of cell material, and the remaining five are used to regenerate three molecules of RuBP. There are four types of rubisco (I–IV). Forms I, II and III are known for carboxylation reaction in different groups of bacteria, while rubisco-like protein or Form IV does not catalyse carboxylation reactions due to lack of amino acid residues required for the reaction but is found in *Archaeoglobus fulgidus*, *Bacillus subtilis* and *Chlorobium tepidum* (Hanson and Tabita 2001; Selesi et al. 2005). Rubisco Form I is composed of large octamer and small octamer and is encoded by *cbb* gene (Badger and Bek 2008). This form is present in photosynthetic and aerobic chemolithotrophs. Based on phylogenetic analysis of *cbbL* gene, there exist green-like and red-like rubisco protein groups. Green-like protein sequences are observed in algae; α-, β- and γ-*proteobacteria*; and plants, while red-like proteins sequences are present in α- and β-*proteobacteria* and nongreen algae (Selesi et al. 2005).

16.2.1.2 Reductive Tricarboxylic Acid cycle (rTCA or Arnon-Buchanan cycle)

The cycle (Fig. 16.3) was first proposed by Evans et al. (1966), supported by the discovery of ferredoxin-dependent pyruvate synthase and 2-oxoglutarate synthase; later the presence of ATP citrate lyase was also confirmed, thus allowing a reverse flux of the well-known TCA (Sintsov et al. 1980). The cycle fixes two molecules of CO_2 to synthesize single acetyl CoA molecule. The acetyl CoA can be converted to pyruvate and phosphoenolpyruvate, which are involved in anaplerotic reactions of the cycle catalysed by pyruvate synthase and phosphoenolpyruvate carboxylase. The cycle consumes ATP and reducing agents (such as NADH and FADH2) and brings CO_2 from the environment into metabolism. It operates in an environment where plenty of CO_2 is available, but there is very little organic carbon. Also plenty of reducing agents are available (such as dissolved H_2 gas) for making reduced cofactors (such as NADH). This cycle requires lesser energy comparable to Calvin cycle. It is found in anaerobic or microaerophilic bacteria and archaea. The photoautotrophic growth of green sulphur bacteria including *Chlorobium limicola* (earlier

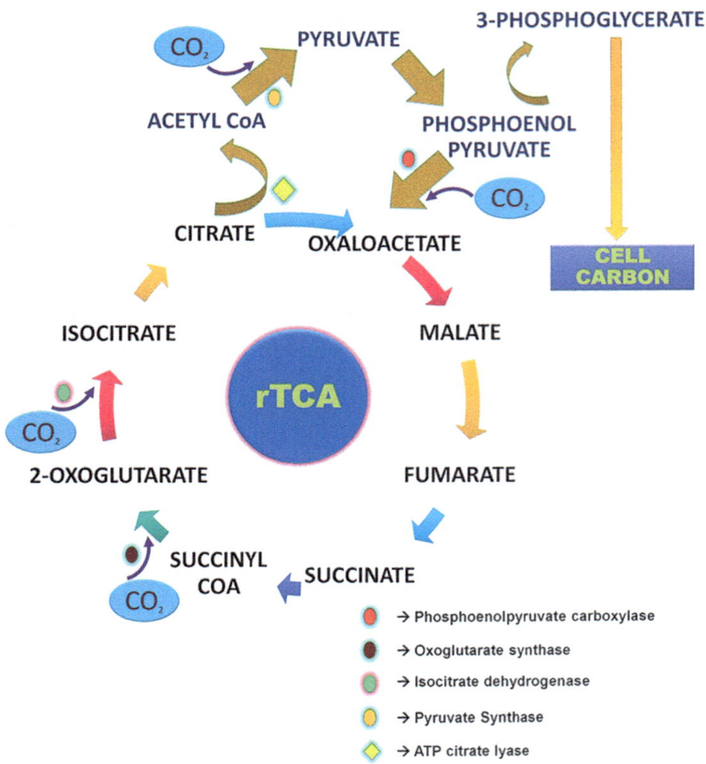

Fig. 16.3 Reductive tricarboxylic acid cycle representing fixation of two molecules of CO_2 and the key enzymes and CO_2-fixing enzymes of the cycle

C. thiosulfatophilum) and *Chlorobaculum tepidum* fixes CO_2 by rTCA (Sintsov et al. 1980; Tang and Blankenship 2010). Members of *ε-proteobacteria* also demonstrate the abundance of rTCA genes (Hügler et al. 2005). The key enzymes of the cycle are 2-oxoglutarate synthase (2-oxoglutarate: ferredoxin oxidoreductase) and the citrate-breaking enzymes, viz. ATP citrate lyase or citryl-CoA lyase (CCL), whereas other enzymes of the cycle are common for both oxidative and reductive TCAs.

(a) *ATP Citrate lyase (ACL)*

It is an enzyme of rTCA that makes the cycle to run in reverse direction in contrast to oxidative TCA, by cleaving citrate into oxaloacetate (Fig. 16.4) and acetyl CoA. ACL is dependent on CoA and MgATP. It is encoded by *aclBA* gene which is composed of two subunits. In some bacteria, this breakdown reaction is catalysed by joint action of citryl-CoA lyase (CCL) and citryl-CoA synthetase (CCS) (Fig. 16.5).

(b) *2-Oxoglutarate: Ferredoxin Oxidoreductase*

It catalysis the carboxylation of succinyl CoA to synthesize 2-oxoglutarate and is encoded by *oorDABC* gene consisting of four subunits (Kanao et al. 2001).

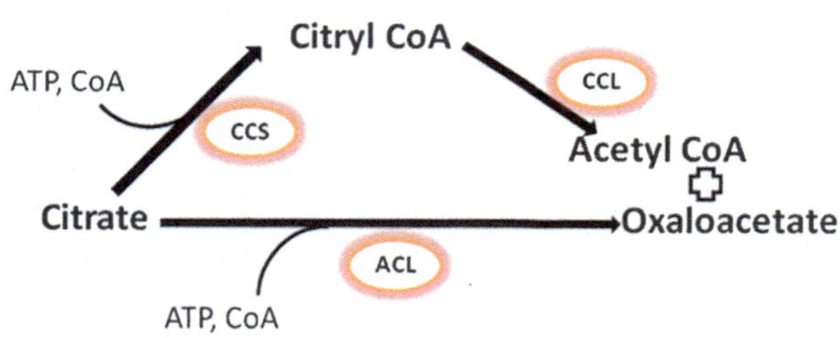

Fig. 16.4 Schematic representation of citrate cleavage reactions within rTCA cycle

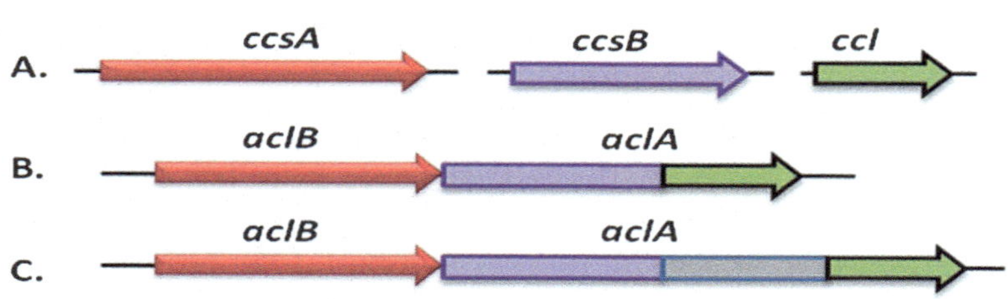

Fig. 16.5 Genetic organization of citrate cleavage gene in different organisms. (**a**) Citryl-CoA synthetase (*ccsA and ccsB*) and citryl-CoA lyase (*ccl*) found in *Aquificaceae* and *Leptospirillum* spp.; (**b**) ATP citrate lyase type I found in *Chlorobiales*, *Epsilonproteobacteria*, *Hydrogenothermaceae*, *Desulfurobacteriaceae* and *Nitrospira*; (**c**) ATP citrate lyase type II found in *Magnetococcus* sp. MC-1 and *Riftia pachyptila* endosymbiont (*CCS* citryl-CoA synthetase, *CCL* citryl-CoA lyase, *ACL* ATP citrate lyase)

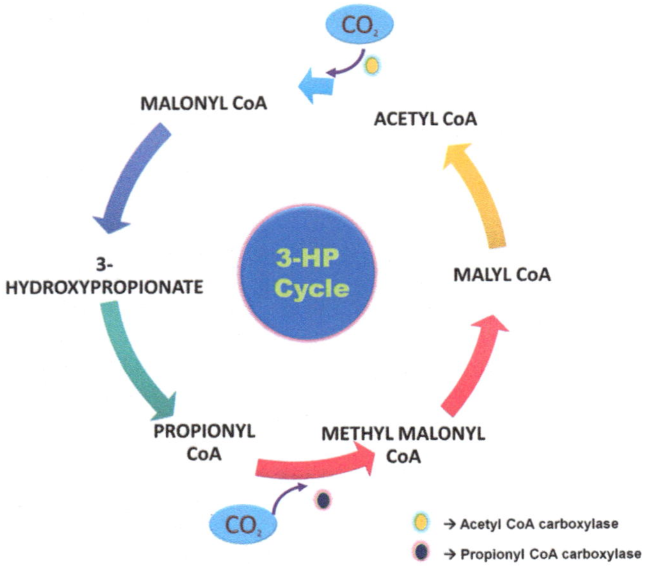

Fig. 16.6 3-hydroxypropionate pathway

16.2.1.3 3-Hydroxypropionate (3-HP) cycle

It is a very old pathway of CO_2 fixation (Fig. 16.6) discovered in a facultative aerobic phototrophic bacterium *Chloroflexus aurantiacus* (Holo 1989). The pathway fixes two molecules of CO_2 to synthesize glyoxylate by the action of acetyl CoA and propionyl-CoA carboxylases and regenerates acetyl CoA. The glyoxylate assimilation into cell material occurs by glyoxylate pathway. The cycle differs in different members of archaea and evolves as 3-HB bicycle in members of the family *Chloroflexaceae*, *hydroxypropionate/hydroxybutyrate cycle* in aerobic Crenarchaeota and *dicarboxylate-hydroxybutyrate cycle* in anaerobic members of Crenarchaeota (Berg et al. 2010).

(a)*Acetyl CoA Carboxylase*

In bacteria and eukaryotes, it serves as a carboxylation enzyme for acetyl CoA to form malonyl-CoA (Fig. 16.7). This reaction serves as the first step of fatty acid synthesis in most of the organisms, while it is a marker gene for CO_2 fixation in archaea (Auguet et al. 2008). It fixes inorganic CO_2 (or bicarbonate) in 3-hydroxypropionate (3-HP) bicycle and 3-hydroxypropionate/4-hydroxybutyrate cycles by *acc* gene. The enzyme is composed of three subunits, viz. biotin carboxylase (*accC gene*), biotin carboxyl carrier protein (*accB* gene) and carboxytransferase (*pccB* gene) (Auguet et al. 2008; Berg et al. 2010).

Fig. 16.7 Partial reaction catalysed by acetyl CoA carboxylase gene

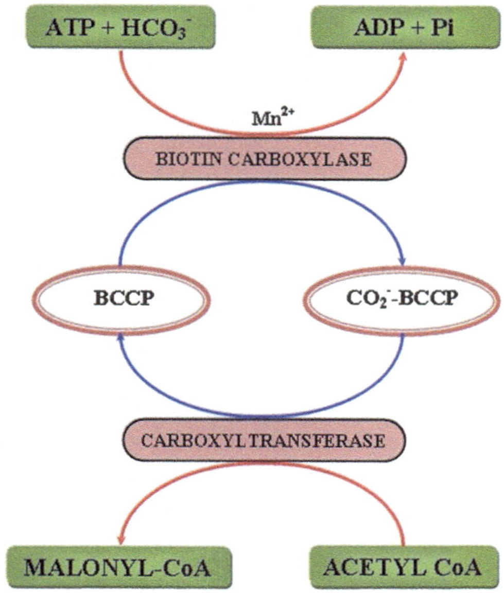

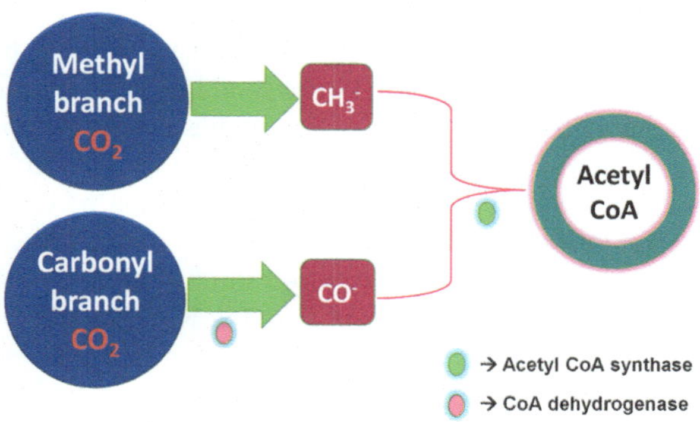

Fig. 16.8 Reductive acetyl CoA (rACo) pathway

16.2.2 Reductive Acetyl CoA Pathway

It a noncyclic pathway of CO_2 fixation (Fig. 16.8), which is restricted to the strict anaerobes, acetogenic bacteria and methanogens (Hügler and Sievert 2011). The CO_2-fixing enzyme of the pathway is carbon monoxide dehydrogenase (CODH)/acetyl CoA synthase. The pathway reduces CO_2 by methyl and carbonyl branch both in bacteria and archaea; the only difference is bacteria produce acetate and archaea produce methane (Borrel et al. 2016). It is also found in Euryarchaeota, viz. denitrifying *Ferroglobus placidus* and sulphate-reducing *Archaeoglobus lithotrophicus* (Berg 2011).

16.3 Carboxylases

About 98% of the biological CO_2 fixation occurs by the action of carboxylases (Thauer et al. 2008), thus playing a critical role in controlling atmospheric CO_2 levels. Carboxylases catalyse the incorporation of CO_2 molecules into an organic substrate. Erb (2011) has defined five different types of carboxylases, viz. (a) autotrophic carboxylases, (b) assimilatory carboxylases, (c) anaplerotic carboxylases, (d) biosynthetic carboxylases and (e) redox-balancing carboxylases. The enzyme phosphoenolpyruvate carboxylase (PEPC) and other carboxylases are major routes of anaplerotic reaction in bacteria. Phosphoenolpyruvate carboxylase is widely distributed amongst plants and bacteria, required for the replenishment of TCA intermediates (Erb 2011). It is a carboxylating enzyme in rTCA and dicarboxylate/4-hydroxybutyrate cycle (Berg et al. 2010). It catalyses the carboxylation of phosphoenolpyruvate to fix one molecule of CO_2. PEPC is a homotetramer with a molecular mass 100 kDa of each subunit. The enzyme phosphoenolpyruvate (PEP) carboxylase is faster than rubisco and has lower Km for bicarbonate.

16.4 Dissecting Microbial Community Existing in Carbon-Starved Environment Like Desert

Desert represent extreme environment for microorganisms due to extreme temperature, water stress, high salinity, nutritional scarcity and elevated UV radiation. These factors contribute to soil sterility as per earlier studies due to the observation of a low number of cultivable or viable microorganisms. But with the advent of molecular tools, the picture was very contrasting. The application of various molecular tools to study microbial community in desert showed a broader range of diversity in desert soil like of Atacama and Antarctica also (Smith et al. 2006; Warren-Rhodes et al. 2006; Pointing et al. 2009; Yergeau et al. 2009; Köberl et al. 2011; Ambardar et al. 2016). Since desert soils are a substantial part of the terrestrial ecosystem and could act as a sink for CO_2. The ecosystems with lesser plant cover like desert favour the existence of diverse photoautotrophic microbes that obtain carbon from the atmosphere (Wood et al. 2008; Lacap et al. 2011). Microbial autotrophy in desert could increase the organic content of the soil thereby supporting life in the near future (Agarwal et al. 2014). However, it is unknown how these microbes obtain carbon and energy required to maintain their life in such an oligotrophic surrounding. Earlier attempts had been made by various researchers to untie indigenous microbial community of cold deserts, viz. Atacama and Antarctica, extensively, while hot desert microbial population still is less explored; also the community knowledge is not correlated with carbon fixation. Here in the following sections, this community knowledge of desert microbes is linked with a functional aspect for CO_2 fixation.

16.5 Microbial Community in Cold Desert

16.5.1 Atacama Desert

Atacama Desert is the most arid regions on earth located in South America. Due to scarcity of liquid water, it represents dry limits for life on earth where Yungay is the hyperarid region at Atacama Desert (Warren-Rhodes et al. 2006). Earlier it was considered as Mars-like lifeless due to the failure of soil DNA isolation from this site (Navarro-González et al. 2003). To challenge these observations, Lester et al. (2007) analysed total organic carbon, phospholipid fatty acids and culturable bacteria count of the soil samples of this region. They have isolated *α-proteobacteria* and firmicute. Both hot and cold deserts represent the occurrence of lithic microbial colonization that principally supports autotrophic microbial communities. Quartz stone in Atacama Desert (Fig. 16.9) was reported to be colonized both red and green hypolithic microbial community (Lacap et al. 2011).

The bacterial diversity of Atacama Desert is unearthed based on 168 nucleotide sequences of 16S rDNA gene retrieved from the finding of the work carried out by Warren-Rhodes et al. (2006), Lester et al. (2007), Dong et al. (2007), Connon et al.

Fig. 16.9 Hypolithic colonization by cyanobacteria in deserts

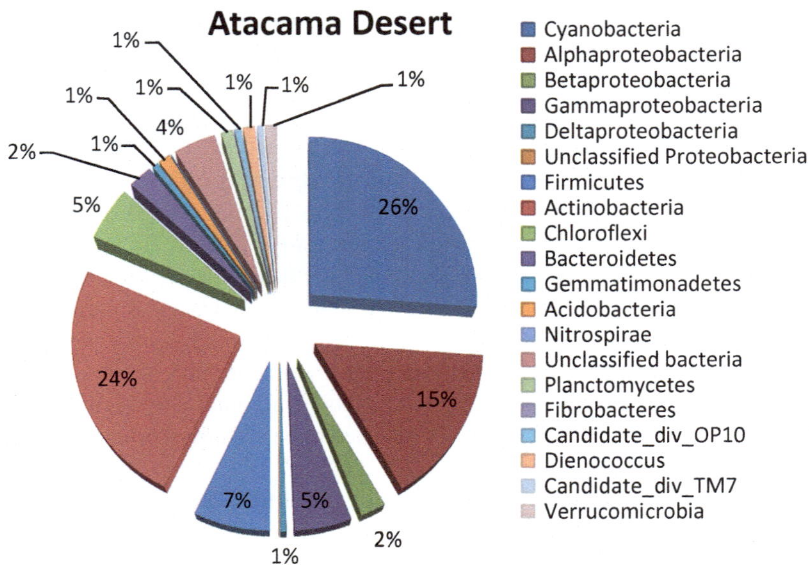

Fig. 16.10 Pictorial representation for bacterial community composition of Atacama Desert, 168 clone data adapted from the published 16S rDNA sequences affiliated to Atacama Desert

(2007) and Lacap et al. (2011). This suggests the dominance of cyanobacteria, actinobacteria and proteobacteria, while (Fig. 16.10) alphaproteobacteria dominates proteobacteria. Atacama being the water-limited habitat that showed the presence of lithic photoautotrophic microbial communities provides an example of very simple structure deprived of plants and higher forms.

The investigation of Connon et al. (2007) reported *Actinobacteria* as a dominant group accounting 94% from the total 16S rRNA gene sequences, and most of them belonged to *Frankia* genus or are representatives of new genera. They showed fewer representatives of group firmicute and proteobacteria. In contrast, the study of Warren-Rhodes et al. (2006) on hypolithic community in Atacama Desert demonstrated the dominance of cyanobacteria members since the rock provides microhabitat which provides comparatively enhanced water availability for phototrophic life forms. In another study, to untie the microbial hypolithic community conducted by Lacap et al. (2011) showed different lithic microbes dominate in different types of hypoliths and give them a visual colour distinction. They reported that the red hypolithic community was populated by diverse group of chloroflexi member, while the green was colonized by cyanobacteria, to reflect the photoautotrophic component (Lacap et al. 2011). The phylotypes of nitrogen-fixing alphaproteobacteria were found to be associated with green hypoliths only, while *Gemmatimonadetes* phylotypes occurred only on red hypoliths. The desiccation-resistant group *Actinobacteria* was found equally distributed in both hypoliths. This study was based on clone library, while DGGE-based analysis was performed by

Warren-Rhodes et al. (2006) to analyse microbial diversity at Atacama. The clone library data increased the information by double or more thus allowing accurate phylogenetic assignment of rRNA gene than the fingerprinting approach (Lacap et al. 2011). This has overcome the restrictions of short sequence data generated by DGGE technique. Another metagenomic study of Atacama Desert from Death Valley showed predominance of microbial community affiliated to the genus *Oceanobacillus*, reflecting the alkaline nature of the habitat (Piubeli et al. 2015). In contrast to other studies from Atacama, authors observed less diverse population; moreover the phylum *Firmicute* is overrepresented.

16.5.2 Antarctica Desert

Antarctica desert is a hyperarid polar cold desert characterized by extreme cold and dry conditions, also elevated soil salinity, nutrient deprived and physical instability due to katabatic winds, all contributing to sterile environment at this niche. Despite this limiting environment, life does exist here. The culture-independent analysis reveals the occurrence of lithic communities and displayed diversity. The microbial community analysis of Antarctica desert is based on 215 nucleotide sequences of 16S rDNA gene retrieved from the studies of Smith et al. (2006) and Pointing et al. (2009). The data of Antarctic dry valley suggests the dominance of photoautotrophic member *Cyanobacteria* (~25%) followed by *Actinobacteria* (~20%) (Fig. 16.11). The 13% of population was affiliated with unclassified bacteria

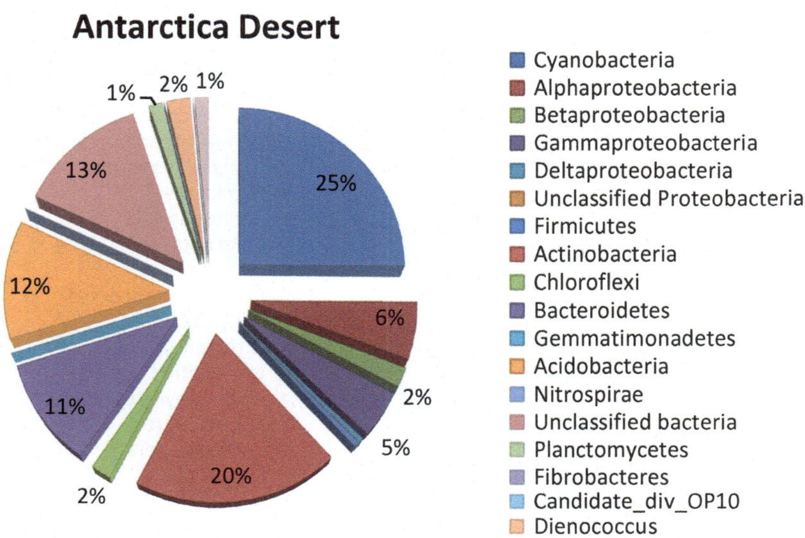

Fig. 16.11 Pictorial representation for bacterial community composition of Antarctica desert based on the 16SrDNA sequence data of 215 clones adapted from the published sequences affiliated with Antarctica desert

suggesting the existence of unreported polar bacteria. The members of group alpha-proteobacteria (~6%) represent nitrogen fixers affiliated to *Sphingomonadales*, *Rhizobiales* and *Rhodospirillales*. The dominant autotroph member in Antarctica is cyanobacteria, but few representative of *Bacteroidetes*, *Planctomycetes* and *Chloroflexi* were also reported but in very few numbers.

16.6 Microbial Community in Hot Desert

In contrast to cold desert, hot desert represents other extreme conditions for the survival of microbes and plants. The microbial community distribution in hot desert like Thar is governed by climatic drivers (rainfall and temperature) and substrate drivers (conductance, pH and salts) (Rao et al. 2016). The presence of hypolithic community is also observed in hot deserts (Dong et al. 2007).

To account the microbial community of hot desert, we included different hot deserts since microbial community in hot desert is not explored extensively. Therefore, the combined data of Jordan desert, Mojave Desert, Sahara desert, Chihuahuan Desert and Thar Desert were included from the reports of Kuhlman et al. (2006), Souza et al. (2006), Chowdhury et al. (2007), Dong et al. (2007), Suárez et al. (2008) and Yousuf et al. (2012) that account for total 344 nucleotide sequences of 16S rDNA gene.

At 97% similarity level, 309 OTUs were identified from the 344 rDNA sequences, which (Fig. 16.12) suggest dominance of proteobacteria (~42.2% classified and

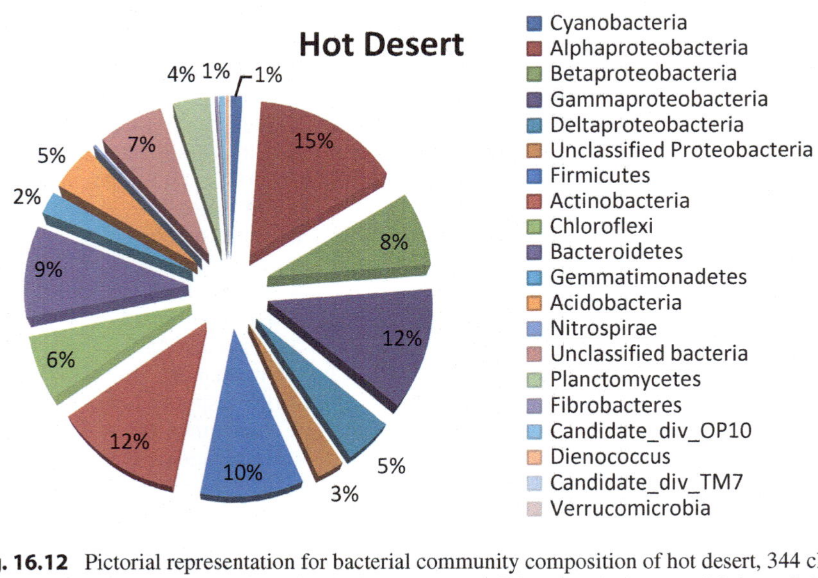

Fig. 16.12 Pictorial representation for bacterial community composition of hot desert, 344 clone data adapted from the published 16S rDNA sequences affiliated to hot desert (Mojave, Sahara, Chihuahuan, Thar, Jordan desert)

~2.6% unclassified proteobacteria); amongst them alphaproteobacteria was dominant followed by gammaproteobacteria (~12%), betaproteobacteria (~8%) and deltaproteobacteria (~5%). The next most abundant members were affiliated to *Actinobacteria* (~12%) and *Firmicute* (~10%), known for their survival in stress. Besides the cyanobacteria and proteobacteria, other members which belonged to chemolithoautotroph were bacteroidetes (~9%) and chloroflexi (~6%). There were ~7% unclassified bacteria representing novel representatives that have not been studied yet. The photoautotrophic member, cyanobacteria, was present in few number (~1%) demonstrating dry conditions in hot desert but also indicating that the lithic communities of hot desert are still unexplored.

The rarefaction curve for Atacama and Antarctica desert reached saturation while of hot desert does not reach asymptote at 20% (phylum level) dissimilarity indicating more species richness in hot desert comparable to cold desert (Fig. 16.13). Hence the surveying effort of cold desert has covered almost complete taxonomic diversity at 80% similarity cut-off, while the microbial community related to hot deserts requires more attention and exploration. The calculated Shannon diversity indices (H′) for hot desert were higher than cold desert, which reflects higher bacterial diversity in hot desert soil (Table 16.2).

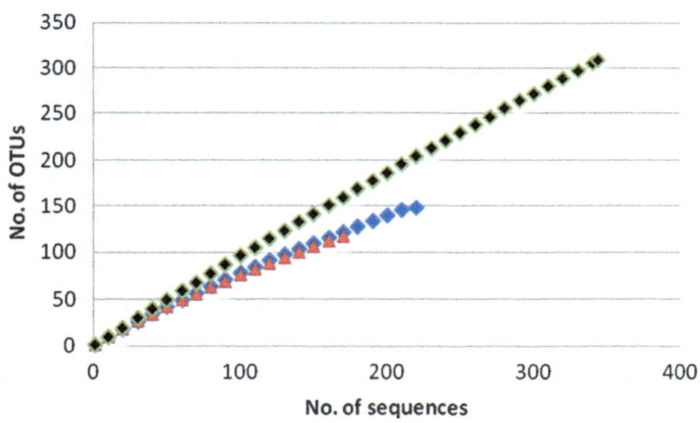

Fig. 16.13 Rarefaction analysis of Atacama (red triangle), Antarctica (blue diamond) and hot desert (black diamond). It indicates that bacterial diversity is higher in hot desert compared to Antarctica and Atacama Desert. The data was calculated employing tools of mothur package, and the OTUs are shown at genetic distance of 3%

Table 16.2 Species richness estimated at 20% genetic dissimilarity of 16S rDNA sequences from different desert types

Desert type	Shannon index (H′)	No. of OTUs	Total no. of 16S rDNA sequences
Antarctica desert	4.7	149	215
Atacama Desert	4.5	117	168
Hot desert	5.6	309	344

The sequence data was retrieved from public database, NCBI

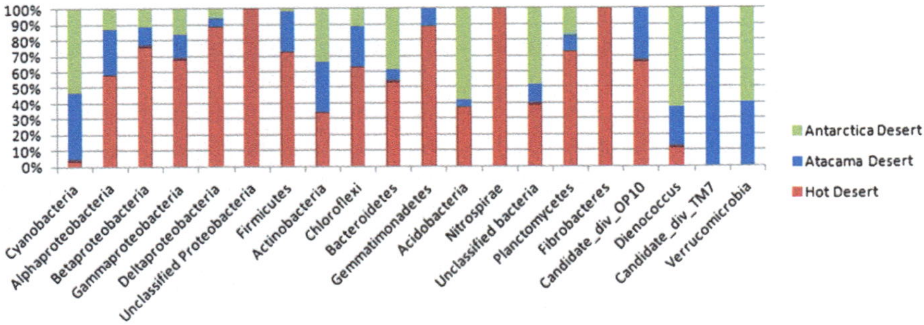

Fig. 16.14 Microbial community dominance (%), representing number of clones phylogenetically assigned to each group in different deserts. Metagenomic data retrieved from the published reports. The taxonomy of 16S rDNA sequences was identified by SILVA database using cut-off 80%

The overall analysis of 727 nucleotide sequences of 16S rDNA gene from microbial community of the three desert type demonstrates the abundance of action bacteria followed by cyanobacteria and alphaproteobacteria (Fig. 16.14). Cyanobacteria were primary photoautotroph in cold desert, while microlithic community in hot desert has not been paid much attention for the photoautotrophs, while chemolithoautotrophs were identified by the occurrence of chloroflexi, bacteroidetes and proteobacteria members in hot desert. The bacteria affiliated with chloroflexi and bacteroidetes/chlorobi and some proteobacteria are known to show autotrophy (Atomi 2002; Yousuf et al. 2012; Yuan et al. 2012). This demonstrates the supremacy of photoautotrophy in cold desert and chemolithoautotrophy in hot desert. Firmicutes were more common to hot desert than cold desert. Also the identification of many unclassified proteobacteria in hot desert and unclassified bacteria in all desert type indicates that desert harbours many bacteria; those were yet not identified.

Desert is considered as one of the unfriendly environments for microbial population due to less precipitation and low organic content. In this study, microbial diversity analysis data revealed ample diversity that could be explored for the isolation of CO_2-fixing bacteria. Eubacterial diversity analysed using culture-independent approach indicates the presence of *Cyanobacteria*, *α-proteobacteria*, *β-proteobacteria*, *ϒ-proteobacteria*, *δ-proteobacteria*, *Planctomycetes*, *Chloroflexi*, *Bacteroides*, *Firmicute*, *Enterobacter*, *Verrucomicrobia*, *Acidobacteria* and *Actinobacteria*. The microbial community of the two cold deserts (Atacama and Antarctica) was explored comparatively more than a particular type of hot desert. The analysis based on 344 sequences of 16S rDNA clones from different hot desert showed comparatively higher diversity, and the rarefaction curve indicates scope for more extensive study to unravel the microbial diversity in hot deserts. Hot deserts were overlooked for its microbial photolithoautotrophic and chemolithoautotrophic community analysis. In desert the endolithic community develops a microhabitat for life, irrespective of hot and cold desert environment. The study reported the association of various microbes in endolithic sample of Kutch that was not reported earlier. The oxygenic phototroph (*Cyanobacteria*) and anoxygenic phototroph (*Chloroflexi*) were found together with many chemolithotrophic bacteria affiliated with purple sulphur and purple non-sulphur bacteria (PSB and PNSB) and green sulphur and green non-sulphur bacteria (GSB and GNSB). The phylogenetic analysis of cyanobacteria from all the desert type revealed that there were some novel cyanobacteria members that were not reported earlier in desert. Therefore, the niche needs more attention to untie the microbial community with advanced molecular tools.

16.7 Oligotrophs: Prospect for CO_2 Fixation

Oligotrophs are the organisms usually present in an environment with a low level of nutrients and are not found in conditions of more plentiful nutrients (Yoshida et al. 2007). There exsists microbes that can grow under extremely low concentrations of carbon and in distilled water also, which were referred as oligotrophs. They have been isolated from various environmental niches, but the mode of nourishment is still an issue. As oligotrophs grow in extremely low nutrient conditions, they are supposed to exclusively use CO_2 as a carbon source for their growth, but their genetic studies with respect to CO_2 fixation are lacking. They might fix CO_2 from the atmosphere following some unreported pathways other than the known pathways involved in autotrophic growth (Yoshida et al. 2011; Agarwal et al. 2014; Yano et al. 2015). Many oligotrophic bacteria require CO_2 for their normal growth at varying concentrations (Ueda et al. 2008) besides the normal nutrient requirement. *Rhodococcus erythropolis* N9T-4 from oil store (Japan) and *Streptomyces* spp. from soil (Japan) were isolated as CO_2-requiring extreme oligotrophs which grow on basal salt medium under CO_2 atmosphere (Yoshida et al. 2007). The mode of CO_2 fixation in these bacteria was unknown, but the proteomic analysis revealed the high activity of NAD-dependent formaldehyde dehydrogenase and N,N′-dimethyl-4-nitrosoaniline-dependent methanol dehydrogenase under oligotrophic growth conditions (Ohhata et al. 2007; Yoshida et al. 2011). The prerequisite of CO_2 atmosphere for the survival of bacteria is known to be a marker for the presence of carbonic anhydrase, an enzyme involved in CO_2 and bicarbonate interconversion and also the acquisition of bicarbonate in non-phototrophic microorganisms. All the bacteria having carbonic anhydrase can produce sufficient bicarbonate from atmospheric CO_2 to support growth and supply it to bicarbonate-dependent enzymes, viz. acetyl-coenzyme A carboxylase, pyruvate carboxylase and phosphoenolpyruvate carboxylase, and hence allow the organism to grow under ambient CO_2 levels (0.035%). Under the absence of carbonic anhydrase, the bacteria can grow when they are either supplied with a high concentration of bicarbonate or under high levels of CO_2. The growth under high levels of CO_2 is actually due to the consumption of bicarbonate that is produced as a consequence to maintain natural equilibrium in the cell.

Desert is a natural oligotrophic environment where microbial activity is expected to govern the soil carbon and nitrogen. Due to alkaline pH of desert soil, CO_2 gets more solubilized. The bacteria inhabiting under oligotrophic conditions are expected to use some CO_2 fixation mechanism by virtue of which they survive (Agarwal and Purohit 2013). The oligotrophic bacteria could be autotrophic but the experimental evidence of such mechanism in oligotrophs has not been elucidated. The terms "autotrophic" and "oligotrophic" are sometimes controversial and to a certain extent confusing; an autotrophic growth is always indicated by the operation of any of the six known CO_2 fixation pathway (Berg 2011), while oligotrophic growth could be any of the pathways other than CO_2 fixation pathways (Yoshida et al. 2011). It has been reported that some bacteria require CO_2 for their growth and those strictly grow under high levels of CO_2 are known as capnophiles. Such bacteria occur widely in nature, but they remain uncharacterized or uncultured since CO_2 is not normally supplied for culturing such isolates from environmental samples. The understanding of growth conditions of autotrophs and oligotrophs requires more attention since CO_2 fixation occurs through several CO_2 fixation pathways and anaplerotic pathways. Some of them are known (six CO_2 fixation pathways), and new pathways could be deciphered through detailed genome studies on such microorganisms. Since facultative chemolithotrophs exhibit mixotrophy, the dependency on CO_2 for growth in such bacteria should be properly investigated. The growth of bacteria supplied with organic carbon source in the presence of CO_2 could yield more biomass, but the absence of CO_2 reduced it significantly. Since the mode of CO_2 fixation and requirement of CO_2 for oligotrophic growth of bacteria are unknown (Yoshida et al. 2007), this could also enable us to find new pathway of CO_2 fixation.

Despite being nutritionally compromised, the microbial community of desert is associated with diverse photoautotrophs and chemolithoautotrophs. It was observed that phototrophic microbial community is generally associated with microlithic population than the bare soil, while chemolithoautotrophic community is present in bare soil too. Exploring the microbial community intelligence of desert microbes linked with autotrophy could enrich soil fertility to maintain green cover (Agarwal et al. 2014). Such knowledge-based studies could be used further to develop strategies and protocols to use CO_2-fixing microbes for sinking carbon in an efficient manner.

Acknowledgement Authors are thankful to the Director, CSIR-National Environmental Engineering Research Institute (NEERI), Nagpur for providing necessary facilities and support. The manuscript has been checked for plagiarism by Knowledge Resource Centre, CSIR-NEERI, Nagpur and assigned KRC No.: CSIR-NEERI/KRC/2017/JULY/EBGD/17.

References

Agarwal L, Purohit HJ (2013) Genome sequence of *Rhizobium lupini* HPC(L) isolated from saline desert soil, Kutch (Gujarat). Genome Announc 1:e00071–e00012. https://doi.org/10.1128/genomeA.00071-12

Agarwal L, Qureshi A, Kalia VC, Kapley A, Purohit HJ, Singh RN (2014) Arid ecosystem: future option for carbon sinks using microbial community intelligence. Curr Sci 106:1357–1363

Ambardar S, Gupta R, Trakroo D, Lal R, Vakhlu J (2016) High throughput sequencing: an overview of sequencing chemistry. Indian J Microbiol 56:394–404. https://doi.org/10.1007/s12088-016-0606-4

Atomi H (2002) Microbial enzymes involved in carbon dioxide fixation. J Biosci Bioeng 94:497–505. https://doi.org/10.1016/S1389-1723(02)80186-4

Auguet J, Borrego CM, Bañeras L, Casamayor EO (2008) Fingerprinting the genetic diversity of the biotin carboxylase gene (*accC*) in aquatic ecosystems as a potential marker for studies of carbon dioxide assimilation in the dark. Environ Microbiol 10:2527–2536. https://doi.org/10.1111/j.1462-2920.2008.01677.x

Badger MR, Bek EJ (2008) Multiple rubisco forms in proteobacteria: their functional significance in relation to CO_2 acquisition by the CBB cycle. J Exp Bot 59:1525–1541. https://doi.org/10.1093/jxb/erm297

Beh M, Strauss G, Huber R, Stetter KO, Fuchs G (1993) Enzymes of the reductive citric acid cycle in the autotrophic eubacterium *Aquifex pyrophilus* and in the archaebacterium *Thermoproteus neutrophilus*. Arch Microbiol 160:306–311. https://doi.org/10.1007/BF00292082

Berberoglu H, Barra N, Pilon L, Jay J (2008) Growth, CO_2 consumption and H_2 production of *Anabaena variabilis* ATCC 29413-U under different irradiances and CO_2 concentrations. J Appl Microbiol 104:105–121. https://doi.org/10.1111/j.1365-2672.2007.03559.x

Berg IA (2011) Ecological aspects of the distribution of different autotrophic CO_2 fixation pathways. Appl Environ Microbiol 77:1925–1936. https://doi.org/10.1128/AEM.02473-10

Berg IA, Kockelkorn D, Ramos-Vera WH, Say RF, Zarzycki J, Hügler M, Alber BE, Fuchs G (2010) Autotrophic carbon fixation in archaea. Nat Rev Microbiol 8(6):447–460. https://doi.org/10.1038/nrmicro2365

Borrel G, Adam PS, Gribaldo S (2016) Methanogenesis and the Wood-Ljungdahl pathway: an ancient, versatile, and fragile association. Gen Biol Evol 8(6):1706–1711. https://doi.org/10.1093/gbe/evw114

Campbell BJ, Cary SC (2004) Abundance of reverse tricarboxylic acid cycle genes in free-living microorganisms at deep-sea hydrothermal vents. Appl Environ Microbiol 70:6282–6289. https://doi.org/10.1128/AEM.70.10.6282-6289.2004

Campbell BJ, Stein JL, Cary SC (2003) Evidence of chemolithoautotrophy in the bacterial community associated with *Alvinella pompejana*, a hydrothermal vent polychaete. Appl Environ Microbiol 69:5070–5078. https://doi.org/10.1128/AEM.69.9.5070-5078.2003

Chowdhury SP, Schmid M, Hartmann A, Tripathi AK (2007) Identification of diazotrophs in the culturable bacterial community associated with roots of *Lasiurus sindicus*, a perennial grass of Thar Desert. India. Microb Ecol 54:82–90. https://doi.org/10.1007/s00248-006-9174-1

Connon SA, Lester ED, Shafaat HS, Obenhuber DC, Ponce A (2007) Bacterial diversity in hyperarid Atacama Desert soils. J Geophys Res 112:1–9. https://doi.org/10.1029/2006JG000311

Ding L, Yokota A (2004) Proposals of *Curvibacter gracilis* gen. nov., sp. nov. and *Herbaspirillum putei* sp. nov. for bacterial strains isolated from well water and reclassification of *[Pseudomonas] huttiensis, [Pseudomonas] lanceolata, [Aquaspirillum] delicatum* and *[Aquaspirillum] autotrophicum* as *Herbaspirillum huttiense* comb. nov., *Curvibacter lanceolatus* comb. nov., *Curvibacter delicatus* comb. nov. and *Herbaspirillum autotrophicum* comb. Nov. Int J Syst Evol Microbiol 54:2223–2230. https://doi.org/10.1099/ijs.0.02975-0

Dong H, Rech JA, Jiang H, Sun H, Buck BJ (2007) Endolithic cyanobacteria in soil gypsum: occurrences in Atacama (Chile), Mojave (United States), and Al-Jafr Basin (Jordan) Deserts. J Geophys Res 112:1–11. https://doi.org/10.1029/2006JG000385

Erb TJ (2011) Carboxylases in natural and synthetic microbial pathways. Appl Environ Microbiol 77:8466–8477. https://doi.org/10.1128/AEM.05702-11

Evans MC, Buchanan BB, Arnon DI (1966) A new ferredoxin-dependent carbon reduction cycle in a photosynthetic bacterium. Proc Natl Acad Sci U S A 55(4):928–934

Ferrera I, Longhorn S, Banta AB, Liu Y, Preston D, Reysenbach AL (2007) Diversity of 16S rRNA gene, ITS region and aclB gene of the aquificales. Extremophiles 11:57–64. https://doi.org/10.1007/s00792-006-0009-2

Fuhrmanna S, Fernera M, Jeffke T, Henne A, Gottschalk G, Meyer O (2003) Complete nucleotide sequence of the circular megaplasmid pHCG3 of *Oligotropha carboxidovorans*: function in the chemolithoautotrophic utilization of CO, H_2 and CO_2. Gene 322:67–75. https://doi.org/10.1016/j.gene.2003.08.027

Hanson TE, Tabita FR (2001) A ribulose-1,5-bisphosphate carboxylase/oxygenase (RubisCO) like protein from *Chlorobium tepidum* that is involved with sulfur metabolism and the response to oxidative stress. Proc Natl Acad Sci U S A 98:4397–4402. https://doi.org/10.1073/pnas.081610398

Holo H (1989) *Chloroflexus aurantiacus* secretes 3-hydroxypropionate, a possible intermediate in the assimilation of CO_2 and acetate. Arch Microbiol 151:252–256. https://doi.org/10.1007/BF00413138

Hu Y, Holden JF (2006) Citric acid cycle in the hyperthermophilic archaeon *Pyrobaculum islandicum* grown autotrophically, heterotrophically, and mixotrophically with acetate. J Bacteriol 188:4350–4355. https://doi.org/10.1128/JB.00138-06

Hügler M, Sievert SM (2011) Beyond the calvin cycle: autotrophic carbon fixation in the ocean. Annu Rev Mar Sci 3:261–289. https://doi.org/10.1146/annurev-marine-120709-142712

Hügler M, Wirsen CO, Fuchs G, Taylor CD, Sievert SM (2005) Evidence for autotrophic CO_2 fixation via the reductive tricarboxylic acid cycle by members of the -subdivision of proteobacteria. J Bacteriol 187:3020–3027. https://doi.org/10.1128/JB.187.9.3020-3027.2005

Hügler M, Huber H, Molyneaux SJ, Vetriani C, Sievert SM (2007) Autotrophic CO_2 fixation via the reductive tricarboxylic acid cycle in different lineages within the phylum Aquificae: evidence for two ways of citrate cleavage. Environ Microbiol 9:81–92. https://doi.org/10.1111/j.1462-2920.2006.01118.x

Kanao T, Fukui T, Atomi H, Imanaka T (2001) ATP-citrate lyase from the green sulfur bacterium *Chlorobium limicola* is a heteromeric enzyme composed of two distinct gene products. Eur J Biochem 268:1670–1678. https://doi.org/10.1046/j.1432-1327.2001.02034.x

Köberl M, Müller H, Ramadan EM, Berg G (2011) Desert farming benefits from microbial potential in arid soils and promotes diversity and plant health. PLoS One 6:e24452. https://doi.org/10.1371/journal.pone.0024452

Kuhlman KR, Fusco WG, La Duc MT, Allenbach LB, Ball CL, Kuhlman GM, Anderson RC, Erickson IK, Stuecker T, Benardini J, Strap JL (2006) Diversity of microorganisms within rock varnish in the Whipple Mountains, California. Appl Environ Microbiol 72:1708–1715. https://doi.org/10.1128/AEM.72.2.1708-1715.2006

Kuznetsov BB, Ivanovsky RN, Keppen OI, Sukhacheva MV, Bumazhkin BK, Patutina EO, Beletsky AV, Mardanov AV, Baslerov RV, Panteleeva AN, Kolganova TV (2011) Draft genome sequence of the anoxygenic filamentous phototrophic bacterium *Oscillochloris trichoides* subsp. DG-6. J Bacteriol 193:321–322. https://doi.org/10.1128/JB.00931-10

Lacap DC, Warren-Rhodes KA, McKay CP, Pointing SB (2011) Cyanobacteria and chloroflexi-dominated hypolithic colonization of quartz at the hyper-arid core of the Atacama Desert, Chile. Extremophiles 15:31–38. https://doi.org/10.1007/s00792-010-0334-3

Lester ED, Satomi M, Ponce A (2007) Microflora of extreme arid Atacama desert soil. Soil Biol Biochem 39:704–708. https://doi.org/10.1016/j.soilbio.2006.09.020

Navarro-González R, Rainey FA, Molina P, Bagaley DR, Hollen BJ, de la Rosa J, Small AM, Quinn RC, Grunthaner FJ, Cáceres L, Gomez-Silva B (2003) Mars-like soils in the Atacama Desert, Chile, and the dry limit of microbial life. Science 302:1018–1021. https://doi.org/10.1126/science.1089143

Ohhata N, Yoshida N, Egami H, Katsuragi T, Tani Y, Takagi H (2007) An extremely oligotrophic bacterium, *Rhodococcus erythropolis* N9T-4, isolated from crude oil. J Bacteriol 189:6824–6831. https://doi.org/10.1128/JB.00872-07

Piubeli F, de Lourdes Moreno M, Kishi LT, Henrique-Silva F, García MT, Mellado E (2015) Phylogenetic profiling and diversity of bacterial communities in the Death Valley, an extreme habitat in the Atacama Desert. Indian J Microbiol 55:392–399. https://doi.org/10.1007/s12088-015-0539-3

Pointing SB, Chan Y, Lacap DC, Lau MCY, Jurgens JA, Farrell RL (2009) Highly specialized microbial diversity in hyper-arid polar desert. Proc Natl Acad Sci U S A 106:19964–19969. https://doi.org/10.1073/pnas.0908274106

Rao S, Chan Y, Bugler-Lacap DC, Bhatnagar A, Bhatnagar M, Pointing SB (2016) Microbial diversity in soil, sand dune and rock substrates of the Thar Monsoon Desert, India. Indian J Microbiol 56:35–45. https://doi.org/10.1007/s12088-015-0549-1

Saini R, Kapoor R, Kumar R, Siddiqi TO, Kumar A (2011) CO_2 utilizing microbes—a comprehensive review. Biotechnol Adv 29:949–960. https://doi.org/10.1016/j.biotechadv.2011.08.009

Schafer S, Gotz M, Elsenreich W, Bacher A, Fuchs G (1989) 13C-NMR study of autotrophic CO_2 fixation in *Thermoproteus neutrophilus*. Eur J Biochem 184:151–156. https://doi.org/10.1111/j.1432-1033.1992.tb16850.x

Selesi D, Schmid M, Hartmann A (2005) Diversity of green-like and red-like ribulose-1,5-bisphosphate carboxylase/oxygenase large-subunit genes (*cbbL*) in differently managed agricultural soils. Appl Environ Microbiol 71:175–184. https://doi.org/10.1128/AEM.71.1.175-184.2005

Shively JM, van Keulen G, Meijer WG (1998) Something from almost nothing: carbon dioxide fixation in chemoautotrophs. Annu Rev Microbiol 52:191–230. https://doi.org/10.1146/annurev.micro.52.1.191

Sintsov NV, Ivanovskii RN, Kondrat'eva EN (1980) ATP-dependent citrate lyase in the green phototrophic bacterium, *Chlorobium limicola*. Mikrobiologiia 49(4):514–516

Smith JJ, Tow LA, Stafford W, Cary C, Cowan DA (2006) Bacterial diversity in three different Antarctic Cold Desert mineral soils. Microb Ecol 51:413–421. https://doi.org/10.1007/s00248-006-9022-3

Souza V, Espinosa-Asuar L, Escalante AE, Eguiarte LE, Farmer J, Forney L, Lloret L, Rodríguez-Martínez JM, Soberón X, Dirzo R, Elser JJ (2006) An endangered oasis of aquatic microbial biodiversity in the Chihuahuan desert. Proc Natl Acad Sci U S A 103:6565–6570. https://doi.org/10.1073/pnas.0601434103

Suárez EB, Matta J, Rolón M, Maldonado L, Detrés Y, De la Motta A, Gelado M, Ramos J, Armstrong R (2008) Molecular identification of the bacterial burden in Sahara Dust samples using a new method to improve the evidence for the effective management of public health measures during an SD event. J Environ Health Res 7:99–106

Tang KH, Blankenship RE (2010) Both forward and reverse TCA cycles operate in green sulfur bacteria. J Biol Chem 285:35848–35854. https://doi.org/10.1074/jbc.M110.157834

Thauer RK, Kaster AK, Seedorf H, Buckel W, Hedderich R (2008) Methanogenic archaea: ecologically relevant differences in energy conservation. Nat Rev Microbiol 6:579–591. https://doi.org/10.1038/nrmicro1931

Ueda K, Tagami Y, Kamihara Y, Shiratori H, Takano H, Beppu T (2008) Isolation of bacteria whose growth is dependent on high levels of CO_2 and implications of their potential diversity. Appl Environ Microbiol 74:4535–4538. https://doi.org/10.1128/AEM.00491-08

Warren-Rhodes KA, Rhodes KL, Pointing SB, Ewing SA, Lacap DC, Gomez-Silva B, Amundson R, Friedmann EI, McKay CP (2006) Hypolithic cyanobacteria, dry limit of photosynthesis, and microbial ecology in the hyperarid Atacama Desert. Microb Ecol 52:389–398. https://doi.org/10.1007/s00248-006-9055-7

Williams TJ, Zhang CL, Scott JH, Bazylinski DA (2006) Evidence for autotrophy *via* the reverse tricarboxylic acid cycle in the marine *Magnetotactic coccus* strain MC-1. Appl Environ Microbiol 72:1322–1329. https://doi.org/10.1128/AEM.72.2.1322-1329.2006

Wood SA, Rueckert A, Cowan DA, Cary SC (2008) Sources of edaphic cyanobacterial diversity in the dry valleys of eastern Antarctica. ISME J 2:308–320. https://doi.org/10.1038/ismej.2007.104

Yang C, Hua Q, Shimizu K (2002) Integration of the information from gene expression and metabolic fluxes for the analysis of the regulatory mechanisms in Synechocystis. Appl Microbiol Biotechnol 58:813–822. https://doi.org/10.1007/s00253-002-0949-0

Yano T, Yoshida N, Yu F, Wakamatsu M, Takagi H (2015) The glyoxylate shunt is essential for CO_2 requiring oligotrophic growth of *Rhodococcus erythropolis* N9T-4. Appl Microbiol Biotechnol 99:5627–5637. https://doi.org/10.1007/s00253-015-6500-x

Yergeau E, Schoondermark-Stolk SA, Brodie EL, Déjean S, DeSantis TZ, Gonçalves O, Piceno YM, Andersen GL, Kowalchuk GA (2009) Environmental microarray analyses of Antarctic soil microbial communities. ISME J 3:340–351. https://doi.org/10.1038/ismej.2008.111

Yoshida N, Ohhata N, Yoshino Y, Katsuragi T, Tani Y, Takagi H (2007) Screening of carbon dioxide extreme oligotrophs from soil. Biosci Biotechnol Biochem 71:2830–2832. https://doi.org/10.1271/bbb.70042

Yoshida N, Hayasaki T, Takagi H (2011) Gene expression analysis of methylotrophic oxidoreductases involved in the oligotrophic growth of *Rhodococcus erythropolis* N9T-4. Biosci Biotechnol Biochem 75:123–127. https://doi.org/10.1271/bbb.100700

Yousuf B, Sanadhya P, Keshri J, Jha B (2012) Comparative molecular analysis of chemolithotrophic bacterial diversity and community structure from coastal saline soils, Gujarat, India. BMC Microbiol 12:150. https://doi.org/10.1186/1471-2180-12-150

Yuan H, Ge T, Chen C, O'Donnell AG, Wu J (2012) Significant role for microbial autotrophy in the sequestration of soil carbon. Appl Environ Microbiol 78:2328–2336. https://doi.org/10.1128/AEM.06881-11

Bioprocess Network for Solid Waste Management

17

Federico Micolucci, Marco Gottardo, and Wanderli Rogério Moreira Leite

Abstract

The anaerobic digestion process is a well-known and still growing technology. It has been implemented in full scale in several waste and wastewater treatment plants.

The key issue is the treatment, stabilization, and waste reduction in order to obtain benefits, especially energy and bioproducts. Efficient separate collection, first of all, is the key for anaerobic digestion success. Separate collection is the prerequisite to have waste streams of good quality at the source for further treatments, and it can be implemented at reasonable costs. We have to change the paradigm in plant pretreatment configuration, thus passing from complicated pretreatment lines to simplified systems. Moreover, urban waste treatment processes should be considered as real productive industries. The future vision of biowaste management leads to consider it as a raw material to produce not only energy but also products (e.g., bioplastics). The new aerobic/anaerobic biorefinery comes from the integration of the cycles that take advantage of the wastewater treatment and the urban organic waste management as a valuable source for the production of products and energy resources. The integrated approach for waste stream treatment gives considerable advantages, which lead this option in the field of the "smart" opportunities for the urban service management.

F. Micolucci (✉)
Department of Chemistry and Bioscience, Sustainable Biotechnology, Aalborg University, A.C Meyers vænge 15, 2450 Copenhagen SV, Denmark
e-mail: fm@bio.aau.dk

M. Gottardo
Department of Environmental Sciences, Informatics and Statistics, University Ca' Foscari of Venice, Via Torino 155, 30172 Venezia Mestre, Italy

W.R.M. Leite
Departmentof Civil Engineering, Federal University of Pernambuco, Av. AcadêmicoHélio Ramos s/n, 50740-530 Recife, PE, Brazil

H.J. Purohit et al. (eds.), *Optimization and Applicability of Bioprocesses*, https://doi.org/10.1007/978-981-10-6863-8_17

Keywords
Anaerobic digestion • Organic waste • Sewage sludge • Pre-treatment • Biogas upgrading • Hydrogen • Volatile fatty acids • Multivariate analysis • Process control

17.1 Introduction

The intense demographic (urban) growth results in greater organic waste production which assigns a crucial importance to improve waste management in a more appropriate way.

Before addressing the specific waste management topic, it is advisable to focus on the substrate: the municipal solid waste (MSW).

It represents one of the most significant solid streams to be disposed, from both the quantitative and qualitative standpoints.

In the European context, the definition of MSW is not univocal among all countries. In fact, it differs according to the management strategies adopted in each member state. Eurostat (http://ec.europa.eu/eurostat) defines municipal waste, the waste produced from domestic and commercial activities, offices, and public places. Waste collection is the primary responsibility and duty of municipal authorities, who should furthermore ensure its appropriate disposal in accordance with their own *waste management* program. The total municipal solid waste production in EU state members earned 252 million tons in 2011 (Eurostat 2011) with an average per capita generation of 541 kg per year, almost 1.5 kg of MSW per day.

Almost 40% of 252 million tons waste/year produced in EU arise as high-quality organic material (high biodegradability and low inert material content) as a consequence of an outstanding separate collection program successfully set up in the municipalities (Bernstad and la Cour Jansen 2011; Cecchi and Cavinato 2015), and in a global perspective, the amount of food wasted is definitely expected to increase (about 44% between 2005 and 2025).

MSW landfilling can enlarge world methane emissions from 31 to 43 million tons.

Figure 17.1 shows the per capita production of waste in EU.

Figure 17.1 highlights the total and per capita decreasing trend of waste production in Europe. Interesting is to determine whether that reduction is related either to international economic crisis that inevitably affects consumption or to prevention and regulation of urban waste in line with EU policies.

In this concern, the yearly edition of Urban Wastes Report (ISPRA 2014a, b) revealed in 2014 an interesting analysis carried out between 2007 and 2012. The relation among data from waste production and spending household consumption was examined. A dissociation tendency of both parameters was noticed, reduction of 0.7% on household consumption, evidently generated by the international economic crisis, in contrast to reductions over 6.5% of waste production. Concerning municipal waste management in the European context (Eurostat 2014), about 33% of all MSW has been landfilled, while 24% has been incinerated. In addition, 28% of this residue has been recycled, whereas over 15% has been treated through biological processes (composting and anaerobic digestion).

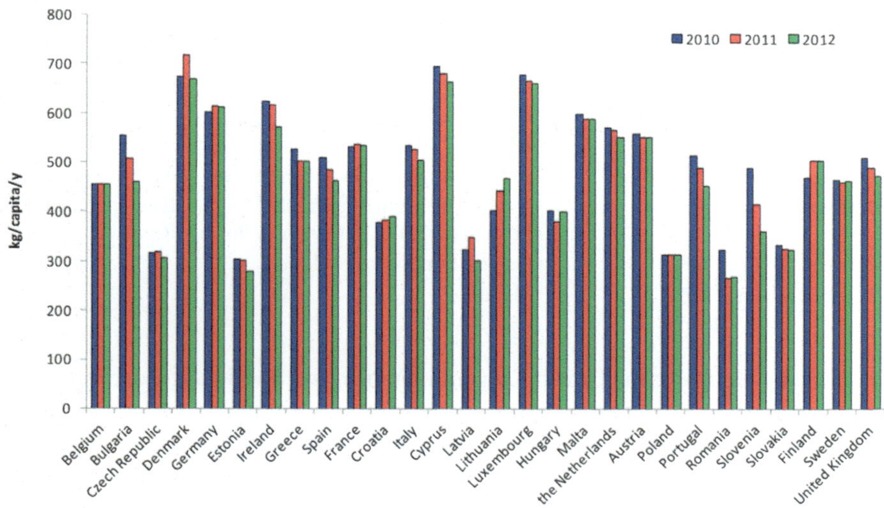

Fig. 17.1 Per capita EU waste production

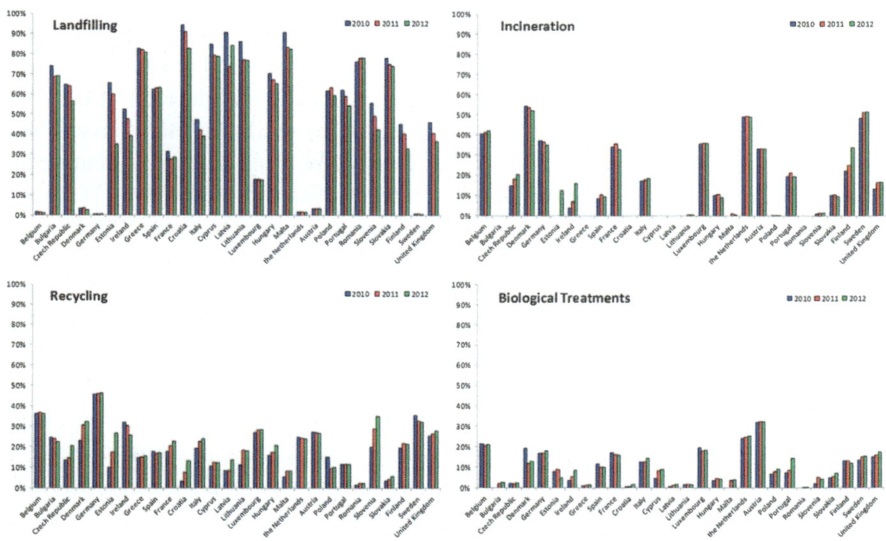

Fig. 17.2 MSW management percentage in Europe (2012). Top left: landfilling. Top right: incineration. Bottom left: recycling. Bottom right: biological treatments

It is essential to underline such a great variability between different approaches to achieve MSW management among the different countries in EU.

Figure 17.2 expresses the most impressive waste management programs are located in countries like Belgium, Denmark, Germany, the Netherlands, Austria, and Sweden. Inter alia, Austria, the Netherlands, Belgium, and Luxembourg are the EU countries that adopted biological treatments to improve waste management,

Table 17.1 Organic fraction of the MSW in Europe (2008–2010)

% of biowaste in total municipal waste	Countries
<20%	Lithuania, Slovenia
Between 20% and 30%	Bulgaria, Denmark, Ireland, Hungary, Latvia
Between 30% and 40%	Italy, Germany, France, Sweden, the United Kingdom
Between 40% and 50%	Austria, Belgium, the Czech Republic, Estonia, Finland, Luxembourg, the Netherlands, Poland, Romania, Spain
Between 50% and 60%	Greece, Portugal, Slovakia
Between 60% and 80%	Malta

followed by Germany, the United Kingdom, France, and Sweden. A closer analysis, however, should also consider the organic fraction content in the solid waste which arises quite uneven among EU countries as shown in Table 17.1 (EEA 2013).

Both Fig. 17.2 and Table 17.1 reveal a tendency regarding the large quantity of organic fraction of MSW (OFMSW); the higher OFMSW is managed by biological treatments. In 2010 the percentage of OFMSW sent to biological treatments was 32% (Austria), 24% (the Netherlands), 22% (Belgium), and 20% (Luxembourg), nations that showed an organic fraction percentage on the waste among 40% and 50%. On the other hand, countries with lower percentage of OFMSW biologically treated were Bulgaria (0%), Lithuania (1%), Latvia (1%), Slovenia (2%), Ireland (4%), and Hungary (4%), wherein nations whose organic fraction is less than 30% of the total MSW.

Germany and France, whose organic fraction of the produced MSW arises 30% and 40%, respectively, send treat over 17% of the MSW using biological treatments, followed by the United Kingdom (15%), Sweden (14%), and Italy (13%). Denmark emerges the only exception since 19% of the overall organic fraction of MSW with a low organic fraction content (20–30%) is sent to biological treatment (2010).

By contrast, less than 10% of the MSW is sent to biological treatment in Romania, the Czech Republic, Estonia, and Poland where the OFMSW is about 40% and 50%. Greece, Portugal, and Slovakia have a worse condition; only 7% of the MSW, at least 50% organic, is sent to biological management. Malta has the worst scenario; more than 50% of the total MSW is organic, but the country does not have biological treatments included on its mainstream solid waste management.

In 2014 the Italian Institute for Environmental Protection and Research (ISPRA) published a report that pointed the national urban waste production in 2013 was about 30 million ton, 400 ton less than the year before (−1.3%).

The next graph shows the Italian waste management between 2009 and 2013 (Fig. 17.3).

Landfilling appears still the main alternative for waste disposal, although in 2013, 800,000 tons less of MSW were landfilled in comparison with the previous year. This is explained by the waste generation decreasing and recycling program improvements and the use of mechanical treatment (MT) to benefit mixed waste handling. MT consists in applying mechanical pretreatment for mixed wastes in order to separate different wastes which are individually disposed of, such the biostabilization of the organic fraction while the dry fraction for incineration or assigned for fuels producing either. Regarding biological treatments, approximately

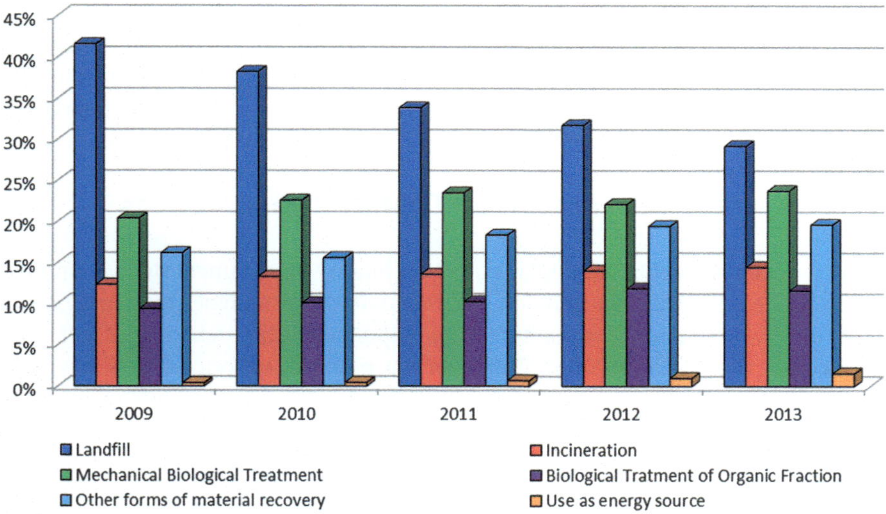

Fig. 17.3 Italian management of MSW in percentage of different treatment and disposal

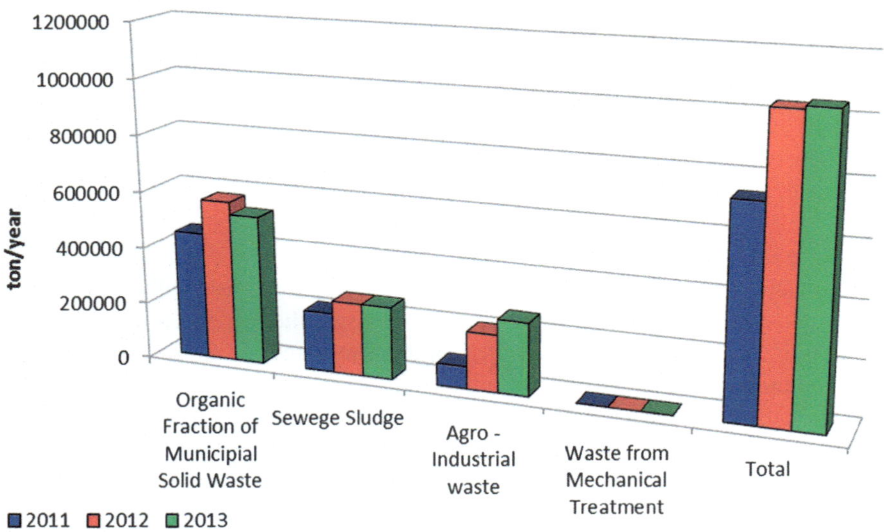

Fig. 17.4 Different organic waste treated by anaerobic digestion (Italy, 2011–2013)

14% of the MSW produced in 2013 was referred for composting treatment and anaerobic digestion. Considering only the fully operating anaerobic treatment plants, at least 1043 million tons of total waste were treated in 2013. Additionally, over 50% (526,000 tons) consisted of OFMSW, and the remaining part consisted of biological sludges and wastes from agro-industrial activities (Fig. 17.4).

Before moving to the next section, an important aspect to discuss as regards the fate of a poor amount of MSW derived from mechanical selection of undifferentiated waste. The next section will depict the low purity of this fraction coming from MT, which does not make it ideally suited for biotreatment, i.e., the methanation process.

17.2 Pretreatment Process for OFMSW in Bioprocesses

Waste collection can be contemplated as the primarily pretreatment step for a proper municipal solid waste management: different collection strategies may bring about quite different organic streams which might require different pretreatment technologies consequently (Pavan et al. 2000). In the past, the unsorted MSW has been mechanically treated, while the sorted OFMSW has been sent to biological treatments. Recently, the yields obtained from both separately and source-sorted OFMSW collection have proven that these approaches arise the most promising way to obtain valuable substrates (Cecchi et al. 2005).

Therefore, two types of organic fraction of MSW are the core of interest, each one presenting different characteristics according to the collection system:

- Undifferentiated collection, supported by mechanical separation as a pretreatment step
- Differentiated collection from large utilities (canteens, markets, etc.) and domestic sources (curbside collection)

These two substrates are discussed below, highlighting their main physical, chemical, and biological differences.

The Organic Fraction from Mechanic Selection

The organic waste, before being sent to biological treatment, either aerobic or anaerobic ones, must be pretreated in order to ensure them adequate for these bioprocesses. Several authors in scientific literature reported the collection as the first and foremost pretreatment for the organic fraction derived from municipal solid waste. Separate collection efficiency determines the complexity of the subsequent waste processing, and the main substrate characteristics influence the choice of waste treatment.

The MT of the organic fraction has been used in a rather extensive way in Europe during the last 30 years, with the dual purpose to obtain an organic fraction with great putrescibility and high calorific material well known as fuel, refuse-derived fuel (RDF) or solid recovered fuel/specified recovered fuel (SRF), which represents a kind of fuel obtained when urban solid waste goes through shredding and dehydrating stages. Meanwhile, RDF is substantially constituted by combustible components reclaimed from some plastics and biodegradable waste found in MSW.

Currently there are different types of mechanical separation plants: simplified-type plants, medium-complex plants, and complex plants.

Complex system presents a complete selection line, comprised of waste size reduction, iron removal machines, multistep sieving, dry fraction shredders, and pelletizing to produce RDF pellets thereafter. Byproducts are far refined that even fit better into biological energy recovering processes. Nowadays, the limited use of the OFMSW-MT processed majority regards its insufficient quality of the end product (waste) compared to differentiated waste.

The following table shows total solid (TS) and volatile solid amounts (VS) determined on different MT-OFMSW samples, coming from undifferentiated waste, curbside collected waste from canteen (Cecchi et al. 1997) and fruit and vegetable shops (Pavan et al. 2000) (Tables 17.2 and 17.3).

Table 17.4 present end product mechanically selected has high solid content than ordinary putrescible pure fractions of urban solid waste. This outcome may be related to the inertial fraction presented in the undifferentiated substrate, neither entirely self-separable in the MT step nor during the compost refining. This aspect is more evident when we consider the percentage of volatile solids (VS): it hardly reaches 50% of total solids (TS). Undifferentiated waste which majority composition is TS (over 40%) is practically not suitable for anaerobic digestion processes. This is checked by Table 17.3 where the maximum potential conversion achieved as methane product per kg of VS fed (B_0) and as biogas produced per kg VS fed (G_0) for three different types of OFMSW is displayed.

Table 17.2 Organic fraction characteristics of different OFMSW samples

Parameters	Units	Average	St. dev.	Organic fraction from mechanical treatment (Cecchi et al. 2005)
TS	g/kg	763	81.3	
VS	% TS	43.9	5.4	
Parameters	Units	Range	Typical value	Organic fraction collection from canteen (Cecchi et al. 1997)
TS	%	21.4–27.4	25.6	
VS	% TS	91.3–99.7	96.5	
Parameters	Units	Average	St. dev.	Organic fraction collection from fruit and vegetable market (Pavan et al. 2000)
TS	g/kg	81.8	15.7	
VS	% TS	81.9	11.3	

Table 17.3 Conversion yields at infinite time for the three types of organic fraction (Cecchi et al. 2005)

Parameters	Units	Organic fraction from mechanical treatment	Organic fraction from source-sorted collection large users	Organic fraction from source-sorted household collection (door-to-door)
B_0	$m^3CH_4/kgVS$	0.16–0.37	0.45–0.49	0.37–0.40
G_0	$m^3biogas/kgVS$	0.29–0.66	0.81–0.89	0.67–0.72

Table 17.4 Organic MSW from kerbside collection of Treviso (Italy 2015)

Substrate	TS g/kg w.w.	VS g/kg w.w.	VS/TS %	COD g/kg TS	TKN g/ kgTS	P g/ kgTS
Organic MSW	298 ± 44	267 ± 32	85 ± 3	1100 ± 454	27 ± 4	4.0 ± 0.2

Table 17.5 Composition of separately collected biowaste in Veneto region, Italy, on wet, dry, and volatile basis (Micolucci et al. 2016)

Fractions	Wet weight %	Dry weight %	Volatile %
Paper	10.75	11.25	12.87
Plastic	0.85	1.20	1.31
Inert	6.88	11.33	8.16
Organic	81.52	76.22	77.66

On the other hand, the organic waste removed from the door-to-door waste collection brings more quality for biological treatments than that one derived from unsorted material. This is assigned by the existence of high organic fraction and few inert aggregates.

In any case, it can be mechanically pretreated in order to remove iron and plastics and to accomplish size reduction. The main characteristics of raw differentiated organic material are reported in Table 17.5.

The results show a dry matter content of 284 gTS/kg w.w. (wet weight), from which 86% are total volatile solid (VS). These values, together with nitrogen (TKN) and phosphorus (P) contents, are similar with those widely reported in literature (Zhang et al. 2014; Bernstad and la Cour Jansen 2011; Cecchi and Cavinato 2015).

Regarding the nutrient content, Table 17.4 highlights the positive influence of curbside collection on the quality of OFMSW since it gathered high biodegradability and an adequate nutrients ratio, particularly COD:N ratio (mean value of 40), more adequate for anaerobic digestion.

From the analysis of waste composition, it is clear how the food fraction accounts for 82% of the entire wet weight food waste, while the rest is made up mainly of paper (10%), still anaerobically biodegradable, and inert material (8%) Table 17.6.

Fruit and vegetable make up 60% of the overall organic waste. Comparing these values with those from Cavinato et al. (2012), in the study of Micolucci et al.(2016), the fraction related to fruit and vegetable was lower; the fraction consisted of other organic waste (e.g., bread, meals, dairy, etc.) was higher instead. According to these differences in the composition, the total solid content found in the OFMSW used in this study (Table 17.3) was higher compared to that used by Cavinato (2012).

Another feature regards the comparison with the municipal organic waste obtained from a Greek study (Malamis et al. 2015), whose organic fraction (fruits, vegetable, and other organic waste) consisted of 86.2%, whereas vegetable and fruits corresponded to about 60.6%. These values are perfectly comparable with other studies. The amount of inert material reported in the Greek study (2.5%) was about 4% less than that revealed in Treviso fraction (Micolucci et al. 2017).

Table 17.6 Composition of the organic fraction municipal solid waste (Treviso, Italy)

Fractions	Wet weight %
Fruit and vegetable	52.3
Other organic waste	31.3
Paper and hardboard	9.6
Inerts and unclassified materials	6.8

The maximum biogas potential conversion on the OFMSW from undifferentiated collection is considerably lower than the one coming from recycling.

Concerning the organic fraction of solid urban waste coming from separate collection prior to the biological treatment, it must be pretreated in order to remove the presence of extra inert fractions (plastics, metals, etc.) and to downsize the organic material.

The most cost-effective and straightforward OFMSW pretreatment is the Soft Wet Refine (SWR), recognized as a power-save technology usually applied both in small and medium installations. This technology can process up to 10,000 tons of OFMSW per year. It is set by a bunch of automatic operations, like primary crushering, bag-breaking machines, designed to effectively open the bags and release the inside materials and reduce the residue size afterward. Following, the iron removal step takes place, in order to remove the magnetic material. Subsequently, a sieving system separates plastics and inert. The final product is finely sieved in a second shredding operation, which ensures a downsized material and a more homogeneous final product.

A variant of the SWR considers the use of a hydropulper, namely, Hard Wet Refine (HWR) Technology. The hydropulper provides the wet processing and pulping of the organic fraction and benefits also the undesirable material removing.

Nowadays, a third alternative approach has included a squeezing step of OFMSW. It can be performed by two different technologies, namely, the extrusion press (EP) and the screw press (SP).

When the organic fraction is subjected to EP process, a high pressure inside a perforated extrusion chamber is set in order to fluidificate the organic fraction (food waste and other putrescible fractions), which thrusts the internal and external chamber pressure difference. This condition benefits the fluidized organic fraction isolation from that fraction more resistant (plastics, glass, rubbers, etc.).

SP is characterized by a perforated metal chamber in which organic waste is forced pressed through a bunch of slits. In this way, the softest and moist part of the fed substrate is withdrawn through the slits, while the hard and dry ones are directed to discharge (paper, lignocellulosic materials, animal bones) (Fig. 17.5) (Hansen et al. 2007).

SP proves beneficial due to its low energy requirement and its applicability for high-purity substrate demands.

Micolucci et al. (2016) carried out the treatment of leachate retrieved from pressed biowaste using mesophilic and thermophilic anaerobic digestion as a

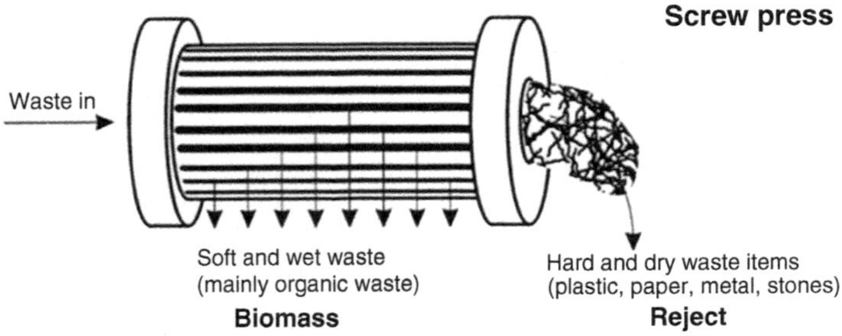

Fig. 17.5 Screw press (Hansen et al. 2007)

treatment to enhance the biogas production and revamp composting plants. Besides, this study underlined how important is a well-established separate collection of the urban organic waste at the source, in order to obtain as a pretreatment pure waste streams with high biodegradability and less inert content. Therefore, the application of a screw press can divide the waste flow into two fractions, one liquid and one solid. Through the liquid fraction of pressed biowaste, energy and byproducts can be obtained. Solid fraction is instead sent to composting.

The research group treated as substrate the pressed organic fraction of MSW from curbside residential pickup services and door-to-door collection. The results after continuous operation periods depicted this leachate was properly treated through AD process under all working conditions, whereas the largest biogas production 0.92 $m^3/kgVS_{fed}$ was obtained under thermophilic temperature when the digester was fed with a biowaste loading rate of 4.7 $kgVS/m^3d$.

Thus, the study demonstrated that biowaste pressing is a new alternative for mechanical treatment plants that needs to be constructed or restored.

"Compositional analysis of food waste from study sites in geographically distinct regions of Europe" of the FP7, EU project "Valorisation of food waste to biogas reported the concept of handling waste streams with a reduced amount of inert material improve the efficiencies avoiding next treatment step failure.

17.3 Process Network for OFMSW

The rapid global population growth associated with the intensive urbanization is rising every year. An inevitable expansion in waste production arises as a consequence as well as increases in the quantity of sludge from wastewater treatment and energy demand.

For this reason, a proper and "smart" management of the waste streams must be accomplished. The key issue is the treatment, stabilization, and waste reduction in order to obtain benefits, especially energy and bioproducts. Despite waste flow management arises as an evident problem, effective ways to produce energy from renewable sources must become a priority and remain in one's mind. Nowadays, energy or other bioproduct recovery from substrates without an economic value added at first glance, such as organic waste or sewage sludge, is quite challenging.

In recent decades, international and scientific communities have shown great interest in renewable energy sources: biomass is certainly the most promising and the most widely studied. Biomass belongs to organic waste and wastewater (sewage sludge) but also agricultural and industrial waste, energy crops, etc.

The OFMSW, the active sludge discharged from the municipal sewage treatment (WAS, waste activated sludge), their main uses and options for treatment, and their energy yield and reuse will be discussed in this paragraph.

In order to integrate the water and waste cycles, the organic fraction of MSW separately collected may be treated in wastewater treatment plants (WWTP) in association with the sludge produced in the waterline. Both substrates may be sent to anaerobic digestion process, which provides stabilization and produces biogas.

The anaerobic digestion process has been widely studied, and it has been implemented in full scale in several waste and wastewater treatment plants. Through this biological process, which occurs in the absence of oxygen, organic substance is degraded by the action of different microbial groups interacting among them: hydrolytic bacteria, acidifying bacteria, and methanogenic Archaea. Starting from heterogeneous and complex substrates composed of proteins, fats, and carbohydrate substances, sugars and amino acids can be obtained by a simple metabolic pathway which ends in the biogas production.

Anaerobic codigestion means the simultaneous anaerobic digestion of two or more organic substrates (Sosnowski et al. 2008), specifically OFMSW and sewage sludge. The use of the OFMSW-WAS codigestion process is a very widely used technique in Europe due to uncountable advantages from both the environmental and economic perspectives.

The disposal of sewage sludge is a hard and actual problem especially because of large mass flows and consequently the economic costs involved. Anaerobic digestion is a valid way to solve these issues since it provides sludge stabilization and a small amount of stabilized solids to be disposed of. Besides, anaerobic digestion benefits energy recovery through the use of biogas. In fact, the biogas obtained can be converted into heat and electricity through cogeneration units (CHP). The waste activated sludge (WAS), however, shows a modest energy yield (250–350 m^3 biogas/tonVS). During the 1980s, Cecchi et al. (1988) had the idea to improve the anaerobic digestion of waste activated sludge with different types of residues (biowaste) as cosubstrates such as the organic residue retrieved from municipal solid waste (OFMSW).

The codigestion WAS-OFMSW is far suitable for WWTP as it provides conditions to treat both the organic fraction of MSW and sewage sludge, gathering greater biogas production which may improve the overall WWTP energy balance.

Several studies (Battistoni et al. 1998; Bolzonella et al. 2006; Astals et al. 2014) have shown several advantages performing codigestion of different organic substrates, besides the incentive for developing integrated cycles among wastewater treatment and organic waste. This strategy permits taking advantages of the synergies available among the two substrates.

The simultaneous WAS-OFMSW processing may also balance nutrient ratio: in general, sewage sludges are characterized by low carbon/nitrogen ratios (COD/

TKN about 11) (Metcalf and Eddy 2006), whereas the OFMSW presents high biodegradability (COD/TKN about 35) and may provide an extra contribution of carbon readily convertible into biogas. Further studies (Pavan et al. 2000; Bolzonella et al. 2006) have also demonstrated how the joint sludge and OFMSW processing enhances biogas yields.

It is important to state the codigestion process can be easily implemented in existing wastewater treatment plants (WWTP), in order to promote the overall operation of existing digester (mud line) and also to improve nutrient removal (waterline).

Summarizing, anaerobic codigestion stands as a well-known process used to stabilize activated sludge and convert the putrescible matter from organic substrates into methane and carbon dioxide, supported by the right nutrient balance present in the feeding mixture.

Another AD technique could be also adopted: the use of a fermenter reactor. The two-stage anaerobic digestion approach physically divides the biochemical step, namely, hydrolysis-acidogenesis in the first reactor (dark fermentation), while acetogenesis-methanogenesis reactions take place in the second reactor. The stages separation gives the optimal growth rates and pH requirements for acidogenic bacteria and methanogens at the same time (5.5–6.5 and 7, respectively). Considering that, diverse working conditions must be assigned in each reactor (De la Rubia et al. 2009). The first fermentation stage in fact constitutes a valuable external carbon source (soluble COD, volatile fatty acids) that can be used as a carrier to further denitrification and biological phosphorous removal (Bolzonella et al. 2001; Leite et al. 2016).

In sum, the codigestion WAS-OFMSW once considered an integrated approach between the water treatment and waste cycles supports harnessing the benefits from a well-known anaerobic digestion process. Proceeding the separated stage, anaerobic digestion process can make feasible the possibility to produce biogas and also improve the waterline nutrient removal capacity in the WWTp.

17.3.1 Biogas Upgrading

The current trend in Europe is to promote the anaerobic digestion as the base to produce a real biofuel optimized through upgrading techniques, useful not only in situ demands but also in the automotive sector.

Natural gas, compressed natural gas, biomethane, and biogas are all sources of thermal energy for combustion, since their mixture with air represents a flammable substance.

The natural gas (NG) is constituted by a mixture composed of about 90% methane (CH_4) and about 10% ethane (C_2H_6), propane (C_3H_8), butane (C_4H_{10}), and several other minor constituents, such as water vapor (H_2O (g)), hydrogen sulfide (H_2S), helium, molecular nitrogen (N_2), and petroleum gas. The latter compounds and molecules are usually removed before the natural gas is sent to the distribution grid. It is possible to distinguish a natural gas with high energy content (H-gas) from a natural gas with lower energy content (L-gas) (Table 17.7).

Table 17.7 Gas compositions

Composition in vol-%		
	Natural gas L	Natural gas H
Methane	82,0	85,4
Ethane	3,3	8,0
Propane	0,6	2,9
Butane	0,3	1,0
Nitrogen	12,6	0,7
Carbon dioxide	1,2	2,0

The compressed natural gas (CNG) consists of natural gas which undergoes a compression treatment at 220 bar. Compression reduces the volume to below 1% at standard conditions (1 atm). The CNG is then stored and distributed in a rigid container, usually in a spherical or cylindrical shape, at a normal pressure of 200–248 bar, and it can be used in vehicles equipped with a conventional internal combustion engine, like the bi-fuel vehicles (gasoline/CNG).

Biogas consists of a mixture of gases produced during the anaerobic digestion process applied to treat different organic substrates. Biogas main composition includes methane, CH_4 (55–65%); carbon dioxide, CO_2 (35 to 45%); 0.02–0.2% hydrogen sulfide (H_2S); and water vapor (saturation). Other trace constituents: ammonia (NH_3), siloxanes, mercaptans.

Biomethane (BM) is instead a gas containing predominantly methane (CH_4) produced from a renewable source. BM can be obtained from biogas previously submitted to a purification process (dehydration, desulfurization, removal of gaseous ammonia, siloxanes, powders) and upgraded (removal of carbon dioxide, CO_2) in order to accomplish standard qualities of natural gas.

Biogas purification starts removing undesirable elements (trace), using gravel filters for dust and drops and through activated carbon filters for siloxanes. Scrubber washing is usually applied to cause ammonia stripping.

Next, water vapor removal can be performed using cooling systems (e.g., underground piping, water traps, appropriate refrigeration machines), compression, absorption in solutions with gycols or resorting to hygroscopic salts, and adsorption on oxide silica (SiO_2) or activated charcoal.

H_2S is typically the most problematic impurity present in biogas, because it is corrosive, it is toxic, and once burned, it turns into SO_2, whose emissions contribute to acid rain formation. Hydrogen sulfide can be removed by dry or liquid technologies.

All trapping processes to dry H_2S usually present an absorbent medium located in a tank, across which the gas stream passes from the bottom upward or vice versa.

The absorbing medium may become saturated with H_2S and inactive in the end. In these cases, a regeneration step must be provided; in general plants are equipped with two towers that work in parallel. Dry technologies to remove sulfuric acid are those historically most used and are able to remove about 200–250 kgS/day. These technologies differ among each other depending on the medium used. For instance, one of the oldest dry technologies to remove H_2S and mercaptans uses iron oxide (e.g. Iron Sponges) in according to the following chemical reaction:

$$Fe_2O_3 + 3H_2S \rightarrow Fe_2S_3 + 3H_2O$$

One kilogram of Fe_2O_3 stoichiometrically removes about 0.64 kg of H_2S.

$$2Fe_2S_3 + 3O_2 \rightarrow 2Fe_2O_3 + 6S$$
$$Fe_2O_3 + 6RSH \rightarrow 2Fe(RS)_3 + 3H_2O$$

In addition, the iron sponge regeneration step may undergo separately since it can be set up in batch; this approach permits to obtain a removal efficiency of 85% (0.56 kg H_2S per kg Fe_2O_3). Advantages: simplicity and efficiency. Disadvantages: the process is extremely expensive.

Another dry technology utilizes insoluble zinc oxide in according to the following chemical reaction.

$$ZnO + H_2S \rightarrow ZS + H_2O$$

The processes with zinc oxide are highly temperature proceeded, between 200 and 400 °C, and the top sulfur load is generally equal to 30–40 kg sulfide per 100 kg of absorbent.

The Puraspec®, patented and commercially sold by the IC Industry (Great Britain), regards the mixing of diverse zinc oxides whose best working efficiency is assigned within the temperature range of 40–200 °C. The zinc sulfide formation is irreversible.

Adsorbents are used to physically absorb a gas phase on a solid surface, instead of being chemically transformed. Molecular sieves (zeolite) are crystalline materials, highly porous, which belong to the class of aluminates.

Polar compounds, such as water, H_2S, SO_2, NH_3, COS, and organosulfur compounds, have been effectively absorbed, and nonpolar substances such as methane can be used to easily remove them.

Not all of the mercaptans are adsorbed by molecular 4A or 5A sieve type due to the limited pore sizing; accordingly, the 13× (10A) molecular sieves are preferred to completely remove sulfur gas.

Either activated carbon adsorbents impregnated with alkaline oxides or solution have been designed to improve the adsorption properties; several ionic compounds (sodium hydroxide, sodium carbonate, potassium hydroxide, potassium iodide) and some metal oxides have been also applied onto absorbents' surface.

Liquid H_2S process removal has partially replaced dry solid technologies because they require less space, lower labor costs, and increase the recovery of elemental sulfur.

Some examples are iron and zinc oxide slurries, quinone process, chelated iron solutions.

The chemical reaction between slurry oxide and H_2S is reported in the following equations:

$$Na_2CO_3 + H_2S \rightarrow NaHS + NaHCO_3$$
$$Fe_2O_3 \cdot 3H_2O + 3NaHS + 3NaHCO_3 \rightarrow Fe_2S_2 \cdot 3H_2O + 3Na_2CO_3 + 3H_2O$$

The regeneration is obtained by aeration, which converts the sulfide into elemental sulfur according to the equation:

$$2Fe_2S_2 \cdot 3H_2O + 3O_2 \rightarrow 2Fe_2O_3 \cdot 3H_2O + 6S$$

The patented product Chemsweet ® (Natco, Inc) has been successfully used as an absorbent in recent years. Once added to water, this substance, made of zinc oxide, zinc acetate, and dispersant, starts to build up some chemical reactions as described below:

$$ZnAc_2 + H_2S \rightarrow 2HAc + ZnS$$
$$ZnO + 2HAc \rightarrow ZnAc_2 + H_2O$$
$$ZnO + H_2S \rightarrow ZnS + H_2O$$

Since the involved reagents cannot be regenerated, this approach results in high maintenance costs.

The redox cycle in the figure shows how the H_2S is converted to elemental sulfur using quinones (Fig. 17.6).

Due to the high operating costs and the formation of thiosulfate as a byproduct, the technologies based on the use of quinone to remove H_2S are generally not advantageous when small gas flow is supposed to be purified.

Technologies that use iron chelate solution (chelated iron) are gaining attention for the H_2S removal. Adsorption and regeneration steps are chemically represented as follows:

$$2Fe^{3+} + H_2S \rightarrow 2Fe^{2+} + S + 2H^+$$
$$2Fe^{2+} + \frac{1}{2}O_2 + H_2O \rightarrow 2Fe^{3+} + 2OH^-$$

The LO-CAT® is a very relevant biogas designed process since it has H_2S removal efficiency of 99%; the catalyst is nontoxic and operates at room temperature, without the need for heating or cooling for the adsorbent purpose.

Energy for pumps and ventilators, chemical products and the replacement of the catalyst due to the formation of thiosulfate and bicarbonate are the main operating costs.

After the purification of biogas, to obtain BioMethane is necessary to remove the carbon dioxide from biogas by processes named Upgrading Techniques. One of the most used upgrading techniques is Pressure Swing Adsorption (PSA).The PSA-type plant operates the CO_2 for CH_4 separation using packed columns with adsorbent material (generally activated carbon or zeolites), which varied pressures are applied along the process.

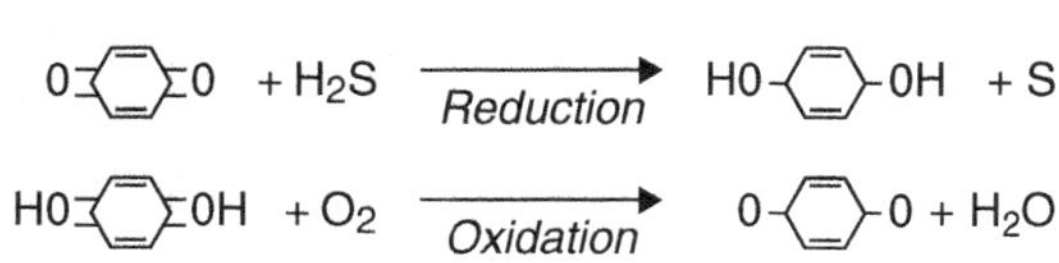

Fig. 17.6 H_2S converted to elemental S using quinones

The main advantages of this technique consist of the simplicity of construction and compact-sized structure that benefits the small-sized plant adoption (up to 250 Nm^3 of raw gas treated per hour).

However, this technology requires a pretreatment step applied in the raw biogas in order to eliminate both H_2S, which may bind irreversibly to the adsorbent material, and H_2O (g), which may damage the structure.

Another common technique is washing with Under Pressure Water (PSW). The basis principle concerns the higher solubility of CO_2 compared with CH_4, in particular at low temperatures (Table 17.8).

H_2S has a higher solubility compared to CO_2, but the costs associated with the selective removal of H_2S that uses the washing water have not still shown sufficient competitiveness with other methods.

In the case of the PWS plants, raw gas is forced through a plastic-packed bed column to increase the surface contact between the gaseous and liquid phase. Inside the column, the gas meets a countercurrent liquid flow. The output liquid is CO_2 "enriched," while the output gas is predominantly constituted of CH_4.

Finally, other upgrading techniques are based on chemical washing with ammine solution (monoethanolamine or dimethylethanolamine) or physical washing with organic solvents.

The liquid processes with amine are the largest used alternative for water purification in order to remove acid gases. Amines arise an interesting scrubber since they can reach high efficiency, and it can regenerate and be designed for either single CO_2 removal or for simultaneous CO_2 and H_2S removal.

$$H_2S \rightarrow HS^- + H^+$$
$$CO_2 + H_2O \rightarrow HCO_3^- + H^+$$
$$RNH_2 + H^+ \rightarrow RNH_3^+$$
$$RNH_2 + CO_2 \rightarrow RNHCOO^- + H^+$$

The plant that operates the physical washing with organic solvents is similar to the PWS plants, with the fundamental difference the CO_2 absorbed by an organic solvent (*glycol as polyethylene*, the family of polyethylene glycols, PEG), namely, Selexol™ or Genosorb®.

The advantage of an organic solvent instead of H_2O resides in the fact that the CO_2 is more soluble in PEG than in water.

Table 17.8 Henry constant of biogas constituents

Henry constant at 25 °C e 1 atm	
	H (M/atm)
Methane	$1.5 \cdot 10^{-4}$
Carbon dioxide	$3.6 \cdot 10^{-2}$
Hydrogen sulfide	$8.7 \cdot 10^{-2}$

At equal upgradeability capacity to upgrade biogas, the liquid phase flow may be lower, and the upgrade installation plant may be lower compared with the PWS plant structure. Even whether physical washing with organic solvents is supposed to be used, the PEG solution should be heated or depressurized to get regenerated. Finally, even though contaminants such as H_2S, H_2O (g), O_2, and N are removed during the upgrading process, they are usually removed in the beginning of the upgrading process.

17.4 Thermophilic Process in OFMSW Treatment (Evolution of the AD Processes)

Anaerobic digestion is a widespread and well-known technology to treat organic waste of diverse stocks (Micolucci et al. 2016). In the past, it was considered a process applied to deal with the organic waste and consequently a management of organic waste flows such as food waste and organic fraction of MSW. Nowadays the energy demands have evolved the AD process that considers the organic waste as a raw material, therefore aimed at a production process (Yoo et al. 2004). Considered this approach as a production process (Gottardo et al. 2017), AD requires its management with new and solid perspectives in order to enhance not only the biogas production and the energy recovery but also the production of raw resources such as volatile fatty acids so as to further obtain, for instance, bioproducts like linear polyesters produced by bacterial fermentation (e.g., fermented organic waste substrate), polyhydroxyalkanoates. PHA-storing bacteria are mainly volatile fatty acid-consuming bacteria. Additionally, AD involves different microorganisms that through synergic way allow not only the production of methane but also other valuable products, hydrogen, and volatile fatty acids (Bolzonella et al. 2005). The ability to extract from the anaerobic process these different products is to divide the anaerobic process itself into two main stages.

Fermentation is the first stage, which mainly includes the step of hydrolysis, acidogenesis, and acetogenesis, and the second stage that substantially optimizes the last step, the methanogenesis (De Falco and Basile 2015). Therefore, through the optimization of the fermentation, hydrogen (gas), volatile fatty acid (liquid), and other low-weight organic compounds such as alcohols and lactic acid (Viturtia et al. 1992) can be obtained.

VFAs, for example, can be then used as external carbon source for biopolymer production, like polyhydroxyalkanoates, altering the conditions of carbon source availability and unavailability (Chua et al. 1997; Dionisi et al. 2004).

Nowadays AD comprises 244 treatment plants in the whole EU, with a handling almost 8 million tons of organic waste capacity (De Baere and Mattheeuws 2015).

Mesophilic digestion is the most adopted, a developed process adopted predominantly to minimize treatment cost due to heating. De Baere and Mattheeuws (2015) highlights the cumulative percentage of mesophilic plants in EU, 67% of the installed plant in 2014. These digesters operate at temperatures between 35 and 40 °C. Only the 33% operate in thermophilic conditions (temperatures between 50 and 55 °C).

The large difference is due to the fact the most of the applications are associated with wastewater treatment plants, with sewage sludge digestion. Mesophilic digestion has been considered as a more stable process and compared to thermophilic with a smaller need for heating. Currently, thermophilic digestion needs to be encouraged. It has been proven at full scale its stability over many years of studies and of plant operation. Treating organic fraction of MSW, biogas production rates can be 30–50% higher than mesophilic digestion; hence, in fact, the issue for further digester heating is a problem absolutely bridgeable. Thermophilic digestion allows also to operate through a digestion process in two separate stages.

Between environmental damage and pick oil, there is not only energy recovery but also the recovery of the waste material as a source for the production of bioproducts. Waste has to be considered from a mere treatment product to a raw precious material for a production system of marketable products, products that may be required in different sectors. In this transition, there is the evolution of the anaerobic digestion processes.

Operate to thermophilic separate stages rather than single stage, so as to obtain a commercial product. Low-cost volatile fatty acid (VFA) production from organic waste by fermentation is an approach that leads to further obtain biodegradable plastics and bioenergy or is utilized in biological nutrient removal processes.

VFAs are short-chain fatty acids usually made up of six carbon atoms at most. Several are the applications related to VFA as substrate, mainly bioenergy production, bioplastic, and biological sewage nutrient depletion. This is why it has increased in the last few years the interest for their production. The VFA production can be chemical or biological routes. Nowadays chemical routes are the main approaches for commercial VFA production; however, in the next year, a growth of interest in the biological production is expected. In this context, the use of organic-rich waste is more indicated than pure sugar for economical and ethical reasons. Particularly, organic fraction of municipal solid waste, with high total COD (150–347 gCOD/kg), is an important source of VFA production.

In general, VFA production from biowaste is an anaerobic process, namely, anaerobic fermentation or dark fermentation: firstly (hydrolysis process), where complex organic substances are broken down into simpler compounds (acidogenesis process), which are fermented into VFAs. The processes involve a complex consortium of anaerobic microorganism, such as *Clostridia*, *Bacteroides*, and *Enterobacteriaceae*, and these processes are conducted concurrently in a mono-stage reactor. Acidic condition supports the production of VFAs from food waste (Lee et al. 2014).

The weak link in the digestion chain is the hydrolysis, in the fermentation step. One of the approaches that help the rate-limiting hydrolysis step is anaerobic digestion at high temperatures, the thermophilic. Thermophilic temperature compared to psychrophilic or mesophilic increases the concentration and the production of VFA (hydrolysis is enhanced at higher temperatures) (Lu and Ahring 2005). Acidogenesis is favorite to thermophilic temperature; hence, it leads to higher VFA yields (Lee et al. 2014) even though some studies are contradictory. However, it is necessary to improve the hydrolysis and acidogenesis process since dark fermentation allows, depending on the process parameters, to produce high value-added products, volatile fatty acids, and/or hydrogen.

17.5 Biohydrogen Production from OFMSW

Hydrogen mixed with biogas can be considered as a substitute fuel to conventional fossil fuels only if the production of hydrogen using renewable energy resources. Currently, more than 95% of the hydrogen produced comes from thermochemical processes, energetically expensive and moreover employing fossil fuels. An example is the "steam-reforming" natural gas (Levin et al. 2004), or the use of high-temperature steam and appropriate catalysts to operate the alkane cleavage to hydrogen and carbon dioxide. Water hydrolysis, after the thermochemical processes, is the most voyaged way for the production of hydrogen (Levin et al. 2004). It could be an optimum hydrogen production process as the necessary electric energy to its use can be derived from renewable energy sources (about 460 kJ/mol bond dissociation energy of the water molecule). Examples are wind, solar, and hydropower (Adamson 2004). However, even now, this technology presents substantial obstacles in terms of cost and low efficiency. Hydrogen can also be produced by biological pathways, through the action of enzyme complexes, the *hydrogenase* and *nitrogenase* (Kótay and Das, 2008). These enzyme complexes are produced by some prokaryotic microorganisms, photosynthetic bacteria and fermentative bacteria, eukaryotic bacteria, and some types of microalgae. Specifically, biohydrogen can be produced from three biological processes: the water biophotolysis, the photo-fermentation, and finally the dark fermentation. Photolysis in the production of hydrogen from water is performed by microalgae. During a normal process of oxygenic photosynthesis, the reduced ferredoxin, the result of the flow of electrons along the thylakoid membrane, is oxidized by a nicotinamide adenine dinucleotide phosphate ($NADP^+$), and NADPH thus formed, together with adenosine triphosphate (ATP), will be used to fix the carbon dioxide.

Since the industrial revolution, fossil fuels were essential elements for the development of human civilization. After about two centuries of exploitation of fossil fuel reserves and increasing air pollution, the evidence of global climate change and the prospect of an imminent achieving of peak oil have become pressing issues for the international community, which must develop systems able to produce energy in an environmentally responsible way, including the exploitation of renewable energy sources. A proposed approach concerns the production of hydrogen as a renewable fuel, looking at the forthcoming "hydrogen economy" which has been widely discussed recently. Nowadays, this approach still has many unsolved issues such as hydrogen storage technologies and its application descriptions to fuel engines. Indeed, there are some advisable procedures regarding hydrogen handling: high-pressure demands, special container materials in order to minimize diffusion and prevent leaks, and huge security precautions. Moreover, it is important to state liquid hydrogen has low volumetric power density (1/3 of compressed natural gas, CNG). Considering the aforementioned and the necessity of developing new infrastructure slow down the realization of this approach.

At present, a feasible scenario is the hydrogen fuel injection (HFI). HFI means to mix a gaseous fuel with hydrogen to obtain a mixture with improved combustion characteristics. Compared to other conventional fuels, hydrogen presents lower minimum ignition energy amount and also greater laminar flame speed than conventional fuels; therefore, it is necessary that just a slight quantity of H_2 be added to the main fuel in order to boost up its combustion propriety. Finally, hydrogen acts as a catalyst for combustion and only secondarily as an energy carrier. In the early 1990s, the Hydrogen Components Inc. (HCI) conducted several studies concerning the feasibility to use a blend of compressed natural gas (CNG) and hydrogen as an improved fuel for internal combustion chamber engines, and they showed that the lean burn of mixture of hydrogen (7% by energy or 20% by volume) and CNG can reduce pollutant emissions (mainly NO_x) into the atmosphere, without causing damages in the CNG usage efficiency. The use of this mixture does not require storage system nor particular changes both in the CNG engines and infrastructure. As a result, HCI patented this mixture and the commercial name of this fuel was Hythane®. Several studies (i.e., in Italy performed by ENEA) were carried out on this fuel confirming the results shown above.

Beyond the benefits, nonrenewable energy sources are used to produce methane and hydrogen, by reforming processes of fossil fuels with the production of syngas, a gaseous mixture of CO and hydrogen, an intermediate in creating the synthetic natural gas (SNG). To cope with this problem, the attention has recently moved toward the biological H_2 production to mix with biogas: a very well-known mixture, namely, biohythane. Many studies have shown that adding a little hydrogen amount (nearly 10% v/v) strongly increases the biogas ignition efficiency of both automotive internal combustion engines (Graham et al. 2008) and also in internal cogeneration mode engines. These results are ascribed to the valuable CO_2 removal during the prior biogas upgrading step.

This paragraph will point out the feasibility of biological hydrogen and biogas production process using biowaste as substrate.

Dark Fermentation

Hydrogen production by dark fermentation (DF) is operated by hydrogenase enzyme, which catalyzes the reduction of protons by electrons provided by a reduced ferredoxin.

Ferredoxin is reduced mainly in the pyruvate conversion to acetyl-coA. It can be also reduced by the reduced nicotinamide adenine dinucleotide, NADH, obtained by glyceraldehyde-3-phosphate to 1,3-diphosphoglycerate conversion. The reduction of ferredoxin by NADH occurs exclusively when the hydrogen partial pressure is very low, below 60 Pa (Kraemer and Bagley, 2007; Mathews and Wang 2009; Valdez-Vazquez and Poggi-Varaldo, 2009).

Fermentation's final byproducts are the acetic acid and hydrogen, presenting a maximum yield of 4 moles of H_2 (expressed as consumed moles of glucose) (Angenet et al. 2004; Levin et al. 2004).

17.5.1 Biohydrogen Production by Dark Fermentation and Biohythane: Basic Concepts

The biological production of hydrogen (biohydrogen) is enhanced microbiological conversion of water and organic substrates in hydrogen fuel, in the case of dark fermentation. This process takes place through the action of hydrogenase and nitrogenase enzymes (Kotay and Das, 2008). The main biological role of hydrogen-producing microorganisms concerns biophotolysis of water, photo-fermentation, and dark fermentation (through heterotrophic bacteria).

Bacteria which carry out dark fermentation transform sugars, starches, and other carbohydrates and organic substrates into H_2 and CO_2, together with organic acids, alcohols, and other byproducts. Particularly these bacteria are *Clostridium* spp., *Thermoanaerobacterium* spp., *Enterobacter*, and *Bacillus* (Reith et al. 2003). Presently, the dark fermentation is the process that arouses greater interest because biohydrogen and renewable biogas (biohythane) can be produced at the same time in consortium with anaerobic digestion.

Biohydrogen generation comes from the action of hydrogenase enzymes. These enzymes reduce the protons present in the water to $H_2(g)$, by the oxidation of a strong reducing agent. In nature, the reducing agent is a reduced electron carrier, such as ferredoxin, the flavodoxin (this is produced by several bacteria as an iron-free alternative to the ferredoxin, where the iron is a limiting factor), or the NADPH (reduced).

The production of molecular hydrogen in the dark fermentation is carried out by the enzyme hydrogenase which catalyzes the reduction of protons by electrons provided by a reduced ferredoxin; the latter is reduced mainly in the conversion reaction of pyruvate to acetyl-CoA, but it can also be reduced by the reduced nicotinamide adenine dinucleotide, NADH, obtained during the conversion of glyceraldehyde-3-phosphate to 1,3-diphosphoglycerate. The reduction of ferredoxin by NADH occurs exclusively when the hydrogen partial pressure is very low, below 60 Pa (Kraemer and Bagley, 2007; Mathews and Wang, 2009; Valdez-Vazquez and Poggi-Varaldo, 2009); in this condition the acetic acid is the final product of the fermentation, and a maximum production of molecular hydrogen is obtained, 4 moles of H_2 per mole of glucose consumed (Fig. 17.7) (Angenet et al. 2004; Levin et al. 2004).

However, this condition is difficult to maintain, particularly at a full-scale production system (Valdez – Vazquez and Poggi-Varaldo, 2009); most of the NADH will be oxidized through reactions that lead to the formation of smaller products with a consequent decrease of the molecular hydrogen production. In fact, when the products that are formed at the end of fermentation are more reduced than acetic acid, fewer electrons are available to reduce the protons and thus generate molecular hydrogen (Fig. 17.8).

Lactic acid, however, is another glucose's fermentation byproduct produced through three metabolic pathways (known as solvatogenic fermentation) which do not lead to the production of hydrogen: the homofermentative way (a), heterofermentative way (b), and the bifidum pathway (c).

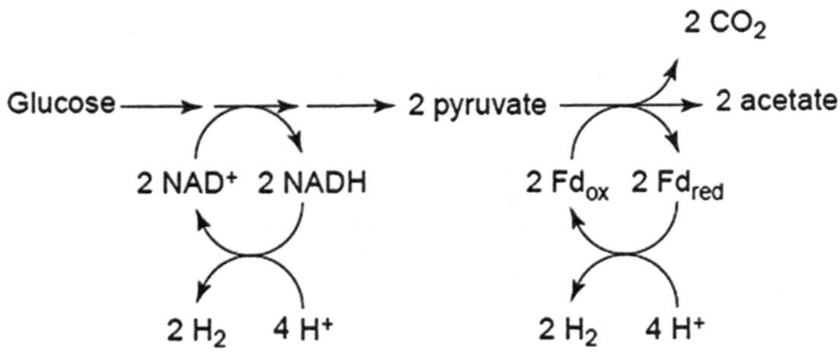

Fig. 17.7 Hydrogenogenic metabolic fermentation with production of acetic acid (Angenet et al. 2004)

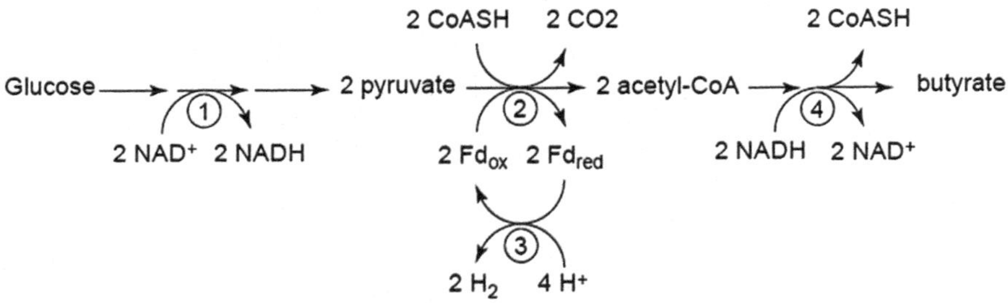

Fig. 17.8 Hydrogenogenic metabolic fermentation with production of butyric acid (Angenet et al. 2004)

$$C_6H_{12}O_6 \rightarrow 2CH_3CHOHCOOH \tag{a}$$

$$C_6H_{12}O_6 \rightarrow CH_3CHOHCOOH + CH_3CH_2OH + CO_2 \tag{b}$$

$$2C_6H_{12}O_6 \rightarrow 2CH_3CHOHCOOH + 3CH_3COOH \tag{c}$$

Literature studies have shown that the highest yields of hydrogen production are obtained when the end product of the fermentation is a mixture of acetic acid and butyric acid, while the lowest hydrogen yields have been related to other reduced by-products, such as alcohols and lactic acid (Levin et al. 2004; Mathews and Wang, 2009).

The modest hydrogen productivity can also be associated with the presence of propionic acid as a final fermentation product (Levin et al. 2004); this can be explained by the action of *Clostridium propionicum* (Cardon and Barker 1947), a bacterium capable of producing propionic acid consuming molecular hydrogen (Hawkes et al. 2002).

$$C_6H_{12}O_6 + 2H_2 \rightarrow 2CH_3CH_2COOH + 2H_2O$$

It has to be considered also that not necessarily an increase of acetic acid as an end product of fermentation is indicative of an improvement in the production yield of molecular hydrogen; some species of *Clostridium*, such as *Clostridium aceticum* (Braun and Gottschalk, 1981), can lower the hydrogen production yield by converting carbon dioxide and molecular hydrogen to acetic acid (Hawkes et al. 2007); and acetic acid with propionic acid can also be a result of the lactic acid fermentation (Antonopoulou et al. 2008).

$$4H_2 + 2CO_2 \rightarrow CH_3COOH + 2H_2O$$

$$3CH_3CHOHCOOH \rightarrow 2CH_3CH_2COOH + CH_3COOH + CO_2 + H_2O$$

The advantages of dark fermentation compared to other biological techniques for hydrogen production are in fact related to the use of waste substrates such as the organic fraction of MSW; in the dark fermentation, in fact, carbohydrates are the preferential substrates for hydrogen production (Okamoto et al. 2000), which constitute, together with protein, the main components of household organic waste (Chong et al. 2009). A greater hydrogen production is provided from the conversion of each substrate mole containing both carbohydrate and organic acids (Reith et al. 2003).

A second advantage is the possible use of the effluent coming from the process to produce biogas completing the process of anaerobic digestion; in this way, it is therefore possible to obtain a mixture biohydrogen-biogas which in effect constitutes a valid alternative to traditional fossil fuels: the biohythane.

Several studies have shown that pH condition and, in a minor way, temperature condition can influence the hydrogenase enzyme activity along the H_2 production (Hallenbeck et al. 2009; Reith. et al. 2003; Valdez-Vazques and Poggi-Varaldo, 2009). Authors emphasized the greatest enzyme activity occurred in thermophilic temperature conditions (55 °C) in the pH range of 5.0–6.0 (optimum value at 5.5). On the other hand, potential inhibitors of the hydrogenase enzyme are the hydrogen partial pressure that causes a metabolic pathway shift to solvatogenic fermentation and free ammonia inhibition.

The obstacle regarding the continuous biohydrogen production via dark fermentation is on one hand how to keep the optimal pH value since the accumulation of organic acids (produced during fermentative metabolism) can decrease the pH value in another hand. In order to accomplish a better pH control, some chemical methods have already been tested (Valdez-Vazques and Poggi-Varaldo, 2009). Recently some authors proposed an interesting strategy for pH control, dark fermentation (first stage) coupled with anaerobic digestion (second stage) process using the enriched buffer compound digestate recirculation to control the pH of the first stage.

Moreover, in this way, it is possible to produce biohythane, an alternative fuel to traditional fossil fuels (Fig. 17.9).

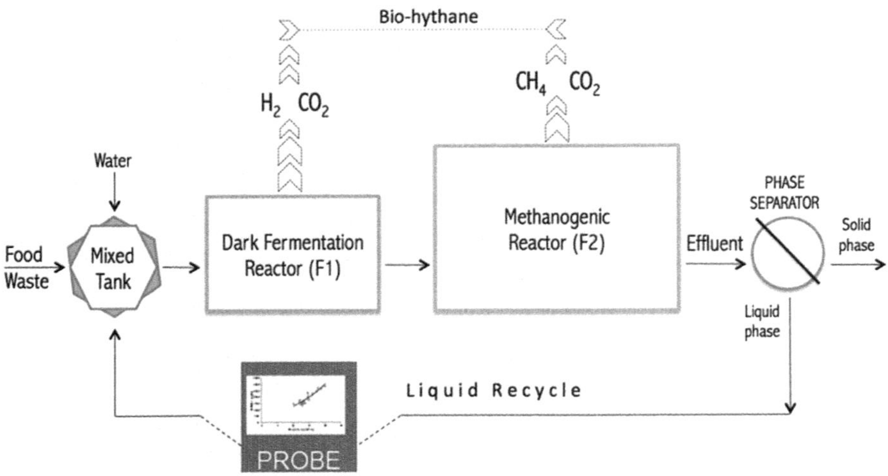

Fig. 17.9 Two-stage anaerobic digestion, process scheme (Micolucci et al. 2014)

17.5.2 State of the Art of Biohydrogen and Biohythane Production from Biowaste: Process Applications

Biohydrogen production by DF can be executed at mesophilic (35 °C–37 °C) or thermophilic (55 °C) conditions, although the thermophilic biohydrogen production is generally applied at laboratory and pilot-scale systems. Biowastes are abundant organic materials recovered by separate collection of municipal wastes or by food industry, and it is a material with a large energy content that must be exploited. These two aspects (temperature and substrates), together with other process parameters, are reported in Table 17.9 with the aim of reviewing the two-stage process for biowaste exploitation.

From the data shown in Table 17.9, some important guidelines can be pointed out:

- Hydraulic retention time (HRT): hydrogen production increases with HRT decreasing; in fact the application of low HRT (from 2 to 5 days) in CSTR systems allows the washout of methanogens (Hawkes et al. 2007), favoring the development of fermentative producing hydrogen bacteria. The HRT, however, must be greater than the specific growth rate of the microorganisms producing hydrogen to prevent their washout (Kongjan and Angelidaki 2010).
- Organic loading rate (OLR): OLR applied varies considerably among 14 and 38.5 kgVS/(m^3_rd). Basically, hydrogen production decreases by increasing the organic load applied. This may be due to the accumulation of hydrolyzed substances that became toxic to the microorganism cells. In fact, high-VFA concentrations can damage H_2 fermentation just like pH does, since molecular acids are

Table 17.9 State of the art of two-stage hydrogen and methane fermentation from biowaste (food/municipal substrates)

Reference	Substrate	H_2 reactor	CH_4 reactor	Working volume	OLR	HRT	Temp	pH	pH control	SHP average	SMP average
	(Biowaste)	Type	Type	L	kgVS/(m^3r^ad)	d	°C			LH_2/kgVS	LCH_4/kgVS
Liu et al. (2006)	Household	CSTR	CSTR	H_2: 0.4	H_2: 37.5	H_2: 2	H_2: 37	H_2: 5–5.5	NA	43	500
	Solid waste	Lab.Scale	Lab.Scale	CH_4: 3	CH_4: 4.1	CH_4: 15	CH_4: 37	CH_4: 7.5			
Chu et al. (2008)	OFMSW	CSTR	Fluid bed reactor	H_2: 10	H_2: 38.4	H_2: 1.3	H_2: 55	H_2: 5.5	R	205	464
		Lab.Scale		CH_4: 40	CH_4: 6.6	CH_4: 5	CH_4: 35	CH_4: 7.5			
Wang and Zhao (2009)	Food waste kitchen W	Rotating Drum	CSTR	H_2: 200	H_2: 22.65	H_2: 6.7	H_2: 40	H_2: 5.2–5.8	NO	65	546
				CH_4: 800	CH_4: 4.61	CH_4: 26.7	CH_4: 40	CH_4: 6.7–7.3			
Chu et al. (2010)	Food waste	CSTR	Fluid bed reactor	H_2: 10	H_2: 38.4	H_2: 1.3	H_2: 55	H_2: 5.5	R	205	464
		Lab.Scale		CH_4: 40	CH_4: 6.6	CH_4: 5	CH_4: 35	CH_4: 7.8			
Lee et al. (2010)	Food waste	CSTR	UASB	H_2: 500	H_2: 39^	H_2: 1.9	H_2: 33	H_2: 5.5	NaOH	83	210
				CH_4: 2300	CH_4: 6.4^	CH_4: 7.7	CH_4: 36	CH_4: 7.4			

(continued)

Table 17.9 (continued)

Reference	Substrate	H_2 reactor	CH_4 reactor	Working volume	OLR	HRT	Temp	pH	pH control	SHP average	SMP average
	(Biowaste)	Type	Type	L	kgVS/(m^3r^ad)	d	°C			LH_2/kgVS	LCH_4/kgVS
Cavinato et al. (2011)	OFMSW	CSTR	CSTR	H_2: 200	H_2: 16/21	H_2: 3	H_2: 55	H_2: 5.5	R	51	416
				CH_4: 760	CH_4: 5/10	CH_4: 12.6	CH_4: 55	CH_4: 7.8			
Elbeshbishy and Nakhla (2011)[b]	Food Waste	SBHR	/	H_2: 2	H_2: 14.5	H_2: 2	H_2: 37	H_2: 5-6	NA	332	/
		Lab.Scale									
Luo et al. (2011)	Organic wastes	CSTR	CSTR	H_2: 1.2	CH_4: 3	H_2: 3	H_2: 55	H_2: 5.3	NaOH	48	320
		Lab.Scale	Lab.Scale	CH_4: 3.5		CH_4: 12	CH_4: 55	CH_4: 7.9			
Nazlina et al. (2011)[c]	Food waste	Stirred batch	/	H_2: 0.2	H_2: 50^	H_2: 0.5–1	H_2: 55	H_2: 5.5	A-B	63	/
Sreela-or et al. (2011)[c]	Food waste	Stirred batch	/	H_2: 0.07	H_2: 2.5	H_2: 3	H_2: 30	H_2: 5-6	NA	104	/
Zhu et al. (2011)[a, c]	OFMSW + sewage sludge	CSTR	CSTR	H_2: 1	H_2: 15.2	H_2: 1.25	H_2: 35	H_2: 5.8	NaOH	47	522
		Lab.Scale	Lab.Scale	CH_4: 5	CH_4: 5.6	CH_4: 12.5	CH_4: 35	CH_4: 6.8–7.2			
Cavinato et al. (2012)	OFMSW	CSTR	CSTR	H_2: 200	H_2: 16.3	H_2: 3.3	H_2: 55	H_2: 5.5–6	R	66.7	483
				CH_4: 760	CH_4: 4.8	CH_4: 12.6	CH_4: 55	CH_4: 8.4			

Chinellato et al. (2013)	Food waste	CSTR	CSTR	H_2: 2	H_2: 20	H_2: 3	H_2: 52	H_2: 5.2	R	117	311
		Lab.Scale	Lab.Scale	CH_4: 5	CH_4: 3–6	CH_4: 12	CH_4: 52	CH_4: 7.7			
Gottard et al. (2013)	OFMSW	CSTR	/	H_2: 4.5	H_2: 16	H_2: 3	H_2: 55	H_2: 5.5	R	30	/
		Lab.Scale									
Giuliano et al. (2014)	OFMSW	CSTR	CSTR	H_2: 200	H_2: 16-18	H_2: 3.3	H_2: 55	H_2: 5.3	R	53	350
				CH_4: 760	CH_4: 4-5	CH_4: 12.6	CH_4: 55	CH_4: 8.2			

[a]Removal methanigen biomass flushing air
[b]Sonication/*NA* not available, *R* recirculation, *A-B* acid-base control. ^ on COD basis
[c]HST (substrate heat treatment)

analog uncoupler agents that benefit protons to penetrate the cell, while causing injury in the natural membrane pH stability. Thus, it has been reported bacteria undergo major metabolic shift from acidogenesis to solventogenesis whenever they need to provide a natural cell detoxification mechanism in order to prevent VFA inhibition effects (Valdez-Vazquez and Poggi-Varaldo 2009).

- Inhibition of hydrogenotrophic activity: most experiments aimed to optimize the fermentation phase and often applied pH monitoring/control (Shin and Youn 2005; Gomez et al. 2006; Lee et al. 2010) and/or substrate pretreatment (Han et al. 2005; Chou et al. 2008; Lee et al. 2010) such as heat treatment (100 °C – 10 min). The newly proposed approach in opposite to the conventional use of chemicals for pH controlling in the DF stage consists of recirculating the digestate coming from the methanogenic reactor to the hydrogen-producing reactor, in order to take advantage from its remained buffer capacity (such as ammonium and bicarbonate compounds) (Cecchi et al. 2005) and also to provide both nutrient recycling and the feedstock dilution just in case (Kataoka et al. 2005; Chu et al. 2008; Lee et al. 2010).
- Biohydrogen and biomethane yields: taking into account the gas yields obtained by pilot-scale CSTR plant, a full-scale simulation of the process could be done. Researches of Cavinato et al. (2011, 2012) and Giuliano et al. (2014) using 200 l pilot-scale reactors have reported hydrogen and methane amounts around 51.0–66.7 $LH_2/kgVS_{fed}$ and 350–483 $LCH_4/kgVS_{feed}$, respectively. Considering the best yields, it is possible to obtain 16 m^3 of hydrogen per ton of biowaste and 110 m^3 of methane per ton of biowaste, which means about 166 m^3 biohythane per ton of biowaste (0.8 $m^3/kgVS_{added}$). The specific energy production that could be obtained is 404 kWh per ton of waste, greater than all the energy demands estimated in 20–40 kWh per ton of waste treated (Cecchi et al. 2005).

17.5.3 Research Trends and Challenges: Automatic Control and Full-Scale Implementation of the Two-Stage Process

Biohydrogen and biohythane processes at a full scale are presently under assessment, both in economic and long-term feasibility viewpoints. In fact, the continuous recirculation can lead to an increase of alkalinity in the system favoring the proliferation of hydrogenotrophic microorganisms with a consequent low hydrogen yields and also cause ammonia accumulation in the system. As previously stated, the digestate recirculation benefits the process control preventing also ammonia inhibition in the system. However, this approach might be feasible because commercial ammonia probes applied in such a heterogeneous media would scarcely provide and register accurate data with high confidential level over time. Micolucci et al. (2014) have modelled and predicted the undissociated ammonia in a two stage anaerobic digestion system through chemical (volatile fatty acids) and physical-chemical parameters (alkalinity and electrical conductivity) carried out during

1 year. In fact, addressing reference set points to those parameters, which could indicate a safe range for the system operation, was a crucial step in order to deliver a precisely defined amount of recycled digestate considering the demands and the reactors' evolution on time. For this reason, among the monitored parameter was also inserted the conductivity, which allows to use simple, resilient, and cheaper online probes. Finally, the real semiautomatic statistical model was a completed setup and validated, being useful to predict ammonia concentration in such biological processes.

A further advantage of a logical controlling based on set point regards the possibility to provide itself the best environmental conditions to microorganisms. As a consequence, more biohydrogen and biomethane can be easily achieved (Micolucci et al. 2014). This implementation is one of the most innovative, bringing pilot plant research to market.

The main results obtained from the aforementioned research can be briefly summarized as follows (Micolucci et al. 2014):

- In steady-state condition, the specific hydrogen production was 66.7 LH_2/kgVS added, while the biogas produced in the methanogenic phase was some 0.72 m^3/kgVS added. In these conditions, a 1 m^3H_2/m^3 reactor can be obtained as HPR (hydrogen production rate).
- The finale biohythane composition characteristics were CH_4 58%, H_2 7%, and CO_2 36%. Hydrogen content in the biohythane mixture was always more than 5% in all the conditions studied being ideal suitability for engine employment.
- The global biogas production obtained was 0.80 m^3/kgVS added, while gas production rate (GPR) was some 2.75 m^3/m^3d.
- Power conversion reaches the amount of 404 kWh per ton of pretreated waste.

It is noticeable that, if we compare the energy spent for the OFMSW treatment (20–40 kWh per ton of waste treated), we can obtain ten times the energy we use to treat it.

An important part of the research carried out concerns the possibility to implement an automatic expert system to control the dark fermentation (DF) process (Gottardo et al. 2017). In fact, the maximum hydrogen and volatile fatty acids productions arise in the first stage reactor, as a result of the effluent recirculation coming from the second stage (methanogenic stage) with no need of external chemicals for this purpose. Such approach, however, does not entirely prevent the methanogenic activity inhibition. Previous studies reported the use of variable recirculation flow as the most recommended approach in order to control the entire system, decreasing the opportunities of ammonia inhibition to occur. The approach used is based on a low-cost, robust probe as the system condition detector. The research demonstrates that, using conductivity, VFA, and alkalinity, ammonia concentration was estimated by means of a multiple linear regression model with statistical software (Fig. 17.10).

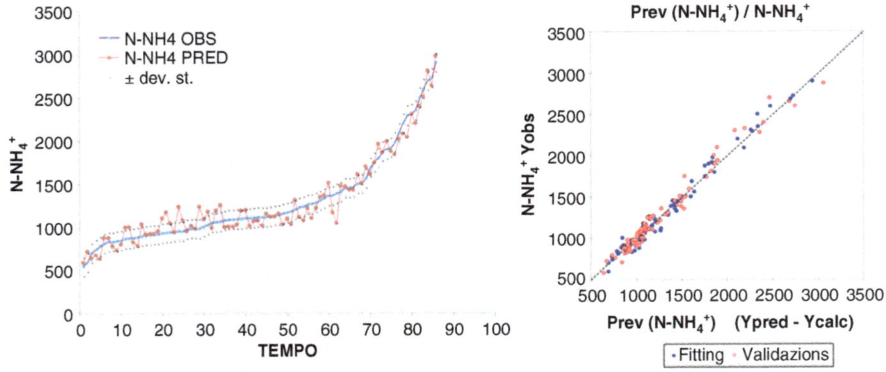

Fig. 17.10 Statistical answer of the control model developed and correlation between predicted and real ammonia value (Micolucci et al. 2014)

As a final conclusion of the work carried out, conductivity alone (thus using a low-cost system) can be considered as the most representative parameter to describe the process conditions and so its automatic control. These results clearly show the possibility to implement an automatic control to improve process operation and monitoring based on some statistics and math concepts (Gottardo et al. 2017).

17.6 Perspectives

Nowadays, energy demands have evolved the anaerobic digestion process that considers the organic waste as a raw material, therefore a production process. Considering this approach as a production process, the biorefinery should require its management and process analysis through analytical multivariate statistical methods. Biological process gets improved when monitoring and fault detection approaches are implemented since they intend to boost benefits on planned operations and on productivity levels.

Today the knowledge level of AD processes is well established, but the technology to control the anaerobic systems is still growing in order to cope the difficulties of process stability and enhance energy and bioproduct production (Leite et al. 2016). A very important target for the current technology will be the overall remote process control of biorefinery plants. The aim of process control in the further cycle integration will be not only to ensure its stability but also to enhance process performances.

Acknowledgments The authors wish to thank the European Union, 7° Framework program 2007–2013, Valorgas Project (ENERGY.2009.3.2.2, Valorization of food waste to biogas), PRIN-MIUR 2007, and LIFE+ ENVIRONMENT POLICY AND GOVERNMENT project "Development and implementation of a demonstration system on Integrated Solid Waste Management for Tinos in line with the Waste Framework Directive" (*LIFE10 ENV/GR/000610*). Part of the work was carried out by the financial support of the project PRIN 2012 WISE "Advanced process to sustainable useful innovative products from organic waste."

References

Adamson KA (2004) Hydrogen from renewable resources—the hundred year commitment. Energ Policy 32:1231–1242. https://doi.org/10.1016/S0301-4215(03)00094-6

Angenent LT, Karim K, Al Dahhan MH, Wrenn BA, Domìguez-Espinosa R (2004) Production of bioenergy and biochemicals from industrial and agricultural wastewater. Trends Biotechnol 22:477–485. https://doi.org/10.1016/j.tibtech.2004.07.001

Antonopoulou G, Gavala HN, Skiadas IV, Angelopoulos K, Lyberatos G (2008) Biofuels generation from sweet sorghum: fermentative hydrogen production and anaerobic digestion of the remaining biomass. Bioresour Technol 99(1):110–119

Astals S, Batstone DJ, Mata-Alvarez J, Jensen PD (2014) Identification of synergistic impacts during anaerobic co-digestion of organic wastes. Bioresour Technol 169:421–427. https://doi.org/10.1016/j.biortech.2014.07.024

Battistoni P, Pavan P, Cecchi F, Mata-Alvarez J, Majone M (1998) Integration of civil wastewater and municipal solid waste treatments. The effect on biological nutrient removal processes. In: Proceedings of the European conference on new advances in biological nitrogen and phosphorus removal for municipal or industrial wastewaters, 12–14 October 1998, Narbonne, France, pp 129–137

Bernstad A, la Cour Jansen J (2011) A life cycle approach to the management of household food waste–a Swedish full-scale case study. Waste Manag 31:1879–1896. https://doi.org/10.1016/j.wasman.2011.02.026

Bolzonella D, Innocenti L, Pavan P, Cecchi F (2001) Denitrification potential enhancement by addition of the anaerobic fermented of the organic fraction of municipal solid waste. Water Sci Technol 44(1):187–194

Bolzonella D, Fatone F, Pavan P, Cecchi F (2005) Anaerobic fermentation of organic municipal solid wastes for the production of soluble organic compounds. Ind Eng Chem Res 44:3412–3418. https://doi.org/10.1021/ie048937m

Bolzonella D, Battistoni P, Susini C, Cecchi F (2006) Anaerobic codigestion of waste activated sludge and OFMSW: the experiences of Viareggio and Treviso plants (Italy). Water Sci Technol 53:203–211. https://doi.org/10.2166/wst.2006.251

Braun K, Gottschalk G (1981) Effect of molecular hydrogen and carbon dioxide on chemo-organotrophic growth of Acetobacterium woodii and clostridium aceticum. Arch Microbiol 128(3):294–298

Cavinato C, Bolzanella D, Fatone F, Cecchi F, Pavan P (2011) Optimization of two-phase thermophilic anaerobic digestion of biowaste for hydrogen and methane production through reject water recirculation. Bioresour Technol 102:8605–8611. https://doi.org/10.1016/j.biortech.2011.03.084

Cavinato C, Giuliano A, Bolzonella D, Pavan P, Cecchi F (2012) Bio-hythane production from food waste by dark fermentation coupled with anaerobic digestion process: a long-term pilot scale experience. Int J Hydrogen Energ 37:11549–11555. https://doi.org/10.1016/j.ijhydene.2012.03.065

Cavinato C, Giuliano A, Bolzonella D, Pavan P, Cecchi F (2012) Bio-hythane production from food waste by dark fermentation coupled with anaerobic digestion process: a long-term pilot scale experience. Int J Hydrogen Energy 37:11549–11555.

Cardon BP, Barker HA (1947) Amino acid fermentations by clostridium propionicum and Diplococcus glycinophilus. Arch Biochem 12:165

Cecchi F, Cavinato C (2015) Anaerobic digestion of bio-waste: a mini-review focusing on territorial and environmental aspects. Waste Manag Res 33:429–438. https://doi.org/10.1177/0734242X14568610

Cecchi F, Traverso PG, Perin G, Vallini G (1988) Comparison of co-digestion performance of two differently collected organic fractions of municipal solid waste with sewage sludges. Environ Technol Lett 9:391–400. https://doi.org/10.1080/09593338809384582

Cecchi F, Pavan P, Mata-Alvarez J (1997) Kinetic study of the thermophilic anaerobic digestion of the fresh and precomposted mechanically selected organic fraction of MSW. J Environ Sci Health A 32:195–213. https://doi.org/10.1080/10934529709376536

Cecchi F, Battistoni P, Pavan, Bolzonella D, Innocenti L (2005) Digestioneanaerobicadellafrazioneorganica dei rifiutisolidi. Aspettifondamentali, progettuali, gestionali, diimpattoambientaleedintegrazioneconladepurazionedelleacquereflue. Edizioni APAT: Agenzia per laProtezionedell'Ambiente e per i ServiziTecnici. Manuali e lineeguida 13/2005, Roma

Chinellato G, Cavinato C, Bolzonella D, Heaven S, Banks CJ (2013) Biohydrogen production from food waste in batch and semi-continuous conditions: evaluation of a two-phase approach with digestate recirculation for pH control. Int J Hydrogen Energ 38:4351–4360. https://doi.org/10.1016/j.ijhydene.2013.01.078

Chong ML, Sabaratnam V, Shirai Y, Hassan MA (2009) Biohydrogen production from biomass and industrial wastes by dark fermentation. Int J HydrogenEnerg 34:3277–3287. https://doi.org/10.1016/j.ijhydene.2009.02.010

Chou CH, Wang CW, Huang CC, Lay JJ (2008) Pilot study on the influence of stirring and pH on anaerobes converting high-solid organic wastes to hydrogen. Int J Hydrogen Energ 33:1550–1558. https://doi.org/10.1016/j.ijhydene.2007.09.031

Chu CF, Li YY, KQ X, Ebie Y, Inamori Y, Kong HN (2008) A pH-temperature-phased two-stage process for hydrogen and methane production from food waste. Int J Hydrogen Energ 33:4739–4746. https://doi.org/10.1016/j.ijhydene.2008.06.060

Chu CF, Ebie YXKQ, Li YY, Inamori Y (2010) Characterization of microbial community in the two-stage process for hydrogen and methane production from food waste. Int J Hydrogen Energ 35:8253–8261. https://doi.org/10.1016/j.ijhydene.2009.12.021

Chua H, PHF Y, Ho LY (1997) Coupling of waste water treatment with storage polymer production. Appl Biochem Biotechnol 63:627–635. https://doi.org/10.1007/BF02920461

De Baere L, Mattheeuws B (2015) State of the art of anaerobic digestion of municipal solid waste in Europe. In: Proceedings of the international conference on solid waste 2011 – moving towards sustainable resource management, p 416

De Falco M, Basile A (eds) (2015) Enriched methane: the first step towards the hydrogen economy. Springer, Cham

De la Rubia MA, Raposo F, Rincón B, Borja R (2009) Evaluation of the hydrolytic–acidogenic step of a two-stage mesophilic anaerobic digestion process of sunflower oil cake. Bioresour Technol 100:4133–4138. https://doi.org/10.1016/j.biortech.2009.04.001

Dionisi D, Majone M, Papa V, Beccari M (2004) Biodegradable polymers from organic acids by using activated sludge enriched by aerobic periodic feeding. Biotechnol Bioeng 85:569–579. https://doi.org/10.1002/bit.10910

EEA (2013) http://www.eea.europa.eu/publications/eea-annual-report-2013

Elbeshbishy E, Nakhla G (2011) Comparative study of the effect of ultrasonication on the anaerobic biodegradability of food waste in single and two-stage systems. Bioresour Technol 102:6449–6457. https://doi.org/10.1016/j.biortech.2011.03.082

EUROSTAT (n.d.) Eurostat 2011 [WWW Document]. http://ec.europa.eu/eurostat, http://ec.europa.eu/eurostat/web/products-datasets/-/ten00110

EUROSTAT 2011–2012-2015a. http://ec.europa.eu/eurostat, http://ec.europa.eu/eurostat/web/products-datasets/–/ten00110

Eurostat (2014) Generation of Waste. http://ec.europa.eu/eurostat/data/database

Eurostat (2015) http://ec.europa.eu/eurostat

Giuliano A, Zanetti L, Micolucci F, Cavinato C (2014) Thermophilic two-phase anaerobic digestion of source sorted organic fraction of municipal solid waste for bio-hythane production: effect of recirculation sludge on process stability and microbiology over a long-term pilot scale experience. Water Sci Technol 69:2200–2209. https://doi.org/10.2166/wst.2014.137

Gomez X, Moran A, Cuetos MJ, Sanchez ME (2006) The production of hydrogen by dark fermentation of municipal solid wastes and slaughterhouse waste: a two-phase process. J Power Sources 157(2):727–732

Gottardo M, Cavinato C, Bolzonella D, Pavan P (2013) Dark fermentation optimization by anaerobic digested sludge recirculation: effects on hydrogen production. Chem Eng 32:997–1002. https://doi.org/10.3303/CET1332167

Gottardo M, Micolucci F, Bolzonella D, Uellendahl H, Pavan P (2017) Pilot scale fermentation coupled with anaerobic digestion of food waste - effect of dynamic digestate recirculation. Renew Energy 114:455–463

Graham LA, Rideout G, Rosenblatt D, Hendren J (2008) Greenhouse gas emissions from heavy-duty vehicles. Atmos Environ 42:4665–4681. https://doi.org/10.1016/j.atmosenv.2008.01.049

Hallenbeck P, Ghosh D, Skonieczny M, Yargeau V (2009) Microbiological and engineering aspects of biohydrogen production. Indian J Microbiol 49:48–59. https://doi.org/10.1007/s12088-009-0010-4

Han SK, Kim SH, Kim HW, Shin HS (2005) Pilot-scale two-stage process: a combination of acidogenic hydrogenesis and methanogenesis. Water Sci Technol 52:131–138

Hansen TL, Jansen JC, Davidsson A, Christensen TH (2007) Effects of pre-treatment technologies on quantity and quality of source-sorted municipal organic waste for biogas recovery. Waste Manag 27:398–405. https://doi.org/10.1016/j.wasman.2006.02.014

Hawkes FR, Dinsdale R, Hawkes DL, Hussy I (2002) Sustainable fermentative hydrogen production: challenges for process optimisation. Int J Hydrogen Energ 27:1339–1347. https://doi.org/10.1016/S0360-3199(02)00090-3

Hawkes F, Hussy I, Kyazze G, Dinsdale R, Hawkes D (2007) Continuous dark fermentative hydrogen production by mesophilic microflora: principles and progress. Int J Hydrogen Energ 32:172–184. https://doi.org/10.1016/j.ijhydene.2006.08.014

ISPRA (2014a) Rapportorifiutiurbani. Edizione 2014. Rapporti 202/2014. ISPRA, 2014: RapportoRifiutiUrbani. ISBN 978-88-448-0680-4

ISPRA (2014b) Italian emission inventory 1990–2012. Informative inventory report 2014. Rapporti 201/2014

Kataoka N, Ayame S, Miya A, Ueno Y, Oshita N, Tsukahara K, Sawayama S, Yokota N (2005) Studies on hydrogen-methane fermentation process for treating garbage and waste paper. ADSW 2005 conference proceedings, 2, process engineering

Kongjan P, Angelidaki I (2010) Extreme thermophilic biohydrogen production from wheat straw hydrolysate using mixed culture fermentation: effect of reactor configuration. Bioresour Technol 101:7789–7796. https://doi.org/10.1016/j.biortech.2010.05.024

Kotay SM, Das D (2008) Biohydrogen as a renewable energy resource. Prospects and potentials. Int J Hydrogen Energ 33:258–263. https://doi.org/10.1016/j.ijhydene.2007.07.031

Kraemer JT, Bagley DM (2007) Improving the yield from fermentative hydrogen production. Biotechnol Lett 29:685–695. https://doi.org/10.1007/s10529-006-9299-9

Lee DY, Ebie Y, KQ X, Li YY, Inamori Y (2010) Continuous H2 and CH4 production from high-solid food waste in the two-stage thermophilic fermentation process with the recirculation of digester sludge. Bioresour Technol 101:S42–S47. https://doi.org/10.1016/j.ijhydene.2007.07.031

Lee WS, Chua ASM, Yeoh HK, Ngoh GC (2014) A review of the production and applications of waste-derived volatile fatty acids. Chem Eng J 235:83–99. https://doi.org/10.1016/j.cej.2013.09.002

Leite WRM, Gottardo M, Pavan P, Belli Filho P, Bolzonella D (2016) Performance and energy aspects of single and two phase thermophilic anaerobic digestion of waste activated sludge. Renew Energy 86:1324–1331

Levin DB, Pitt L, Love M (2004) Biohydrogen production: prospects and limitations to practical application. Int J Hydrogen Energ 29:173–185. https://doi.org/10.1016/S0360-3199(03)00094-6

Liu D, Liu D, Zeng RJ, Angelidaki I (2006) Hydrogen and methane production from household solid waste in the two-stage fermentation process. Water Res 40:2230–2236

Lu J, Ahring BK (2005) Effects of temperature and hydraulic retention time on thermophilic anaerobic pretreatment of sewage sludge. In: Proceedings anaerobic digestion of solid waste, Copenhagen, pp 159–164

Luo G, Xie L, Zhou Q, Angelidaki I (2011) Enhancement of bioenergy production from organic wastes by two-stage anaerobic hydrogen and methane production process. Bioresour Technol 102:8700–8706. https://doi.org/10.1016/j.biortech.2011.02.012

Malamis D, Moustakas K, Bourka A, Valta K, Papadaskalopoulou C, Panaretou V, Skiadi O, Sotiropoulos A (2015) Compositional analysis of biowaste from study sites in Greek municipalities. Waste Biomass Valoriz 6:637–646. https://doi.org/10.1007/s12649-015-9406-z

Mathews J, Wang G (2009) Metabolic pathway engineering for enhanced biohydrogen production. Int J Hydrogen Energ 34:7404–7416. https://doi.org/10.1016/j.ijhydene.2007.07.031

Metcalf, Eddy (2006) Ingegneriadelleacquereflue. Trattamento e riuso. Mc Graw Hill, Milano. ISBN: 9788838661884

Micolucci F, Gottardo M, Bolzonella D, Pavan P (2014) Automatic process control for stable biohythane production in two-phase thermophilic anaerobic digestion of food waste. Int J Hydrogen Energ 39:17563–17572. https://doi.org/10.1016/j.ijhydene.2014.08.136

Micolucci F, Gottardo M, Cavinato C, Pavan P, Bolzonella D (2016) Mesophilic and thermophilic anaerobic digestion of the liquid fraction of pressed biowaste for high energy yields recovery. Waste Manag 48:227–235. https://doi.org/10.1016/j.wasman.2015.09.031

Micolucci F, Gottardo M, Pavan P, Cavinato C, Bolzonella D (2017) Pilot scale comparison of single and double-stage thermophilic anaerobic digestion of food waste. Article in press. doi:https://doi.org/10.1016/j.jclepro.2017.10.080

Nazlina HMYNH, Nor'Aini AR, Man HC, Yusoff MZM, Hassan MA (2011) Microbial characterization of hydrogen-producing bacteria in fermented food waste at different pH values. Int J Hydrogen Energ 36:9571–9580. https://doi.org/10.1016/j.ijhydene.2011.05.048

Okamoto M, Miyahara T, Mizuno O, Noike T (2000) Biological hydrogen potential of materials characteristic of the organic fraction of municipal solid wastes. Water Sci Technol 41:25–32. ISSN:02731223

Pavan P, Battistoni P, Cecchi F, Mata-Alvarez J (2000) Two-phase anaerobic digestion of source sorted OFMSW (organic fraction of municipal solid waste): performance and kinetic study. Water Sci Technol 41:111–118. ISSN:02731223

Reith JH, Wijffels RH, Barten H (2003) Bio-methane & bio-hydrogen, status and perspectives of biological methane and hydrogen production. Dutch Biological Hydrogen Foundation, Petten

Shin HS, Youn JH (2005) Conversion into hydrogen by thermophilic acidogenesis. Biodegradation 16:33–44. https://doi.org/10.1007/s10531-004-0377-9

Sosnowski P, Klepacz-Smolka A, Kaczorek K, Ledakowicz S (2008) Kinetic investigations of methane co-fermentation of sewage sludge and organic fraction of municipal solid wastes. Bioresour Technol 99:5731–5737. https://doi.org/10.1016/j.biortech.2007.10.019

Sreela-or C, Imai T, Plangklang P, Reungsang A (2011) Optimization of key factors affecting hydrogen production from food waste by anaerobic mixed cultures. Int J Hydrogen Energ 36:14120–14133. https://doi.org/10.1016/j.ijhydene.2011.04.136

Valdez-Vazquez I, Poggi-Varaldo HM (2009) Hydrogen production by fermentative consortia. Renew Sust Energ Rev 13:1000–1113. https://doi.org/10.1016/j.rser.2008.03.003

Viturtia AM, Mata-Alvarez J, Sans C, Costa J, Cecchi F (1992) Chemicals production from wastes. Environ Technol 13:1033–1041. https://doi.org/10.1080/09593339209385240

Wang X, Zhao YC (2009) A bench scale study of fermentative hydrogen and methane production from food waste in integrated two-stage process. Int J Hydrogen Energ 34:245–254. https://doi.org/10.1016/j.ijhydene.2008.09.100

Yoo CK, Lee JM, Lee IB, Vanrolleghem PA (2004) Dynamic monitoring system for full-scale wastewater treatment plants. Water Sci Technol 50:163–171. ISSN: 0273-1223

Zhang C, Su H, Baeyens J, Tan T (2014) Reviewing the anaerobic digestion of food waste for biogas production. Renew Sust Energ Rev 38:383–392. https://doi.org/10.1016/j.rser.2014.05.038

Zhu H, Parker W, Conidi D, Basnar R, Seto P (2011) Eliminating methanogenic activity in hydrogen reactor to improve biogas production in a two-stage anaerobic digestion process co-digesting municipal food waste and sewage sludge. Bioresour Technol 102:7086–7092. https://doi.org/10.1016/j.biortech.2011.04.047

The Application of Computer Image Analysis in Water Toxicity Tests

18

Piotr Lewicki, Robert Mazur, and Dawid Bedla

Abstract

Laboratory bioassay belongs to very important tools in environmental monitoring; biological assessment of water quality plays a fundamental role in whole system of water monitoring. The WFD requires the use of bioassay in assessing the threat of aquatic ecosystems from certain pollutants runoff into the aquatic environment. Classical methods of acute and chronic toxicity biotests are based on classical visual evaluation and observation of the negative effects of the selected toxicants impact. The usage of image analysis tool such as computer in water biomonitoring increases their sensitivity both in qualitative and quantitative assessment of negative effects from the selected toxicants. They make new bioassay very useful for the optimisation bioprocesses involved in the protection and control of aquatic ecosystems.

Keywords

Computer image analysis • Bioassay • Water pollutants • Biomonitoring • Environmental biotechnology

P. Lewicki (✉)
Faculty of Tourism and Recreation, Department of Statistics and Computer Science, University of Physical Education, Al. Jana Pawła II 78, 31-571 Kraków, Poland
e-mail: piotr.lewicki@awf.krakow.pl

R. Mazur
AGH University of Science and Technology, Kraków, Poland
e-mail: dydaktyka.mazur@gmail.com

D. Bedla
University of Agriculture in Krakow, Krakow, Poland
e-mail: d.bedla@ur.krakow.pl

H.J. Purohit et al. (eds.), *Optimization and Applicability of Bioprocesses*,
https://doi.org/10.1007/978-981-10-6863-8_18

18.1 Introduction

The methods of water bioassays play a very important role in water monitoring systems, being necessary for efficient protection of biodiversity and ecological balance of aquatic ecosystems. Introduction of more sensitive embryological criteria for the assessment of impact of various pollutants on reproduction of different species of invertebrates and vertebrates was a starting point for experimental study in different countries. Juvenile stages were recommended as especially useful for the optimisation bioprocesses involved in the protection of water environment (Dobrowolski 1976). Comparison of efficiency of different biotests to various pollutants of water is very useful for the selection of adequate and economical tests in the assessment of different methods of wastewater treatment and in ecotoxicological survey of the quality of lakes and rivers in different countries (Pardos et al. 1999). Many authors publish results of their studies based on classical methods of biotests (Kirsanov et al. 2014; Gophen 2016). Nevertheless, classical methods of bioassays depended on traditional visual assessment, requiring the investigator's attention, which can contribute to a risk of making wrong conclusions. Many of these methods are time-consuming and introduce the errors due to researchers' subjective assessment.

Critical evaluation of classical biotest methods indicates the need to develop innovative methods based on the application of computer image analyses for water bioassays. Methods of computer imaging and data analysis are very popular in numerous fields of science and industry, but in ecotoxicology there is still much to be done in this field.

The main aim of this experimental study was the creation of new tools allowing a very fast, highly precise quantitative and qualitative assessment in the field of water bioassay methods. The authors attempted the automation of the toxicity test with the application of computer image analysis methods (Mazur and Lewicki 2008). Four bioassays, crucial for ecotoxicology, were modified using automatic image analysis method.

In the first stage, the bioassays were selected and analysed in detail to select the most appropriate algorithms for the analysis of morphometric parameters (fixed and variable).

The following bioassays were selected:

1. *Hydra attenuata* assay
2. *Lemna minor* growth inhibition test
3. Determination of the inhibition of the mobility of *Daphnia magna* Straus (Cladocera, Crustacea) – acute toxicity test
4. *Tubifex tubifex*, *Lymnaea stagnalis*, *Xenopus laevis* – toxicity embryo test (Dobrowolski 1976)

All of these bioassays play an important role in biomonitoring of water quality and can be used both separately and in a battery of bioassays in toxicological assessment of water samples (water pollution, chemical substances, various water stressors and eco-toxins). There are many reports in literature of study results with the application of the above-mentioned methods, and they are included in international standards (protocols) (Trottier et al. 1997; OECD 2004; Bandow and Weltje 2012; Gopalapillai et al. 2014; Agathokleous et al. 2016). Classical methods in these and many other tests are very labour-consuming and burdened with the possibility of errors resulting from subjective perceptions. The application suitable algorithm of image analysis made these bioassays more accurate, eliminated the above-mentioned errors, accelerated the process of assessment and fully automated them. These new possibilities made bioassays easy to operate by technical service, without the required continuous presence of scientists (Sozzani et al. 2014; Rahaman et al. 2015).

The use of computer image analysis as a tool in bioassays was introduced by Mazur and Lewicki (2008), who applied the macro-programmes in Aphelion 3.02 software. It allowed the researchers to track the changes in shrinking of tentacles in *Hydra vulgaris* (Mazur and Lewicki 2008), to measure the frond area for a bioassay on duckweed *Lemna minor* and to detect the motion in embryos of *Lymnaea stagnalis* in cocoons exposed to specific toxins (Mazur and Lewicki 2008).

18.2 Traditional Methods vs Computer Image Analysis Application, Focus on *Hydra* sp. Bioassay

18.2.1 *Hydra attenuata* Assay

To eliminate any other effects such as impact of bacteria and pathogenic microorganisms for bioassay, cultural stock of any *Hydra* species should be selected from axenic cultures and maintained in sterile conditions. Rahat and Reich established the proper techniques for preparation of *Hydra* culture stock for metabolic, cell and developmental study (Muscatine and Lenhoff 1965; Rahat and Reich 1983). Other researchers also developed the method of *Hydra* colonies maintenance, feeding and care for bioassay purposes (Trottier et al. 1997; Pollino and Holdway 1999) (Tables 18.1 and 18.2).

Standard methods of *Hydra* sp. bioassay, based on the very time-consuming procedure described in detail by many authors (Trottier et al. 1997), were adopted by other authors to various species of *Hydra* (Pardos et al. 1999; Pollino and Holdway 1999; Arkhipchuk et al. 2006; Brown et al. 2014).

Table 18.1 *Hydra* medium according to Trottier et al. (1997) (for axenic culture *Hydra attenuata*)

Chemical substances	Sample weight/volume
Calcium chloride ($CaCl_2$)	2.94 g
N-tris (hydroxyl-methyl) methyl 1-2-aminoethanesulfonic acid (TES buffer)	0.08 g
Ethylene-di-amine-tetra-acetic acid (EDTA)	2.2 g
Ultrapure water	20 dm^3

Table 18.2 *Hydra* medium according to Muscatine and Lenhoff (1965) (for culture *Hydra* sp.)

Chemical substances	Sample weight/volume
Sodium chloride (NaCl)	1 mM
Calcium chloride ($CaCl_2$)	1 mM
Potassium chloride (KCl)	0.1 mM
Magnesium sulphur dioxide	0.1 mM
Tris pH 7.6 (pH correction – with concentrated hydrochloric acid HCl)	1 mM
Ultrapure water	1 dm^3

There were studies on the application of *Hydra attenuata* according to Trottier's procedure (Trottier et al. 1997) and *Hydra vulgaris* according to Trottier's procedure with Muscatine and Lenhoff (1965) medium.

The invertebrates were maintained at room temperature (20–24 °C) under 16–8 h of light and dark photoperiod. The feeding procedure was based on early hatched sterilised brain shrimps *Artemia salina*. The organisms were not fed in 96 h prior to the acute toxicity test.

18.2.1.1 Classical Method

Observation of changes in *Hydra* morphology is provided by microscope at six to ten times magnification (in periods 0, 24, 48, 72 and 96 h) of toxic exposure. The control group should maintain right shape of tentacles for the duration of bioassay.

18.2.1.1.1 Assessment Endpoints

Experimental organisms undergo deep changes in morphology with the increase in the concentration of toxins. The toxic effect can manifest as in Fig. 18.1 (Lewicki 2006).

- Normal shape of *Hydra* body (Fig. 18.2a)
- Slight changes – clubbed tentacles (Fig. 18.2b)
- Deeper changes – short-ended tentacles (Fig. 18.2c)
- Deep severe intoxication – tulip phase (which leads directly to the death of the tested organism) (Fig. 18.2d)
- Last stage – post-tulip phase finally leads to disintegration of *Hydra* body (which is identical with the lethal effect) (Fig. 18.2e)

The following elements belong to the measurement of the endpoints: EC50 and LC50, threshold concentrations (TC) from NOEC and LOEC – which are essential to calculate TC = (NOEC × LOEC).

According to Wilby's scale, the toxic effect can be divided into ten possible stages (Table 18.3), which are quite difficult to determine precisely.

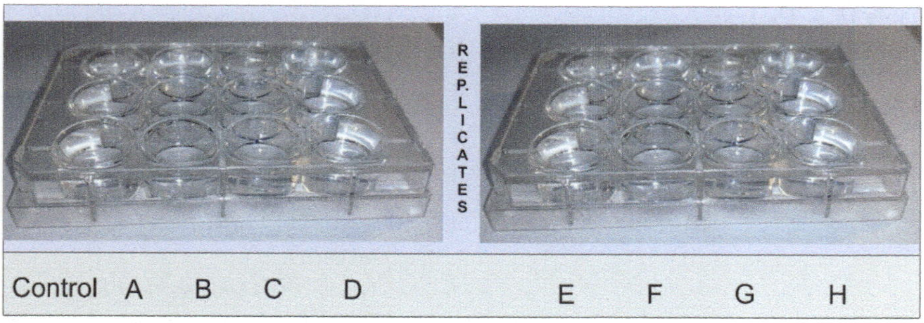

Fig. 18.1 *Hydra* sp. bioassay in chambers of microplates

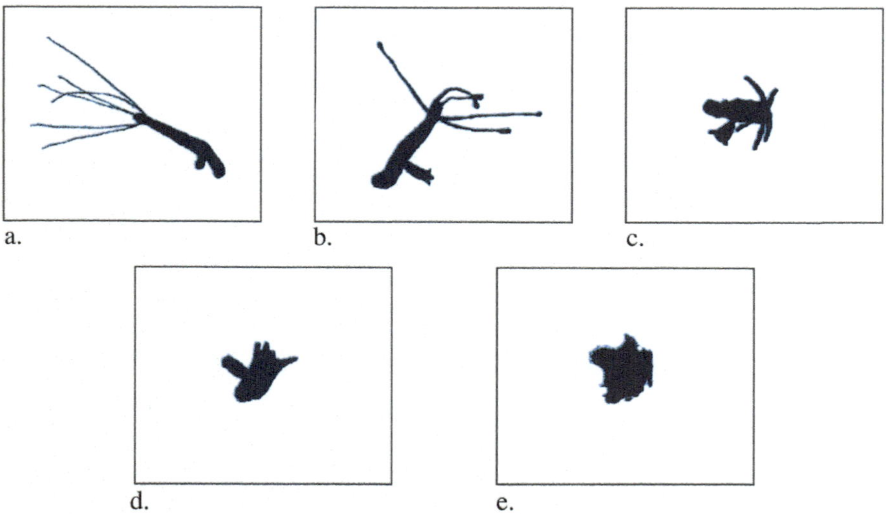

Fig. 18.2 The changes in *Hydra* sp. body morphology exposed to waterborne eco-toxins (Lewicki 2006)

18.2.2 Application of Computer Image Analysis: New Challenges

Lewicki studied the possibility of using computer analysis methods in the chronic toxicity for *Hydra vulgaris* Pallas, on the basis of the morphological changes analysis caused by water stressors. He created a new tool for more advanced bioassays, characterised by higher precision in the detection of changes and leading to a very objective assessment in relation to the adopted criteria of morphometric changes.

In classical assays, the morphological changes in *Hydra* bodies (five levels of pathological changes, Fig. 18.2) were visually (i.e. subjectively) assessed by the experimenter. The proposed new macro-programme for image analysis was applied due to better assessment parameter of the changes in the morphometry of *Hydra's* body exposed on various water toxins. Analysed features are determined in automatic mode, and such an assessment is much more objective compared to the traditional method.

Table 18.3 The key assessment criteria for manifesting of progressive toxicity effects. Introduced by Wilby (1988)

Score	Morphological changes
10	Extended tentacles and body reactive (normal metabolism)
9	Partially contracted, slow reactions (first sing of intoxication)
8	Clubbed tentacles, body slightly contracted
7	Shortened tentacles, body slightly contracted
6	Tentacles and body shortened
5	Totally contracted, tentacles visible
4	Totally contracted, no visible tentacles
3	Expanded, tentacles visible
2	Expanded, no visible tentacles
1	Dead but intact

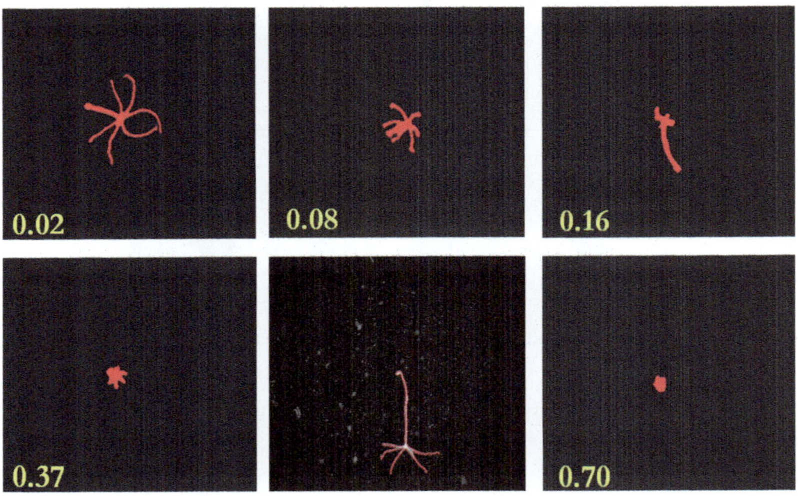

Fig. 18.3 The computer image analysis of *Hydra vulgaris* body exposed to water pollutants (Lewicki 2006)

Tested groups of *Hydra* during the exposure on eco-toxins show the same pathogenic changes as in the five stages mentioned above (Fig. 18.2). In a final step of intoxication process, body shapes take spherical form (view from the top) converging to the circle (Fig. 18.3).

The algorithm detection used shape factor – (mathematical operator) "compactness", as a digital determinant of these features. This factor takes the value 1 for ideal circle shape and decreases to 0 for irregular and elongated shapes (in plain view) (Fig. 18.4).

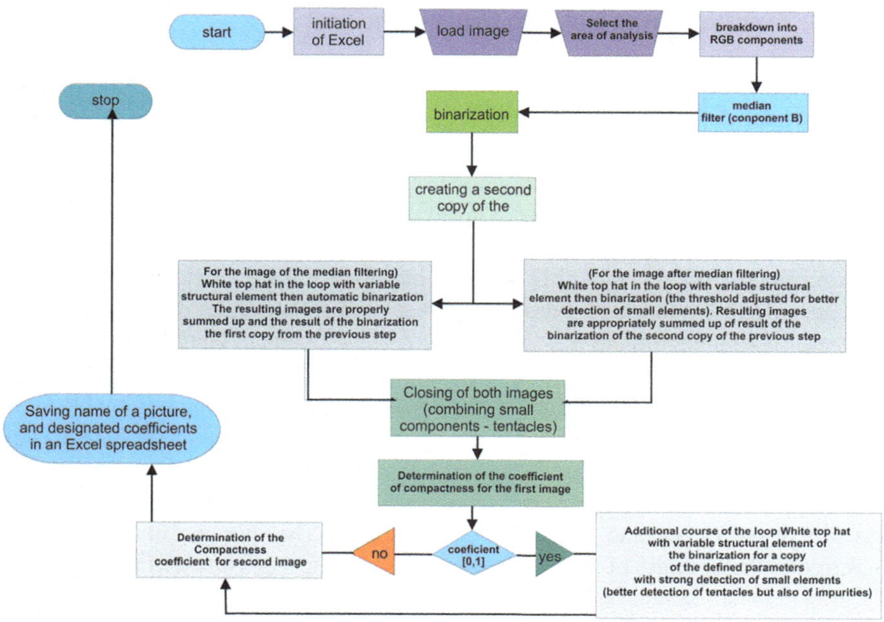

Fig. 18.4 The algorithm of image analysis in Aphelion software for *Hydra* sp. assay (Lewicki 2006)

Mathematical formula for this factor:

$$\text{COMPACTENE } S = \frac{16S}{L^2}$$

L – the circuit of the object in the image points
S – the number of points belonging to the object

This macro-programme seems to be crucial in *online* biomonitoring. It is characterised by a statistically acceptable repeatability. Classic methods were developed into an innovative tool, allowing us to determine toxic levels of tested water stressors.

The assay results (based on image analysis) can be easily plotted and assigned to the *compactness* value to define five stages of morphological changes. The newly developed method can be used as an automated system for early detection of potential toxicity of pollutants in the water environment.

18.3 *Lemna minor* Growth Inhibition Test

Many authors report about advanced imaging techniques used in research of plant growth and development (Nowakowski et al. 2011; Sozzani et al. 2014). Morphometric studies in plant physiology are based on the use of a variety of tools, from the microscope, by laser scanners, CT scanners and high-resolution cameras (using variable spectra of light, etc.) (Mazur and Lewicki 2008). In order to increase the precision of the measurements, the researchers have been adopting technologies making it possible to create 3D models that allow further advanced research of structure. Taking into account the changes between 3D models, i.e. the growth rate, increase in volume, direction and character of development changes of the investigated plants, etc. (Sozzani et al. 2014), it is possible to obtain 4D model in final results (dynamic model) (Sozzani et al. 2014). Computer programmes enable automation of measurement methods and their father statistical analysis (Rahaman et al. 2015; Agathokleous et al. 2016).

Morphometric analysis of *Lemna minor* is based on 2D, due to the nature of the growth of duckweed and flat shape of frond area. The increase in surface area, referred to the biomass of tested organisms, allows the assessment of the dynamics of growth under certain boundary conditions.

18.3.1 Classical *Lemna minor* and *Spirodela polyrhiza* Bioassays

Lemna minor organisms have been applied in toxicity bioassay since the 1930s and, with other plant species, have been used to investigate the effects of the phenoxy herbicides in plants (Moody and Miller 2005). Nowadays duckweed is recommended by various authorities in phytotoxicity testing as a part of biomonitoring methods in water quality assessment (Moody and Miller 2005).

Most of water ecotoxicological studies are based on classical *Lemna minor* bioassays:

1. ISO 20079: 2005: Water quality – determination of the toxic effect of water constituents and waste water on duckweed (*Lemna minor*) – duckweed growth inhibition test
2. ISO/DIS 20227: Water quality – determination of the growth inhibition effects of waste waters, natural waters and chemicals on the duckweed *Spirodela polyrhiza* – the method using a stock culture independent microbiotest
3. **OCSPP 850.4400: Aquatic Plant Toxicity Test using *Lemna* spp. [EPA 712-C-008]**

Bioassay protocols include very specific procedures for the preparation of test and culturing conditions:

- Culturing procedures
- Preparation of stock solution and solvents (nutrient media and diluents)

- Guidelines for the exposure technique with treatment levels
- Laboratory facilities and whole monitoring equipment
- Environmental conditions (and necessary boundary conditions)
- Methods of observation and measurement endpoints and statistical analysis

In classical methods of *Lemna minor* (Fig. 18.5) and *Spirodela polyrhiza*, in toxicity assessment, investigators based on measures of effects such as:

1. Frond number

   The number of fronds in each container during the bioassay time is calculated at least every 3 days.
2. Dry weight

   Destructive method based on classical weighing of dry matter after drying in 60 °C in 24 h.
3. Frond area

   It is determined by the application of classical planimetric methods or computer image analysis.
4. Determination of crowding:

$$w_{\text{frond}} = \frac{w_{\text{total}}}{n}$$

where

w_{total} = total dry weight plant mass in the container (after test period)
n = the number of frond the test group (after test period)

5. Changes in fronds view (appearance or condition)

   Phytostatic and phytocidal effects recorded during test period, necrosis and other negative visual changes on the frond area.
6. Yield calculations:

Fig. 18.5 *Lemna minor* toxicity bioassay (classical methods)

$$Y = b_1 - b_0$$

where

Y = yield of tested population duckweed biomass (given as frond number, dry weight or the average frond area)
b_0 = biomass (dry weight, average frond area and *Lemna* leaves number) at the beginning of test
b_1 = biomass (dry weight, average frond area and *Lemna* leaves number) at the end of test

7. Average growth rate calculations:

$$\overline{r}_{i-j} = \frac{\ln(b_j) - \ln(b_i)}{\Delta t}$$

where

$\overline{r}_{i-j}$ = mean specific growth rate per day (given as: biomass, frond number, dry weight or average frond area)
b_0 = observed (biomass, frond number, dry weight or average frond area) at the beginning of test (in time "*j*")
b_1 = observed (biomass, frond number, dry weight or average frond area) at the end of test (in time "*i*")
Δt = time period from i to j in days

8. Percentage of inhibition:
   Based on yield or specific growth rate, for each response variable, the percent of inhibition – I % can be calculated.

$$\%I = \frac{(C - X)\cdot 100}{C}$$

where

C = the value of control response
X = the treatment average response (given as biomass, frond number, dry weight or average frond area) et the end of test

9. Toxicity measure:
   - LOEC, NOEC, LC_{50}, LC_{100}, other statistical analyses, regression curve, SD, AVG, 95% confidence interval, etc.

In classical method of bioassay, all of these endpoints are time-consuming and need to engage the full attention of the researcher.

In the above-mentioned biotest, the increase of mass is measured comparing the duckweed mass after drying it in the control and experimental sample, which is invasive method and does not allow the continuation of the experiment. More reliable and more accurate method of defining the specific growth rate of tested organisms is the measurement of the area of fronds in the control and experimental samples by application of computer image analysis. This is a non-invasive method allowing the continuation of the experiment (Mazur and Lewicki 2008; Janczak et al. 2013).

18.3.2 Computer Image Analysis in *Lemna* sp. Bioassay

Owing to the application of computer image analysis methods, our research can be more reliable and accurate. The prices of electronic devices for digital data acquisition have been dropping down, therefore construction of laboratory stand for automatic image analysis in toxicological bioassay is available for each ecological or ecotoxicological research department.

Based on the digital data, the process of image analysis can be carried out in a very advanced way with high precision.

Authors presented an example of algorithm for automatic image analysis in Aphelion software.

The procedure of toxicity test preparing is following standard guidelines, but the endpoint response to toxic effects is based on frond surface area of *Lemna* or *Spirodela*. For this purpose, advanced mathematical formula of image analysis was employed in Aphelion programme.

The algorithm scheme (Fig. 18.6):

The analysis begins with the opening of the picture to be analysed (Fig. 18.7). The next stage contains a median filter to reduce "noise" from the background. In the next operation, the colour of picture is converted to monochrome. After that, the image is broken down into RGB components, and the blue component is used in the further analysis because it provides the best contrasts of duckweed objects. The Black Top Hat filter is used in the next step for better mining of the objects. The most important stage is the binarisation process (corrected by Entropy Threshold), with removing of unwanted small objects (erosion filter + picture reconstruction). Now the areas with *Lemna* objects are marked (image acquisition filter). The beaker top view with duckweed is outlined in black, and in this defined region, frond surface area is measured. The last stage is marking the analysed objects by the programme, and the results of measurement are shown in the Excel or other programmes (Lewicki 2006) (Fig. 18.6).

On the Fig. 18.8 are presented main stages of computer image analysis, with final frond area measurements.

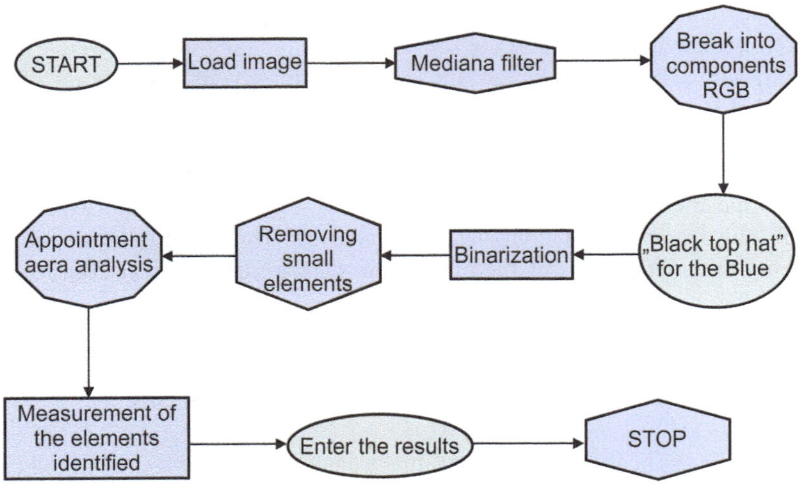

Fig. 18.6 The algorithm of the duckweed detection and surface area measurement

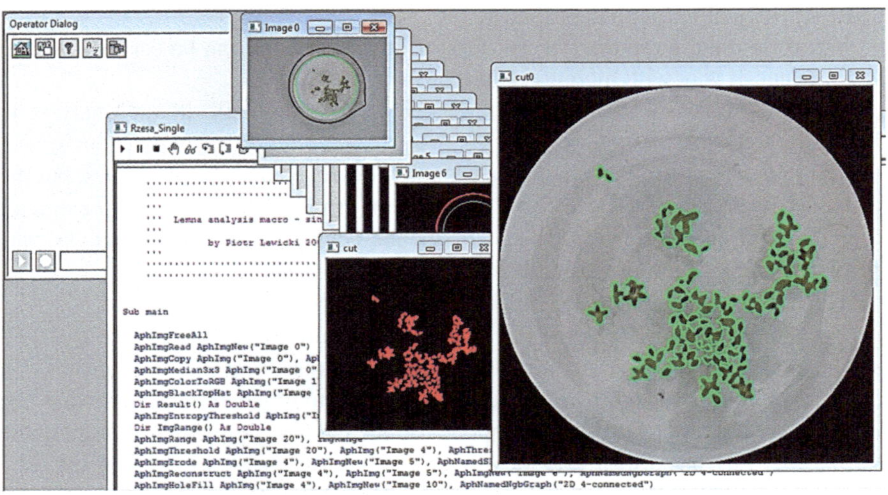

Fig. 18.7 The *Lemna minor* fronds area analysis process in Aphelion

18.3.2.1 An Example of *Lemna* Bioassay (Toxicity Test on Non-ionic Detergent Brij 72)

Table 18.4 presents the results of chronic toxicity test, based on the duckweed frond area. The values in each record were presented as the sum of all the frond area (average area for the whole tested population – based on three repetitions) (Fig. 18.9).

Changes in average frond area of *Lemna minor* can be used for later calculations of specific growth rate in time. Acute toxicity effect or chronic toxicity effect can be estimated.

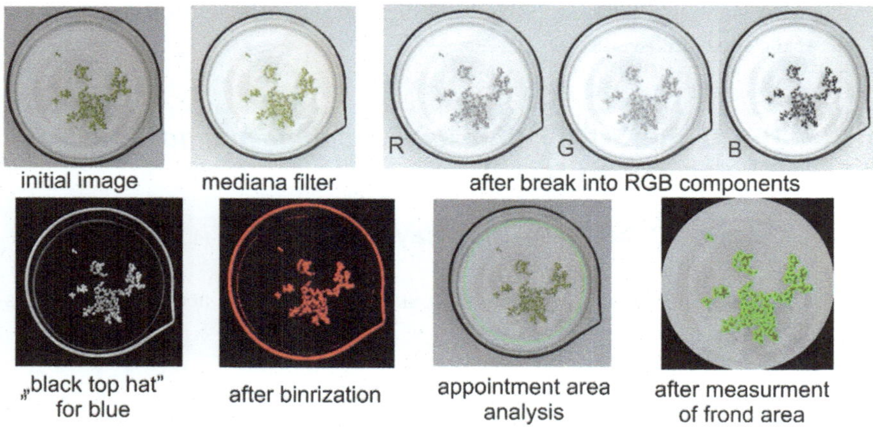

Fig. 18.8 Main stages in computer image analysis of duckweed frond area

Table 18.4 The bioassay results for the mean surface area of test populations (from three repetitions) of *Lemna minor* exposed to different concentrations of Brij 72 (average number of pixels ±SD)

Surfactants con. mg•dm^{-3}	1 week	2 weeks	3 weeks	4 weeks
Control	19.5 ± 0.2	22.5 ± 0.3	25.7 ± 0.1	28.2 ± 0.6
262.5	22.9 ± 0.9	27.1 ± 1.2	30.6 ± 1.5	30.7 ± 2.0
327.5	22.3 ± 1.0	23.5 ± 1.6	25.7 ± 1.7	25.2 ± 1.9
410	21.2 ± 0.9	21.0 ± 2.6	24.0 ± 3.9	24.1 ± 3.0
512.5	21.1 ± 1.5	19.0 ± 1.3	21.1 ± 2.8	21.0 ± 3.1
640	22.7 ± 0.9	21.6 ± 0.6	22.4 ± 4.6	21.4 ± 4.0
800	18.9 ± 2.2	16.8 ± 0.7	15.9 ± 1.3	14.2 ± 2.6
1000	20.3 ± 2.5	20.8 ± 2.2	18.3 ± 4.7	14.4 ± 2.2

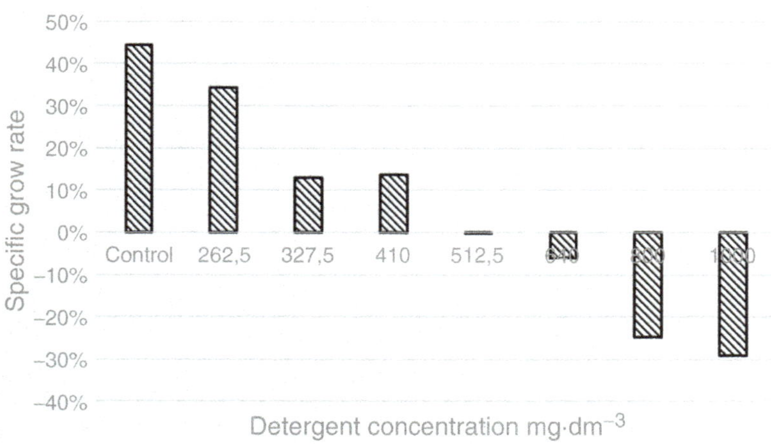

Fig. 18.9 The growth dynamics of *Lemna minor* exposed to Brij 35 solutions in each test group after 4 weeks of experiment

Detailed statistical analysis proved that the measurement error did not exceed 4.4% of the measured value.

18.4 The Acute Toxicity Biotests on *Daphnia magna* Straus (Cladocera, Crustacea)

18.4.1 Classical *Daphnia* sp. Bioassays

Daphnia sp. toxicity bioassay – the main standard protocols accepted by professional laboratories:

- ISO 6341: 2012 Water quality – determination of the inhibition of the mobility of *Daphnia magna* Straus (Cladocera, Crustacea) – acute toxicity test
- OECD (2004), Test No. 202: *Daphnia* sp. Acute Immobilisation Test, OECD Guidelines for the Testing of Chemicals, Section 2, OECD Publishing, Paris.
- OECD 211, Reg. (EG) Nr. 440/2008, C.20, ISO 10706 *Daphnia magna* Reproduction Test
- OECD (2012), Test No. 211: *Daphnia magna* Reproduction Test, OECD Guidelines for the Testing of Chemicals, Section 2, OECD Publishing, Paris.
- U.S. EPA (United States Environmental Protection Agency) (2002a) Short-term Methods for Estimating the Chronic Toxicity of Effluents and Receiving Waters to Freshwater Organisms. Fourth Edition. Office of Water (4303 T), Washington, DC 20460, Doc. EPA-821-R-02-013.
- Environment Canada (EC) (2000) Biological Test Method: Reference Method for Determining Acute Lethality of Effluents to *Daphnia magna*, Environmental Technology Centre, Ottawa, Report EPS/RM/14 Second Edition.
- PN-EN ISO 6341: 2002, Water quality – determination of the inhibition of the mobility of *Daphnia magna* S. (Cladocera, Crustacea) – acute toxicity test (ISO 6341:1996). Warsaw, PKN.

Daphnia sp. (Fig. 18.10) belongs to the most recognised bioindicators, regarding various toxic substances (Jonczyk and Gilron 2005).There is an enormous number of literature reports based on the biomonitoring results with application bioassay on *Daphnia* sp. (Dominik et al. 2000; Bernot et al. 2005; Toropov and Benfenati 2006; Fulton and Meyer 2014; Roy et al. 2015; Beasley et al. 2015). Different versions of

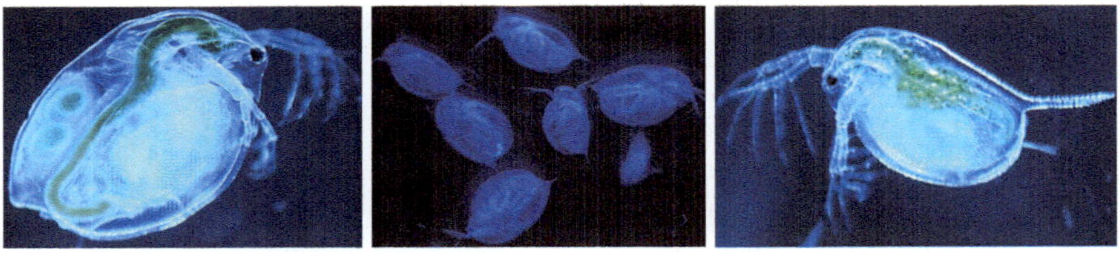

Fig. 18.10 *Daphnia* sp. during acute toxicity bioassay (type: Fluo-tox)

the bioassay have the protocols and guidelines for researchers (ISO 2012; OECD 2004, 2012; U.S. EPA 2002). The standard methods have been developed due to the toxicity assessment and were implemented by national and international institution and agency for environmental protection and conducting toxicity research. The main standards are available in the ISO, OECD and EPA database and must be followed by professional toxicology laboratories during toxicity bioassay with application of *Daphnia* sp. (Jonczyk and Gilron 2005).

In standard procedures of bioassay, fundamental states are preserved:

1. Conditions for culturing and feeding of *Daphnia* sp. (temperature, pH, light, nutrients, water hardness and quality, dissolved oxygen, etc.)
2. Organism selection and acclimation for laboratory conditions
3. Reagents, test organisms and media
4. Assessment criteria for mass culture
5. Applied equipment and laboratory glassware, culture vessel and other materials
6. Testing procedures and conditions
7. Procedure of samples preparation and dilution
8. Monitoring of physico-chemical parameters and endpoints
9. Test validity criteria and interpretation
10. Statistical analysis and presentation of results

The most popular toxicity test endpoints:

- Immobilisation (mortality) (24–48 h)
- Inhibition of reproduction (7–20 days)
- Inhibition of fluorescence using (MUG: 4-Methylumbelliferyl β-D-Galactopyranoside) (Fluo-tox), (45 min)

Mobility of *Daphnia* organisms strongly depends on water quality, and the changes in average velocities of tested water fleas movement can be used as the toxicity endpoint. Classical observation of speed of tested organism is out of quantity evaluation. Application of computer image analysis brings new opportunities in more objective quantity assessment of these parameters.

18.4.2 Computer Image Analysis in *Daphnia* sp. Bioassay

Kinetic disturbances by xenobiotic and other water pollutants can be precisely presented on the velocities plots and subjected to advanced statistical analysis. Authors presented an advanced algorithm and procedure of image analysis in Aphelion software. The standard bioassay procedure must be adopted to make available good condition for automatic image acquisitions.

The tests were conducted in the Petri dishes. In order to obtain flat and wide surface area, camera settings must be strictly defined (Fig. 18.11).

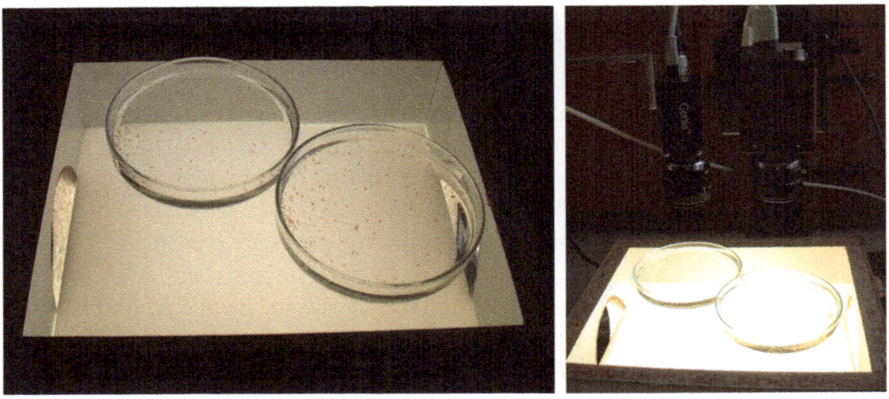

Fig. 18.11 *Daphnia magna* bioassay with application of computer image analysis in movement detection (laboratory experimental stand)

Relevant measurements were performed from the sequences of images that had been recorded in sequence of certain specified fixed intervals. It does not really matter whether they were registered in the form of video film, which was subdivided into film frames, or as a series of separate images. The only important parameter is the speed of acquisition and therefore the interval at which the individual images were recorded. The most important parameter was resolution of recorded images; the authors chose scale factor for such settings k = 0.0714 mm/pix 1 mm, which means that the object presented on the Petri dish with a length of 1 mm had a length 14 pix in the recorded image. The minimum time of video recording is 30 frames per second. The average length of the *Daphnia* selected for test was in the range from 0.6 to 1.2 mm and a surface area from 0.15 to 0.65 mm^2 (Kawa 2008).

Authors developed a suitable algorithm for *Daphnia* detection and determining their movement trajectory. Based on this data, proper measurement can be calculated such as velocities of *Daphnias* and other dependent variables (Mazur et al. 2006; Kawa 2008).

There are two versions of algorithm. The first one is based on the detection of gravity centre in all analysed objects and calculates the vectors of displacements of these points (Fig. 18.12). The results are the basis for further calculations of derivatives of kinetic parameters. The second algorithm uses the method of determining the motion trajectory of objects along with a curve of displacement traces, which gives the velocity value and all derivatives (Fig. 18.13). The scheme of second algorithm is very similar to the first one, but the differences are in the stages of the objects measurements (Fig. 18.14).

Authors applied a developed tool in *Daphnia* bioassay; the results present differences in kinetic parameters of tested population of daphnia in water toxins such KCl. This new tool can be easily adopted to laboratory practice of toxicity tests, based on *Daphnia* sp. The new version of the biotest will be available for very sensitive and quantitative assessment of water quality.

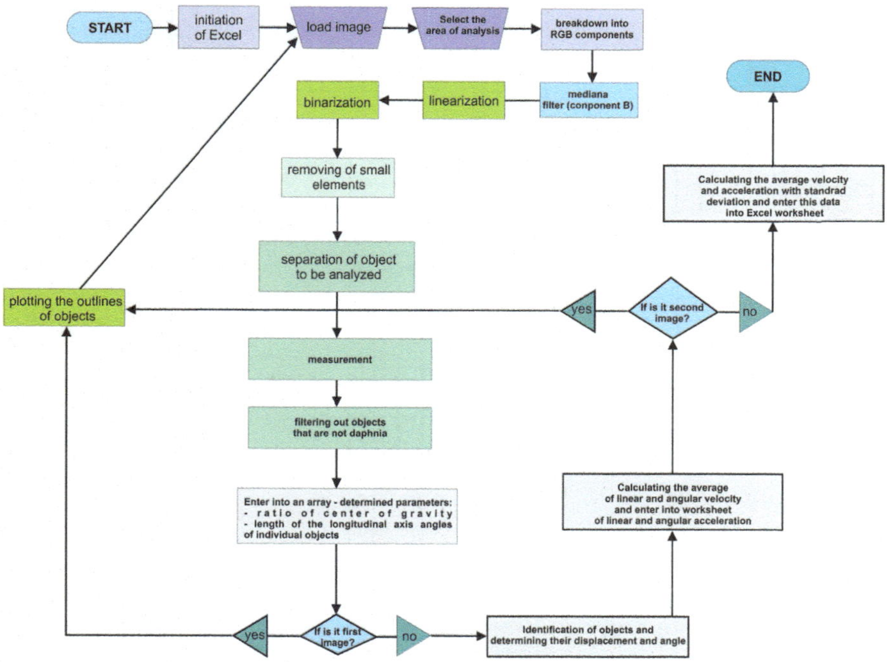

Fig. 18.12 The algorithm of the *Daphnia* detection with velocity measurement of their movements based on Kawa's research (Kawa 2008)

18.5 *Lymnaea stagnalis*: Toxicity Embryo Test

18.5.1 *Lymnaea stagnalis*: Bioassay

With its shell height of up to 7 cm, *Lymnaea stagnalis* L.(the great pond snail) is the largest aquatic snail in temperate zones of Europe and Northern America. The great pond snail feeds on fragments of plants, carrion, algae and other microorganisms.

Lymnaea stagnalis is one of the snail species being a good candidate for chronic toxicity test. The *Lymnaea stagnalis* is a widely spread species, living in many types of freshwater reservoirs (Jura 2005). Moreover, *L. stagnalis* is halotolerant (up to 0.7%). Thus it is also common in the waters of the Baltic Sea (Bandow and Weltje 2012).

Full cycle of embryonic development of this snail takes place in eggs laid in cocoons. The cocoons are usually attached to frond areas of hydrophytes. Embryo capsule of eggs is transparent, which is helpful during the observation of developmental stages.

In the process of eggs development, four different stages can be distinguished (Lalah et al. 2007):

- Morula: 0–3 days, yellowish appearance, slow movements.
- Trochophora: 3–5 days, grainy appearance, uninterrupted movement.

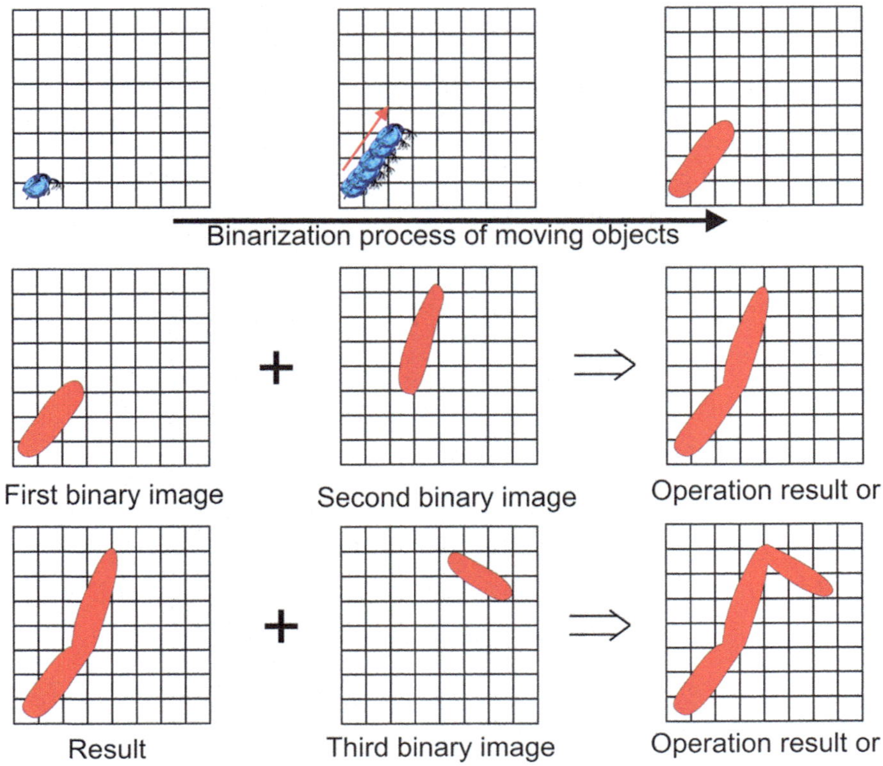

Fig. 18.13 Image analysis in Aphelion, object detection and determining the trajectory movement after binarisation process (Lewicki 2006)

- Veliger: 5–7 days, the embryo loses its round shape, foot and shell can be distinguished.
- Hippo: 7 days and above, the embryo is fully developed; shell, foot and eyes are differentiated; heart beat is visible.

Lymnaea stagnalis is very sensitive for changes in water chemical parameters. Especially developing embryos are particularly relevant as model organism for describing of adverse effects of tested contamination on critical developmental processes stages (Bandow and Weltje 2012).

Embryo toxicity bioassay based on snail eggs is very well known (Grosel et al. 2006; Lewicki 2006; Lalah et al. 2007; Mazur et al. 2013). Eggs should be collected with cocoon envelopes, and then the envelopes should be macerated for better penetration of waterborne toxicants mixture to the inside. Then the embryos were put in the toxicants for 24 h. Not only snail eggs but also young individuals after hatching are useful to toxicity assessment bioassay (Mazur et al. 2013).

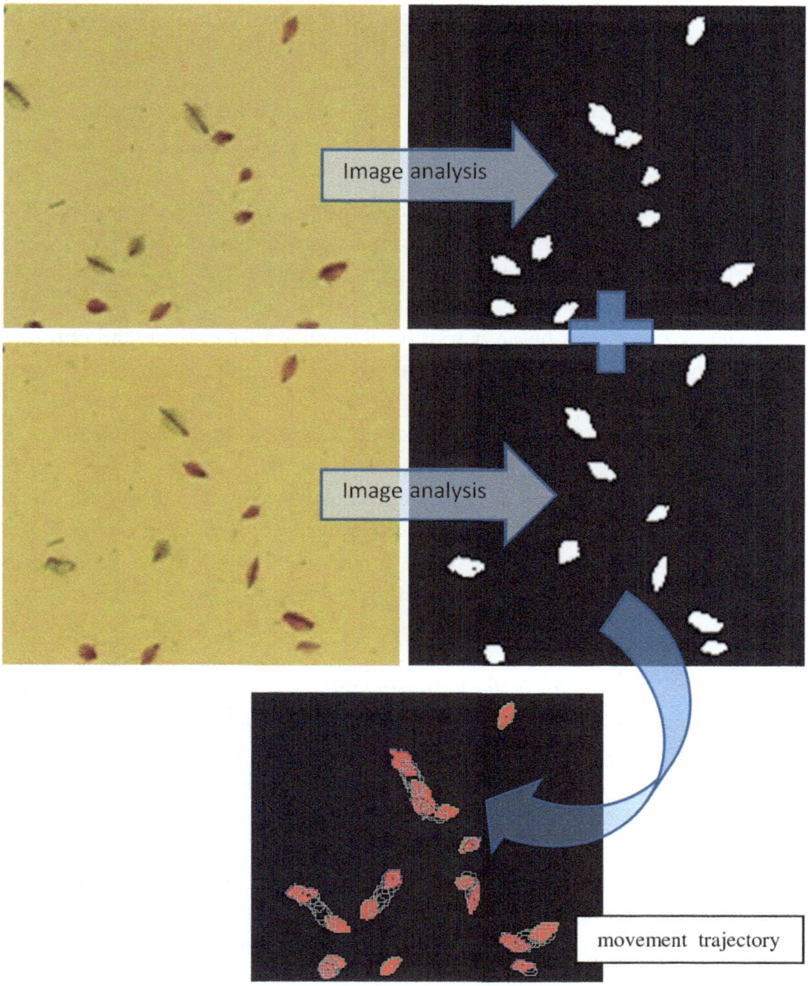

Fig. 18.14 *Daphnia* movement detection in Aphelion (Kawa 2008)

In traditional toxicity bioassay tests, the response of the species during 24 h is conducted. In the research by Lewicki (2006), acidification and ammonia were measured. Sensitivity of the great pond snail to the stress of all the tested water toxicants was observed. The most toxic of the three was ammonia, followed by sulphuric acid and nitric acid.

Mortality is one of the most popular factors to assess toxicity of substances in traditional tests. Stress can result in disorders in functioning of exposed organism and consequently their death.

The LC50 level confirmed that ammonia belongs to the most dangerous ecotoxicants for great pond snails; the LC50/24 h was 24.27 mg·dm^{-3} for embryos and 24.72 for juveniles. Embryonic forms were more sensitive than juvenile ones. The acute toxicity bioassay results reviled that LC50/24 h for *Lymnaea stagnalis* embryos was on level 81.17 mg·dm¯3 for the sulphuric acid test solution and 105.19 mg·dm^{-3} for the nitric acid test solution.

The bigger differences are recorded in acute toxicity results for juvenile *Lymnaea stagnalis*. The organisms after hatching are characterised by LC50/24 h levels, respectively, 126.49 mg·dm¯3 in the sulphuric acid and 170.47 mg·dm^{-3} in the nitric acid test solutions (Mazur et al. 2010).

The *Lymnaea stagnalis* belongs to model organism for actuate and chronic bioassay purposes especially in toxicity evaluation of samples from water reservoirs contaminated by acidification and alkalinisation.

The classical protocol of *Lymnaea stagnalis* embryo test is still at the stage of development; there are no accepted standard protocols as in the matter of *Daphnia magna, Hydra* sp. and *Lemna* sp. Different ecotoxicological laboratories work on the environmental condition for toxicity bioassay. The same as in the case of other bioindicators applied in toxicity test, the main issues of protocols that need standardising and specifying to the international standards for this species are the following:

- Culturing procedures still differ between laboratories.
- Preparation of stock solution and solvents (nutrient media and diluents).
- Guidelines for exposure technique with treatment levels.
- Laboratory facilities and whole monitoring equipment.
- Environmental conditions (and necessary boundary conditions).

As many authors reported widespread occurrence of this species, its sensitivity and susceptibility to a wide range of environmental stress factors make it an excellent bioindicator for laboratory tests and environmental investigations (in situ) (Coeurdassier et al. 2003; Grosell et al. 2006; Mazur et al. 2013; Ducrot et al. 2014; Tufi et al. 2016).

18.5.2 Computer Image Analysis in the Development of *Lymnaea* Embryo Bioassay

Mazur and Lewicki (2008) have already made an attempt to apply computer image analysis in toxicity bioassay with embryos of great pond snails. Image analysis is a novelty in this bioassay. Toxicity symptoms can be recognised as the morphological and physiological changes in the tested snail body. The image analysis allows us to not only detect the fact of disorders or damages but also distinguish between different effects of pollutants (e.g. acids, ammonia, detergents, etc.). Laboratory studies (Mazur et al. 2013) reviled that the great pond snail can be used as a proper and sensitive bioindicator for bioassay purposes. Biotests with embryonic and juvenile forms are especially desired because this young stage of organisms is fulfilling all the 5R criterion stated by Walker et al. (2002).

Detection algorithm (Fig. 18.15) requires the image series from the movie time-lapse of *Lymnaea stagnalis* embryos (Fig. 18.16). The final picture (Fig. 18.17) presents results of the analysis, and these results are implemented to excel worksheet for further calculations and statistical analyses.

Precision of image analysis is very high, and the measurement error is 0.044. The macro-programme can detect all embryos and specify those with the movements. Based on this analysis, we can obtain toxicity endpoint and subject them to further statistical analysis.

18.6 Conclusions

Many advantages were brought owing to the usage of image analysis by sophisticated computer programmes in the field of water bioassay (Sozzani et al. 2014; Agathokleous et al. 2016).

We obtained new very advanced tools for water biomonitoring, with possibility of their use as an early warning system against the negative impact of different toxicants in fresh water systems in online monitoring of surface waters.

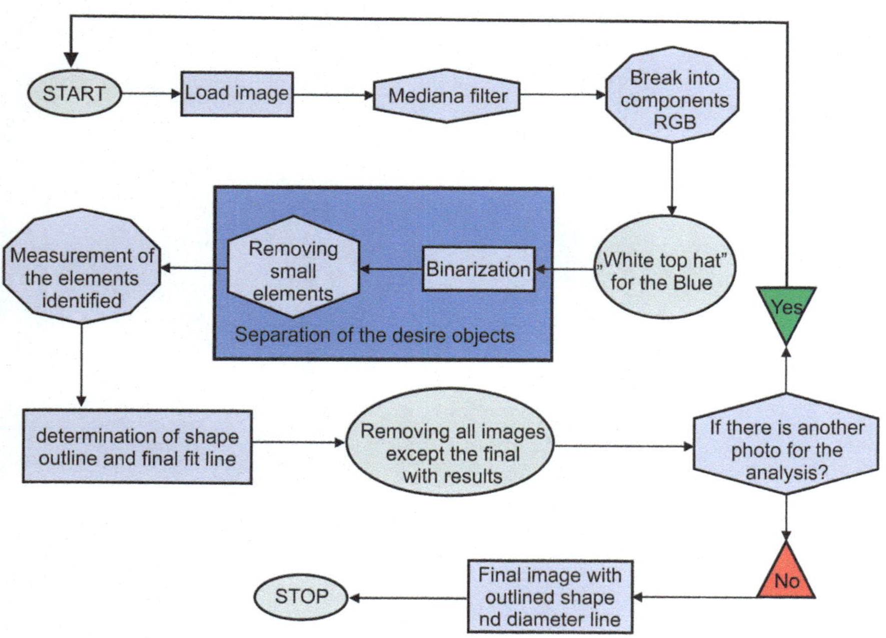

Fig. 18.15 The scheme of image analysis algorithm of investigated embryos during toxicity test

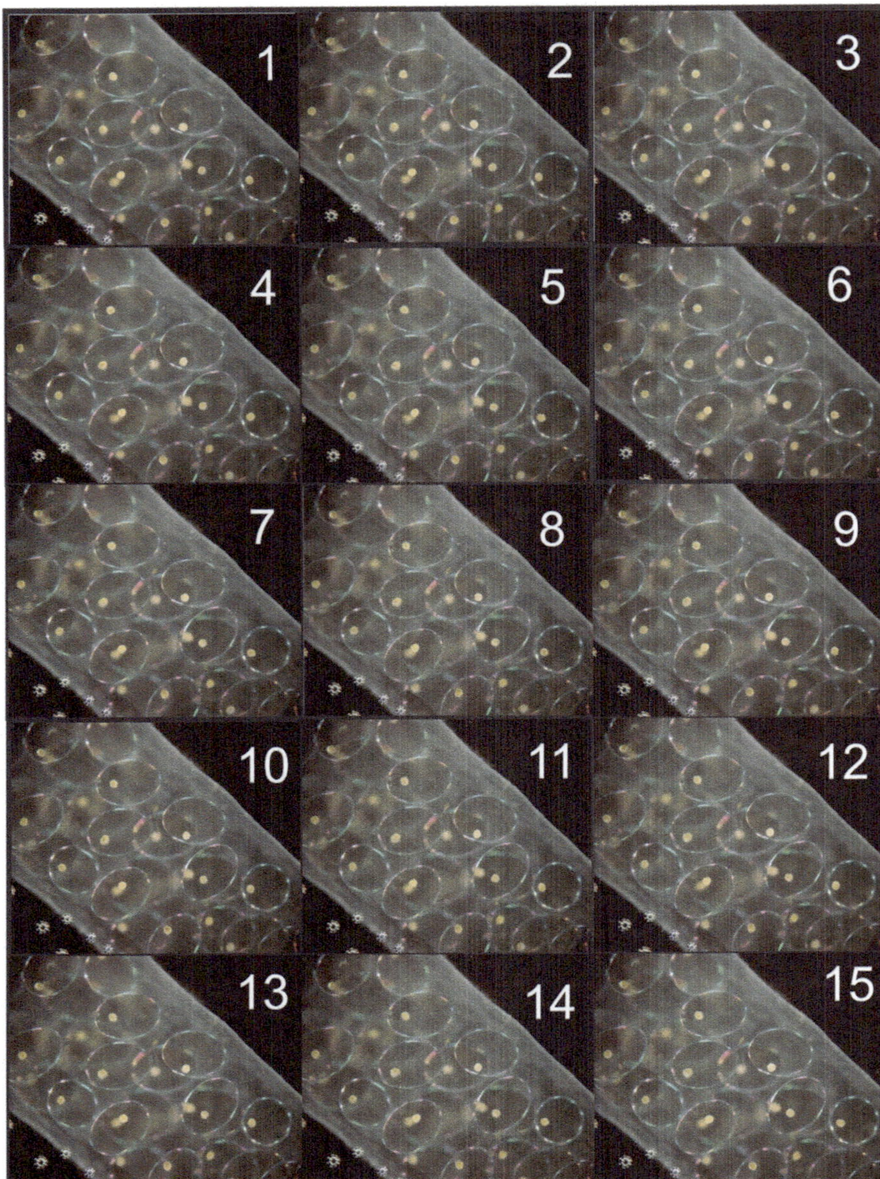

Fig. 18.16 A series of images from the movie time-lapse of *Lymnaea stagnalis* embryos in a cocoon exposed on water toxins

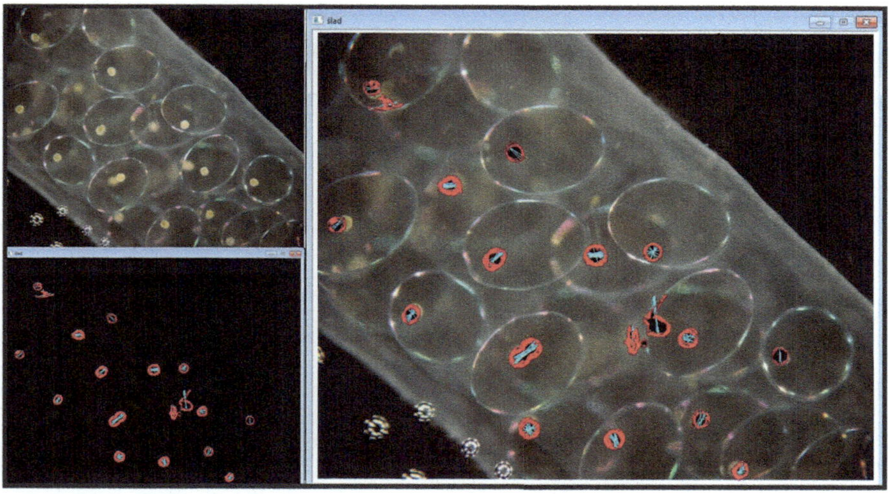

Fig. 18.17 Movement detection of *Lymnaea* embryos during bioassay by Aphelion algorithm detection

New bioassay can be applied in the automated mode; the time of operation is accelerated. The use of bioassay automation is very promising in the area of environmental engineering and nature protection.

References

Agathokleous E, Mouzaki-Paxinou AC, Saitanis CJ, Paoletti E, Manning WJ (2016) The first toxicological study of the antiozonant and research tool ethylene diurea (EDU) using a *Lemna minor* L. bioassay: hints to its mode of action. Environ Pollut 213:996–1006. https://doi.org/10.1016/j.envpol.2015.12.051

Arkhipchuk VV, Blaise C, Malinovskaya MV (2006) Use of hydra for chronic toxicity assessment of waters intended for human consumption. Environ Pollut 142:200–211. https://doi.org/10.1016/j.envpol.2005.10.012

Bandow C, Weltje L (2012) Development of an embryo toxicity test with the pond snail *Lymnaea stagnalis* using the model substance tributyltin and common solvents. Sci Total Environ 435:90–95. https://doi.org/10.1016/j.scitotenv.2012.07.005

Beasley A, Belanger SE, Brill JL, Otter RR (2015) Evaluation and comparison of the relationship between NOEC and EC10 or EC20 values in chronic *Daphnia* toxicity testing. Environ Toxicol Chem 34:2378–2384. https://doi.org/10.1002/etc.3086

Bernot RJ, Brueseke MA, Evans-White MA, Lamberti GA (2005) Acute and chronic toxicity of imidazolium-based ionic liquids on *Daphnia magna*. Environ Toxicol Chem 24:87–92. https://doi.org/10.1897/03-635.1

Brown KA, Mays T, Romoser A, Marroquin-Cardona A, Mitchell NJ, Elmore SE, Phillips TD (2014) Modified hydra bioassay to evaluate the toxicity of multiple mycotoxins and predict the detoxification efficacy of a clay-based sorbent. J Appl Toxicol 34:40–48. https://doi.org/10.1002/jat.2824

Coeurdassier M, De Vaufleury A, Badot PM (2003) Bioconcentration of cadmium and toxic effects on life-history traits of pond snails (*Lymnaea* palustris and *Lymnaea* stagnalis) in laboratory bioassays. Arch Environ Cont Tox 45:0102–0109

Dobrowolski JW (1976) Embryological aspects of environmental toxicology environmental quality and safety, vol 5. Academic, New York, pp 8–81

Dominik J, Benninghoff C, Dobrowolski IJ, Gueguen C, Pardos M, Thomas L, Wagner A (2000) Water quality assessment with combined chemical and biological methods in the Vistula River and waste waters in the region of Kraków (Poland). In: Fifth international symposium and exhibition on environmental contamination in Central and Eastern Europe, 12–14 September. Prague, Symposium Programme

Ducrot V, Askem C, Azam D, Brettschneider D, Brown R, Charles S, Holbech H (2014) Development and validation of an OECD reproductive toxicity test guideline with the pond snail *Lymnaea stagnalis* (Mollusca, Gastropoda). Regul Toxicol Pharmacol 70:605–614. https://doi.org/10.1016/j.yrtph.2014.09.004

Environment Canada (EC) (2000) Biological test method: reference method for determining acute lethality of effluents to *Daphnia magna*. Environmental Technology Centre, Ottawa, Report EPS/RM/14 Second Edition

Fulton BA, Meyer JS (2014) Development of a regression model to predict copper toxicity to *Daphnia magna* and site-specific copper criteria across multiple surface-water drainages in an arid landscape. Environ Toxicol Chem 33:1865–1873. https://doi.org/10.1002/etc.2631

Gopalapillai Y, Vigneault B, Hale BA (2014) Root length of aquatic plant, *Lemna minor* l., as an optimal toxicity endpoint for biomonitoring of mining effluents. Integr Environ Assess Manag 10:493–497. https://doi.org/10.1002/ieam.1558

Gophen M (2016) Bioassay indication of Nemagon toxicity on *Daphnia magna* (Straus 1820). Int J Bioassays 5:4739–4741

Grosell M, Gerdes RM, Brix KV (2006) Chronic toxicity of lead to three freshwater invertebrates—Brachionus calyciflorus, *Chironomus tentans* and *Lymnaea stagnalis*. Environ Toxicol Chem 25:97–104. https://doi.org/10.1897/04-654R.1

ISO 20079 (2005) Water quality – determination of the toxic effect of water constituents and waste water on duckweed (*Lemna minor*) – duckweed growth inhibition test

ISO 6341 (2012) Water quality – determination of the inhibition of the mobility of *Daphnia magna* Straus (*Cladocera, Crustacea*) – acute toxicity test

ISO norm 20079 (2007) Dissemination of the toxic effect of water constituents and wastewater on duckweed (Lemna minor) – Duckweed growth inhibition test. Accessed on November 5th

Janczak D, Lewicki P, Mazur R, Boniecki P, Dach J, Przybył J, Pawlak M, Pilarski K, Czekała W (2013) The selected examples of the application of computer image analysis in the assessment of environmental quality. In: Proceedings of SPIE 8878, fifth international conference on digital image processing (ICDIP). https://doi.org/10.1117/12.2031069

Jonczyk E, Gilron G (2005) Chapter 10: Acute and chronic toxicity testing with Daphnia sp. In: Blaise C, Férard JF (eds) Small-scale freshwater toxicity investigations, pp 337–394

Jura Cz (2005) Invertebrates: fundamentals of functional morphology, taxonomy and phylogeny (Bezkręgowce: podstawy morfologii funkcjonalnej, systematyki i filogenezy). PWN

Kawa R (2008) The new method for the assessment of the water toxicity using a computer image analysis. PhD thesis, PK – Krakow

Kirsanov D, Legin E, Zagrebin A, Ignatieva N, Rybakin V, Legin A (2014) Mimicking *Daphnia magna* bioassay performance by an electronic tongue for urban water quality control. Anal Chim Acta 824:64–70. https://doi.org/10.1016/j.aca.2014.03.021

Lalah JO, Severin GF, Schramm KW, Lenoir D, Behechti A, Guenther K (2007) Effects of a branched p-nonylphenol isomer (4(3′,6′-dimethyl-3′-heptyl)-phenol) on embryogenesis in *Lymnaea stagnalis* L. Arch Environ Contam Toxicol 52:104–112. https://doi.org/10.1007/s00244-004-0228-4

Lewicki P (2006) Computer image analysis in selected bioassays for the assessment of water quality. PhD thesis, AGH – Dep. GGiIŚ. Krakow

Mazur R, Lewicki P (2008) Application of the computer image analyses in the embryotests as a new method of water biomonitoring. Pol J Environ Stud 17:409–412. Olsztyn

Mazur R, Kawa R, Lewicki P (2006) The monitoring of selected pollutants of waters with the application of the selected eco-toxicological criteria. Pol J Environ Stud 16:334–338. Olsztyn
Mazur R, Lewicki P, Kasztalski K (2010) The application of the computer image analysis methods in environmental biomonitoring. Tech Trans 2–M/2010:167–174. E. PK, Krakow
Mazur R, Wagner A, Zhou M (2013) The application of the *Lymnaea stagnalis* embryo-test in the toxicity bioindication of surfactants in fresh waters. Ecol Indic 30:190–195. https://doi.org/10.1016/j.ecolind.2013.02.002
Moody M, Miller J (2005) Chapter 7: Lemma minor growth inhibition test. In: Blaise C, Férard JF (eds) Small-scale freshwater toxicity investigations. Springer, Dordrecht, pp 277–290
Muscatine L, Lenhoff HM (1965) Symbiosis of hydra and algae. I. Effects of some environmental cations on growth of symbiotic and aposymbiotic hydra. Biol Bull 128(3):415–424
Nowakowski K, Boniecki P, Raba B (2011) Image analysis and neural networks in the process of identifying of selected mechanical damage maize caryopses. J Res Appl Agric Eng 56:100–102. https://doi.org/10.1109/ICDIP.2009.78
OECD (2004) Test no. 202: *Daphnia sp.* acute immobilization test, OECD guidelines for the testing of chemicals, section 2. OECD Publishing, Paris
OECD (2012) Test no. 211: *Daphnia magna* reproduction test, OECD guidelines for the testing of chemicals, section 2. OECD Publishing, Paris
Pardos M, Benninghoff C, Guéguen C, Thomas R, Dobrowolski J, Dominik J (1999) Acute toxicity assessment of Polish (waste) water with a microplate-based *Hydra attenuata* assay: a comparison with the Microtox® test. Sci Environ Total Environ 243:141–148
PN-EN ISO 6341 (2002) Water quality – determination of the inhibition of the mobility of *Daphnia magna* Straus (*Cladocera, Crustacea*) – acute toxicity test (ISO 6341:1996). PKN, Warsaw
Pollino CA, Holdway DA (1999) Potential of two hydra species as standard toxicity test animals. Ecotox Environ Safe 43:309–316
Rahaman MM, Chen D, Gillani Z, Klukas C, Chen M (2015) Advanced phenotyping and phenotype data analysis for the study of plant growth and development. Front Plant Sci 6:619. https://doi.org/10.3389/fpls.2015.00619
Rahat M, Reich V (1983) Preparing axenic hydra. In: Hydra: research methods. Springer, New York, pp 79–83
Roy K, Das RN, Popelier PL (2015) Predictive QSAR modelling of algal toxicity of ionic liquids and its interspecies correlation with *Daphnia* toxicity. Environ Sci Pollut Res 22:6634–6641. https://doi.org/10.1007/s11356-014-3845-0
Sozzani R, Busch W, Spalding EP, Benfey PN (2014) Advanced imaging techniques for the study of plant growth and development. Trends Plant Sci 19(304–310):55–61. https://doi.org/10.1016/j.tplants.2013.12.003
Toropov AA, Benfenati E (2006) QSAR models for *Daphnia* toxicity of pesticides based on combinations of topological parameters of molecular structures. Bioorg Med Chem 14:2779–2788. https://doi.org/10.1016/j.bmc.2005.11.060
Trottier S, Blaise C, Kusui T, Johnson EM (1997) Acute toxicity assessment of aqueous samples using a microplate-based *Hydra* attenuata assay. Environ Toxic Water 12:265–271
Tufi S, Wassenaar PN, Osorio V, De Boer J, Leonards PE, Lamoree MH (2016) Pesticide mixture toxicity in surface water extracts in snails (*Lymnaea stagnalis*) by an in vitro acetylcholinesterase inhibition assay and metabolomics. Environ Sci Technol 50:3937–3944. https://doi.org/10.1021/acs.est.5b045777
U.S. EPA (United States Environmental Protection Agency) (2002) Short-term methods for estimating the chronic toxicity of effluents and receiving waters to freshwater organisms, 4th edn. Office of Water (4303T), Washington, DC. 20460, Document EPA-821-R-02-013
Walker CH, Hopkin SP, Sibly RM, Peakall DB (2002) Principles of ecotoxicology, 2nd edn. PWN, Warsaw
Wilby OK (1988) The hydra regeneration assay. In: Proceedings of workshop organized by Association Francaise de Teratologie, pp 108–124

Mining the Microbial Community for Redefining the Bioprocesses in the Future

19

Hemant J. Purohit, Anshuman A. Khardenavis, Atul N. Vaidya, and Vipin Chandra Kalia

Abstract

Biological systems operate efficiently under defined set of conditions to ensure their survival. In order to ensure maximum benefit to human beings, bioprocess optimization is necessary. At present, the emphasis is on exploiting natural resources without causing any damage to the environment and ecosystem. Naturally occurring biological systems are quite efficient; however, in comparison to chemical processes, they are slow and vulnerable. Hence, the focus of research is on optimization of these bioprocesses, with emphasis on efficient microbial communities by understanding biological pathways and their specific requirements. This needs development of efficient and innovative reactor systems accompanied by efficient downstream processing.

Keywords

Bioprocess • Bioreactors • Downstream • Microbes • Pathways • Upscaling

H.J. Purohit (✉) • A.A. Khardenavis
Environmental Biotechnology and Genomics Division, CSIR – National Environmental Engineering and Research Institute (CSIR-NEERI), Nehru Marg, Nagpur 440020, Maharashtra, India
e-mail: hj_purohit@neeri.res.in; hemantdrd@hotmail.com

A.N. Vaidya
Solid and Hazardous Waste Management Division, CSIR – National Environmental Engineering and Research Institute (CSIR-NEERI), Nehru Marg, Nagpur 440020, Maharashtra, India

V.C. Kalia
Microbial Biotechnology and Genomics, CSIR – Institute of Genomics and Integrative Biology, New Delhi, India

H.J. Purohit et al. (eds.), *Optimization and Applicability of Bioprocesses*,
https://doi.org/10.1007/978-981-10-6863-8_19

19.1 Introduction

Under defined conditions, biological systems for their survival carry out a series of biochemical steps. Biotechnological solutions device the rules where without compromising with basic physiological processes spins out a desired expected outcome. To achieve the maximum output, balancing of overall physiology is bioprocess optimization (Purohit et al. 2016). In the current scenario of increasing energy demand, depleting natural resources, and ever-increasing environmental awareness, bioprocesses occupy a unique position as sole means of converting hitherto untapped resources into energy and useful chemicals in an environmentally sound manner (Arasu et al. 2015; Balakrishnan et al. 2015; Bandyopadhyay et al. 2015; Go et al. 2015; Park et al. 2015; Yin et al. 2015; Ambardar et al. 2016; Hernández-Saldaña et al. 2016; Peña-Yam et al. 2016; Parmar et al. 2017; Sanchart et al. 2017; Sharma and Lal 2017). One of the most addressed areas is depleting fossil fuels and surge of biomass-based bioprocesses. They are befitting examples underlining the inevitability of application of bioprocesses (Patel et al. 2017). Bioprocesses are mostly eco-friendly, efficient, and targeted. Bioprocesses are abundantly operative in nature, and there is a need to critically observe them, study them, and mimic them. All the naturally occurring bioprocesses are the most energy-efficient and are associated with minimal entropy increase. However, bioprocesses have certain disadvantages when compared to competing chemical processes, major ones being slower rates and vulnerability to climatic conditions. Therefore, all the research efforts are to be directed towards optimization of bioprocesses (Kumar and Prasad 2011). Employing bioprocesses for practical and large-scale applications need:

- Efficient microbial community
- Thorough understanding of microbial pathways and specific requirements
- Efficient and innovative reactor systems
- Efficient downstream processing

19.2 Prerequisites for Practical and Large-Scale Applications of Bioprocesses

19.2.1 Role of Microbial Community

Microbial products and microbial process-driven outcome are different scenarios. In controlled condition using pure culture, atypical output can be envisaged and for which designing the bioprocess rules is rather simple. However, in the environmental conditions with so-called mix population, the understanding of bioprocesses is very difficult. Depending upon the availability of resources, a typical community emerges with defined ratio of key microbes (Pandit et al. 2016; Gulhane et al. 2017). Still there are no absolute tools available to understand this consortium and associated bioprocesses. Development of efficient microbial communities needs continual research in isolation and identification of useful species, their enrichment, genetic modifications, and developing protocol for maintenance (He et al. 2016). For this purpose, many modern tools, particularly omic tools, are available. In addition, many modern equipments are developed for extremely precise studies in molecular biology. Among the various tools, application of laser stimulation stands out as the most innovative approach which could find application in optimization of natural processes such as in sewage and wastewater treatment, bioremediation of pollutants, and in improving the growth rate, biomass productivity, and adaptability of plants to pollutants such as heavy metals (Jakubiak and Gdowska 2013; Tursunov and Dobrowolski 2015).

In the environmental processes, the composition of the microbial community is decided by the availability of resources and overall stress in terms of oxidizing/reducing agents. After dissecting such association and understanding, the possible roles of each member are always different than the acquired physiology under the influence of community organization. If a defined community is perturbed by introducing an "invader" strain, the overall performance changes due to additional intelligence provided by new genetic information to that system (De Roy et al. 2013). However, this assembled new intelligence is also prone to environmental effectors and will shift if a new variable is introduced to the system. The thermodynamic phenomenon could be difficult to comprehend using even a single-bacterium system with either change in redox or complication with nutrients (Puranik and Purohit 2015). In environmental scenarios such as contaminated sites or wastewater treatment facilities, there is a selection of microbial population over a period of time with the available pollutant as nutrients (Khardenavis et al. 2008; Kapley et al. 2007). This community with their dominating player and key biotransforming functions survives with stress shocks. However, to understand the community evolution, an approach of the community by design could provide more insight into this phenomenon (Fredrickson 2015). As the process condition changes, the same community can acquire new architecture of sharing nutrient, and with the topology of the organization in space, they symbiotically survive with acquired flow of metabolites, such as shift in granulation of community from aerobic to the anaerobic environment (Sarma et al. 2017). With this background, research efforts needed for the improvement in the bioprocess applications could be delineated and implemented. For example, bioremediation of contaminated sites offers a great potential for the bioprocesses; however, we need to develop communities that would efficiently work in the environmental matrix available at the specific site (Sharma et al. 2014; Rao et al. 2017). Secondly we need efficient means to spread and proliferate desired communities in the nonhomogeneous soil mass. In case of remediation of water bodies, problems envisaged are retention of the microbial species in flowing systems and exposure to dissolved organics and metals.

19.2.2 Microbial Pathways

The discoveries in microbial pathway will find a mega shift due to interventions from metagenomics (Jadeja et al. 2014; More et al. 2014). Research should focus on isolation and identification of microbial communities having abilities to perform multiple tasks, rather than working in association with other species. The species should be robust and efficient and sustain natural variations in the systems (Deshmukh et al. 2016). With the intervention of omics tool, the understanding of microbial community requires a holistic approach. Thorough understanding of the microbial pathways is necessary to understand the possibilities and opportunities for the production of value-added chemicals from abundant natural substrates (Azman et al. 2017). This would create a spectrum of chemicals that would replace synthetically produced commercial chemicals (Ray and Kalia 2017). Understanding of pathway should lead to identification of precursors, cofactors, nutritional requirements, process conditions, etc. Modern tools could be used along with tracer techniques to establish unknown pathways (Antoniewicz 2015). This would also lead to identification of useful enzymes. The new pathway discoveries will lead to new enzymatic options (Jadeja et al. 2014; Ghosh et al. 2017).

The performance of a community has to be analysed with possible metabolic plasticity required for symbiotic survival of the different bacteria (Gulhane et al. 2017). In most of the design microbial community with given pollutant load and with possible intermediates and/or product, the overall pathway could be predicted, and rules driving such pathway could be validated (Tikariha et al. 2016). There is a need for exploration of understanding new unexplored biotransformation pathways. Metagenomics approach would yield information of such biochemical events which could be the basis of new strategies for controlled biotransformation of molecules (Comolli 2014). The new drug discovery strategies and new molecule synthesis/precursors for various intermediates will be identified through the exploration of metagenomics (Kalia and Purohit 2011; Kapley et al. 2015).

New options are emerging for metabolomic engineering where functional protein chips are being used (Jung and Stephanopoulos 2004), which is now getting extended to controlled enzymatic reaction on chip (Mahler 2014). The systems are further extended to design microelectronic devices, where the electrical signals are generated via biochemical reactions. The system relies on natural redox system to generate signals (Gordonov et al. 2014). The new models are developed for in silico prediction of the metabolic pathways to design the desired product; software such as COBRA is applied for simulation generation (King et al. 2015). Alternatively, algorithms are designed for the strain to get desired metabolic output (Long et al. 2015). The flexibilities associated with microbial metabolism have opportunities for the future green industry, where the gene will be engineered based on in silico design to provide a biochemical pathway for a variety of biotechnological processes of industrial importance (Thodey et al. 2014; Santero et al. 2016). Similar options of new plant-based products are emerging with Cas-9 nuclease system, where by controlling RNA expression, a phenotype can be decided (Raitskin and Patron 2016; Ehrenworth and Peralta-Yahya 2017).

Metabolic engineering has a lot to learn starting from the biggest limitation of interpretation of dynamic system to freeze a desired biochemical pathway for desired output. For a single system, it is still a challenge; therefore in case of environmental or mix population system, we are limited by both wet lab and computational prediction. The DNA/RNA database provides little leads with BLAST and another homology-based search algorithm for predicting the closest function associated with unknown DNA/RNA fragments (Shim et al. 2017). Similarly, for different organic molecules from C_1 to most complex ones, an MS/MS database will be required. The library of such MS/MS fragments correlated with transcriptome data might suggest for that particular condition how a biological system responds or could be exploited. This will propose new intermediates and identify the new metabolic options which cells or community acquires. In environmental conditions, microbial community intelligence drives the biotransformation pathways that work in favour of bioremediation (Yadav et al. 2014, 2015).

19.2.3 Efficient and Innovative Reactor System

Reactors or bioreactors are needed when bioprocesses are put to large-scale applications. Efficient contact between substrates, appropriate microbial community, and electron acceptor are the basic needs of any bioreactor (Cassarini et al. 2015; Zhang et al. 2015). Owing to the very nature of microorganisms, suspended culture reactor or fixed-film reactors can be developed. Fixed-film reactors are typically plug-flow kind of reactors and offer more robust designs that can sustain shock loadings as well as adverse conditions (Robles Martinez et al. 2015). Fixed-film reactor systems usually adopt multiple microbial communities and as such are better choice for waste management systems, wherein many organic pollutants are encountered in different concentrations. On the contrary, fixed-film systems may not be a good choice when bioprocess is targeted to produce one particular type of product. Here suspended growth reactors, which are of CSTR type, are more suitable as we can maintain a desired microbial community (Arya et al. 2016). Typical limitations can be related to the maximum product concentration attainable. The overall choice of reactor system would depend upon the microbial community to be used, targeted application, and desired results. Reactor systems should always be looked into with possible downstream processing options. Bioprocesses are known to produce the desired chemicals in low concentrations as compared to chemical processes. Therefore, when these are applied for production of value-added chemicals, downstream processing for the purification of product can be very cumbersome and prohibitive (Jungbauer 2013). Therefore, there is a need to develop efficient downstream processing using membrane, ion exchange, adsorption, and other advanced physicochemical techniques. Depending upon the option of reactor design which has been selected with the same composition of organic loading, the system selects the different microbial communities. In biofilm conditions, even the most toxic molecules get biotransformed due to the survival options acquired by the community which protect them (Ghosh et al. 2017).

19.3 Application Domain for the Bioprocesses

If these idealities are achieved, bioprocesses sure would replace chemical processes. Accordingly, bioprocesses would find abundant applications in:

- Conversion of biomass to fuels/value-added chemicals
- Reclamation of waste resources like water and land
- Waste management, typically wastewater and solid waste

Problems envisaged in the biomass to fuel applications are pertaining to drastic variations in the biomass type and concentrations thereof, reactor volumes involved, and product purification requirements (Kalia et al. 2016). These issues sometimes render bioprocesses cost prohibitive. It should be remembered, however, that cost benefit is temporal concept and would favour similar bioprocesses cost-effective if crude oil prices are dramatically shooting up. Nevertheless, there is a need to make waste to fuel bioprocesses more efficient, more precise, and more robust (Kumar et al. 2014).

Waste management appears to be the perfect domain for the application of bioprocesses, owing to the fact that chemical processes have limitations while dealing with relatively low concentrations of wastes and are therefore cost prohibitive in many instances. Wastewater reclamation can resolve water crisis likely to be faced in the near future, and bioprocesses would play a key role in this (Goh et al. 2013). Of course the research needs delineated earlier are still valid, and we have to develop very efficient and robust systems. Emphasis should be laid on the reuse of water rather than merely meeting the discharge standards. The innovation in applying different reactors and ensuring possible pretreatment could create a situation where the industrial grade water recovery could be achieved (Yadav et al. 2015).

Solid waste management offers a great potential as well as great challenge for application of bioprocesses. Solid waste management costs are very high, and any resource recovery is always advantageous (Allesch and Brunner 2014). Solid wastes comprise many natural resources in scattered and nonhomogeneous structure, which pose a great challenge for the application of bioprocesses. Here development of multi-tasking microbial communities can give desired breakthrough; however, there is an equal need for development of robust reactor system that would take care of non-homogeneity of solid waste. Strategies to utilize segregated solid waste as raw material will provide a base for bioprocesses where different organic synthesis applying microbial capacities will be possible (Gulhane et al. 2016).

Acknowledgement The authors are thankful to Director, CSIR-National Environmental Engineering Research Institute and Director, CSIR-Institute of Genomics and Integrative Biology for providing necessary facilities for this work [CSIR-NEERI/KRC/2017/JULY/EBGD/13].

References

Allesch A, Brunner PH (2014) Assessment methods for solid waste management: a literature review. Waste Manag Res 32:461–473. https://doi.org/10.1177/0734242X14535653

Ambardar S, Gupta R, Trakroo D, Lal R, Vakhlu J (2016) High throughput sequencing: an overview of sequencing chemistry. Indian J Microbiol 56:394–404. https://doi.org/10.1007/s12088-016-0606-4

Antoniewicz MR (2015) Parallel labeling experiments for pathway elucidation and 13 C metabolic flux analysis. Curr Opin Biotechnol 36:91–97. https://doi.org/10.1016/j.copbio.2015.08.014

Arasu MV, Al-Dhabi NA, Rejiniemon TS, Lee KD, Huxley VAJ, Kim DH, Duraipandiyan V, Karuppiah P, Choi KC (2015) Identification and characterization of *Lactobacillus brevis* P68 with antifungal, antioxidant and probiotic functional properties. Indian J Microbiol 55:19–28. https://doi.org/10.1007/s12088-014-0495-3

Arya V, Philip L, Bhallamudi SM (2016) Performance of suspended and attached growth bioreactors for the removal of cationic and anionic pharmaceuticals. Chem Eng J 15(284):1295–1307. https://doi.org/10.1016/j.cej.2015.09.070

Azman AS, Othman I, Fang CM, Chan KG, Goh BH, Lee LH (2017) Antibacterial, anticancer and neuroprotective activities of rare *Actinobacteria* from mangrove forest soils. Indian J Microbiol 56:177–187. https://doi.org/10.1007/s12088-016-0627-z

Balakrishnan D, Bibiana AS, Vijayakumar A, Santhosh RS, Dhevendaran K, Nithyanand P (2015) Antioxidant activity of bacteria associated with the marine sponge *Tedania anhelans*. Indian J Microbiol 55:13–18. https://doi.org/10.1007/s12088-014-0490-8

Bandyopadhyay P, Mishra S, Sarkar B, Swain SK, Pal A, Tripathy PP, Ojha SK (2015) Dietary *Saccharomyces cerevisiae* boosts growth and immunity of IMC *Labeo rohita* (Ham.) juveniles. Indian J Microbiol 55:81–87. https://doi.org/10.1007/s12088-014-0500-x

Cassarini C, Bhattarai S, Gonzalez-Gil G, Rene ER, Vogt C, Musat N, Lens PN (2015) Anaerobic oxidation of methane using different sulphur compounds as electron acceptors in a bioreactor. In: Book of abstracts, June 29, p 47. doi:https://doi.org/10.13140/RG.2.1.1564.7842

Comolli LR (2014) Intra-and inter-species interactions in microbial communities. Front Microbiol 5:629. https://doi.org/10.3389/fmicb.2014.00629

De Roy K, Marzorati M, Negroni A, Thas O, Balloi A, Fava F, Verstraete W, Daffonchio D, Boon N (2013) Environmental conditions and community evenness determine the outcome of biological invasion. Nat Commun 4:1383. https://doi.org/10.1038/ncomms2392

Deshmukh R, Khardenavis AA, Purohit HJ (2016) Diverse metabolic capacities of fungi for bioremediation. Indian J Microbiol 56:247–264. https://doi.org/10.1007/s12088-016-0584-6

Ehrenworth AM, Peralta-Yahya P (2017) Accelerating the semisynthesis of alkaloid-based drugs through metabolic engineering. Nat Chem Biol 13:249–258. https://doi.org/10.1038/nchembio.2308

Fredrickson JK (2015) Ecological communities by design. Science 348:1425–1427. https://doi.org/10.1126/science.aab0946

Ghosh S, Qureshi A, Purohit HJ (2017) Enhanced expression of catechol 1,2 dioxygenase gene in biofilm forming *Pseudomonas mendocina* EGD-AQ5 under increasing benzoate stress. Int Biodeterior Biodegrad 118:57–65. https://doi.org/10.1016/j.ibiod.2017.01.019

Go T-H, Cho K-S, Lee S-M, Lee O-M, Son H-J (2015) Simultaneous production of antifungal and keratinolytic activities by feather-degrading *Bacillus subtilis* S8. Indian J Microbiol 55:66–73. https://doi.org/10.1007/s12088-014-0502-8

Goh S, Zhang J, Liu Y, Fane AG (2013) Fouling and wetting in membrane distillation (MD) and MD-bioreactor (MDBR) for wastewater reclamation. Desalination 15(323):39–47. https://doi.org/10.1016/j.desal.2012.12.001

Gordonov T, Kim E, Cheng Y, Ben-Yoav H, Ghodssi R, Rubloff G, Yin JJ, Payne GF, Bentley WE (2014) Electronic modulation of biochemical signal generation. Nat Nanotechnol 9:605–610. https://doi.org/10.1038/nnano.2014.151

Gulhane M, Khardenavis AA, Karia S, Pandit P, Kanade GS, Lokhande S, Vaidya A, Purohit HJ (2016) Biomethanation of vegetable market waste in an anaerobic reactor: effect of effluent recirculation and carbon mass balance analysis. Bioresour Technol 215:100–109. https://doi.org/10.1016/j.biortech.2016.04.039

Gulhane M, Pandit P, Khardenavis A, Singh D, Purohit H (2017) Study of microbial community plasticity for anaerobic digestion of vegetable waste in Anaerobic Baffled Reactor. Renew Energy 101:59–66. https://doi.org/10.1016/j.renene.2016.08.021

He G, Wang X, Liao G, Huang S, Wu J (2016) Isolation, identification and characterization of two aluminum-tolerant fungi from acidic red soil. Indian J Microbiol 56:344–352. https://doi.org/10.1007/s12088-016-0586-4

Hernández-Saldaña OF, Valencia-Posadas M, de la Fuente-Salcido NM, Bideshi DK, Barboza-Corona JE (2016) Bacteriocinogenic bacteria isolated from raw goat milk and goat cheese produced in the Center of México. Indian J Microbiol 56:301–308. https://doi.org/10.1007/s12088-016-0587-3

Jadeja NB, More RP, Purohit HJ, Kapley A (2014) Metagenomic analysis of oxygenases from activated sludge. Bioresour Technol 165:250–256. https://doi.org/10.1016/j.biortech.2014.02.045

Jakubiak M, Gdowska K (2013) Innovative environmental technology applications of laser light stimulation. Енергетика і автоматика 3:14–21

Jung GY, Stephanopoulos G (2004) A functional protein chip for pathway optimization and in vitro metabolic engineering. Science 304:428–431. https://doi.org/10.1126/science.1096920

Jungbauer A (2013) Continuous downstream processing of biopharmaceuticals. Trends Biotechnol 31:479–492. https://doi.org/10.1016/j.tibtech.2013.05.011

Kalia VC, Purohit HJ (2011) Quenching the quorum sensing system: potential antibacterial drug targets. Crit Rev Microbiol 37:121–140. https://doi.org/10.3109/1040841X.2010.532479

Kalia VC, Prakash J, Koul S (2016) Biorefinery for glycerol rich biodiesel industry waste. Indian J Microbiol 56:113–125. https://doi.org/10.1007/s12088- 016-0583-7

Kapley A, Thierry B, Purohit HJ (2007) Eubacterial diversity of activated biomass from a CETP. Res Microbiol 158:494–500. https://doi.org/10.1016/j.resmic.2007.04.004

Kapley A, Tanksale H, Sagarkar S, Prasad AR, Rathod AK, Sharma N, Qureshi A, Purohit HJ (2015) Antimicrobial activity of *Alcaligenes* sp. HPC 1271 against multidrug resistant bacteria. Funct Integr. Gen 16:57–65. https://doi.org/10.1007/s10142-015-0466-8

Khardenavis AA, Kapley A, Purohit HJ (2008) Phenol-mediated improved performance of active biomass for treatment of distillery wastewater. Int Biodeterior Biodegrad 62:38–45. https://doi.org/10.1016/j.ibiod.2007.06.016

King ZA, Lloyd CJ, Feist AM, Palsson BO (2015) Next-generation genome-scale models for metabolic engineering. Curr Opin Biotechnol 35:23–29. https://doi.org/10.1016/j.copbio.2014.12.016

Kumar RR, Prasad S (2011) Metabolic engineering of bacteria. Indian J Microbiol 51:403–409. https://doi.org/10.1007/s12088-011-0172-8

Kumar P, Pant DC, Mehariya S, Sharma R, Kansal A, Kalia VC (2014) Ecobiotechnological strategy to enhance efficiency of bioconversion of wastes into hydrogen and methane. Indian J Microbiol 54:262–267. https://doi.org/10.1007/s12088-014-0467-7

Long MR, Ong WK, Reed JL (2015) Computational methods in metabolic engineering for strain design. Curr Opin Biotechnol 34:135–141. https://doi.org/10.1016/j.copbio.2014.12.019

Mahler G (2014) Metabolic engineering: enzyme control on a chip. Nat Nanotechnol 9:571–572. https://doi.org/10.1038/nnano.2014.160

More PR, Mitra S, Raju SC, Kapley A, Purohit HJ (2014) Mining and assessmentof catabolic pathways in the metagenome of a common effluent treatment plant to induce the degradative capacity of biomass. Bioresour Technol 153:137–146. https://doi.org/10.1016/j.biortech.2013.11.065

Pandit PD, Gulhane MK, Khardenavis AA, Purohit HJ (2016) Mining of hemicellulose and lignin degrading genes from differentially enriched methane producing microbial community. Bioresour Technol 216:923–930. https://doi.org/10.1016/j.biortech.2016.06.021

Park J-M, Radhakrishnan R, Kang S-M, Lee I-J (2015) IAA producing *Enterobacter* sp. I-3 as a potent bio-herbicide candidate for weed control: a special reference with lettuce growth inhibition. Indian J Microbiol 55:207–211. https://doi.org/10.1007/s12088-015-0515-y

Parmar NR, Pandit PD, Purohit HJ, Nirmal Kumar JI, Joshi CG (2017) Influence of diet composition on cattle rumen methanogenesis: a comparative metagenomic analysis in Indian and exotic cattle. Indian J Microbiol 57:226–234. https://doi.org/10.1007/s12088-016-0635-z

Patel SK, Lee JK, Kalia VC (2017) Dark-fermentative biological hydrogen production from mixed biowastes using defined mixed cultures. Indian J Microbiol 57:171–176. https://doi.org/10.1007/s12088-017-0643-7

Peña-Yam LP, Esaú Ruíz-Sánchez E, Barboza-Corona JE, Reyes-Ramírez A (2016) Isolation of Mexican Bacillus species and their effects in promoting growth of chili pepper (*Capsicum annuum L.* cv Jalapeño). Indian J Microbiol 56:375–378. https://doi.org/10.1007/s12088-016-0582-8

Puranik S, Purohit HJ (2015) Dynamic interactive events in gene regulation using *E. coli* dehydrogenase as a model. Funct Integr Genomics 15:175–188. https://doi.org/10.1007/s10142-014-0418-8

Purohit HJ, Kapley A, Khardenavis A, Qureshi A, Dafale NA (2016) Insights in waste management bioprocesses using genomic tools. In: Sariaslani S, Gadd GM (eds) Advances in Applied Microbiology, vol 97, pp 121–170. https://doi.org/10.1016/bs.aambs.2016.09.002

Raitskin O, Patron NJ (2016) Multi-gene engineering in plants with RNA-guided Cas9 nuclease. Curr Opin Biotechnol 37:69–75. https://doi.org/10.1016/j.copbio.2015.11.008

Rao STH, Papathoti NK, Gundeboina R, Mohamed YK, Mudhole GR, Bee H (2017) Hexavalent chromium reduction from pollutant samples by *Achromobacter xylosoxidans* SHB 204 and its kinetics study. Indian J Microbiol 57:292–298. https://doi.org/10.1007/s12088-017-0654-4

Ray S, Kalia VC (2017) Microbial cometabolism and polyhydroxyalkanoate co-polymers. Indian J Microbiol 57:39–47. https://doi.org/10.1007/s12088-016-0622-4

Robles Martinez A, Latrille E, Ruano MV, Steyer JP (2015) A robust fuzzy-logic-based controller for bio-methane production in anaerobic fixedfilm reactors. World Congress on Anaerobic Digestion (AD14), Viña del Mar, CHL, 2015-11-15-2015-11-18, 2015

Sanchart C, Rattanaporn O, Haltrich D, Phukpattaranont P, Maneerat S (2017) *Lactobacillus futsaii* CS3, a new GABA-producing strain isolated from Thai fermented shrimp (Kung–Som). Indian J Microbiol 57:211–217. https://doi.org/10.1007/s12088-016-0632-2

Santero E, Floriano B, Govantes F (2016) Harnessing the power of microbial metabolism. Curr Opin Microbiol 31:63–69. https://doi.org/10.1016/j.mib.2016.03.003

Sarma SJ, Tay JH, Chu A (2017) Finding knowledge gaps in aerobic granulation technology. Trends Biotechnol 35:66–78. https://doi.org/10.1016/j.tibtech.2016.07.003

Sharma A, Lal R (2017) Survey of (meta)genomic approaches for understanding microbial community dynamics. Indian J Microbiol 57:23–38. https://doi.org/10.1007/s12088-016-0629-x

Sharma JK, Gautam RK, Misra RR, Kashyap SM, Singh SK, Juwarkar AA (2014) Degradation of di-through hepta-chlorobiphenyls in clophen oil using microorganisms isolated from long term PCBs contaminated soil. Indian J Microbiol 54:337–342. https://doi.org/10.1007/s12088-014-0459-7

Shim JE, Lee T, Lee I (2017) From sequencing data to gene functions: co-functional network approaches. Anim Cells Syst 21:77–83. https://doi.org/10.1080/19768354.2017.1284156

Thodey K, Galanie S, Smolke CD (2014) A microbial biomanufacturing platform for natural and semisynthetic opioids. Nat Chem Biol 10:837–844. https://doi.org/10.1038/nchembio.1613

Tikariha H, Pal RR, Qureshi A, Kapley A, Purohit HJ (2016) In silico analysis for prediction of degradative capacity of Pseudomonas putida SF1. Gene 591:382–392. https://doi.org/10.1016/j.gene.2016.06.028

Tursunov O, Dobrowolski JW (2015) A brief review of application of laser biotechnology as an efficient mechanism for the increase of biomass for bio-energy production via clean thermo-technologies. Am J Renew Sust Energ 1:66–71

Yadav TC, Khardenavis AA, Kapley A (2014) Shifts in microbial community in response to dissolved oxygen levels in activated sludge. Bioresour Technol 165:257–264. https://doi.org/10.1016/j.biortech.2014.03.007

Yadav TC, Pal RR, Shastri S, Jadeja NB, Kapley A (2015) Comparative metagenomics demonstrating different degradative capacity of activated biomass treating hydrocarbon contaminated wastewater. Bioresour Technol 188:24–32. https://doi.org/10.1016/j.biortech.2015.01.141

Yin X, Ma L, Pei X, Du P, Li C, Xie T, Yu L, Yu L, Wang Q (2015) Creation of functionally diverse chimerical α-glucosidase enzymes by swapping homologous gene fragments retrieved from soil DNA. Indian J Microbiol 55:114–117. https://doi.org/10.1007/s12088-014-0493-5

Zhang Z, Wang Y, Hu J, Wu Q, Zhang Q (2015) Influence of mixing method and hydraulic retention time on hydrogen production through photofermentation with mixed strains. Int J Hydrog Energy 40:6521–6529. https://doi.org/10.1016/j.ijhydene.2015.03.118

Printed by Printforce, the Netherlands